THERMODYNAMICS – FUNDAMENTALS AND ITS APPLICATION IN SCIENCE

Edited by **Ricardo Morales-Rodriguez**

Thermodynamics – Fundamentals and Its Application in Science

http://dx.doi.org/10.5772/2615
Edited by Ricardo Morales-Rodriguez

Contributors

Ahmet Gürses, Mehtap Ejder-Korucu, Nikolai Bazhin, Yi Fang, Bohdan Hejna, A. Plastino, Evaldo M. F. Curado, M. Casas, Zdeňka Kolská, Milan Zábranský, Alena Randová, Ronald J. Bakker, Elisabeth Blanquet, Ioana Nuta, Lin Li, Rıza Atav, L.E. Panin, Yu Liu, Kui Wang, Philippe Vieillard, Nong-Moon Hwang, Jae-Soo Jung, Dong-Kwon Lee, Vasiliy Fefelov, Vitaly Gorbunov, Alexander Myshlyavtsev, Marta Myshlyavtseva, Adela Ionescu, Paiguy Armand Ngouateu Wouagfack, Réné Tchinda, Raul Măluțan, Pedro Gómez Vilda, Xuejing Hou, Harvey J.M. Hou, C.A.S. Silva, Hui-Zhen Fu, Yuh-Shan Ho

CBS, Edition **2017**

Published by InTech

Janeza Trdine 9, 51000 Rijeka, Croatia

Notice

Publishing Process Manager Marina Jozipovic
Technical Editor InTech Prepress, Novi Sad
Cover InTech Design team

Additional hard copies can be obtained from orders@intechopen.com

Thermodynamics – Fundamentals and Its Application in Science,
Edited by Ricardo Morales-Rodriguez
p. cm.
ISBN 978-953-51-0779-8

INTECH
open science | open minds

Contents

Preface

This book is a result of a careful selection of scientific contributions involved in the thermodynamic area and it is titled "Thermodynamics - Fundamentals and Its Application in Science". Thermodynamics is very important for the description of phenomena in different fields on science. Therefore, this book contains chapters describing the fundamentals and the diverse applications in different areas under development, which allow the access of different kind of readers; for instance, bachelor and postgraduate students, researchers, etc.

The book is divided in six sections and the classification was done according to the purpose, relevance and approaches employed in the development of the contributions.

The first section describes the classical thermodynamics, where firstly an overview about classical thermodynamics considering diverse fundamental concepts of the area is described. This section also has a contribution presenting the mechanism of useful work and heat production in reversible systems.

The second section includes some chapters based on statistical mechanics, for instance, one of the chapters described the protein folding phenomena based on Gibbs free energy through the use of quantum mechanics, topic of high importance currently. On the other hand, another of the contributions in this section describes the information capacity of quantum transfer channels and thermodynamics analogies. The last chapter of this section introduces some axioms which allow one to derive the MaxEnt equations and viceversa, giving an alternative foundation for equilibriums statistical mechanics.

The property prediction in thermodynamics is presented in the following section. A chapter explaining the use and implementation of group contribution methods for property prediction of organic compounds is firstly described. The description of pure gases and multi/component fluid systems is presented in another chapter, which in fact used a modified version of the Van der Waals equation. The last chapter is this section illustrates the interest area of macroscopic modelling on the thermodynamics simulation and gives some interesting examples in different domains in the material and product design areas employing some predicted properties.

The fourth section contains the application of some thermodynamic insights in the material and products area. One of the chapters introduces some computational

Section 1

Classical Thermodynamics

Chapter 1

A View from the Conservation of Energy to Chemical Thermodynamics

Ahmet Gürses and Mehtap Ejder-Korucu

Additional information is available at the end of the chapter

http://dx.doi.org/10.5772/51383

1. Introduction

According to the **conservation of energy law**, **energy**, which is the capacity to do work or to supply heat, can be neither created nor destroyed; it can only be converted from one form into another. For example, the water in a reservoir of dam has potential energy because of its height above the outlet stream but has no kinetic energy because it is not moving. As the water starts to fall over the dam, its height and **potential energy**, *(Ep)* is energy due to position or any other form of stored energy, decrease while its velocity and **kinetic energy**, *(E_K)* is the energy related to the motion of an object with mass m and velocity v, increase. The total of potential energy plus kinetic energy always remains constant. When the water reaches the bottom and dashes against the rocks or drives the turbine of a generator, its kinetic energy is converted to other forms of energy-perhaps into heat that raises the temperature of the water or into electrical energy [6]. If any fuel is burned in an open medium, its energy is lost almost entirely as heat, whereas if it is burned in a car engine; a portion of the energy is lost as work to move the car, and less is lost as heat. These are also typical examples of the existence of the law.

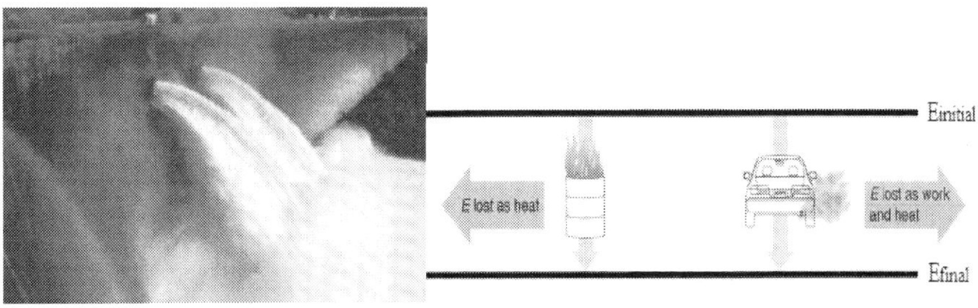

Figure 1. Some examples showing the existence of the conservation of energy law

It is thought that the summation of the introduction as a detailed concept map related with the conservation of energy would be better. This map in Fig.2 presents a concise view for many concepts of thermodynamics and their relations.

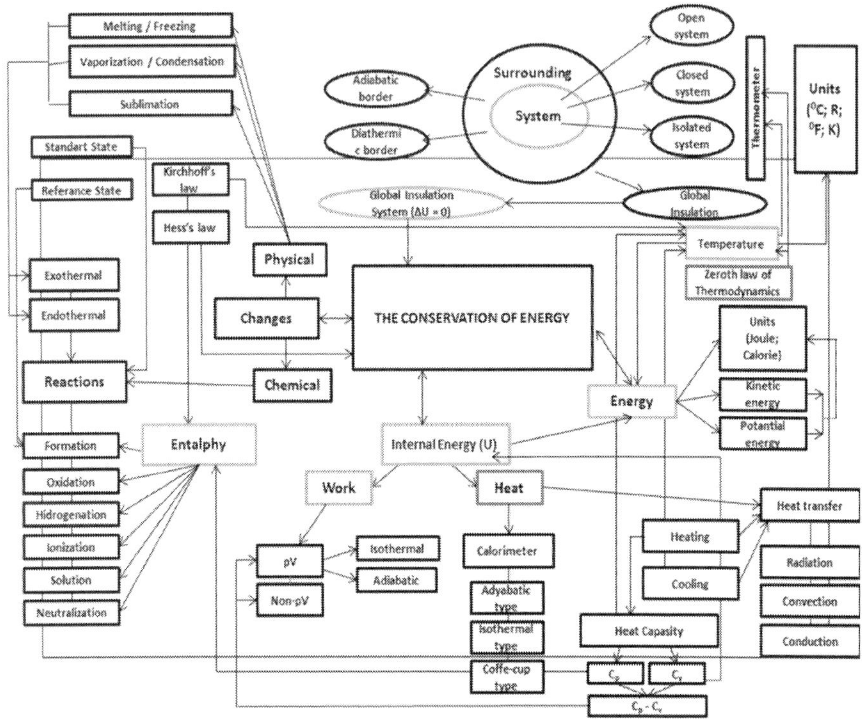

Figure 2. A concept map for the conservation of energy law

The macroscopic part of universe under study in thermodynamics is called the system. The parts of the universe that can interact with the system are called the surroundings [5]. In order to describe the thermodynamic behavior of a physical system, the interaction between the system and its surroundings must be understood. Thermodynamic systems are thus classified into three main types according to the way they interact with the surroundings (Fig.3); **open system:** matter and energy can be exchanged with the surroundings; **closed system:** can exchange energy but not matter with the surroundings and **isolated system:** cannot exchange matter or energy with the surroundings [4, 5]. Addition, a boundary that does permit energy transfer as heat (such as steel and glass) is called **diathermic** and a boundary that does not permit energy transfer as heat is called **adiabatic** [1;4].

The system may be homogeneous or heterogeneous. An exact definition is difficult, but it is convenient to define a **homogeneous system** as one whose properties are the same in all parts, or at least their spatial variation is continuous. A **heterogeneous system** consists of two or more distinct homogeneous regions or **phases**, which are separated from one another by surfaces of discontinuity [4].

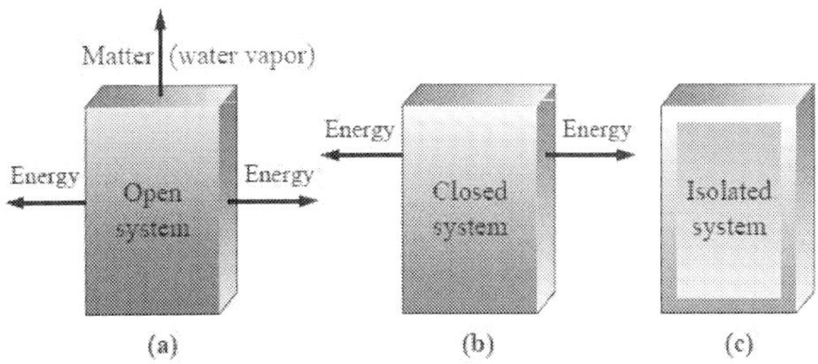

Figure 3. Systems and their surroundings; (a) open system (b) closed system and (c) isolated system.

Heating is a process in which the temperature of system, separated with diathermic border from its surrounding, is increased. This process leads to passing system from a state of lower energy to higher one. Heating process based on the energy difference between system and its surrounding provides identify of an important property which indicates the flow direction of energy. This property is called **temperature.** Temperature is a physical property of matter that quantitatively expresses the common notions of hot and cold. Objects of low temperature are cold, while various degrees of higher temperatures are referred to as warm or hot. Heat spontaneously flows from bodies of a higher temperature to bodies of lower temperature, at a rate that increases with the temperature difference and the thermal conductivity. No heat will be exchanged between bodies at same temperature; such bodies are said to be in "**equilibrium**". On the other hand, kinetic energy associated with the random motion of particles is called thermal energy, and the thermal energy of a given material is proportional to temperature. However, the magnitude of thermal energy in a sample also depends on the number of particles in the sample and so it is an **extensive property**. The water in a swimming pool and a cup of water taken from the pool has the same temperature, so their particles have the same average kinetic energy. The water in the pool has much more thermal energy than the water in the cup, simply because there are a larger number of molecules in the pool. A large number of particles at a given temperature have a higher total energy than a small number of particles at the same temperature [7]. Quantitatively, temperature which is an **intensive property** is measured with thermometers, which may be calibrated to a variety of temperature scales [9].

If two thermodynamic systems, A and B, each of which is in **thermal equilibrium** independently, are brought into thermal contact, one of two things will take place: either a flow of heat from one system to the other or no thermodynamic process will result. In the latter case the two systems are said to be in thermal equilibrium with respect to each other [11]. When same treatment has been repeated other system, C, if there is thermal equilibrium between B and C; the condition of thermodynamic equilibrium between them may be symbolically represented as follows,

$$\text{If } A = B \text{ and } B = C \quad (1)$$

This observation has also been schematically shown in Fig.4

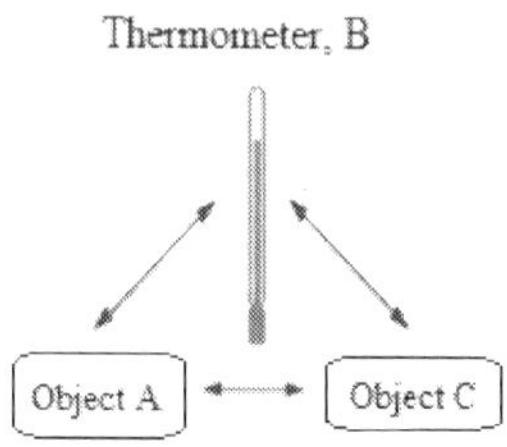

Figure 4. A schematic presentation of the zeroth law of thermodynamics.

Based on preceding observations, some of the physical properties of the system B (Thermometer) can be used as a measure of the temperature, such as the volume of a fixed amount of a liquid merqury or any alcohol under standard atmospheric pressure. The **zeroth law of thermodynamics** is the assurance of the existence of a property called the temperature [11].

The zeroth law allows us to assert the existence of temperature as a state function. Having defined temperature, how do we measure it? Of course, you are familiar with the process of putting a liquid-mercury thermometer in contact with a system, waiting until the volume change of the mercury has ceased, indicating that thermal equilibrium between the thermometer and the system has been reached [5].

It is necessary to know four common different temperature scales, namely Fahrenheit (^{0}F), Celsius (^{0}C), Kelvin (K) and Rankine (^{0}R). To convert these scales one another can be used the following equations [9,10].

$$T / K = \theta / {}^0C + 273.15 \quad (2)$$

$$[oR] = [oF] + 459.67 [oF] = 1.8 [oC] + 32 [oR] = 1.8 [K] \quad (3)$$

2. The first law of thermodynamic (the conservation of energy)

In thermodynamics, the total energy of a system is called its **internal energy**, U. The internal energy is the total kinetic and potential energies of the particles in the system. It is denoted by ΔU the change in internal energy when a system changes from an initial state i with internal energy U_i to a final state of internal energy U_f :

$$\Delta U = U_f - U_i \quad (4)$$

The internal energy is a state function in the sense that its value depends only on the current state of the system and is independent of how that state has been prepared. In other words, internal energy is a function of the properties as variables that determine the current state of the system. Changing any one of the state variables, such as the pressure and temperature, results in a change in internal energy. The internal energy is a state function that has consequences of the greatest importance. The internal energy is an extensive property of a system and is measured in joules (1 J = 1 kg m^2 s^{-2}). The molar internal energy, U_m, is the internal energy divided by the amount of substance in a system, $U_m = U/n$; it is an intensive property and commonly reported in kilojoules per mole (kJ mol^{-1}) [12].

A particle has a certain number of motional degrees of freedom, such as the ability to translate (the motion of its centre of mass through space), rotate around its centre of mass, or vibrate (as its bond lengths and angles change, leaving its centre of mass unmoved). Many physical and chemical properties depend on the energy associated with each of these modes of motion. For example, a chemical bond might break if a lot of energy becomes concentrated in it, for instance as vigorous vibration. According to it, the average energy of each quadratic contribution to the energy is 1/2 kT. The mean energy of the atoms free to move in three dimensions is kT and the total energy of a monatomic perfect gas is NkT, or nRT (because $N = nNA$ and $R = NAk$, NA: Avogadro's number and k: Boltzman's constant). I t can therefore be written;

$$U_m(T) = U_m(0) + RT(\textit{monatomic gas; translation only}) \tag{5}$$

where $U_m(0)$ is the molar internal energy at $T = 0$, when all translational motions have ceased and the sole contribution to the internal energy arises from the internal structure of the atoms. This equation shows that the internal energy of a perfect gas increases linearly with temperature. At 25°C, 3/2 RT = 3.7 kJ mol^{-1}, so translational motion contributes about 4 kJ mol^{-1} to the molar internal energy of a gaseous sample of atoms or molecules.

When the gas consists of molecules, we need to take into account the effect of rotation and vibration. A linear molecule, such as N_2 and CO_2, can rotate around two axes perpendicular to the line of the atoms, so it has two rotational modes of motion, each contributing a term 1/2 kT to the internal energy. Therefore, the mean rotational energy is kT and the rotational contribution to the molar internal energy is RT.

$$U_m(T) = U_m(0) + 5/2RT(\textit{linear molecule; translation and rotation only}) \tag{6}$$

A nonlinear molecule, such as CH_4 or H_2O, can rotate around three axes and, again, each mode of motion contributes a term 1/2 kT to the internal energy. Therefore, the mean rotational energy is 3/2 kT and there is a rotational contribution of 3/2 RT to the molar internal energy. That is,

$$U_m(T) = U_m(0) + 3RT(\textit{nonlinear molecule; translation and rotation only}) \tag{7}$$

The internal energy now increases twice as rapidly with temperature compared with the monatomic gas. Another way: for a gas consisting of 1 mol of nonlinear molecules to undergo the same rise in temperature as 1 mol of monatomic gas, twice as more energy must be supplied. Molecules do not vibrate significantly at room temperature and, as a first approximation; the contribution of molecular vibrations to the internal energy is negligible except for very large molecules such as polymers and biological macromolecules. None of the expressions we have been derived depends on the volume occupied by the molecules: there are no intermolecular interactions in a perfect gas, so the distance between the molecules has no effect on the energy. That is, the internal energy of a perfect gas is independent of the volume it occupies. The internal energy of interacting molecules in condensed phases also has a contribution from the potential energy of their interaction. However, no simple expressions can be written down in general. Nevertheless, the crucial molecular point is that, as the temperature of a system is raised, the internal energy increases as the various modes of motion become more highly excited [12].

By considering how the internal energy varies with temperature when the pressure of the system is kept constant; many useful results and also some unfamiliar quantities can be obtained. If it is divided both sides of eqn ($(dU = (\partial U/\partial V)_T\, dV + (\partial U/\partial T)_v\, dT$) ($(\partial U/\partial V)_T = \pi_T$ and π_T called as **the internal pressure** ; $(\partial U/\partial T)_v = C_v$ and it is called as **heat capacity at constant volume**)) by dT and impose the condition of constant pressure on the resulting differentials, so that dU/dT on the left becomes $(\partial U/\partial T)_p$, So;

$$\left(\partial U / \partial T\right)_p = \pi_T \left(\partial V / \partial T\right)_p + C_v \tag{8}$$

It is usually sensible in thermodynamics to inspect the output of a manipulation like this to see if it contains any recognizable physical quantity. The differential coefficient on the right in this expression is the slope of the plot of volume against temperature (at constant pressure). This property is normally identified as thermal expansion coefficient, α, of a substance, which is defined as

$$\alpha = 1/V \left(\partial V / \partial T\right)_p \tag{9}$$

When it is introduced the definition of α into the equation for $(\partial U/\partial T)_p = \alpha\pi_T V + C_v$,this equation is entirely general (provided the system is closed and its composition is constant). It expresses the dependence of the internal energy on the temperature at constant pressure in terms of C_v, which can be measured in one experiment, in terms of α, which can be measured in another, and in terms of the quantity π_T, for a perfect gas, $\pi_T = 0$, so

$$\left(\partial U / \partial T\right)_p = C_v \tag{10}$$

That is, the constant-volume heat capacity of a perfect gas is equal to the slope of a plot of internal energy against temperature at constant pressure as well as (by definition) to the slope at constant volume. It can also be predicted that the change of internal energy with temperature at constant pressure means a total energy change raised from increase in both energy of kinetics and potential energy of particles in higher temperature. The translational motion of particles against constant external pressure will lead to expansion, but thermal expansion characteristics of substance control its magnitude. Thus, we can see that heating in constant volume only changes internal energy as q_v ($\Delta U = q_v$), whereas its change in constant pressure additionally includes changing of potential energy of particles due to translation motion. That is, changes in constant pressure require a different definition of the transferred energy [12].

3. From conservation of energy to heat and work

It has been found experimentally that the internal energy of a system may be changed either by doing work on the system or by heating it. Whereas we may know how the energy transfer has occurred (because we can see if a weight has been raised or lowered in the surroundings, indicating transfer of energy by doing work, or if ice has melted in the surroundings, indicating transfer of energy as heat), the system is blind to the mode employed. Heat and work are equivalent ways of changing a system's internal energy. It is also found experimentally that, if a system is isolated from its surroundings, then no change in internal energy takes place. This summary of observations is now known as the **First Law of thermodynamics** or the **Conservation of Energy** and is expressed as follows [1;12].

The internal energy of an isolated system is constant.

$$\Delta U = 0 \tag{11}$$

A system cannot be used to do work, leave it isolated, and then come back expecting to find it restored to its original state with the same capacity for doing work. The experimental evidence for this observation is that no 'perpetual motion machine', a machine that does work without consuming fuel or using some other source of energy, has ever been built. These remarks may be summarized as follows. If we write w for the work done on a system, q for the energy transferred as heat to a system, and ΔU for the resulting change in internal energy, and then it follows that

$$\Delta U = q + w \tag{12}$$

Equation summarizes the equivalence of heat and work and the fact that the internal energy is constant in an isolated system (for which $q = 0$ and $w = 0$). The equation states that the change in internal energy of a closed system is equal to the energy that passes through its boundary as heat or work. It employs the 'acquisitive convention', in which w and q are positive if energy is transferred to the system as work or heat and are negative if energy is lost from the system. In other words, we view the flow of energy as work or heat from the system's perspective [1;12].

Heat (q)

Heat flows by virtue of a temperature difference. Heat will flow until temperature gradients disappear [8]. When a heater is immersed in a beaker of water (the system), the capacity of the system to do work increases because hot water can be used to do more work than cold water [1].

An **exothermic process** is a process that releases energy as heat. All combustion reactions are exothermic. An **endothermic process** is a process that absorbs energy as heat. An example of an endothermic process is the vaporization of water. An endothermic process in a diathermic container results in energy flowing into the system as heat. An exothermic process in a similar diathermic container results in a release of energy as heat into the surroundings. If an endothermic process in nature was taken place by a system divided an adiabatic boundary from its surroundings, a lowering of temperature of the system results; conversely if the process is exothermic, temperature rises [1]. These are the expected results of conversation of energy.

Work (*w*)

Energy is the essence of our existence as individuals and as a society. Just as energy is important to our society on a macroscopic scale, it is critically important to each living organism on a microscopic scale. The living cell is a miniature chemical factory powered by energy from chemical reactions. The process of cellular respiration extracts the energy stored in sugars and other nutrients to drive the various tasks of the cell. Although the extraction process is more complex and more subtle, the energy obtained from "fuel" molecules by the cell is the same as would be obtained from burning the fuel to power an internal combustion engine [3].

The fundamental physical property in thermodynamics is work is done when an object is moved against an opposing force. Doing work is equivalent into raising a weight somewhere in the surrounding. An example of doing work is the expansion of a gas that pushes out a piston and raises a weight. A chemical reaction that derives an electric current through a resistance also does work, because the same current could be driven through a motor and used to raise a weight [1].

Work is the transferred energy by virtue of a difference in mechanical properties from a boundary between the system and the surroundings. There are many types of work; such as mechanical work, electrical work, magnetic work, and surface tension [8].

The SI unit of both heat and work (kg m^2/s^2) is given the name joule (J), after the English physicist James Prescott Joule (1818-1889) [6].

$$1\ J = 1\ kg\ m^2/s^2$$

In addition to the SI unit joule, some chemist's still use the unit calorie (cal). Originally defined as the amount of energy necessary to raise the temperature of 1 g of water by 1°C (specially, from 14.5 °C to 15.5 °C), one calorie is now defined as exactly 4,184 J [6].

The work–energy theorem

In mechanics, the work–energy theorem demonstrates that the total work done on a system is transformed into kinetic energy. This is represented in a very simple and meaningful equation as follows:

$$W_{total} = \Delta E_k \tag{13}$$

in which, W_{total} is the total work done on the system, including the work carried out by all the external forces (Wex), as well as the work developed by the internal forces within the system (W_{in}). Thus;

$$W_{total} = W_{ex} + W_{in} \tag{14}$$

Now, if the external work is separated into two terms, namely the work done by the external conservative forces ($W_{ex,c}$), which are associated with an external potential energy ($Ep_{,ex}$), and the non-conservative external work ($W_{non\text{-}cex}$), the whole external work can be written as:

$$W_{ex} = W_{ex,c} + W_{non-cex} = -\Delta E_{p,ex} + W_{non-cex} \tag{15}$$

Similarly, the work developed by the internal forces within the system can be also expressed as the sum of a conservative work term plus the non-conservative internal work. Thus;

$$W_{in} = W_{in,c} + W_{non-c'in} = -\Delta E_{p,in} + W_{non-cin} \tag{16}$$

Where, $\Delta E_{p,in}$ is the internal potential energy of the system.

As for the kinetic energy of a system, mechanics shows that it can be considered as consisting of two terms, as follows:

$$E_k = 1/2Mv^2_{CM} + E_{k,CM} \tag{17}$$

M being the total mass of the system,v_{CM} the velocity of its center of mass, and $E_{k,CM}$ the kinetic energy of the system with respect to its center of mass. The first term on the right-hand side of equation above represents the kinetic energy of the center of mass of the system, as if it had the mass of the whole system. Thus, as the velocity v_{CM} is taken with respect to an external reference frame, this first term can be called the external kinetic energy of the system ($E_{k,ex}$), whereas the second term would be its internal kinetic energy ($E_{k,in}$). Accordingly, the increase of the kinetic energy of a system can be written in the following way:

$$\Delta E_k = \Delta E_{k,ex} + \Delta E_{k,in} \tag{18}$$

Then, the substitution of equations 14, 15, 16 and 18 into equation 13 allows us to order terms as follows:

$$W_{ex} = \Delta E_{p,ex} + \Delta E_{k,ex} + \Delta E_{p,in} + \Delta E_{k,in} - W_{in} \tag{19}$$

Equation 19 is a general developed expression of the work–energy theorem derived from mechanics. It should be noticed that, though it does not describe the details of the energetic terms, each of them is explicitly stated, which will be of great help both to define and to understand the contribution of thermodynamics when establishing the first law [17].

Reversibility and Reversible changes

A reversible change in thermodynamics is a change that can be reversed by an infinitesimal modification of a variable. One example of reversibility that we have encountered already is the thermal equilibrium of two systems with the same temperature. The transfer of energy as heat between the two is reversible because, if the temperature of either system is lowered infinitesimally, then energy flows into the system with the lower temperature. If the temperature of either system at thermal equilibrium is raised infinitesimally, then energy flows out of the hotter system. There is obviously a very close relationship between reversibility and equilibrium: systems at equilibrium are poised to undergo reversible change. Suppose a gas is confined by a piston and that the external pressure, p_{ex}, is set equal to the pressure, p, of the confined gas. Such a system is in mechanical equilibrium with its surroundings because an infinitesimal change in the external pressure in either direction causes changes in volume in opposite directions. If the external pressure is reduced infinitesimally, the gas expands slightly. If the external pressure is increased infinitesimally, the gas contracts slightly. In either case the change is reversible in the thermodynamic sense. If, on the other hand, the external pressure differs measurably from the internal pressure, then changing p_{ex} infinitesimally will not decrease it below the pressure of the gas, so will not change the direction of the process. Such a system is not in mechanical equilibrium with its surroundings and the expansion is thermodynamically irreversible [12].

To achieve reversible expansion we set p_{ex} equal to p at each stage of the expansion. In practice, this equalization could be achieved by gradually removing weights from the piston so that the downward force due to the weights always matches the changing upward force due to the pressure of the gas. When we set $p_{ex} = p$, eqn ($dw = -p_{ex}dV$) becomes

$$dw = -p_{ex}dV = -pdV \qquad \textit{reversible expansion work} \qquad (20)$$

(Equations valid only for reversible processes are labeled with a subscript rev.) Although the pressure inside the system appears in this expression for the work, it does so only because p_{ex} has been set equal to p to ensure reversibility. The total work of reversible expansion from an initial volume V_i to a final volume V_f is therefore

$$w = -\int_{Vi}^{Vf} pdV \qquad (21)$$

The integral can be evaluated once it is known how the pressure of the confined gas depends on its volume [12].

At the isothermal, reversible expansion of a perfect gas, the work made by keeping the system in thermal contact with its surroundings can be stated as follows;

$$w = -nRT\int_{Vi}^{Vf} dV / V = -nRT\ln V_f / V_i \qquad (22)$$

When the final volume is greater than the initial volume, as in an expansion, the logarithm in Eqn. 22 is positive and hence $w < 0$. In this case, the system has done work on the surroundings and there is a corresponding reduction in its internal energy, but due to there is a compensating influx of energy as heat, overall the internal energy is constant for the isothermal expansion of a perfect gas, which clearly indicates **the Conservation of Energy Law.** The equations also show that more work is done for a given change of volume when the temperature is increased: at a higher temperature the greater pressure of the confined gas needs a higher opposing pressure to ensure reversibility and the work done is correspondingly greater (Fig. 5) [12].

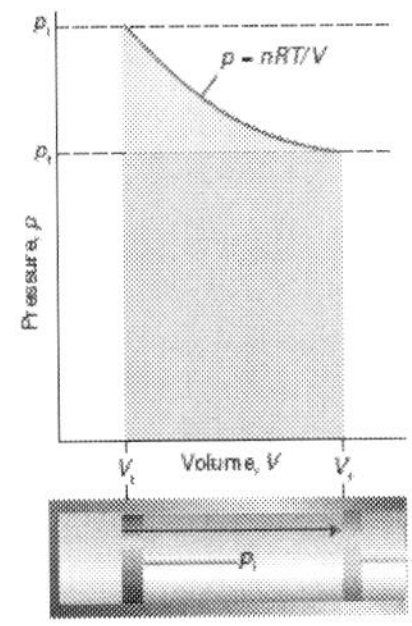

Figure 5. The work done by a perfect gas when it expands reversibly and isothermally is equal to the area under the isotherm $p = nRT/V$. The work done during the irreversible expansion against the same final pressure is equal to the rectangular area shown slightly darker. It can be seen that the reversible work is greater than the irreversible work [12].

Adiabatic changes

We are now equipped to deal with the changes that occur when a perfect gas expands adiabatically. A decrease in temperature should be expected: because work is done but no heat enters the system, the internal energy decrease, and therefore the temperature of the working gas also decrease. In molecular terms, the kinetic energy of the particles decrease as work is done, so their average speed decreases, and hence the temperature decrease. This means that in the case of perfect gas, change in the distance between particles cannot be responsible for the changing of internal energy but **the Conservation of Energy Law** requires a measurable reduction in kinetic energy of particles, i.e. a reduction in their velocities. The change in internal energy of a perfect gas when the temperature is changed from T_i to T_f and the volume is changed from V_i to V_f can be expressed as the sum of two steps (Fig. 6).

In the first step, only the volume changes and the temperature is held constant at its initial value. However, because the internal energy of a perfect gas is independent of the volume the molecules occupy, the overall change in internal energy arises solely from the second step, the change in temperature at constant volume. Provided the heat capacity is independent of temperature, this change is

$$\Delta U = C_V\left(T_f - T_i\right) = C_V \Delta T \tag{23}$$

Because the expansion is adiabatic, we know that $q = 0$; because $\Delta U = q + w$, it then follows that ΔU = ad. The subscript 'ad' denotes an adiabatic process. Therefore, by equating the two expressions we have obtained for ΔU, we obtain

$$W_{ad} = C_V \Delta T \tag{24}$$

Since internal energy, pressure, and volume are all state functions, enthalpy is also a state function. But what exactly is enthalpy? To help answer this question, consider a process carried out at constant pressure and where the only work allowed is pressure-volume work ($w = - p\Delta V$). Under this conditions, the expression

$$\Delta U = q_P + w \tag{33}$$

becomes

$$\Delta U = q_p - p\Delta V \tag{34}$$

or

$$q_p = \Delta U + p\Delta V \tag{35}$$

where q_P, ΔU and $p\Delta V$ are the transferred energy as heat to the system heated at constant pressure, change in internal energy and change in pV, respectively. Since p is constant; the change in pV is due only to a change in volume. Thus

$$(pV) = p\,\Delta V \tag{36}$$

$$\Delta H = q_p \tag{37}$$

Heat capacity in terms of derivative at constant pressure, i.e. changes in the energy of a system heated at constant pressure is also expressed as follows:

$$C_p = (\partial H / \partial T)_p \tag{38}$$

The slope of a plot of internal energy against temperature at constant volume, for a perfect gas C_V is also the slope at constant pressure. In order to obtain an easy way to derive the relation between Cp and C_V for a perfect gas, both heat capacities can be used in terms of derivatives at constant pressure:

$$Cp - Cv = (\partial H / \partial T)p - (\partial U / \partial T)v = (\partial H / \partial T)p - (\partial U / \partial T)p \tag{39}$$

Then, if ($H = U + pV = U + nRT$) is introduced into the first term, which results in

$$Cp - Cv = (\partial U / \partial T)p + nR - (\partial U / \partial T)p \quad \text{and}\left(C_p - C_v = nR\right) \tag{40}$$

This means that in the case of a perfect gas, R, universal gas constant may be considered as the work done to push back the atmosphere per unit increase in temperature. However, the general relation between the two heat capacities for any pure substance is demonstrated as follows;

$$C_p - C_v = \alpha^2 TV / \kappa_T \tag{41}$$

This formula is a thermodynamic expression, which means that it applies to any substance (that is, it is universally true). It reduces to previous equation for a perfect gas when it was set as $\alpha = 1/T$ and $\kappa_T = 1/p$

Because thermal expansion coefficients, α, of liquids and solids are small, it is tempting to deduce from last equation that for them $Cp \approx Cv$. But this is not always so, because the compressibility κ_T might also be small, so α^2/κ_T might be large. That is, although only a little work need be done to push back the atmosphere, a great deal of work may have to be done to pull atoms apart from one another as the solid expands. As an illustration, for water at 25 ^{0}C, Eqn (41) gives $C_p = 75.3$ J/Kmol compared with $C_v = 74.8$ J/Kmol. In some cases, the two heat capacities differ by as much as 30 per cent [21]. The constant-pressure heat capacity C_p differs from the constant-volume heat capacity C_v by the work needed to change the volume of the system to maintain constant pressure. This work arises in two ways. One is the work of driving back the atmosphere (external work); the other is the work of stretching the bonds in the material, including any weak intermolecular interactions (internal work). In the case of a perfect gas, the second makes no contribution. This suggests that the difference between two heat capacities is related to both internal work and external work done by the particles of a substance as an expended result of the conservation of energy.

By considering the variation of H with temperature at constant volume, the validity of the enthalpy function can be differently verified. Firstly; for a closed system of constant composition, H is expressed in the total differential of T and p;

$$dH = ((\partial H / \partial p)_T dp + (\partial H / \partial T)_p dT \tag{42}$$

and then, divided this equation though by dT;

$$(\partial H / \partial T)_v = (\partial H / \partial p)_T (\partial p / \partial T)_v + C_p \tag{43}$$

The manipulation of this expression provides more involved equation which can be applied to any substance. Because all the quantities that appear in it can be measured in suitable experiments.

$$(\partial H / \partial T)_v = (1 - \alpha\mu / \kappa_T)\, C_p \tag{44}$$

where the isothermal compressibility, κ_T, is defined as

$$\kappa_T = -1 / V\ (\partial V / \partial p)_T \tag{45}$$

and the Joule – Thomson coefficient, μ, is defined as

$$\mu = (\partial T / \partial p)_H \tag{46}$$

This expression derived for the changing of the enthalpy with temperature at constant volume suggests that change in H with increased temperature at constant V is lower than that at constant p and the difference between them depend on some characteristic properties of particles, such as κ_T, μ and α,, indicating its relation with the absence of external work.

The measurement of an enthalpy change

An enthalpy change can be measured calorimetrically by monitoring the temperature change that accompanies a physical or chemical change occurring at constant pressure. A calorimeter for studying processes at constant pressure is called an isobaric calorimeter. A simple example is a thermally insulated vessel open to the atmosphere: the heat released in the reaction is monitored by measuring the change in temperature of the contents. For a combustion reaction an adiabatic flame calorimeter may be used to measure ΔT when a given amount of substance burns in a supply of oxygen. Another route to ΔH is to measure the internal energy change using a bomb calorimeter, and then to convert ΔU to ΔH. Because solids and liquids have small molar volumes, for them both pV_m , external work and internal work is so small that the molar enthalpy and molar internal energy are almost identical ($H_m = U_m + pV_m \approx U_m$). Consequently, if a process involves only solids or liquids, the values of ΔH and ΔU are almost identical. Physically, such processes are accompanied by a very small change in volume; the system does negligible work on the surroundings when the process occurs, so the energy supplied as heat stays entirely within the system, as a expended result of the conversation of energy.

Calorimeters

The heat that is given out or taken in, when a chemical reaction occurs can be measured using a calorimeter. A simple, constant-pressure calorimeter (Coffee-cup calorimeter) for measuring heat for reactions in solution is shown in Fig. 8. This figure also shows a bomb calorimeter. The container is an expanded polystyrene cup with a lid. This material provides insulation which ensures that heat loss to, or gains from the surroundings is minimized; the outer cup in Fig. 8 provides additional insulation. As the reaction takes place, the thermometer records any change in temperature.

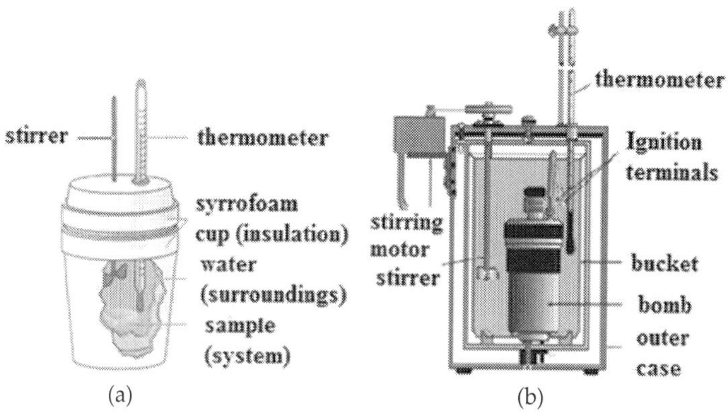

Figure 8. Calorimeter types; (a) Coffee-cup calorimeter [18] and (b) Bomb calorimeter [22].

The relationship between the temperature change and the energy is given as follows;

$$\textit{Energy change in } J = m\,(g).C\,(J/gK).\,\Delta T\,(K) \tag{47}$$

where C is the specific heat capacity of the solution. Since the reaction is carried out at constant pressure, the energy is equal to the enthalpy change. For dilute aqueous solutions, it is usually sufficient to assume that the specific heat capacity of the solution is the same as for water: C_{water} 4.18 J/gK. It is assumed that no heat is used to change the temperature of the calorimeter itself. Where a calorimeter is made from expanded polystyrene cups, this is a reasonable assumption because the specific heat capacity of the calorimeter material is so small. However, the approximation is not valid for many types of calorimeter and such pieces of apparatus must be calibrated before use. Measurements made in the crude apparatus shown together with in Figure 8 are not accurate, and more specialized calorimeters, such as **bomb calorimeter** must be used if accurate results are required [18;19]. A bomb calorimeter is ideally suited for measuring the heat evolved in a combustion reaction. The system is everything within the double-walled outer jacket of the calorimeter. This includes the bomb and its contents, the water in which the bomb is immersed, the thermometer, the stirrer, and so on. Before using above equation, it must be emphasized that a rise in temperature of a system insulated with its surrounding does not occur by any heat transferred from the surroundings to the system because of the temperature difference between them. The difference in temperature observed during the measuring may be arisen from the change in the composition of system and considered as if the involved reaction occurs as exothermic or endothermic in diathermic condition. At constant pressure (where only pV work is allowed), the change in enthalpy ΔH of the system is equal to the energy flow as heat. This means that for a reaction studied at constant pressure, the flow of heat is a measure of the change in enthalpy for the system. For this reason, the terms heat of reaction and change in enthalpy are used interchangeably for reactions studied at constant pressure. For a chemical reaction, the enthalpy change is given by the equation

$$\Delta H = H_{products} - H_{reactants} \tag{48}$$

In a case in which the products of a reaction have a greater enthalpy than the reactants, ΔH will be positive. Thus heat will be absorbed by the system, and the reaction is **endothermic**. On the other hand, if the enthalpy of the products is less than that of the reactants, ΔH will be negative. In this case the overall decrease in enthalpy is achieved by the generation of heat, and the reaction is **exothermic** [3]. Energy diagrams for exothermic and endothermic reactions are shown in Figure 9.

Standard enthalpy change

The standard enthalpy change of a reaction refers to the enthalpy change when all the reactants and products are in their **standard states**. The notation for this thermochemical quantity is $\Delta_r H^0$ (T) where the subscript 'r' stands for 'reaction', the superscript 'o' means 'standard state conditions', and (T) means 'at temperature T'. This type of notation is found for other thermodynamic functions that we meet later on.

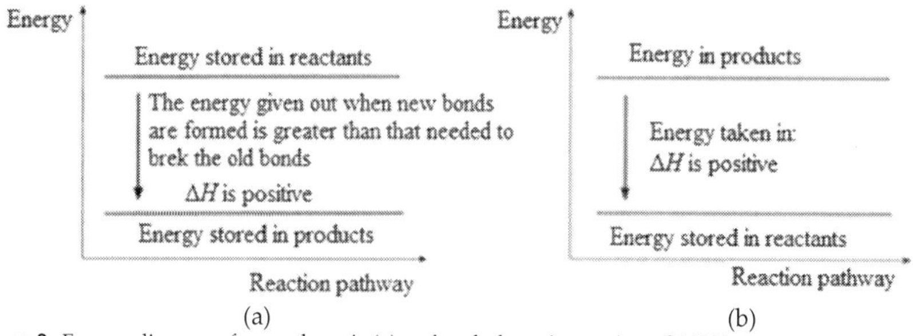

Figure 9. Energy diagrams for exothermic (a) and endothermic reactions (b) [20].

The standard state of a substance is its most thermodynamically stable state under a pressure of 1 bar (1.00 x 10^5 Pa) and at a specified temperature, T. Most commonly, T = 298.15 K, and the notation for the standard enthalpy change of a reaction at 298.15K is then $\Delta_r H^0$ (298.15 K). It is usually sufficient to write $\Delta_r H^0$ (298 K) [19].

Standard enthalpies of formation

The standard enthalpy of formation (ΔH_f^0) of a compound is defined as the change in enthalpy that accompanies the formation of one mole of a compound from its elements with all substances in their reference states. The reference state of a element is its most thermodynamically stable state under a pressure of 1 bar (1.00 x 10^5 Pa) and at a specified temperature, T.

A degree symbol on a thermodynamic function, for example, ΔH, indicates that the corresponding process has been carried out under standard conditions. The standard state for a substance is a precisely defined reference state. Because thermodynamic functions often depend on the concentrations (or pressures) of the substances involved, it must be used a common reference state to properly compare the thermodynamic properties of two substances. This is especially important because, for most thermodynamic properties, it can be measured only *changes* in the property

Enthalpy is a state function, so it can be chosen *any* convenient pathway from reactants to products and then sums the enthalpy changes along the chosen pathway. A convenient pathway, shown in Fig. 10, involves taking the reactants apart to the respective elements in their reference states in reactions (a) and (b) and then forming the products from these elements in reactions (c) and (d). This general pathway will work for any reaction, since atoms are conserved in a chemical reaction [3].

From Fig. 10, it can be seen that reaction (a), where methane is taken apart into its elements,

$$CH_4(g) \rightarrow C(s) + 2H_2(g) \tag{49}$$

is just the reverse of the formation reaction for methane:

$$C(s) + 2H_2(g) \rightarrow CH_4(g) \; \Delta H^0{}_f = -75 \; kJ/mol \tag{50}$$

Since reversing a reaction means changing the sign of ΔH, but keeping the magnitude the same, ΔH for reaction (a) is $-\Delta H^0{}_f$, *or 75 kJ. Thus* $\Delta H^0\,(a) = 75$ *kJ.*

It can be secondarily considered reaction (b). Here oxygen is already an element in its reference state, so no change is needed. Thus $\Delta H^0\,(b) = 0$.

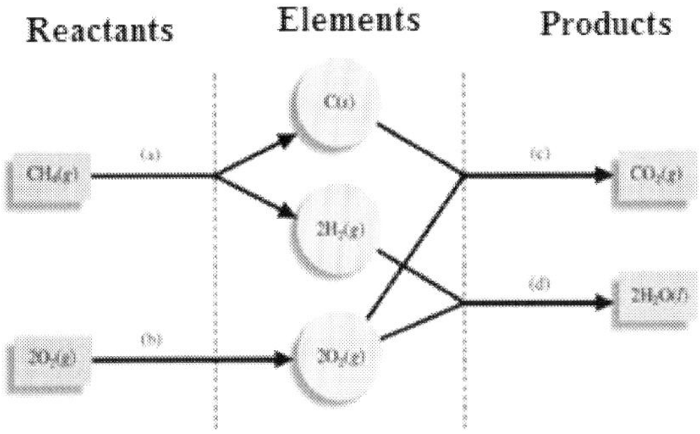

Figure 10. In this pathway for the combustion of methane, the reactants are first taken apart in reactions (a) and (b) to form the constituent elements in their reference states, which are then used to assemble the products in reactions (c) and (d) [3].

The next steps, reactions (c) and (d), use the elements formed in reactions (a) and (b) to form the products. That is, reaction (c) is simply the formation reaction for carbon dioxide:

$$C(s) + O_2(g) \rightarrow CO_2(g) \quad \Delta H^0{}_f = -394\ kJ\,/\,mol \tag{51}$$

and

$$\Delta H^0{}_{(c)} = \Delta H^0{}_f \ for\ CO_2(g) = -394\ kJ\,/\,mol \tag{52}$$

Reaction (d) is the formation reaction for water:

$$H_2(g) + 1/2O_2(g) \rightarrow H_2O\ (l) \quad \Delta H^0{}_f = -286\ kJ\,/\,mol \tag{53}$$

However, since 2 moles of water are required in the balanced equation, it must be formed 2 moles of water from the elements:

$$2H_2(g) + O_2(g) \rightarrow 2H_2O \tag{54}$$

Thus

$$\Delta H^0{}_{(d)} = 2\ x\ \Delta H^0{}_f \ for\ H_2O\ (l) = 2(-286\ kJ\,/) = -572\ kJ\,/\,mol \tag{55}$$

It has now been completed the pathway from the reactants to the products. The change in enthalpy for the reaction is the sum of the ΔH values (including their signs) for the steps:

$$\Delta H^0{}_{reaction} = \Delta H^0{}_{(a)} + \Delta H^0{}_{(b)} + \Delta H^0{}_{(c)} + \Delta H^0{}_{(d)} \tag{56}$$

$$\begin{aligned} &= \left[-\Delta H^0{}_f \text{ for } CH_4(g)\right] + 0 + \left[\Delta H^0{}_f \text{ for } CO_2(g)\right] + \left[2\,x\,\Delta H^0{}_f \text{ for } H_2O\,(l)\right] \\ &= -\left(-75\ kJ\right) + 0 + \left(-394kJ\right) + \left(-572\ kJ\right) = -891\ kJ \end{aligned} \tag{57}$$

This process is diagramed in Fig. 11. It can be seen that the reactants are taken apart and converted to elements [not necessary for O_2 $_{(g)}$] that are then used to form products. It can be seen that this is a very exothermic reaction because very little energy is required to convert the reactants to the respective elements but a great deal of energy is released when these elements form the products. This is why this reaction is so useful for producing heat to warm homes and offices. If it is examined carefully the pathway used in this example, it can be understood that first, the reactants were broken down into the elements in their reference states and then the products were then constructed from these elements. This involved formation reactions and thus enthalpies of formation. The entire process can be summarized as follows:

$$\Delta H^0{}_{reaction} = \sum \nu_p \Delta H^0{}_{f\ (products)} - \sum \nu_p \Delta H^0{}_{f\ (reactants)} \tag{58}$$

where, ν is stoichiometric coefficients for both reactants and products. The enthalpy change for a given reaction can be calculated by subtracting the enthalpies of formation of the reactants from the enthalpies of formation of the products.

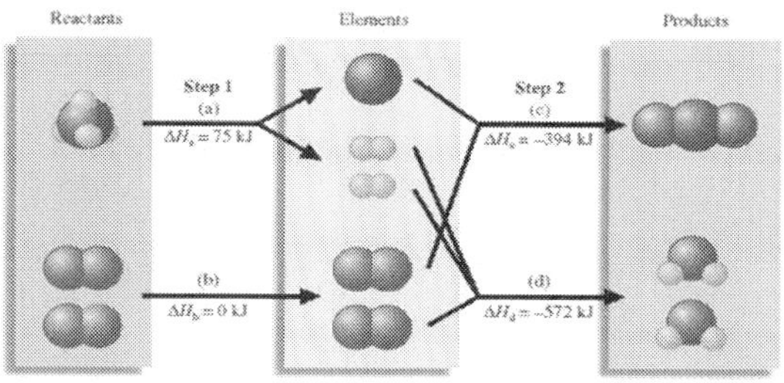

Figure 11. A schematic diagram of the energy changes for the reaction $CH_4\,(g) + 2O_2\,(g) \rightarrow CO_2\,(g) + 2H_2O\,(l)$ [3].

Hess's law

Another way to calculate values of ΔH for reactions involves manipulating equations for other reactions with known ΔH values. When chemical equations are added to yield a different chemical equation, the corresponding ΔH values are added to get the ΔH for the desired equation. This principle is called **Hess's law** and it is an application of the **first law of thermodynamic** or the **conservation of energy**. For example, it can be calculated the ΔH for the reaction of carbon with oxygen gas to yield carbon dioxide from the values for the reaction of carbon with oxygen to yield carbon monoxide and that of carbon monoxide plus oxygen to yield carbon dioxide [9].

Desired

$$C(s) + O_2(g) \rightarrow CO_2(g) \tag{59}$$

Given

$$C(s) + 1/2\, O_2(g) \rightarrow CO(g) \qquad \Delta H = -110\,kJ \tag{60}$$

$$CO(g) + 1/2\, O_2(g) \rightarrow CO_2(g) \quad \Delta H = -283\,kJ \tag{61}$$

Adding the two chemical equations given:

$$C_{(s)} + CO_{(g)} + O_{2(g)} \rightarrow CO_{2(g)} + CO_{(g)} \tag{62}$$

Eliminating CO from both sides results in the desired equation:

$$C(s) + O_2(g) \rightarrow CO_2(g) \tag{63}$$

Therefore, adding these two ΔH values will give the ΔH desired:

$$\Delta H = (-110\ kJ) + (-283\ kJ) = -393\ kJ \tag{64}$$

It must be noticed that enthalpies of formation have not been used explicitly in this process [9]. The Hess's law is shown schematically in Fig 12 [3].

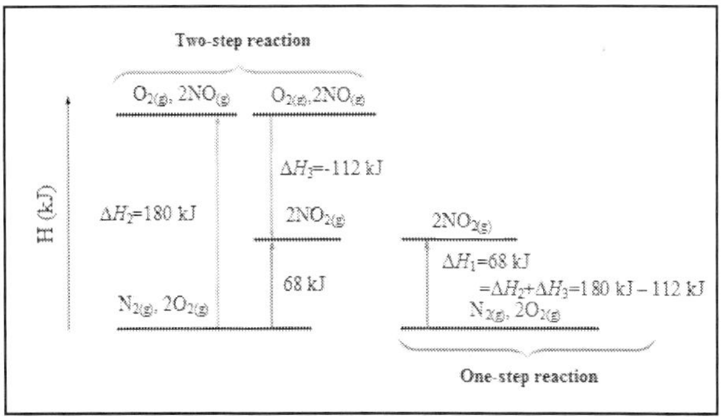

Figure 12. A schematic diagram of Hess's law. The same change in enthalpy occurs when nitrogen and oxygen react to form nitrogen dioxide, regardless of whether the reaction occurs in one (red) or two (blue) steps [3].

The overall reaction can be written in one step, where the enthalpy change is represented by ΔH_1 [3].

$$N_2(g) + 2O_2(g) \rightarrow 2NO_2(g) \quad \Delta H_1 = 68\ kJ \tag{65}$$

This reaction also can be carried out in two distinct steps, with enthalpy changes designated by ΔH_2 and ΔH_3:

$$N_2(g) + O_2(g) \rightarrow 2NO\,(g) \quad \Delta H_2 = 180\ kJ \tag{66}$$

$$2NO\,(g) + O_2(g) \rightarrow 2NO_2(g) \quad \Delta H_3 = -112\ kJ \tag{67}$$

Net reaction:

$$N_2(g) + 2O_2(g) \rightarrow 2NO_2(g) \quad \Delta H_2 + \Delta H_3 = 68\ kJ \tag{68}$$

The sum of the two steps gives the net, or overall, reaction and that

$$\Delta H_1 = \Delta H_2 + \Delta H_3 = 68\ kJ \tag{69}$$

The temperature dependence of reaction enthalpies

The standard enthalpies of many important reactions have been measured at different temperatures. However, in the absence of this information, standard reaction enthalpies at different temperatures may be calculated from heat capacities and the reaction enthalpy at some other temperature (Fig. 13). In many cases heat capacity data are more accurate than reaction enthalpies. It follows from eqn ($dH = C_p dT$ at constant pressure) that, when a substance is heated from T_1 to T_2, its enthalpy changes from

$H\,(T_1)$ to

$$H(T_2) = H(T_1) + \int_{T1}^{T2} Cp dT \tag{70}$$

(It has been assumed that no phase transition takes place in the temperature range of interest.) Because this equation applies to each substance in the reaction, the Standard reaction enthalpy changes from $\Delta H^0_r\,(T_1)$ to

$$\Delta H_r^{\,0}(T_2) = \Delta H_r^{\,0}(T_1) + \int_{T1}^{T2} \Delta Cp dT \tag{71}$$

where ΔCp is the difference of the molar heat capacities of products and reactants under standard conditions weighted by the stoichiometric coefficients that appear in the chemical equation [12];

$$\Delta C_p = \sum \nu C_{p(Products)} - \sum \nu C_{p(Reactants)} \tag{72}$$

Equation 71 is known as **Kirchhoff's law**. It is normally a good approximation to assume that ΔCp_r is independent of the temperature, at least over reasonably limited ranges. Although the individual heat capacities may vary, their difference varies less significantly. In some cases the temperature dependence of heat capacities is taken into account by using equation below.

$$C_p = a + bT + c / T^2 \tag{73}$$

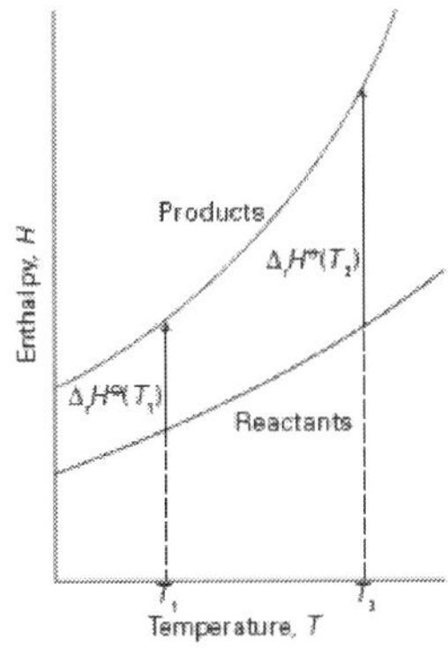

Figure 13. An illustration of the content of Kirchhoff's law. When the temperature is increased, the enthalpy of both the products and the reactants increases. In each case, the change in enthalpy depends on the heat capacities of the substances. The change in reaction enthalpy reflects the difference in the changes of the enthalpies [12].

Author details

Ahmet Gürses

Ataturk University, K.K. Education Faculty, Department of Chemistry, Erzurum, Turkey

Mehtap Ejder-Korucu

Kafkas University, Faculty of Science and Literature, Department of Chemistry, Kars, Turkey

5. References

[1] Atkins. P., and de Paula. J., Atkıns' Physical Chemistry. Oxford Universty Press. Seventh Edition. 2002.

[2] Petrucci. R. H., Harwood. W. S., Genaral Chemistry Principles and Modern Applications. Sıxth Edition. MacMillan Publishing Company. New York. 1993

[3] Zumdahl. S. S and Zumdahl S. A., Chemistry.Houghton Mifflin Company. Seventg Edition. 2007.

[4] Svein Stolen and Tor Grande. Chemical Thermodynamics of Materials. 2004. John Wiley & Sons, Ltd ISBN 0 471 492320 2.

[5] Levine. I. N., Physical Chemistry. McGRAW-HILL International Edition. Third Edition 1988.

[6] McMURRAY. F., Chemistry. Fourth edition.

[7] Gilbert. T. R., Kirss. R. V., Foster. N., and Davies. G., Chemistry the science in context. Second edition. 2009. W. W. Norton & Company, Inc.

[8] Lee. H-G., Chemical Thermodynamics for Metals and Materials. Imperial College Press. 1999.

[9] Goldberg. D. E., Fundamentals of Chemistry. Fifth edition. The McGraw-Hill. 2007

[10] Powers. J. M., Lecture Notes on Thermodynamics. Notre Dame, Indiana; USA. 2012

[11] Tanaka. T., The Laws of Thermodynamics. Cambridge Universty Press.

[12] Atkins. P., and de Paula. J., Atkins' Physical Chemistry. Oxford Universty Press 2009.

[13] Laider, Keith, J. (1993). *The World of Physical Chemistry* (http:/ / books. google. com) Oxford University Press. ISBN 0-19-855919-4.

[14] International Union of Pure and Applied Chemistry, Physical Chemistry Division. "Quantities, Units and Symbols in Physical Chemistry" (http:/ / old. iupac. org/ publications/ books/ gbook/ gren book). Blackwell Sciences. p. 7. "The adjective specific before the name of an extensive quantity is often used to mean divided by mass."

[15] International Bureau of Weights and Measures (2006), *The International System of Units (SI)* (http:/ / www. bipm. org) (8th ed.), ISBN 92-822-2213-6,

[16] Gilbert. T. R., 7.chemistry-the science in context-second edition. 2008.

[17] Minguez. J. M., The work-energy theorem and the first law of thermodynamics. International Journal of Mechanical engineering Education 33/1.

[18] Silberg. M. S., Principles of General Chemistry. McGraw-Hill Higher Education. 2007.

[19] Housecroft. C. E and Constable. E. C., Chemistry: An Introduction to Organic, Inorganic and Physical Chemistry. 3rd Edition. Pearson Prentice Hall. 2006.

[20] Jones. A., Clement. M., Higton. A., and Golding. E., Access to Chemistry. Royal Society of Chemistry. 1999.

[21] Atkins. P. W., Physical Chemistry. Oxford University Press. Sixth Edition. 1998.

[22] Mortimer. R. G., Physical Chemistry. Third edition. Academic Press in an imprint of Elsevier.2008.

Chapter 2

Useful Work and Gibbs Energy

Nikolai Bazhin

Additional information is available at the end of the chapter

http://dx.doi.org/10.5772/50119

1. Introduction

Devices for performing chemical reactions are widely used to produce heat and work. Heat, in turn, produces work, e.g., in the form of electric energy in the so-called heat engines. It is the well-known fact that the efficiency of heat engines is restricted by Carnot principle. Therefore, it is generally recognized that heat cannot be fully converted to work. The efficiency of electric energy production due to the burning of fossil fuel of various kinds varies from 30 to 50 %.

On the other hand, there are galvanic and fuel cells whose efficiency can reach theoretically unity if it implies the ratio between the electric energy produced and the value of a change in Gibbs energy during chemical reactions occurring in a cell. These devices operate at constant temperature and pressure. It is concluded then that the devices, similar to a galvanic cell, cannot work at constant and uniform temperature according to the principle of heat engine. These devices assumed to operate only due to the direct transformation of chemical reaction energy, described by a change in the Gibbs energy, into work [1]. This viewpoint causes, however, numerous contradictions. The goal of this work is to analyze in detail the mechanism of useful work and heat production in chemical systems functioning at constant temperature and pressure.

2. Fundamental functions

In this Section, the fundamental functions of thermodynamics will be characterized, as the fundamental functions play the important role in the description of the process of the energy transformation. This notion includes four functions, i.e., internal energy, enthalpy, Helmholtz energy, and Gibbs energy. All of them are the state functions of energy dimension. It is generally believed that the value with energy dimension describes energy but this is by no means always the case. Below, the fundamental functions will elucidate whether these are energy values or not.

2.1. Internal energy

According to IUPAC [2], "**internal energy** U is the quantity the change in which is equal to the sum of heat, q, brought to the system and work, w, performed on it, $\Delta U = q + w$ ". Because of various transformations, the internal energy can be converted to other kinds of energy. However, the initial quantity of internal energy should be conserved due to the law of energy conservation. Conservation is the most important characteristic of energy. Hence U is energy.

2.2. Enthalpy

According to IUPAC, "**enthalpy,** $H = U + pV$ *is the internal energy* of a system plus the product of pressure and volume. Its change in a system is equal to the heat, brought to the system at constant pressure" [2]. It is worth noting that the change in the value describing energy always corresponds to the change in energy and not only in some special cases. Generally, a change in enthalpy may be inconsistent with the change in real physical values. Let us consider, e.g., the process of ideal gas heating at constant volume. The quantity of heat, taken in by ideal gas from the heat bath, q, equivalently changes only the internal gas energy, $\Delta U = q$, and simultaneously causes changes in gas enthalpy, $\Delta H = \Delta U + \Delta pV > q$. However, this change fails to reflect the changes in any physically significant values. Thus, the enthalpy is not energy but a function of state having the dimension of energy. It is easier and more correct to assume that the enthalpy is the part of a calculating means used to describe thermodynamic processes.

2.3. Helmholtz energy

According to IUPAC, "**Helmholtz energy (function),** *A, is the internal energy* minus the product of thermodynamic temperature and entropy: $A = U - TS$. It was formerly called free energy" [2]. Let us see why this quantity was called free energy. According to the first law [3]

$$\Delta U = q + w, \tag{1}$$

where q is the heat, entering the system, and w is the total work performed on the system by the surroundings. Usually, the total work is given as the sum of two terms: expansion work $(-p\Delta V)$ and so-called useful work (w_{useful})

$$w = -p\Delta V + w_{\text{useful}}. \tag{2}$$

If the process occurs at constant volume, the expansion work is absent and w describes the useful work. In the case of reversible process [3]

$$q = T\Delta S. \tag{3}$$

Thus, eq. (1) is of the form in the case of reversible process

$$w_{\text{useful}} = \Delta U - T\Delta S = \Delta(U - TS) = \Delta A, \tag{4}$$

where ΔA is the change in Helmholtz energy at constant temperature and volume. Since the term "energy" means the capacity of the system to perform work, from eq. (4) it is formally concluded that A is the energy (but in this case eq. (4) can have the second explanation – A is the numerical characteristic of work and not the work itself). Further, from eq. (4) it was concluded that only the part of internal energy U minus TS can be used to produce work. Therefore, the TS quantity was called "bound energy" and (U - TS) – "free energy". The meaning of these notions will be considered in more detail using the Gibbs energy as an example because it is more often used in chemical applications.

2.4. Gibbs energy

According to IUPAC [2], "**Gibbs energy (function),** $G = (H - TS)$, is the enthalpy minus the product of thermodynamic temperature and entropy. It was formerly called free energy or free enthalpy". The reasons for the appearance of the term "Gibbs energy" are similar to those discussed when considering the Helmholtz energy except for the fact that the Gibbs energy describes the reversible useful work performed at constant temperature and pressure. This is readily observed by substituting eq. (3) and eq. (2) into eq. (1) with regard to $V\Delta p = 0$

$$w_{\text{useful}} = \Delta H - T\Delta S = \Delta(H - TS) = \Delta G, \tag{5}$$

where ΔG is the change in Gibbs energy at constant temperature and pressure in the reversible process.

Unfortunately, the word "energy" as defined by IUPAC for a Gibbs energy (and also for a Helmholtz energy), causes great confusion. The Gibbs energy $G = H - TS$ consists of two terms, the enthalpy and the entropy one. The origin of both of the terms is quite different despite the same dimension. The enthalpy, considered above, is not energy.

Consider now the problem of TS nature. In the case of the reversible process $T\Delta S = q$, but in the case of the irreversible process $T\Delta S > q$ and additional contributions to $T\Delta S$ can arise without change in energy. For example, it is well known the increase of the entropy in the process of ideal gas expansion in vacuum without heat consumption ($q = 0$).

Let us consider another example. Let the ideal gas-phase system involves a spontaneous process of the monomolecular transformation of substance A into B. As suggested a change in enthalpy tends to zero in this reaction. Thus, the internal energy, temperature, pressure, and volume will not undergo changes in this process. However, the entropy will increase due to the entropy of the mixing, because the entropy is a function of state. The value of the TS product will increase accordingly. However, the energy and even the bound energy cannot arise from nothing, whereas the entropy, being a function of state, can increase thus reflecting a change in system state without any changes in the internal system energy. Therefore, the TS term is not the energy, which also implies the absence of the term "bound" energy.

Since neither enthalpy nor *TS* are the energy quantities, the difference between them cannot represent energy. Thus, *G* cannot represent energy precisely in terms of this notion. Note that in the irreversible process, occurring at constant temperature and pressure, the Gibbs energy decreases and thus, is not conserved. This is readily demonstrated by e.g., the aforementioned example of a monomolecular transformation of substance A into B. Thus, conservation, as the most important criterion for energy quantity, fails for the Gibbs energy. It is concluded then that the Gibbs energy is not energy [4] but a function of state. In this regard, the Gibbs energy does not differ from heat capacity. The notions of the non-existing in reality quantities of "free energy" and "bound energy" cause only confusion and are, at present, obsolete [5].

Nevertheless, the notions that the Gibbs energy is the energy and thus, obeys conservation laws, prove to be long-lived, which causes erroneous interpretation of a number of processes some of which are of paramount importance.

Let us consider now the reaction of adenosine triphosphate (ATP) hydrolysis in water solution which is of great concern in biochemistry

$$ATP + H_2O = ADP + Pi.$$

Under the standard conditions [6] $\Delta_r G^\circ = -7$, $\Delta_r H^\circ = -4$ kcal·mol^{-1}. According to D. Haynie [6, p. 143], "measurement of the enthalpy change of ATP hydrolysis shows that $\Delta H^\circ = -4$ kcal·mol^{-1}. That is, the hydrolysis of one mole of ATP at 25 ºC results in about 4 kcal·mol^{-1} being transferred to a solution in the form of heat and about 3 kcal·mol^{-1} remaining with ATP and Pi in the form of increased random motion."

The heat of 4 kcal·mol^{-1} is actually released into solution due to hydrolysis. Unfortunately, it is then assumed that the Gibbs energy is conserved which makes his difference in $\Delta_r H^\circ$ and $\Delta_r G^\circ$ of 3 кcal·mol^{-1} be located on the degrees of freedom of product molecules. However, in this connection, the product molecules could appear in the non-equilibrium excited states and transfer this energy to solvent molecules which would result in the emission of 7 kcal·mol^{-1} rather than 4 kcal·mol^{-1} which contradicts the experiment. There are no additional 3 кcal·mol^{-1} on the degrees of freedom of ATP and Pi because the Gibbs energy is not conserved.

2.5. Conclusions

Thus, among the four quantities, that claims to be called energy quantity, only the internal energy deserves this name. The other functions, i.e., enthalpy *H*, Helmholtz energy *A*, and Gibbs energy *G* are the parts of a mathematical apparatus for calculating various quantities, such as useful work, equilibrium constants, etc. This also means that the useful work is only calculated by using functions *A* and *G*, but cannot arise from the change in either the Helmholtz or the Gibbs energy. The physical nature of the work performed should be considered separately. Since the Helmholtz energy and the Gibbs energy are not energies, then, to avoid misunderstanding, it is better to exclude the word "energy" from the name of corresponding functions and to use the second variant of the name of these functions according to IUPAC: a Helmholtz function and a Gibbs function [7].

3. Direct conversion of chemical reaction energy to useful work

This Section is devoted to the discussion of the generally accepted theory of the direct conversion of energy [1, 8] produced by chemical reaction to useful work. For simplicity, hereafter exergonic $(\Delta_r G < 0)$ and exothermic $(\Delta_r H < 0)$ reactions will be considered.

The useful work of the chemical reaction occurring at constant temperature and pressure in the reversible conditions can be calculated through the change in the Gibbs function (5). When the interest is the useful work performed by the system in the environment $(-w_{\text{useful}})$, then

$$-w_{\text{useful}} = -\Delta_r H + T\Delta_r S = -\Delta_r G. \tag{6}$$

From eq. (6) it follows formally that the useful work of the reversible system in the environment is the sum of the enthalpy member $(-\Delta_r H)$ and the entropy member $T\Delta_r S$. In this connection it is interesting to discuss the various situations which arise in dependence on the relation between $(-\Delta_r H)$ and $T\Delta_r S$.

From eq. (6) it follows, that for $\Delta_r S > 0$ the useful work in the environment exceeds $(-\Delta_r H)$: $-w_{\text{useful}} > -\Delta_r H$. Therefore, the system must drag the thermal energy from the environment in the volume $T\Delta_r S$ to perform useful work. What is the physical reason for thermal energy consumption? Why does the system consume thermal energy of volume $T\Delta_r S$ neither more or less? How can two different contributions produce the same useful work?

The second case of $\Delta_r S < 0$ is also of interest. In this case $-w_{\text{useful}} < -\Delta_r H$ and the system must evolve the part of reaction heat to the environment. Why cannot the system use the total reaction heat for useful work production if this energy is at its disposal? Why can the system transfer energy of volume $T\Delta_r S$ and neither more or less to the environment?

The third case is $\Delta_r H = 0$. Here the system can use only the thermal energy of the environment to produce useful work.

In the fourth case $\Delta_r S = 0$ and the system performs work formally due to the reaction heat $(-\Delta_r H)$ without exchanging thermal energy with the environment. But it is not the case.

As mentioned in the Introduction, at present, it is generally accepted that the high efficiency of reversible devices is inconsistent with the notions that heat can be used to produce work and that such devices realize the direct conversion of chemical system energy into work. But below it will be shown that in all the cases, the useful work results from the heat of volume $\Delta_r G$ dragged from the environment.

4. A mechanism of useful work production - A Van't Hoff Equilibrium Box

In this Section, it will demonstrate the mechanism of producing useful reversible work which involves no notions of the direct conversion of reaction energy into useful work. To this end, let us consider a Van't Hoff equilibrium box (VHEB) [9 – 11]. A thermodynamic system must provide realization of the reversible process. This means that all changes in the system are infinitely slow at infinitely minor deviation from equilibrium.

It is assumed that in the system the following reaction occurs

$$\sum_i \nu_i A_i = \sum_j \nu_j B_j, \tag{7}$$

where A_i are the reagents and B_j are the products. The reaction takes place in the reactor (Fig. 1) where the reagents and products are in equilibrium. The chemical process is afforded by reservoirs with reagents and products contained in the system. For simplicity, the reagents and products are assumed to be in standard states. The system should have instruments to transport both the reagents from standard vessels to reactor and the products from reactor to standard vessels. In addition, the system should have tools to perform work, because the reversible process must be followed by reversible work production. The instruments and tools for performing work can be used together. The reactor, transporting instruments, and tools for performing work can be placed either separately or together. To provide constant and uniform temperature, it is necessary to locate the system, including reactor, standard vessels, transporting instruments, and tools in a thermostat, which can also imply the environment of constant temperature.

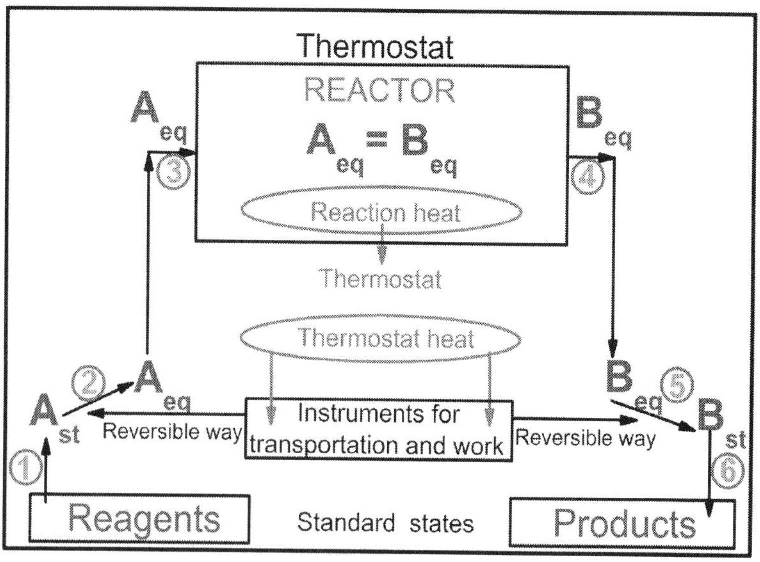

Figure 1. Production of useful work and heat in a closed reversible system. A_{st} (B_{st}) indicate the reagents A_i (products B_j) in the standard states; A_{eq} (B_{eq}) indicate the reagents A_i (products B_j) in the equilibrium states, which correspond to the equilibrium at the reactor; the green figures in circles indicate the step numbers (see text)

Let us consider reversible chemical process in a closed system (Fig. 1). The realization of the reversible chemical process consists in reversible transformation of the reagents to products via chemical reaction. Let us consider the closed system.

4.1. Closed system

The process of reversible work production includes six stages.

Step 1. A small amount of substance A_i is removed from the vessel with reagent A_i in the standard state upon reversible process. Gaseous substances can be removed from a standard vessel and put into a portable cylinder with pistons [11]; solid or liquid substances are placed in lock chambers.

Then, the change in the Gibbs function and the work are zero $(\Delta G_{1i} = 0, w_{1i} = 0)$ because a minor portion of substance A_i is in the same standard state as the residual reagent.

Step 2. Reagent A_i is transformed reversibly from the standard into the equilibrium state in the reactor. For example, for ideal gas, the gas pressure will vary from a standard value to the equilibrium partial value in the reactor. In this stage, the reversible work w_{2i} is produced and $\Delta G_{2i} \neq 0$. The work w_{2i} can be done only due to thermostat heat because there are no other energy sources in the system (reaction is in the equilibrium). This work depends on the difference in the physical states of reagent A_i in the initial and equilibrium states. All reagents A_i participate in all stages in quantities proportional to ν_i. For ideal gas, the useful work is

$$w_{2i} = \nu_i RT \ln(p_{i,\text{eq}} / p_{i,\text{st}}), \tag{8}$$

where $p_{i,\text{eq}}$ is the equilibrium pressure of i-th gas in the reactor, and $p_{i,\text{st}}$ is the pressure of i-th gas in a standard vessel. For gaseous components, e.g., the process of reversible gas expansion (compression) in a portable cylinder for producing the maximum useful work, must proceed to the value $p_{i,\text{eq}}$. If expansion (compression) stops at $p_i > p_{i,\text{eq}}$, then the inlet of gas into the reactor causes irreversible gas expansion and thus, the useful work will be less than the maximum one. When due to expansion (compression) the final pressure is less than $p_{i,\text{eq}}$, then the inlet of gas into the reactor causes the irreversible inlet of the i-th gas from an equilibrium mixture in the reactor to the portable cylinder, which also leads to a decrease in useful work. The solid and liquid substances can be transported by lock chambers. The pressure above either solid or liquid substances is varied from 1 bar to the value of the total equilibrium pressure in the reactor. The pressure is created using a minor portion of equilibrium reaction mixture.

The thermostat enthalpy varies as follows: $\Delta H_{2i,\text{thermostat}} = w_{2i}$.

Step 3. Reagent A_i is reversibly introduced into the reactor. Gaseous components are introduced into the reactor through semipermeable membranes using portable cylinders [11]; the solid or liquid ones – by means of lock chambers. Hence, $\Delta G_{3i} = 0,\ w_{3i} = 0$.

The useful work production and the change in thermostat enthalpy ($\Delta H_{2,\text{thermostat}}$) take place only at step 2:

$$w_2 = \Delta H_{2,\text{thermostat}}, \tag{9}$$

where $w_2 = \sum_i w_{2i}$ and $\Delta H_{2,\text{thermostat}} = \sum_i \Delta H_{2i,\text{thermostat}}$.

The same procedure is used to bring products from the standard vessels to reactor.

Step 4. An equivalent amount of product B_j is reversibly removed from the reactor. After this step is $\Delta G_{4j} = 0, w_{4j} = 0$.

Step 5. Product B_j, removed from the reactor, is reversibly transformed from the equilibrium state into the standard one to perform work w_{5j}. The change in the Gibbs function is not zero, $\Delta G_{5j} \neq 0$. The change in the thermostat enthalpy is $\Delta H_{5j,\text{thermostat}} = w_{5j}$.

Step 6. Product B_j, removed from the reactor, is reversibly introduced into the standard vessel. In this case is $\Delta G_{6j} = 0$ and $w_{6j} = 0$.

The change in the thermostat enthalpy upon thermal energy conversion into useful work at step 5 is

$$w_5 = \Delta H_{5,\text{thermostat}}, \tag{10}$$

where and $w_5 = \sum_j w_{5j}$, $\Delta H_{5,\text{thermostat}} = \sum_j \Delta H_{5j,\text{thermostat}}$

The change in the thermostat enthalpy at the second and fifth steps obeys the equation

$$\Delta H_{2,\text{thermostat}} + \Delta H_{5,\text{thermostat}} = -Q_{\text{dragged}}, \tag{11}$$

where Q_{dragged} is the heat dragged by tools.

For the reversible process, the maximal useful work is numerically equal to $\Delta_r G$, eq. (5) and, hence, the heat dragged by tools from thermostat in the volume

$$w_{\text{useful}} = w_2 + w_5 = \Delta_r G = \Delta_r H - T\Delta_r S = -Q_{\text{dragged}}. \tag{12}$$

The process has resulted in the useful work of the reaction, $\Delta_r G$, but the reaction did not occur yet. To put it otherwise, reaction work was performed without reaction. Only the thermal energy of the thermostat (environment) may be the source of work. This means that the process of useful work production and the reaction itself may be temporally and spatially separated. Thus, eq. (6) numerically connects reaction parameters and the magnitude of the work. However, no reaction energy is needed to produce the work. There is no need to subdivide energy sources into reaction source ($-\Delta_r H$) and thermal $T\Delta_r S$, because there is only one energy source: the thermal energy of thermostat (environment).

For $(\Delta_r S > 0)$, the thermal energy dragged by the tools exceeds $-\Delta_r H$, $(Q_{\text{dragged}} > -\Delta_r H)$; in the case of $(\Delta_r S < 0)$, the dragged thermal energy is less than $-\Delta_r H$ $(Q_{\text{dragged}} < -\Delta_r H)$; in the case of $(\Delta_r S = 0)$ the energy extracted is equal to $-\Delta_r H$ $(Q_{\text{dragged}} = -\Delta_r H)$ and for $(\Delta_r H = 0)$ the dragged thermal energy is of volume $T\Delta_r S$ $(Q_{\text{dragged}} = T\Delta_r S)$. The volume of extracted thermal energy is controlled by chemical equilibrium via $\Delta_r G$.

Thus, the mixture in the reactor is moved off balance to be restored later. As a result, the reaction heat is emitted into the thermostat. Indeed, because of the elementary chemical act in the reactor, the energy released concentrates at the degrees of freedom of the product molecules. As the reactor temperature is constant, this energy is dissipated in the reactor and transferred to the thermostat which causes a $-\Delta_r H$ change in thermostat enthalpy. The total change in thermostat enthalpy is

$$\Delta H_{\text{thermostat}} = w_2 + w_5 - \Delta_r H. \tag{13}$$

The cycle is over. Equations (12, 13) can be used to calculate the total change in thermostat enthalpy

$$\Delta H_{\text{thermostat}} = \Delta_r G - \Delta_r H = -T\Delta_r S. \tag{14}$$

The change in thermostat enthalpy is controlled only by reaction entropy [11].

4.2. The main principles of reversible device functioning in useful work production

This consideration demonstrates the main characteristics of the reversible process of useful work production at constant temperature and pressure in closed systems:

1. The useful work arises from the stage of the reversible transport of reagents from reservoir to reactor and from the stage of the reversible transport of products from the reactor.
2. The only energy source of useful work is the thermal energy of thermostat (or environment).
3. The heat released by chemical reactions is dissipated to the thermostat; the reaction heat is infinitely small in comparison with the volume of the thermostat thermal energy; therefore no reaction heat is really needed to produce work.
4. Although the useful work is produced by the cooling of one body (thermostat), the second law of thermodynamics is not violated, because the process is followed by a change in the amount of reagents and products.
5. The useful work is produced by heat exchange with thermostat (environment) according to the scheme

   reaction heat → thermal thermostat energy → useful work (scheme I)

6. The maximal useful work is equal the heat dragged by tools from thermostat.
7. Useful work depends on the difference in the concentrations of standard and equilibrium states of reagents and products. Therefore, the amount of extracted energy can be calculated via the change in the Gibbs function.

8. There is no direct conversion of the Gibbs energy into useful work. Gibbs energy is equal numerically the thermal energy dragged by systems from the environment for doing work.

4.3. The energy limit of chemical reactions - Open systems

Usually, the total energy which can be produced by chemical system, is $-\Delta_r H$. However, this holds for closed systems only. For open systems, the case is quite different [11].

The open system is depicted in Fig. 2. As compared with the closed one (Fig. 1), the open system consists of two thermostats: the first one contains a reactor and the second one contains standard vessels with substances A and B and tools. The second thermostat can be replaced by the environment. In the process, steps 1, 2, 5, and 6 occur in the second thermostat (environment); steps 3 and 4 take place in the first one. Thus, the heat is released in the first thermostat only and the work is performed by thermal energy of the second thermostat (environment). The processes of heat and work production are spatially separated! The energy potential of the open system obeys the equation

$$q + w_{\text{useful}} = \Delta_r H + \Delta_r G \tag{15}$$

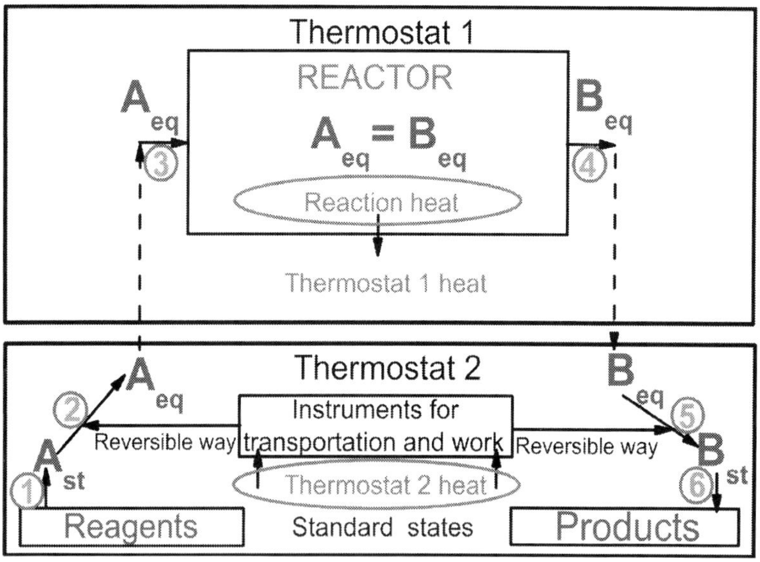

Figure 2. Production of useful work and heat in the open reversible system. The designations see in the subscription to Fig. 1

In the case of coal burning, it is possible to obtain the double total energy [11]. Thus, understanding the mechanism of useful work production in the reversible process allows us to predict an increase in the energy potential of chemical reactions in the open system.

It is worth noting that the open system under study is not a heat pump. The heat pump consumes energy to transfer heat from a cold body to the warm one. The open system studied does not consume external energy and produces heat due to chemical reactions in one thermostat and performs work by extraction of thermal energy from the second one.

4.4. Conclusions

The chemical reaction heat is always released in the reversible chemical processes and passes to the environment independent of the fact whether the system produces work or not, whether it is closed or open. The discussed mechanism of useful work production in the reversible systems did not use such notions as "free energy", "bound energy", "direct Gibbs energy conversion". The useful work arises only due to the heat exchange with a thermal basin in the process, described by the scheme I. The total energy of chemical system can be high and equal to $\Delta_r H + \Delta_r G$.

5. The mechanism of electric work production in a galvanic cell

The current theory of galvanic cells [1, 3, 8] is based on a direct transformation of the energy ($\Delta_r G$) of oxidation-reduction reactions into electric work. However, using VHEB as an example, It is clear that the energy of chemical reactions is first converted into the thermal energy of thermostat (environment) and then the thermal energy is extracted from the thermostat and transformed into work by means of special devices. It is assumed then that in galvanic cells, useful work is produced via the mechanism similar to the VHEB one [12, 13]. The $\Delta_r G$ value is used to calculate electric work which does not, however, mean that the electric work is performed at the expense of the Gibbs energy, all the more it was shown that the Gibbs energy is not energy. The electric work of a galvanic cell results from the electrodes discharged. Electric charging of electrodes is caused by chemical reactions in electrodes.

The mechanism of electric energy production in galvanic cells will be solved by analyzing the behavior of one ion. But it does not denote that thermodynamics will be applied to a real single ion: thermodynamic parameters of one ion imply the averaged parameters of many ions.

5.1. A galvanic cell

For simplicity a Daniell cell will be considered, consisting of zinc (№1) and copper (№2) electrodes (Fig. 3). The activity of salts in solutions is denoted by a_1 and a_2, respectively. Let the cell with an open, external circuit be in equilibrium. Close now the external circuit for the moment and two electrons will transfer from the zinc to the copper electrode. The balance of the cell is distorted. Consider now the establishment of equilibrium on the zinc electrode (Fig 3). To this end, the zinc ion must leave a metallic plate and escape into the bulk. The dissolving of zinc ions is described by the change in a Gibbs function

$$\Delta_r G_1 = \Delta_r G_1^{\circ} + RT \ln a_1, \tag{16}$$

where $\Delta_r G_1^o$ is a standard change in the Gibbs function upon dissolving. The ion penetrates further into solution with execution of the work (w_{1g}) in the electric field

$$w_{1g} = nF\Delta_{\text{Met1}}^{\text{sol}}\varphi_1, \tag{17}$$

where n is the number of electrons, participating in the reactions, F is the Faraday constant, and $\Delta_{\text{Met1}}^{\text{sol}}\varphi_1$ is the difference in potentials of solution and metal. The work described in equation (17), is the electric work spent to charge an electrode. It is performed at the extraction of the thermal energy of solution due to the absence of other energy sources in the system. Since in equilibrium, the chemical potential of ions in solution equals the chemical potential of metal, it is possible to derive the equation for electrochemical equilibrium

$$\Delta_r G_1^o + RT\ln a_1 + nF\Delta_{\text{Met1}}^{\text{sol}}\varphi_1 = 0, \tag{18}$$

which readily gives the expression for both the work performed on the first electrode and its galvanic potential [3]

$$w_{1g} = nF\Delta_{\text{Met1}}^{\text{sol}}\varphi_1 = -\Delta_r G_1^o - RT\ln a_1, \tag{19}$$

$$\Delta_{\text{Met1}}^{\text{sol}}\varphi_1 = -\frac{\Delta_r G_1^o}{nF} - \frac{RT\ln a_1}{nF}. \tag{20}$$

The latter is the potential for a half-cell. Thus, the approach, based on the consideration of the behavior of one ion, provides a common expression for electrode potential.

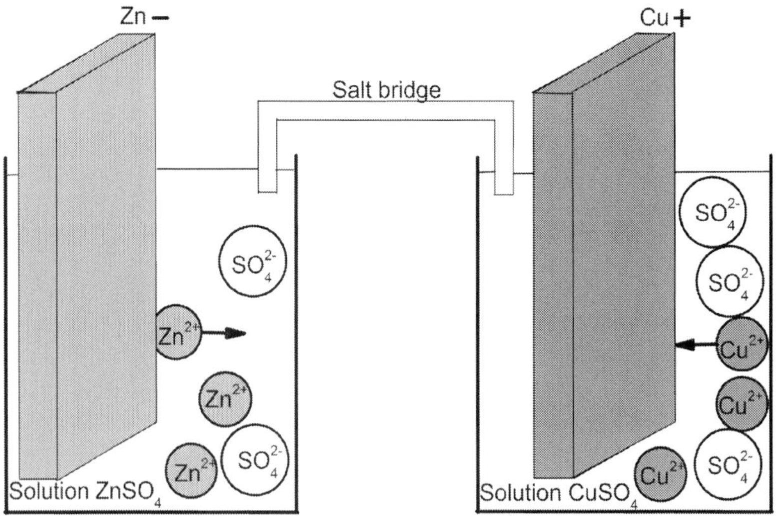

Figure 3. Establishment of equilibrium on the electrodes

The change in electrode enthalpy involves dissolution enthalpy and thermal energy consumption upon ion transport into solution. The equation for dissolution enthalpy is readily obtained from eq. (16)

$$\Delta_r H_1 = \Delta_r H_1^o - RT^2 \frac{\partial \ln a_1}{\partial T}, \tag{21}$$

where $\Delta_r H_1^o$ is a standard change in enthalpy during ion dissolving. The total change in the enthalpy of the first electrode ($\Delta H_{1,\text{thermostat}}$) is the sum of the expressions $-\Delta_r H_1$ and $-w_{g1}$

$$\Delta H_{1,\text{thermostat}} = -\Delta_r H_1 - w_{g1} = -T\Delta_r S_1^o + RT\ln a_1 + RT^2 \frac{\partial \ln a_1}{\partial T} = -T\Delta_r S_1, \tag{22}$$

where $\Delta_r S_1^o$ is a standard change in entropy upon ion dissolving and $\Delta_r S_1$ is the change in entropy upon ion dissolving on the first electrode which amounts to

$$\Delta_r S_1 = \Delta_r S_1^o - R\ln a_1 - RT \frac{\partial \ln a_1}{\partial T}. \tag{23}$$

As follows from eq. (22), the change in enthalpy, related to the first electrode, is independent of the processes, occurring on the second one. Therefore, studying either release or absorption of heat on a separate electrode, one may calculate the change in entropy due to the escape of the ions of the same type into the bulk.

A corresponding expression for the second electrode is of the same form but index "1" should be substituted by index "2":

$$\Delta_r G_2 = \Delta_r G_2^o + RT\ln a_2, \tag{24}$$

$$w_{2g} = nF\Delta_{\text{Met2}}^{\text{sol}}\varphi_2 = -\Delta_r G_2^o - RT\ln a_2 \tag{25}$$

$$\Delta_{\text{Met2}}^{\text{sol}}\varphi_2 = -\frac{\Delta_r G_2^o}{nF} - \frac{RT\ln a_2}{nF}. \tag{26}$$

$$\Delta_r H_2 = \Delta_r H_2^o - RT^2 \frac{\partial \ln a_2}{\partial T}, \tag{27}$$

$$\Delta_r S_2 = \Delta_r S_2^o - R\ln a_2 - RT \frac{\partial \ln a_2}{\partial T}, \tag{28}$$

$$\Delta H_{2,\text{thermostat}} = -\Delta_r H_2 - w_{2g} = -T\Delta_r S_2^o + RT\ln a_2 + RT^2 \frac{\partial \ln a_2}{\partial T} = -T\Delta_r S_2. \tag{29}$$

In the operation of the galvanic cell, the processes on the second electrode are oppositely directed which should be taken into account in consideration of the thermodynamic cell parameters.

Equations (20) and (26) allow to get the Nernst equation for the potential of the cell [3]

$$E = -\Delta_{\text{Met2}}^{\text{sol}}\varphi_2 + \Delta_{\text{Met1}}^{\text{sol}}\varphi_1 = -\frac{\Delta_r G^o}{nF} - \frac{RT}{nF}\ln\frac{a_1}{a_2}, \tag{30}$$

where E – the cell potential, $\Delta_r G^o = \Delta_r G_1^o - \Delta_r G_2^o$.

The electric work of the galvanic cell (w_{el}) results from the transformation of the potential energy of the charged electrodes into electric energy. The potential energy arises from the thermal energy of both of the electrodes upon ions transport into solution and equals

$$w_{el} = w_{1g} - w_{2g} = -\Delta_r G^o - RT\ln\frac{a_1}{a_2} = -w_{\text{useful}}. \tag{31}$$

The change in thermostat enthalpy is of the form

$$\Delta H_{\text{thermostat}} = \Delta H_{1,\text{thermostat}} - \Delta H_{2,\text{thermostat}} = -T\Delta_r S, \tag{32}$$

which is in fair agreement with similar expression, described by eq. (14), for the VHEB. The detailed equation for $\Delta H_{\text{thermostat}}$ can be get after substitution corresponding expressions (22) and (29) into (32)

$$\Delta H_{\text{thermostat}} = -T(\Delta_r S_1^o - \Delta_r S_2^o) - RT\ln(a_2 / a_1) - RT^2(\frac{\partial \ln a_2}{\partial T} - \frac{\partial \ln a_1}{\partial T}). \tag{33}$$

The sum of eqs. (31) and (33) gives the total energy (electric work + heat), produced by the galvanic cell

$$w_{el} + \Delta H_{\text{thermostat}} = -\Delta_r H^o - RT^2(\frac{\partial \ln a_2}{\partial T} - \frac{\partial \ln a_1}{\partial T}). \tag{34}$$

From eq.(34) it follows that the total energy produced by the galvanic cell is equal to the heat emitted by oxidation-reduction reaction.

Thus, the approach, based on the analysis of the behavior of one ion gives the same results as the present-day theory. However, it uses not a mysterious, direct transformation of the chemical energy ($\Delta_r G$) into electric work, but the concept of chemical energy conversion into the thermal one, and then, the thermal energy of thermostat (environment) is converted into the potential energy of charged electrodes [12, 13]. The electric energy of the galvanic cell arises according to the scheme:

reaction heat → thermal thermostat energy → potential energy of charged electrodes → electric energy.

Thus, in various systems with uniform temperature, useful work is produced by the same mechanism through the exchange of thermal energy with thermostat (environment). No direct conversion of chemical energy into useful work is observed. Unfortunately, in the galvanic cell, the processes of heat release and useful work production cannot be spatially separated, because both of them occur in a double layer. Therefore, galvanic cells are unpromising in production of a double amount of energy.

5.2. A concentration cell

Consider now a concentration cell, consisting of two electrodes, e.g., the zinc ones of different solution activity. Standard changes in the Gibbs function, enthalpy, and entropy for the concentration cell tend to zero due to the same chemical nature of both of the electrodes. By definition, it has been considered that $a_2 > a_1$. From eq. (31) it follows

$$w_{\text{el}} = RT\ln(a_2 / a_1) = -w_{\text{useful}}, \tag{35}$$

which is a usual expression for the electric energy of the concentration cell. From eq. (34) it follows

$$w_{\text{el}} + \Delta H_{\text{thermostat}} = -RT^2\left(\frac{\partial \ln a_2}{\partial T} - \frac{\partial \ln a_1}{\partial T}\right). \tag{36}$$

For the system in which the activities are temperature-independent, the electric energy arises from the thermal thermostat energy (environment)

$$w_{\text{el}} = -\Delta H_{\text{thermostat}} = -w_{\text{useful}}, \tag{37}$$

which is in fair agreement with conventional concepts.

6. Useful work of the systems with concentration gradient

An ideal system with concentration gradient has no potential energy because the mixing does not result in heat release and $\Delta H = 0$. Nevertheless, the system with concentration gradient can be used, as any non-equilibrium system, to produce useful work if this system is supplied with special tools for extracting heat from the environment with a simultaneous transformation of the extracted thermal energy into work upon the system approaches to the equilibrium. The volume of useful work is equal to the volume of the heat extracted from the environment

$$w_{\text{useful}} = \Delta H_{\text{thermostat}} \tag{38}$$

The useful work of the system with a concentration gradient w_{useful} obeys expressions (35) and (38). The concentration cell is a good example of such a system.

7. General conclusions

Any non-equilibrium state can serve as an energy source. The thermostat (environment) is an active participant of the process of reversible useful work production in devices operating at constant temperature. The heat released by chemical reactions, always dissipates in the thermostat (environment). The useful work is produced by special tools that provide the extraction of the thermal energy of the thermostat (environment) and the transformation of thermal energy into work at the process of the restoration of the chemical equilibrium. The volume of the useful work is equal, in reversible conditions, to the change in Gibbs function. A spatial separation of reactor and tools can lead to a substantial increase in the energy produced. The direct conversion of the Gibbs energy into useful work does not exist. The concepts of free and bound energy become unnecessary.

Author details

Nikolai Bazhin
Institute of Chemical Kinetics and Combustion, Novosibirsk State University, Institutskaya 3, Novosibirsk, Russia

8. References

[1] Denbigh K (1971) The Principles of Chemical Equilibrium, 3rd ed., The University Press: Cambridge, 494 p.
[2] IUPAC Green Book (1996) Quantities, Units and Symbols in Physical Chemistry 48 p.
[3] Atkins P (2001) The Elements of Physical Chemistry, 3rd ed., Oxford University Press, Oxford,.549 p.
[4] Lower S. (2010) Chemistry. Available http://www.chem1.com/acad/webtext/virtualtextbook.html
[5] Gokcen NA, Reddy RG (1996) Thermodynamics, Second edition, Springer, 416 p.
[6] Haynie DN (2008), Biological Thermodynamics, 2nd ed., Cambridge University Press, 422 p.
[7] Haywood RW (1980) Equilibrium Thermodynamics for Engineers and Scientists, J. Wiley, N.Y., 456 p.
[8] Adam NK (1956) Physical Chemistry, Oxford, Clarendon Press, 658 p.
[9] Physical Chemistry (1980) Gerasimov YaI, Ed., Chimiya, Moscow (in Russian), 1279 p.
[10] Steiner LE (1948) Introduction to Chemical Thermodynamic, 2nd ed.; McGraw-Hill Book Company, New York, 516 p.
[11] Bazhin NM, Parmon VN (2007) Conversion of the Chemical Reaction Energy into Useful Work in the Van't Hoff Equilibrium Box. J. Chem. Ed. 84: 1053 – 1055
[12] Bazhin NM, Parmon VN (2007) Ways of Energy Conversion in Electrochemical Cells. Doklady Physical Chemistry 417: 335-336
[13] Bazhin NM (2011) Mechanism of electric energy production in galvanic and concentration cells. J. Eng. Thermophysics 20: 302-307

Section 2

Statistical Thermodynamics

Chapter 3

Gibbs Free Energy Formula for Protein Folding

Yi Fang

Additional information is available at the end of the chapter

http://dx.doi.org/10.5772/52410

1. Introduction

Proteins are life's working horses and nature's robots. They participate in every life process. They form supporting structures of cell, fibre, tissue, and organs; they are catalysts, speed up various life critical chemical reactions; they transfer signals so that we can see, hear, and smell; they protect us against intruders such as bacteria and virus; they regulate life cycles to keep that everything is in order; etc., just mention only a few of their functions.

The first thing drawing our attention of proteins are their size. Proteins are macromolecules, that is, large molecules. Non-organic molecules usually are small, consisting of from a couple of atoms to a couple of dozen atoms. A small protein will have thousands of atoms, large ones have over ten thousand atoms. With their huge number of atoms, one can imaging that how complicated should be of a protein molecule. Fortunately, there are some regularities in these huge molecules, i.e., proteins are polymers building up by monomers or smaller building blocks. The monomers of proteins are amino acids, life employs 20 different amino acids to form proteins. In cell, a series of amino acids joined one by one into amino acids sequences. The order and length of this amino acid sequence is translated from DNA sequences by the universal genetic code. The bond joining one amino acid to the next one in sequence is peptide bond (a covalent bond) with quite regular specific geometric pattern. Thus amino acids sequences are also called peptide chains. But the easy translation and geometric regularities stop here. The peptide chain has everything required to a molecule, all covalent bonds are correctly formed. But to perform a protein's biological function, the peptide chain has to form a specific shape, called the protein's **native structure**. Only in this native structure a protein performs its biological function. Proteins fall to wrong shapes not only will not perform its function, but also will cause disasters. Many disease are known to be caused by some proteins taking wrong structure.

How the peptide chain take its native structure? Is there another genetic code to guide the process of taking to the native structure? In fact, at this stage, life's most remarkable drama takes stage. Once synthesized, the peptide chain of a protein spontaneously (some need the help of other proteins and molecules) fold to its native structure. This process is called **protein folding**. At this stage, everything is governed by simple but fundamental physical laws.

The **protein folding problem** then can be roughly divided into three aspects: 1. folding process: such that how fast a peptide chain folds, what are the intermediate structures between the initial shape and the native structure. 2. the mechanics of the folding, such as what is the deriving force. 3. the most direct application to biological study is the prediction of the native structure of a protein from its peptide chain. All three parts of the protein folding problem can have a unified treatment: writing down the Gibbs free energy formula $G(\mathbf{X})$ for any conformation $\mathbf{X} = (\mathbf{x}_1, \cdots, \mathbf{x}_i, \cdots, \mathbf{x}_M) \in \mathbb{R}^{3M}$ of protein, where $\mathbf{x}_i \in \mathbb{R}^3$ is the atom $\mathbf{a}_i$'s atomic center.

The fundamental law for protein folding is the **Thermodynamic Principle**: the amino acid sequence of a protein determines its native structure and the native structure of the protein has the minimum Gibbs free energy among all possible conformations as stated in Anfinsen (1973). Let $\mathbf{X}$ be a conformation of a protein, is there a natural Gibbs free energy function $G(\mathbf{X})$? The answer must be positive, as G. N. Lewis said in 1933: "There can be no doubt but that in quantum mechanics one has the complete solution to the problems of chemistry." (quoted from Bader (1990), page 130.) Protein folding is a problem in biochemistry, why such a formula $G(\mathbf{X})$ has not been found and what is the formula? This chapter is trying to give the answers.

First, the Gibbs free energy formula is given, it has two versions, the chemical balance version (1) and the geometric version (2).

1.1. The formula

Atoms in a protein are classified into classes H_i, $1 \leq i \leq H$, according to their levels of hydrophobicity. The formula has two versions, the chemical balance version is:

$$G(\mathbf{X}) = \mu_e N_e(\mathbf{X}) + \sum_{i=1}^{H} \mu_i N_i(\mathbf{X}), \tag{1}$$

where $N_e(\mathbf{X})$ is the mean number of electrons in the space included by the first hydration shell of $\mathbf{X}$, μ_e is its chemical potential. $N_i(\mathbf{X})$ is the mean number of water molecules in the first hydration layer that directly contact to the atoms in H_i, μ_i is the chemical potential.

Let $M_{\mathbf{X}}$ (see FIGURE 3) be the molecular surface for the conformation $\mathbf{X}$, defining $M_{\mathbf{X}\,i} \subset M_{\mathbf{X}}$ as the set of points in $M_{\mathbf{X}}$ that are closer to atoms in H_i than to any atoms in H_j, $j \neq i$. Then the geometric version of $G(\mathbf{X})$ is:

$$G(\mathbf{X}) = \nu_e \mu_e V(\Omega_{\mathbf{X}}) + d_w \nu_e \mu_e A(M_{\mathbf{X}}) + \sum_{i=1}^{H} \nu_i \mu_i A(M_{\mathbf{X}_i}), \quad \nu_e, \nu_i > 0, \tag{2}$$

where $V(\Omega_{\mathbf{X}})$ is the volume of the domain $\Omega_{\mathbf{X}}$ enclosed by $M_{\mathbf{X}}$, d_w the diameter of a water molecule, and $A(M_{\mathbf{X}})$ and $A(M_{\mathbf{X}_i})$ the areas of $M_{\mathbf{X}}$ and $M_{\mathbf{X}_i}$, $\nu_e[V(\Omega_{\mathbf{X}}) + d_w A(M_{\mathbf{X}})] = N_e$, $\nu_i A(M_{\mathbf{X}_i}) = N_i(\mathbf{X})$, $1 \leq i \leq H$. The ν_e and ν_i are independent of $\mathbf{X}$, they are the average numbers of particles per unit volume and area.

Before the actual derivation is given, some basic facts should be stated, such as hydrophobicity, protein structures, and the environment in which the protein folds. Brief description of the methods in the experimental measurements and theoretical derivation of the Gibbs free energy of the protein folding is introduced to give the motivation and idea of the derivation. By making critics on the previous derivation, the necessary concepts would be clarified, what are important in the derivation would be identified, and would set the thermodynamic system that most fit the reality currently known about the protein folding process. Then both classical and quantum statistical derivations were given, the only difference is that in the classical statistically derived formula, the volume and the whole surface area terms in formula (2) are missing. Thus it is that only quantum statistical method gives us the volume and whole surface terms in formula (2). After the derivations, some remarks are made. A direct application of the Gibbs free energy formula (2) is the *ab initio* prediction of proteins' natives structures. Gradient formulas of $G(\mathbf{X})$ are given to be able to apply the Newton's fastest descending method. Finally, it should be emphasized that the gradient $\bigtriangledown G(\mathbf{X})$ not only can be used to predict the native structure, it is actually the force that forces the proteins to fold as stated in Ben-Naim (2012). In Appendix, integrated gradient formulas of $G(\mathbf{X})$ on the molecular surface are given.

2. Proteins

2.1. Amino acids

There are 20 different amino acids that appear in natural proteins. All amino acids have a common part, or the **back bone** consisting of 9 atoms in FIGURE 1 (except the R).

NH_2 is the amino group and COOH is the carboxyl group of the back bone. Single amino acid is in polar state, so the amino group gains one more hydrogen from the carboxyl group, or perhaps the amino group losses one electron to the carboxyl group. Geometrically it is irrelevant since after forming peptide bonds the amino group will loss one H to become NH and the carboxyl group will loss one OH to become CO. Thus an amino acid in the sequence is also called a **residue**.

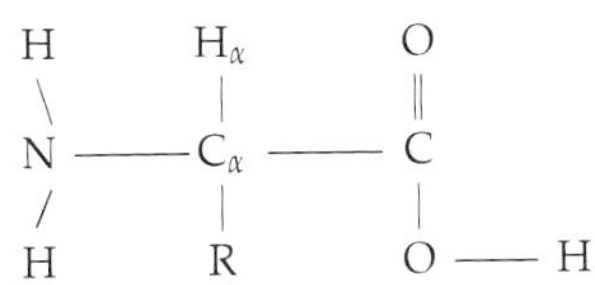

Figure 1. An generic amino acid.

The group R in FIGURE 1 is called **side chain**, it distinguishes the 20 different amino acids. A side chain can be as small as a single hydrogen atom as in Glycine, or as large as consisting of 18 atoms including two rings as in Tryptophan. 15 amino acids have side chains that contain more than 7 atoms, i.e., more atoms than that of the back bone in an amino acid sequence. Except Glycine, a C_β carbon in a side chain forms a covalent bond with the **central carbon** C_α of the back bone.

2.2. Hydrogen bonds

A hydrogen bond is the attractive interaction of a hydrogen atom with an electronegative atom (the **accepter**), like nitrogen, oxygen or fluorine (thus the name "hydrogen bond", which must not be confused with a covalent bond to hydrogen). The hydrogen must be covalently bonded to another electronegative atom (forming a **donor group**) to create the hydrogen bond. These bonds can occur between molecules (intermolecular), or within different parts of a single molecule (intramolecular). The hydrogen bond is stronger than the van der Waals interaction, but weaker than covalent or ionic bond. Hydrogen bond occurs in both inorganic molecules such as water and organic molecules such as DNA, RNA, and proteins.

Some amino acids' side chains contain hydrogen bond donors or acceptors that can form hydrogen bond with either other side chains in the same protein (intramolecular hydrogen bond) or with surrounding water molecules (inter-molecular hydrogen bond). Those amino acids whose side chains do not contain either donors or acceptors of hydrogen bond are classified as hydrophobic.

2.3. Hydrophobicity levels

Every atom in a protein belongs to a moiety or atom group, according to the moiety's level of ability to form hydrogen bond, the atom is assigned a hydrophobicity level. All the hydrophobicity scales are tested or theorized in some aspects of individual amino acid, either as a independent molecule or as a residue in a protein, in various environments such as solvent, PH value, temperature, pressure, etc. That is just like taking a snap shot of an object with complicated shape. All snap shots are different if taking from different angles of view. Therefore, there are many different classifications of hydrophobicity, for example, in Eisenberg and McLachlan (1986) there are five classes, C, O/N, O^-, N^+, S. Let a protein have M atoms $\{\mathbf{a}_1, \cdots, \mathbf{a}_i, \cdots, \mathbf{a}_M\}$. One can assume that there are H hydrophobic classes, such that $\{\mathbf{a}_1, \cdots, \mathbf{a}_i, \cdots, \mathbf{a}_M\} = \cup_{i=1}^{H} H_i$.

2.4. Protein structures

Let a molecule have M atoms, listed as $(\mathbf{a}_1, \cdots, \mathbf{a}_i, \cdots, \mathbf{a}_M)$. A presentation of a structure $\mathbf{X}$ of this molecule is a series atomic centers (nuclear centers) of the atoms $\mathbf{a}_i$, $\mathbf{x}_i \in \mathbb{R}^3$. Hence it can be written as a point in $\mathbb{R}^{3M}$, $\mathbf{X} = (\mathbf{x}_1, \cdots, \mathbf{x}_i, \cdots, \mathbf{x}_M)$. The space $\mathbb{R}^{3M}$ then is called the *control space*. The real shape of the structure $\mathbf{X}$ is realized in $\mathbb{R}^3$, called the *behavior space* as defined in Bader (1990), it is a bunch of overlapping balls (spheres), $P_{\mathbf{X}} = \cup_{i=1}^{M} B(\mathbf{x}_i, r_i)$, where r_i is the van der Waals radius of the atom $\mathbf{a}_i$ and $B(\mathbf{x}, r)$ is the closed ball $\{\mathbf{y} : |\mathbf{y} - \mathbf{x}| \leq r\} \subset \mathbb{R}^3$, of center $\mathbf{x}$ and radius r.

Protein native structures are complicated. Unlike the famous double-helix structure of DNA structure, the only general pattern for protein structure is no pattern at all. To study the native structures of proteins people divide the structures in different levels and make structure classifications.

The amino acid sequence of a protein is called its **primary structure**. Regular patterns of local (along the sequences) structures such as helix, strand, and turn are called the **secondary structure** which contain many intramolecular hydrogen bonds in regular patterns. The global assembly of these secondary structures, connected by turns and irregular loops, is called the **tertiary structure**. For proteins having multiple amino acid sequences or structurally associated with other molecules there are also **quaternary structures**, see Branden and Tooze (1999) and Finkelstein and Ptitsyn (2002).

The secondary structures are local structures, they are usually in helix, strand, and turn. A common feature of them is that they have regular geometric arrangement of their main chain atoms, such that there are good opportunities to form hydrogen bonds between different residues. Several strands may form sheet, stabilized by regular pattern of hydrogen bonds. Turns and loops are necessary for the extended long chain to transfer to a sphere like shape. Turns are short, 3 or 4 residues long. Loops involves many residues, but without any regular pattern of hydrogen bonds. Loops often form the working place of the protein, therefore appear on the out surface of the native structure.

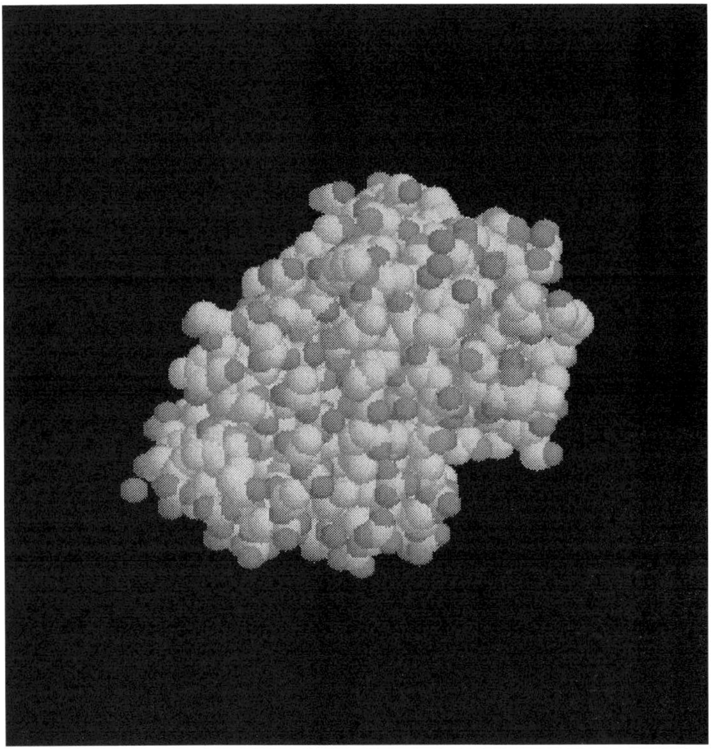

Figure 2. P_X is a bunch of overlapping balls, called the space-filling model, or CPK model.

3. Some functions in thermodynamics

A thermodynamic system consists of particles in a region $\Gamma \subset \mathbb{R}^3$ and a bath or environment surrounding it. A wall, usually the boundary $\partial\Gamma$ separating the system with its surrounding.

If no energy and matter can be exchanged through the wall, the system is an **isolated system**. If only energy can be exchanged, the system is a **closed system**. If both energy and matter can be exchanged with the surrounding, the system is an **open system**.

For an open system Γ of variable particles contacting with surrounding thermal and particle bath, let U, T, S, P, V, μ and N be the inner energy, temperature, entropy, pressure, volume, chemical potential, and the number of particles of the system Γ respectively, then

$$dU = TdS - PdV + \mu dN, \tag{3}$$

By Legendre transformations various extensive quantities can be derived,

$$F = U - TS, \quad G = U - TS + PV, \quad \phi = F - \mu N = U - TS - \mu N \tag{4}$$

where F, G, and ϕ are Helmholtz, Gibbs free energies, and thermodynamic potential respectively. Then

$$dF = -SdT - PdV + \mu dN, \quad dG = -SdT + VdP + \mu dN, \quad d\phi = -SdT - PdV - Nd\mu. \tag{5}$$

Which shows that $U = U(S,V,N)$, $F = F(T,V,N)$, $G = G(T,P,N)$, $\phi = \phi(T,V,\mu)$. All **extensive** quantities satisfy a linear homogeneous relation, i.e., consider a scaling transformation which enlarges the actual amount of matter by a factor λ, then all extensive quantities are multiplied by a factor λ. U, S, V, N, F, G, ϕ are extensive, while T, P, μ are **intensive**. Thus

$$\lambda U = U(\lambda S, \lambda V, \lambda N), \quad \lambda F = F(T, \lambda V, \lambda N), \quad \lambda G = G(T, P, \lambda N), \quad \lambda\phi(T,V,\mu) = \phi(T, \lambda V, \mu). \tag{6}$$

From equations in (5) $(\frac{\partial \phi}{\partial V})_{T,\mu} = -P$. By equations in (6)

$$\phi = \frac{d(\lambda\phi)}{d\lambda} = V\left(\frac{\partial\phi}{\partial V}\right)_{T,\mu} = -PV \tag{7}$$

and

$$\phi(T,V,\mu) = -PV. \tag{8}$$

Equation (8) is true for any open thermodynamics system.

4. Statistical mechanics

Thermodynamics is a phenomenological theory of macroscopic phenomena that neglects the individual properties of particles in a system. Statistical mechanics is the bridge between the macroscopic and microscopic behavior. In statistical mechanics, the particles in a system obey either classical or quantum dynamic laws, and the macroscopic quantities are statistical averages of the corresponding microscopic quantities. If the particles obey classical dynamical law, it is the classical statistical mechanics. If the particles obey quantum dynamical law, it is the quantum statistical mechanics. But the averaging to get macroscopic quantities from microscopic ones are in the same principle and formality.

Protein folding studies the structure of the protein molecule, what is the native structure and why and how the protein folds to it. All these aspects are specific properties of a particle, the protein molecule. To get the Gibbs free energy formula $G(\mathbf{X})$ for each conformation $\mathbf{X}$, statistical mechanics is needed with careful specification of the thermodynamic system.

4.1. The canonic ensemble

Statistical mechanics uses ensembles of all microscopic states under the same macroscopic character, for example, all microscopic states corresponding to the same energy E. The probability of this ensemble then is proportional to

$$p_E \propto \exp(-\beta E), \tag{9}$$

where $\beta = 1/kT$, k the Boltzmann constant and T the temperature. If there are only a series energy levels $E_1, E_2, \cdots$, then the probability distribution for canonic ensemble is

$$p_i = \frac{\exp(-\beta E_i)}{\sum_{n=1}^{\infty} \exp(-\beta E_n)}. \tag{10}$$

Various of thermodynamic quantities, such as the inner energy of the system, can be put as the means:

$$U = \langle E_i \rangle = \frac{\sum_{i=1}^{\infty} E_i \exp(-\beta E_i)}{\sum_{n=1}^{\infty} \exp(-\beta E_n)}. \tag{11}$$

If only the Halminltonian $H(\mathbf{q},\mathbf{p})$ is known, where $\mathbf{q} = (\mathbf{q}_1, \cdots, \mathbf{q}_i, \cdots, \mathbf{q}_N) \in \Gamma^N$ is the position of the N particles in the thermodynamic system $\Gamma \subset \mathbb{R}^3$ under study, and $\mathbf{p} = (\mathbf{p}_1, \cdots, \mathbf{p}_i, \cdots, \mathbf{p}_N)$ momentums of these particles, the *canonical phase-space density* of the system then is

$$p_c(\mathbf{q},\mathbf{p}) = \frac{\exp[-\beta H(\mathbf{q},\mathbf{p})]}{\frac{1}{N!h^{3N}} \int_{\Gamma^N} d\mathbf{q}^N \int_{\mathbb{R}^{3N}} \exp[-\beta H(\mathbf{q},\mathbf{p})] d\mathbf{p}^N} = \frac{\exp[-\beta H(\mathbf{q},\mathbf{p})]}{\mathcal{Z}(T,V,N)}. \tag{12}$$

where $N!$ is the Gibbs corrector because that the particles in the system is indistinguishable. $\mathcal{Z}(T,V,N)$ is called the *canonic partition function*, it depends on the system's temperature T, volume V, and particle number N. Note that under the assumption of the canonic ensemble, they are all fixed for the fixed thermodynamical system Γ. Especially, $V = V(\Gamma) = \int_\Gamma d\mathbf{q}$ implicitly set that $\Gamma \subset \mathbb{R}^3$ has a volume.

Then the entropy S is

$$\begin{aligned} S = S(\Gamma) = \langle -k \ln p_c \rangle &= \frac{k}{N!h^{3N}} \int_{\Gamma^N} d\mathbf{q}^N \int_{\mathbb{R}^{3N}} [\beta H(\mathbf{q},\mathbf{p}) + \ln \mathcal{Z}(T,V,N)] p_c(\mathbf{q},\mathbf{p}) d\mathbf{p}^N \\ &= \frac{1}{T} [\langle H \rangle + kT \ln \mathcal{Z}(T,V,N)]. \end{aligned} \tag{13}$$

From which the Helmholtz free energy $F = F(\Gamma)$ and the Gibbs free energy $G = G(\Gamma)$ are obtained,

$$F = U - TS = -kT \ln \mathcal{Z}(T,V,N), \quad G = PV + F = PV - kT \ln \mathcal{Z}(T,V,N). \tag{14}$$

Therefore, to obtain the Gibbs free energy one has to really calculate $\ln \mathcal{Z}(T, V, N)$, a task that often cannot be done.

4.2. The grand canonic ensemble

The grand canonic ensemble or macroscopic ensemble deals with an open thermodynamic system Γ, i.e., not only energy can be exchanged, matter particles can also be exchanged between Γ and environment. Therefore, the particle number N in Γ is variable.

In classical mechanics, suppose that the phase space is $(\mathbf{q}, \mathbf{p}) \in \Gamma^N \times \mathbb{R}^{3N}$. Let H be the Hamiltonian, the grand canonic phase-space density is

$$p_{gc}(\mathbf{q}, \mathbf{p}, N) = \frac{\exp[-\beta(H - \mu N)]}{\sum_{N=0}^{\infty} \frac{1}{N! h^{3N}} \int_{\Gamma^N} \mathrm{d}\mathbf{q}^N \int_{\mathbb{R}^{3N}} \exp[-\beta(H(\mathbf{q}, \mathbf{p}) - \mu N] \mathrm{d}\mathbf{p}^N} = \frac{\exp[-\beta(H - \mu N)]}{\mathcal{Z}(T, V, \mu)}, \tag{15}$$

where $V = V(\Gamma)$ is the volume of the system. By definition the entropy is

$$\begin{aligned} S(\Gamma) &= \langle -k \ln p_{gc} \rangle = k \sum_{N=0}^{\infty} \int_{\Gamma^N} \mathrm{d}\mathbf{q}^N \int_{\mathbb{R}^{3N}} \{\beta[H(\mathbf{q}, \mathbf{p}) - \mu N] + \ln \mathcal{Z}\} p_{gc}(\mathbf{q}, \mathbf{p}) \mathrm{d}\mathbf{p}^N \\ &= \frac{1}{T} \left[\langle H \rangle - \mu \langle N \rangle + kT \ln \mathcal{Z} \right]. \end{aligned} \tag{16}$$

Here $\langle H \rangle = U$ is the inner energy of the system Γ, $\langle N \rangle = N(\Gamma)$ is the mean number of particles in Γ. More importantly, the function $-kT \ln \mathcal{Z}(T, V, \mu)$ is nothing but the grand canonic potential ϕ, from equation (8) it is just $-PV$. Thus

$$G = U + PV - TS = \mu \langle N \rangle. \tag{17}$$

5. Experimental measuring and theoretical derivation of the Gibbs free energy of protein folding

The newly synthesized peptide chain of a protein automatically folds to its native structure in the physiological environment. Change of environment will make a protein denatured, i.e., the protein no longer performs its biological function. The facts that denaturation does not change the protein molecule, that the only thing changed is its structure, was first theorized by Hisen Wu based on his own extensive experiments, Hisen Wu (1931). It was found that after removing the agents that caused the change of environment, some protein can automatically retake its native structure, this is called renaturation or refolding. After many experiments in denaturation and renaturation, Anfinsen summarized the Thermodynamic Principle as the fundamental law of the protein folding, Anfinsen (1973). Anfinsen's work actually show that protein refolds spontaneously after removing denaturation agents. Therefore, in the physiological or similar environment, the native structure has the minimum Gibbs free energy; and in a changed environment, the denatured structure(s) will have the smaller Gibbs free energy. The Thermodynamic Principle of protein folding then is the general thermodynamics law, if a change happens spontaneously, then the end state will have smaller Gibbs free energy than the initial state.

To apply the Thermodynamic Principle in the research of protein folding, it is necessary to know the Gibbs free energy formula $G(\mathbf{X})$ for each conformation $\mathbf{X}$. Until now, theoretical derivation of $G(\mathbf{X})$ is unsuccessful and rarely being tried. Most knowledge of the Gibbs free energy of protein folding comes from experiment observations.

5.1. Experimental measuring of $\triangle G$

The basic principle of experimentally measuring $\triangle G$, the difference in Gibbs free energy between the native and the denatured structures of a protein is as follows. For protein molecules in a solution, the criterion of the protein is in the native structure is that it performes its biological function, otherwise the protein is denatured or not in the native structure. The level of biological function indicates the degree of the denaturation. Let B be the native structure, denote its molar concentration as $[B]$. Denote A as an non-native structure of the same protein in the solution and $[A]$ its molar concentration.

Three things to be borne in mind: 1. the environment is the physiological environment or similar one such that the protein can spontaneously fold; 2: individual molecule cannot be directly measured, so the measuring is in per mole term, $R = N_A k$ instead of k should be used, where N_A is the Avogadro's number; 3: the environment in reality has constant pressure P, hence the enthalpy $H = U + PV$ can replace the inner energy U, where V is the volume of the system (it is a subset of the whole $\mathbb{R}^3$).

As expressed in (9), the probabilities of the protein takes the conformations A and B are

$$p_A \propto W_A \exp\left(-\frac{H_A}{RT}\right), \quad p_B \propto W_B \exp\left(-\frac{H_B}{RT}\right), \tag{18}$$

where $H_A = U_A + PV$ and $H_B = U_B + PV$ are the enthalpy per mole for A and B, W_A (W_B) is the number of ways of the enthalpy H_A (H_B) can be achieved by microscopic states. The quantities $[A]$ and $[B]$ are assumed to be measurable in experiment. Therefore their ration $K = [A]/[B]$ is also measurable. Then

$$\triangle G^o = -RT \ln K. \tag{19}$$

To see that equation (19) is true, note that the ratio K is equal to the ratio p_A/p_B and the entropies per mole are $S_A = R \ln W_A$, $S_B = R \ln W_B$, therefore

$$\begin{aligned} -RT \ln K &= -RT \ln \frac{p_A}{p_B} = -RT\left(\frac{H_B}{RT} - \frac{H_A}{RT}\right) - RT(\ln W_A - \ln W_B) \\ &= H_A - H_B - T(R\ln W_A - R\ln W_B) = H_A - H_B - T(S_A - S_B) \\ &= H_A - TS_A - (H_B - TS_B) = G_A - G_B = \triangle G^o. \end{aligned} \tag{20}$$

But in reality, the ratio K is measurable in experiment is only theoretical, since in physiological environment $K \cong 0$, i.e., almost all protein molecules take the native structure B.

There is no way to change the native structure B to A while keeping the environment unchanged. In experiments, one has to change the environment to get the protein denatured, that is, to change its shape from the native structure B to another conformation A. Heating the solution is a simple way to change the environment, during the heating, the system absorbs an amount of heat H, the system's temperature increased from T_0 to T_1. Then

$$G(A, T_1) - G(B, T_0) = f(H), \tag{21}$$

where $f(H)$ is a function depending on H and its value is obtained from experiment. What really needed is

$$\triangle G = G(A, T_0) - G(B, T_0). \tag{22}$$

To get $\triangle G$, interpolation to equation (21) is used to estimate the value in T_0. Other methods of changing environment face the same problem, i.e., interpolation has neither theoretical nor observation basis.

Equation (19) may give the reason why $\triangle G$ is used whenever referring the Gibbs free energy. For experiment, only $\triangle G$ can be got. In theoretical derivation, this rule no longer to be followed and moreover, without a base structure to compare to, the notation $\triangle G$ will look strange.

More importantly, it should be emphasized again, that the Thermodynamic Principle really says that in the physiological environment the native structure has the minimum Gibbs free energy; and in other environment, the native structure no longer has the minimum Gibbs free energy. Summarizing, it is

$$G(B, T_0) < G(A, T_0), \quad G(A, T_1) < G(B, T_1). \tag{23}$$

It should always keep in mind that before comparison, first clarify the environment.

When deriving the Gibbs free energy formula, the first thing is also to make clear what is the environment. Another reality that should be borne in mind is that during the protein folding process, the environment does not change.

Remember that after removing the denaturation agent some proteins will spontaneously refold to their native structure, this is called the refolding or renaturation. Distinguish the original protein folding problem and protein refolding problem is another important issue. Only in the refolding case, a theoretical derivation can make the environment change, for example, lower the temperature to the room temperature (around 300K). Some discussions on protein folding are really talking about refolding, because they start from changing the environment from nonphysological to physiological.

While experiment has no way to change the native structure without disturbing the environment, theory can play a role instead. Formulas (1) and (2) give us the chance to compare $\triangle G$, as long as the accurate chemical potentials' values are known.

5.2. Theoretical consideration of the protein folding problem

Protein folding is a highly practical field. Very few attention was paid to its theoretical part. For example, almost nobody has seriously considered the Gibbs free energy formula. Instead, all kinds of empirical models are tried in computer simulation, without any justification in fundamental principle.

One attempt to theoretically get the Gibbs free energy formula from canonic ensemble is summarized by Lazaridis and Karplus (2003), the theoretical part of it is reported below and why it is not successful will be briefly pointed out. Their notations such as $\mathbf{R} = \mathbf{X}$ as conformation, $A = F$ as the Helmholtz free energy, $Q = \mathcal{Z}$ as the partition function, $\Lambda = h$, etc., will be kept in this section.

Treating the protein folding system as the set of all conformations plus surrounding water molecules with a phase point $(\mathbf{R}, \mathbf{r})$, where $\mathbf{r}$ are coordinates of N water molecules plus their orientations. The Hamiltonian H can be decomposed as

$$H = H_{mm} + H_{mw} + H_{ww}, \tag{24}$$

where mm means interactions inside the protein, mw between protein and water molecules, and ww water to water, all in the atomic level. Triplet interactions mmm, mmw, etc., can also be considered, but for simplicity only take the pairwise atomic interactions.

Applying the canonic ensemble, the canonic partition function is

$$Q = \frac{\int \exp(-\beta H) d\mathbf{r}^N d\mathbf{R}^M}{N! \Lambda^{3M} \Lambda^{3N}} = \frac{Z}{N! \Lambda^{3M} \Lambda^{3N}},$$

and the Helmholtz free energy is given by

$$A = -kT \ln Q = -kT \ln \left[\int \exp(-\beta H) d\mathbf{r}^N d\mathbf{R}^M \right] + kT \ln(N! \Lambda^{3M} \Lambda^{3N}). \tag{25}$$

To separate the contributions made by water molecules and the conformations, the *effective energy* W is defined,

$$\exp(-\beta W) = \exp(-\beta H_{mm}) \frac{\int \exp(-\beta H_{mw} - \beta H_{ww}) d\mathbf{r}^N}{\int \exp(-\beta H_{ww}) d\mathbf{r}^N} = \exp(-\beta H_{mm}) \exp(-\beta X), \tag{26}$$

Define

$$\langle \exp(-\beta H_{mw}) \rangle_o = \frac{\int \exp(-\beta H_{mw}) \exp(-\beta H_{ww}) d\mathbf{r}^N}{\int \exp(-\beta H_{ww}) d\mathbf{r}^N}. \tag{27}$$

The effective energy $W(\mathbf{R})$ is:

$$W(\mathbf{R}) = H_{mm}(\mathbf{R}) + X(\mathbf{R}) = H_{mm}(\mathbf{R}) - kT \ln \langle \exp(-\beta H_{mw}) \rangle_o \equiv H_{mm}(\mathbf{R}) + \triangle G^{\text{slv}}(\mathbf{R}). \tag{28}$$

The term $\triangle G^{\text{slv}}(\mathbf{R})$ is called the *solvation free energy* while H_{mm} is the *intra-macromolecular energy*.

After changing $\mathbf{R}$ to interior coordinates $\mathbf{q}$, it is stated that

$$Z = V 8\pi^2 \int \exp(-\beta H_{ww}) d\mathbf{r}^N \int \exp(-\beta W) d\mathbf{q}, \tag{29}$$

because the interior coordinates has only $3M - 6$ dimension, the integration of the remaining 6 dimension over the system getting the value $V8\pi^2$, implying that each $\mathbf{x}_i$ in $\mathbf{R}$ can be any point in the system that has volume V. As usual, the probability of finding the system at the configuration $(\mathbf{q})$ is:

$$p(\mathbf{q}) = \frac{\exp[-\beta W(\mathbf{q})]}{\int \exp[-\beta W(\mathbf{q})]d\mathbf{q}}. \tag{30}$$

Consequently,

$$\int p(\mathbf{q}) \ln p(\mathbf{q}) d\mathbf{q} = -\ln Z + \ln \int \exp(-\beta H_{ww}) d\mathbf{r}^N + \ln(V8\pi^2) - \beta \int p(\mathbf{q}) W(\mathbf{q}) d\mathbf{q}, \tag{31}$$

From equation (25),

$$\begin{aligned} A &= -kT \int \exp(-\beta H_{ww}) d\mathbf{r}^N + kT \ln\left(\frac{\Lambda^{3M}}{V8\pi^2}\right) + \int p(\mathbf{q}) W(\mathbf{q}) d\mathbf{q} + kT \int p(\mathbf{q}) \ln p(\mathbf{q}) d\mathbf{q} \\ &= A^o + kT \ln\left(\frac{\Lambda^{3M}}{V8\pi^2}\right) + \langle W \rangle - TS^{\text{conf}}, \end{aligned} \tag{32}$$

where $A^o = -kT \int \exp(-\beta H_{ww}) d\mathbf{r}^N$ is the pure Helmholtz free energy of pure solvent; the term $-TS^{\text{conf}} = kT \int p(\mathbf{q}) \ln p(\mathbf{q}) d\mathbf{q}$ is the contribution of the configurational entropy of the macromolecule to the free energy.

The Gibbs free energy is $G = A + PV$. Since the volume is thought negligible under ambient conditions so Gibbs and Helmholtz free energies are considered identical.

Now for any subset of $A \subset \Gamma$, integrals restricted on A gives the Helmholtz energy A_A, i.e.,

$$A_A = A^o + kT \ln\left(\frac{\Lambda^{3M}}{V8\pi^2}\right) + \langle W \rangle_A - TS_A^{\text{conf}}. \tag{33}$$

Thus for two different subsets A and B, the difference in the Helmholtz free energy is

$$\begin{aligned} \triangle A &= A_B - A_A = \langle W \rangle_B - \langle W \rangle_A - T(S_B^{\text{conf}} - S_A^{\text{conf}}) \\ &= \triangle \langle H_{mm} \rangle + \triangle \langle \triangle G^{\text{slv}} \rangle - T \triangle S^{\text{conf}}. \end{aligned} \tag{34}$$

Especially, "If A is the denatured state and B the native state, both of which have to be defined in some way and both of which include many configurations, Eq. (34) gives the free energy of folding."

5.3. Critics of the derivation in Lazaridis and Karplus (2003)

Protein folding is considered a very practical research field, dominating activities are computer simulations with empirical models. There are very few theoretical discussions about protein folding. This derivation in Lazaridis and Karplus (2003) is a rare example deserving an analysis to see why for decades there has been no theoretic progress in this field. Many lessons can be learned from this example.

One important lesson from the derivation Lazaridis and Karplus (2003) is that when dealing with thermodynamics and statistical mechanics, the thermodynamic system must be clearly defined. The system will occupy a space in $\mathbb{R}^3$, what is it? How to delimit it?

More importantly, it is not just one conformation **R**, but all conformations of a single protein are considered in the derivation. As a single point $\mathbf{R} \in \mathbb{R}^{3M}$, no structural features of the conformation **R** are considered, i.e., this particle is structureless. Remember that the research object is the conformation of the protein, we cannot treat them as structureless particles. Yes, classical derivations such as the ideal gas system are defined this way, that is because that the interest is not in the individual particle's structure but the macroscopic properties of the idea gas. The lesson then is that instead of considering all conformations together in a system, specific thermodynamic system has to be tailored for each individual conformation **R**. And such a system contains only one conformation **R**, with its structure geometry, and other particles such as water molecules, thus the Gibbs free energy of such a system will be indexed by **R**, $G = G(\mathbf{R})$.

Perhaps the biggest lesson to be learned is that when solving a problem, one should concentrate on the specific features of the problem to design the ways to attack it, not just imitate successful classical examples.

The derivation of Lazaridis and Karplus (2003) gives the effective energy $W(\mathbf{R})$ as some substitute of the Gibbs free energy without theoretic basis for its relation to the Thermodynamic Principle. Moreover, the formula $W(\mathbf{R})$ tells us nothing of how to calculate it, all are buried in multiple-integrations without clear delimitation. Being the only function for individual conformation **R**, it was pointed out in Lazaridis and Karplus (2003) that "The function W defines a hyper-surface in the conformation space of the macromolecule in the presence of equilibrated solvent and, therefore, includes the solvation entropy. This hyper-surface is now often called an 'energy landscape'. It determines the thermodynamics and kinetics of macromolecular conformational transitions." From this comment it can be seen that the authors are not against individual quantities such as $W(\mathbf{R})$ and think they are important to the study of protein folding. Changing the "effective energy" $W(\mathbf{R})$ to the Gibbs free energy $G(\mathbf{R})$, the comment really makes sense. The lesson should be learned is that never invent theoretical concepts without firm theoretical basis. Another one is that always keep in mind that useful Gibbs free energy formula should be calculable.

From now on, the notation $\mathbf{X} = \mathbf{R}$ will be used to represent a conformation. To put the Thermodynamic Principle in practice, not merely as a talking show, what really needed is $G(\mathbf{X})$, the Gibbs free energy of each individual conformation **X**, not the effective energy $W(\mathbf{R})$. One hopes that the formula $G(\mathbf{X})$ should be calculable, not buried in multiple integrations. To get such a formula, the grand canonic ensemble and eventually the quantum statistics have to be applied.

6. Necessary preparations for the derivation of the Gibbs free energy formula

Summarizing what have learned from the critics of the derivation in Lazaridis and Karplus (2003), in any attempt of derivation of the Gibbs free energy formula one has to: 1. clearly state all assumptions used in the derivation; and 2. for each conformation **X**, set a thermodynamic system $\mathcal{T}_{\mathbf{X}}$ associated with **X**; 3. use the grand canonic ensemble.

6.1. The assumptions

All assumptions here are based on well-known facts of consensus among protein folding students. Let $\mathfrak{U}$ be a protein with M atoms $(\mathbf{a}_1, \cdots, \mathbf{a}_i, \cdots, \mathbf{a}_M)$. A structure (conformation) of $\mathfrak{U}$ is a point $\mathbf{X} = (\mathbf{x}_1, \cdots, \mathbf{x}_i, \cdots, \mathbf{x}_M) \in \mathbb{R}^{3M}$, $\mathbf{x}_i \in \mathbb{R}^3$ is the atomic center (nuclear) position of $\mathbf{a}_i$. Alternatively, the conformation $\mathbf{X}$ corresponds to a subset in $\mathbb{R}^3$, $P_\mathbf{X} = \cup_{i=1}^M B(\mathbf{x}_i, r_i) \subset \mathbb{R}^3$ where r_i's are van der Waals radii.

1. The proteins discussed here are monomeric, single domain, self folding globular proteins.
2. Therefore, in the case of our selected proteins, the environment of the protein folding, the physiological environment, is pure water, there are no other elements in the environment, no chaperonins, no co-factors, etc. This is a rational simplification, at least when one considers the environment as only the first hydration shell of a conformation, as in our derivation of the $G(\mathbf{X})$.
3. During the folding, the environment does not change.
4. Anfinsen (1973) showed that before folding, the polypeptide chain already has its main chain's and each residue's covalent bonds correctly formed. Hence, our conformations should satisfy the following steric conditions set in Fang (2005) and Fang and Jing (2010): there are $\epsilon_{ij} > 0$, $1 \le i < j \le M$ such that for any two atoms $\mathbf{a}_i$ and $\mathbf{a}_j$ in $P_\mathbf{X} = \cup_{k=1}^M B(\mathbf{x}_k, r_k)$,

$$\begin{array}{ll} \epsilon_{ij} \le |\mathbf{x}_i - \mathbf{x}_j|, & \text{no covalent bond between } \mathbf{a}_i \text{ and } \mathbf{a}_j; \\ d_{ij} - \epsilon_{ij} \le |\mathbf{x}_i - \mathbf{x}_j| \le d_{ij} + \epsilon_{ij}, & d_{ij} \text{ is the standard bond length between } \mathbf{a}_i \text{ and } \mathbf{a}_j. \end{array} \quad (35)$$

   All conformations satisfying the steric conditions (35) will be denoted as $\mathfrak{X}$ and in this chapter only $\mathbf{X} \in \mathfrak{X}$ will be considered.
5. A water molecule is treated as a single particle centered at the oxygen nuclear position $\mathbf{w} \in \mathbb{R}^3$, and the covalent bonds in it are fixed. In the Born-Oppenheimer approximation, only the conformation $\mathbf{X}$ is fixed, all particles, water molecules or electrons in the first hydration shell of $P_\mathbf{X}$, are moving.
6. As in section 2.3, there are H hydrophobic levels H_i, $i = 1, \cdots, H$, such that $\cup_{i=1}^H H_i = (\mathbf{a}_1, \cdots, \mathbf{a}_i, \cdots, \mathbf{a}_M)$.

6.2. The thermodynamic system $\mathcal{T}_\mathbf{X}$

Let d_w be the diameter of a water molecule and $M_\mathbf{X}$ be the molecular surface of $P_\mathbf{X}$ as defined in Richards (1977) with the probe radius $d_w/2$, see FIGURE 3. Define

$$\mathcal{R}_\mathbf{X} = \{\mathbf{x} \in \mathbb{R}^3 : \text{dist}(\mathbf{x}, M_\mathbf{X}) \le d_w\} \setminus P_\mathbf{X} \quad (36)$$

as the first hydration shell surrounding $P_\mathbf{X}$, where $\text{dist}(\mathbf{x}, S) = \inf_{\mathbf{y} \in S} |\mathbf{x} - \mathbf{y}|$. Then $\mathcal{T}_\mathbf{X} = P_\mathbf{X} \cup \mathcal{R}_\mathbf{X}$ will be our thermodynamic system of protein folding at the conformation $\mathbf{X}$.

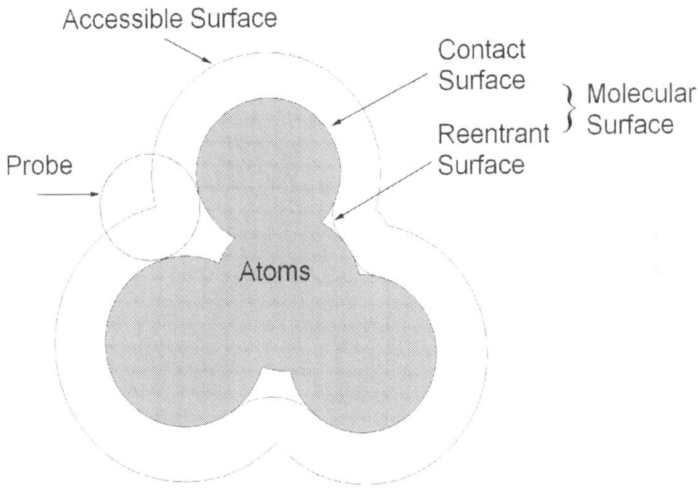

Figure 3. Two dimensional presenting of molecular surface Richards (1977) and solvent accessible surface Lee and Richards (1971). This figure was originally in Fang and Jing (2010).

Let $I_i \subset \{1, 2, \cdots, M\}$ be the subset such that $\mathbf{a}_j \in H_i$ if and only if $j \in I_i$. Define $P_{\mathbf{X}i} = \cup_{j \in I_i} B(\mathbf{x}_j, r_j) \subset P_{\mathbf{X}}$ and as shown in FIGURE 4,

$$\mathcal{R}_{\mathbf{X}i} = \{\mathbf{x} \in \mathcal{R}_{\mathbf{X}} : \text{dist}(\mathbf{x}, P_{\mathbf{X}_i}) \leq \text{dist}(\mathbf{x}, P_{\mathbf{X}} \backslash P_{\mathbf{X}i})\}, \quad 1 \leq i \leq H. \tag{37}$$

Let $V(\Omega)$ be the volume of $\Omega \subset \mathbb{R}^3$, then

$$\mathcal{R}_{\mathbf{X}} = \cup_{i=1}^{H} \mathcal{R}_{\mathbf{X}i}, \quad V(\mathcal{R}_{\mathbf{X}}) = \sum_{i=1}^{H} V(\mathcal{R}_{\mathbf{X}i}), \text{ and for } i \neq j, \quad V(\mathcal{R}_{\mathbf{X}i} \cap \mathcal{R}_{\mathbf{X}j}) = 0. \tag{38}$$

Since $M_{\mathbf{X}}$ is a closed surface, it divides $\mathbb{R}^3$ into two regions $\Omega_{\mathbf{X}}$ and $\Omega'_{\mathbf{X}}$ such that $\partial\Omega_{\mathbf{X}} = \partial\Omega'_{\mathbf{X}} = M_{\mathbf{X}}$ and $\mathbb{R}^3 = \Omega_{\mathbf{X}} \cup M_{\mathbf{X}} \cup \Omega'_{\mathbf{X}}$. Note that $P_{\mathbf{X}} \subset \Omega_{\mathbf{X}}$ and all nuclear centers of atoms in the water molecules in $\mathcal{R}_{\mathbf{X}}$ are contained in $\Omega'_{\mathbf{X}}$. Moreover, $\Omega_{\mathbf{X}}$ is bounded, therefore, has a volume $V(\Omega_{\mathbf{X}})$. For $S \subset \mathbb{R}^3$, denote $\overline{S}$ as the closure of S. Define the hydrophobicity subsurface $M_{\mathbf{X}i}, 1 \leq i \leq H$, as

$$M_{\mathbf{X}i} = M_{\mathbf{X}} \cap \overline{\mathcal{R}_{\mathbf{X}i}}. \tag{39}$$

Let $A(S)$ be the area of a surface $S \subset \mathbb{R}^3$, then

$$M_{\mathbf{X}} = \cup_{i=1}^{H} M_{\mathbf{X}i}, \quad A(M_{\mathbf{X}}) = \sum_{i=1}^{H} A(M_{\mathbf{X}i}), \text{ and if } i \neq j, \text{ then } A(M_{\mathbf{X}i} \cap M_{\mathbf{X}j}) = 0. \tag{40}$$

Although the shape of each atom in a molecule is well defined by the theory of atoms in molecules as in Bader (1990) and Popelier (2000), what concerning us here is the overall

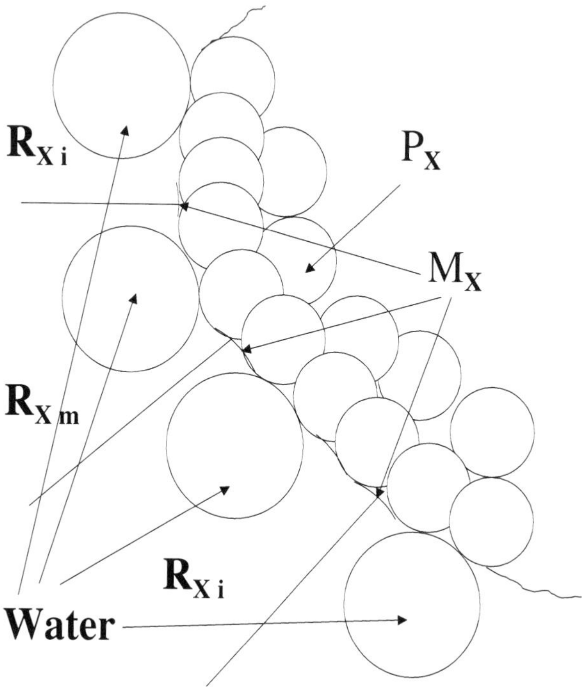

Figure 4. Note that $\mathcal{R}_{Xi}$ generally are not connected, i.e., having more than one block.

shape of the structure P_X. The cutoff of electron density $\rho \geq 0.001$au in Bader (1990) and Popelier (2000) gives the overall shape of a molecular structure that is just like P_X, a bunch of overlapping balls. Moreover, the boundary of the $\rho \geq 0.001$au cutoff is very similar to the molecular surface M_X which was defined by Richards (1977) and was shown has more physical meaning as the boundary surface of the conformation P_X in Tuñón *et. al.* (1992) and Jackson and Sternberg (1993).

7. Gibbs free energy formula: Classical statistical mechanics derivation

The grand canonic ensemble or macroscopic ensemble will be applied to derive the desired Gibbs free energy formula $G(\mathbf{X})$. In addition to let the number of water molecules vary, the assumptions is that the chemical potential μ will be different for water molecules contacting to different hydrophobicity levels H_i (or falling in $\mathcal{R}_{\mathbf{X}i}$). Counting the numbers N_i of water molecules that contact to atoms in H_i, the N and μ in equation (15) should be modified to $N = (N_1, \cdots, N_i, \cdots, N_H)$, $\mu = (\mu_1, \cdots, \mu_i, \cdots, \mu_H)$. Let $(\mathbf{q}, \mathbf{p}) \subset \mathcal{R}_{\mathbf{X}}^M \times \mathbb{R}^{3M}$ be the water molecules' phase space for a fixed N, where $M = \sum_{i=1}^H N_i$. Let $H_{\mathbf{X}} = H_{\mathbf{X}}(\mathbf{q}, \mathbf{p})$ be the Hamiltonian. The grand canonic phase density function will be

$$p_{\mathbf{X}}(\mathbf{q},\mathbf{p},N) = \frac{\exp\{-\beta[H_{\mathbf{X}}(\mathbf{q},\mathbf{p}) - \sum_{i=1}^{H}\mu_i N_i]\}}{\sum_{M=0}^{\infty}\frac{1}{M!h^{3M}}\sum_{\sum N_i=M}\prod_{i=1}^{H}\int_{\mathcal{R}_{\mathbf{X}i}^{N_i}}\mathbf{dq}^{N_i}\int_{\mathbb{R}^{3M}}\exp\{-\beta[H_{\mathbf{X}} - \sum_{i=1}^{H}\mu_i N_i]\}\mathbf{dp}^{3M}}$$

$$= \frac{\exp\{-\beta[H_{\mathbf{X}}(\mathbf{q},\mathbf{p}) - \sum_{i=1}^{H}\mu_i N_i]\}}{\mathcal{Z}(T,V,\mu)}. \tag{41}$$

The entropy $S(\mathbf{X}) = S(\mathcal{T}_{\mathbf{X}})$ is

$$S = \langle -k\ln p_{\mathbf{X}}\rangle = -k\sum_{M=0}^{\infty}\sum_{N_1+\cdots+N_H=M}\prod_{i=1}^{H}\int_{\mathcal{R}_{\mathbf{X}i}^{N_i}}\mathbf{dq}^{N_i}\int_{\mathbb{R}^{3M}}\ln p_{\mathbf{X}}p_{\mathbf{X}}\mathbf{dp}^{M}$$

$$= -k\sum_{M=0}^{\infty}\sum_{N_1+\cdots+N_H=M}\prod_{i=1}^{H}\int_{\mathcal{R}_{\mathbf{X}i}^{N_i}}\mathbf{dq}^{N_i}\left[\beta\sum_{i=1}^{H}\mu_i N_i - \beta H_{\mathbf{X}}(\mathbf{q},\mathbf{p}) - \ln\mathcal{Z}\right]p_{\mathbf{X}}\mathbf{dp}^{M} \tag{42}$$

$$= \frac{1}{T}\left[\langle H\rangle - \sum_{i=1}^{H}\mu_i\langle N_i\rangle + kT\ln\mathcal{Z}(T,V,\mu)\right] \tag{43}$$

$$= \frac{1}{T}\left[U(\mathcal{T}_{\mathbf{X}}) - \sum_{i=1}^{H}\mu_i N_i(\mathcal{T}_{\mathbf{X}}) + kT\ln\mathcal{Z}(T,V,\mu)\right]$$

where $U(\mathbf{X}) = U(\mathcal{T}_{\mathbf{X}}) = \langle H\rangle$ is the inner energy, $N_i(\mathbf{X}) = \langle N_i\rangle$ the mean number of water molecules in $\mathcal{R}_{\mathbf{X}i}$. By equation (8), $kT\ln\mathcal{Z}(T,V,\mu) = -\phi(T,V,\mu) = PV(\mathcal{T}_{\mathbf{X}})$. Therefore, from $G = U + PV - TS$,

$$G(\mathbf{X}) = G(\mathcal{T}_{\mathbf{X}}) = U(\mathbf{X}) + PV(\mathcal{T}_{\mathbf{X}}) - TS(\mathcal{T}_{\mathbf{X}}) = \sum_{i=1}^{H}\mu N_i(\mathbf{X}). \tag{44}$$

The Gibbs free energy given in formula (44) does not involve any integration at all, just counting the number of water molecules contacting atoms in H_i. Furthermore, against the effective energy, potential function H_{mm} plays no role at all, a surprise indeed. But formula (44) also is not easy to calculate, counting the number of water molecules actually need more knowledge of the conformation's boundary, the molecular surface $M_{\mathbf{X}}$. Formula (44) can be directly transfered into a geometric version.

7.1. Converting formula (44) to a geometric version

Since every water molecule in $\mathcal{R}_{\mathbf{X}i}$ has contact with the surface $M_{\mathbf{X}i}$, $N_i(\mathbf{X})$ is proportional to the area $A(M_{\mathbf{X}i})$. Therefore, there are $\nu_i > 0$, such that

$$\nu_i A(M_{\mathbf{X}i}) = N_i(\mathbf{X}), \quad 1 \le i \le H. \tag{45}$$

Substitute in (44),

$$G(\mathbf{X}) = G(\mathcal{T}_{\mathbf{X}}) = \sum_{i=1}^{H}\mu_i\nu_i A(M_{\mathbf{X}i}). \tag{46}$$

For each conformation $\mathbf{X}$, the molecular surface $M_{\mathbf{X}}$ is calculable, see Connolly (1983). The areas $A(M_{\mathbf{X}})$ and $A(M_{\mathbf{X}\,i})$ are also calculable. Therefore, unlike the formula given in (34), this formula is calculable. Moreover, our derivation theoretically justified the surface area models that will be discussed later, only difference is that the molecular surface area is used here instead of the solvent accessible surface area.

But still something is missing. That is, the volume $V(\mathcal{T}_{\mathbf{X}})$, an important thermodynamic quantity, does not show here at all. It seems that no way to put the $V(\mathcal{T}_{\mathbf{X}})$ here in the classical statistical mechanics. To resolve this, the quantum statistical mechanics is necessary.

8. A quantum statistical theory of protein folding

In 1929 Dirac wrote: "The underlying physical laws necessary for the mathematical theory of ... the whole of chemistry are thus completely known, and the difficulty is only that the exact application of these laws leads to equations much too complicated to be soluble." (quoted from Bader (1990), page 132). Yes, the multidimensional Shrödinger equation for protein folding is beyond our ability to solve, no matter how fast and how powerful our computers are. But mathematical theory guarantees that there is a complete set of eigenvalues (energy levels) and eigenfunctions to the Shrödinger equation in the Born-Oppenheimer approximation. Then consider that in the statistical mechanics, ensembles collect all (energy) states of the same system. Although one cannot have exact solutions to the Shrödinger equation, the eigenvalues of it are theoretically known. Thus one can apply the grand canonical ensemble to obtain the desired Gibbs free energy formula $G(\mathbf{X})$. This is the main idea of the derivation.

8.1. The Shrödinger equation

For any conformation $\mathbf{X} \in \mathfrak{X}$, let $\mathbf{W} = (\mathbf{w}_1, \cdots, \mathbf{w}_i, \cdots, \mathbf{w}_N) \in \mathbb{R}^{3N}$ be the nuclear centers of water molecules in $\mathcal{R}_{\mathbf{X}}$ and $\mathbf{E} = (\mathbf{e}_1, \cdots, \mathbf{e}_i, \cdots, \mathbf{e}_L) \in \mathbb{R}^{3L}$ be electronic positions of all electrons in $\mathcal{T}_{\mathbf{X}}$. Then the Hamiltonian for the system $\mathcal{T}_{\mathbf{X}}$ is

$$\hat{H} = \hat{T} + \hat{V} = -\sum_{i=1}^{M} \frac{\hbar^2}{2m_i} \nabla_i^2 - \frac{\hbar^2}{2m_w} \sum_{i=1}^{N} \nabla_i^2 - \frac{\hbar^2}{2m_e} \sum_{i=1}^{L} \nabla_i^2 + \hat{V}(\mathbf{X}, \mathbf{W}, \mathbf{E}), \tag{47}$$

where m_i is the nuclear mass of atom $\mathbf{a}_i$, m_w and m_e are the masses of water molecule and electron; ∇_i^2 is Laplacian in corresponding $\mathbb{R}^3$; and V the potential.

8.2. The first step of the Born-Oppenheimer approximation

Depending on the shape of $P_{\mathbf{X}}$, for each i, $1 \leq i \leq H$, the maximum numbers $N_{\mathbf{X}\,i}$ of water molecules contained in $\mathcal{R}_{\mathbf{X}\,i}$ vary. Theoretically all cases are considered, i.e., there are $0 \leq N_i \leq N_{\mathbf{X}\,i}$ water molecules in $\mathcal{R}_{\mathbf{X}\,i}$, $1 \leq i \leq H$. Let $M_0 = 0$ and $M_i = \sum_{j \leq i} N_j$ and $\mathbf{W}_i = (\mathbf{w}_{M_{i-1}+1}, \cdots, \mathbf{w}_{M_{i-1}+j}, \cdots, \mathbf{w}_{M_i}) \in \mathcal{R}_{\mathbf{X}\,i}^{N_i}$, $1 \leq i \leq H$, and $\mathbf{W} = (\mathbf{W}_1, \mathbf{W}_2, \cdots, \mathbf{W}_{M_H}) \in \prod_{i=1}^{H} \mathcal{R}_{\mathbf{X}\,i}^{N_i}$ denote the nuclear positions of water molecules in $\mathcal{R}_{\mathbf{X}}$. As well, there will be all possible numbers $0 \leq N_e < \infty$ of electrons in $\mathcal{T}_{\mathbf{X}}$. Let $\mathbf{E} = (\mathbf{e}_1, \mathbf{e}_2, \cdots, \mathbf{e}_{N_e}) \in \mathbb{R}^{3N_e}$ denote their nuclear positions. For each fixed $\mathbf{X} \in \mathfrak{X}$ and $N = (N_1, \cdots, N_H, N_e)$, the Born-Oppenheimer approximation has the Hamiltonian

$$\hat{H}_X = -\frac{\hbar^2}{2}\left\{\frac{1}{m_w}\sum_{j=1}^{M_H}\nabla_j^2 + \frac{1}{m_e}\sum_{v=1}^{N_e}\nabla_v^2\right\} + \hat{V}(\mathbf{X},\mathbf{W},\mathbf{E}). \tag{48}$$

The eigenfunctions $\psi_i^{\mathbf{X},N}(\mathbf{W},\mathbf{E}) \in L_0^2(\prod_{i=1}^{H}\mathcal{R}_{\mathbf{X}i}^{N_i}\times\mathcal{T}_{\mathbf{X}}^{N_e}) = \mathcal{H}_{\mathbf{X},N}$, $1\le i<\infty$, comprise an orthonormal basis of $\mathcal{H}_{\mathbf{X},N}$. Denote their eigenvalues (energy levels) as $E_{\mathbf{X},N}^i$, then $\hat{H}_{\mathbf{X}}\psi_i^{\mathbf{X},N} = E_{\mathbf{X},N}^i\psi_i^{\mathbf{X},N}$.

8.3. Grand partition function and grand canonic density operator

Since the numbers N_i and N_e vary, the grand canonic ensemble should be adopted. Let μ_i be the chemical potentials, that is, the Gibbs free energy per water molecule in $\mathcal{R}_{\mathbf{X}i}$. Let μ_e be electron chemical potential. The grand canonic density operator is like in equation (15), or see Greiner *et. al.* (1994) and Dai (2007)

$$\hat{\rho}_{\mathbf{X}} = \exp\left\{-\beta\left[\hat{H}_{\mathbf{X}} - \sum_{i=1}^{H}\mu_i\hat{N}_i - \mu_e\hat{N}_e - \phi(\mathbf{X})\right]\right\}. \tag{49}$$

where $\phi(\mathbf{X})$ is the grand canonic potential ϕ in equation (8) with the index $\mathbf{X}$ and the grand partition function is

$$\begin{aligned}\exp[-\beta\phi(\mathbf{X})] &= \mathrm{Trace}\left\{\exp\left[-\beta\left(\hat{H}_{\mathbf{X}} - \sum_{i=1}^{H}\mu_i\hat{N}_i - \mu_e\hat{N}_e\right)\right]\right\}\\ &= \sum_{i,N}\exp\left\{-\beta\left[E_{\mathbf{X},N}^i - \sum_{i=1}^{H}\mu_i N_i - \mu_e N_e\right]\right\}.\end{aligned} \tag{50}$$

8.4. The Gibbs free energy $G(\mathbf{X})$

As in equation (16), under the grand canonic ensemble the entropy $S(\mathbf{X}) = S(\mathcal{T}_{\mathbf{X}})$ of the system $\mathcal{T}_{\mathbf{X}}$ is

$$\begin{aligned}S(\mathbf{X}) &= -k\mathrm{Trace}(\hat{\rho}_{\mathbf{X}}\ln\hat{\rho}_{\mathbf{X}}) = -k\langle\ln\hat{\rho}_{\mathbf{X}}\rangle = k\beta\left\langle\hat{H}_{\mathbf{X}} - \phi(\mathbf{X}) - \sum_{i=1}^{H}\mu_i\hat{N}_i - \mu_e\hat{N}_e\right\rangle\\ &= \frac{1}{T}\left[\langle\hat{H}_{\mathbf{X}}\rangle - \langle\phi(\mathbf{X})\rangle - \sum_{i=1}^{H}\mu_i\langle\hat{N}_i\rangle - \mu_e\langle\hat{N}_e\rangle\right]\\ &= \frac{1}{T}\left[U(\mathbf{X}) - \phi(\mathbf{X}) - \sum_{i=1}^{H}\mu_i N_i(\mathbf{X}) - \mu_e N_e(\mathbf{X})\right].\end{aligned} \tag{51}$$

Denote $\langle \hat{N}_i \rangle = N_i(\mathbf{X})$ as the mean number of water molecules in $\mathcal{R}_{\mathbf{X}\,i}$, $1 \le i \le H$, and $\langle \hat{N}_e \rangle = N_e(\mathbf{X})$ the mean number of electrons in $\mathcal{T}_{\mathbf{X}}$. The inner energy $\langle \hat{H}_{\mathbf{X}} \rangle$ of the system $\mathcal{T}_{\mathbf{X}}$ is denoted as $U(\mathbf{X}) = U(\mathcal{T}_{\mathbf{X}})$. By equation (8) and the remark after it $\phi(\mathbf{X})(T, V, \mu_1, \cdots, \mu_H, \mu_e) = -PV(\mathbf{X})$, where $V(\mathbf{X}) = V(\mathcal{T}_{\mathbf{X}})$ is the volume of the thermodynamic system $\mathcal{T}_{\mathbf{X}}$. Thus by equation (51) the Gibbs free energy $G(\mathbf{X}) = G(\mathcal{T}_{\mathbf{X}})$ in formula (1) is obtained:

$$G(\mathbf{X}) = G(\mathcal{T}_{\mathbf{X}}) = PV(\mathbf{X}) + U(\mathbf{X}) - TS(\mathbf{X}) = \sum_{i=1}^{H} \mu_i N_i(\mathbf{X}) + \mu_e N_e(\mathbf{X}). \tag{52}$$

8.5. Converting formula (1) to geometric form (2)

As in the classical statistical mechanics case,

$$\nu_i A(M_{\mathbf{X}\,i}) = N_i(\mathbf{X}), \quad 1 \le i \le H. \tag{53}$$

Similarly, there will be a $\nu_e > 0$ such that $\nu_e V(\mathcal{T}_{\mathbf{X}}) = N_e(\mathbf{X})$. By the definition of $\mathcal{T}_{\mathbf{X}}$ and $\Omega_{\mathbf{X}}$, it is roughly $V(\mathcal{T}_{\mathbf{X}} \backslash \Omega_{\mathbf{X}}) = d_w A(M_{\mathbf{X}})$. Thus

$$N_e(\mathbf{X}) = \nu_e V(\mathcal{T}_{\mathbf{X}}) = \nu_e [V(\Omega_{\mathbf{X}}) + V(\mathcal{T}_{\mathbf{X}} \backslash \Omega_{\mathbf{X}})] = \nu_e V(\Omega_{\mathbf{X}}) + d_w \nu_e A(M_{\mathbf{X}}). \tag{54}$$

Substitute equations (45) and (54) into formula (1), formula (2) is obtained.

9. Some remarks

The question to applying fundamental physical laws directly to the protein folding problem is, can it be done? It should be checked that how rigorous is the derivation and be asked that are there any fundamental errors? Possible ways to modify the formula or the derivation will also be discussed.

By applying quantum statistics the protein folding problem is theoretically treated. A theory is useful only if it can make explanations to the observed facts and if it can simplify and improve research methods as well as clarify concepts. It will be shown that $G(\mathbf{X})$ can do exactly these.

If the same theoretical result can be derived from two different disciplines, it is often not just by chance. An early phenomenological mathematical model Fang (2005), starting from purely geometric reasoning, has achieved formula (2), with just two hydrophobic levels, hydrophobic and hydrophilic.

A theory also has to be falsifiable, that is making a prediction to be checked. The fundamental prediction is that minimizing formula (1) or (2) the native structures will be obtained for the amino acid sequences of proteins considered in the assumptions of the formulas. That can only be done after the actual values of the chemical potentials appear in the formulas, for the physiological environment, are determined.

9.1. How rigorous is the derivation?

Two common tools in physics, the first step of the Born-Oppenheimer approximation in quantum mechanics and the grand canonic ensemble in statistical physics, are applied to obtain formula (1).

9.1.1. The Born-Oppenheimer approximation

The Born-Oppenheimer approximation "treats the electrons as if they are moving in the field of fixed nuclei. This is a good approximation because, loosely speaking, electrons move much faster than nuclei and will almost instantly adjust themselves to a change in nuclear position." Popelier (2000). Since the mass of a water molecule is much less than the mass of a protein, this approximation can be extended to the case of when **X** changes the other particles, electrons and water molecules, will quickly adjust themselves to the change as well.

9.1.2. The statistical physics in general and the grand canonic ensemble in particular

"Up to now there is no evidence to show that statistical physics itself is responsible for any mistakes," the Preface of Dai (2007). Via the ensemble theory of statistical mechanics only one protein molecule and particles in its immediate environment are considered, it is justified since as pointed out in Dai (2007) page 10, "When the duration of measurement is short, or the number of particles is not large enough, the concept of ensemble theory is still valid." And among different ensembles, "Generally speaking, the grand canonic ensemble, with the least restrictions, is the most convenient in the mathematical treatment." Dai (2007) page 16. In fact, the canonic ensemble has been tried and ended with a result that the eigenvalues of the quantum mechanics system have to be really calculated, to do it accurately is impossible.

The derivations in this chapter only puts together the two very common and sound practices: the Born-Oppenheimer approximation (only the first step) and the grand canonic ensemble, and apply them to the protein folding problem. As long as protein folding obeys the fundamental physical laws, there should not be any serious error with the derivation.

9.2. Equilibrium and quasi-equilibrium

A protein's structure will never be in equilibrium, in fact, even the native structure is only a snapshot of the constant vibration state of the structure. The best description of conformation **X** is given in Chapter 3 of Bader (1990). Simply speaking, a conformation **X** actually is any point **Y** such that all $\mathbf{y}_i$ are contained in a union of tiny balls centered at $\mathbf{x}_i$, $i = 1, \cdots, M$. In this sense, it can only be anticipated that a quasi-equilibrium description (such as the heat engine, Bailyn (1994) page 94) of the thermodynamic states of the protein folding. This has been built-in in the Thermodynamical Principle of Protein Folding. So the quantities such as $S(\mathbf{X})$, $\phi(\mathbf{X})$, and $G(\mathbf{X})$ can only be understood in this sense. That is, observing a concrete folding process one will see a series of conformations $\mathbf{X}_i$, $i = 1, 2, 3, \cdots$. The Thermodynamic Principle then says that measuring the Gibbs free energy $G(\mathbf{X}_i)$ one will observe that eventually $G(\mathbf{X}_i)$ will converge to a minimum value and the $\mathbf{X}_i$ will eventually approach to the native structure. While all the time, no conformation $\mathbf{X}_i$ and thermodynamic system $\mathcal{T}_{\mathbf{X}_i}$ are really in equilibrium state.

9.3. Potential energy plays no role in protein folding

Formulas (1) and (2) theoretically show that hydrophobic effect is the driving force of protein folding, it is not just solvent free energy besides the pairwise interactions such as the Coulombs, etc., as all force fields assumed. Only in the physiological environment the hydrophobic effect works towards to native structure, otherwise it will push denaturation as discussed in explanation of folding and unfolding. Formulas (1) and (2) show that the Gibbs free energy is actually independent of the potential energy, against one's intuition and a bit of surprising. The explanation is that during the folding process, all covalent bonds in the main chain and each side chain are kept invariant, the potential energy has already played its role in the synthesis process of forming the peptide chain, which of course can also be described by quantum mechanics. According to Anfinsen (1973), protein folding is after the synthesis of the whole peptide chain, so the synthesis process can be skipped and the concentration can be focused on the folding process.

The steric conditions (35) will just keep this early synthesis result, not any $\mathbf{X} = (\mathbf{x}_1, \cdots, \mathbf{x}_i, \cdots, \mathbf{x}_M)$ is eligible to be a conformation, it has to satisfy the steric conditions (35). The steric conditions not only pay respect to the bond length, it also reflect a lot of physic-chemical properties of a conformation: They are defined via the allowed minimal atomic distances, such that for non-bonding atoms, the allowed minimal distances are: shorter between differently charged or polarized atoms; a little longer between non-polar ones; and much longer (generally greater than the sum of their radii) between the same charged ones, etc. For example, minimal distance between sulfur atoms in Cysteine residues to form disulfide bonds is allowed. And for any newly found intramolecular covalent bond between side chains, such as the isopeptide bonds in Kang and Baker (2011), the steric conditions can be easily modified to allow the newly found phenomenon.

The drawback of the steric conditions is that the minimization in equation (57) becomes a constrained minimization.

9.4. Unified explanation of folding and denaturation

Protein denaturation is easy to happen, even if the environment is slightly changed, as described by Hsien Wu (1931). (Hsien Wu (1931) is the 13th article that theorizes the results of a series experiments, and a preliminary report was read before the Xlllth International Congress of Physiology at Boston, August 19-24, 1929, and published in the *Am. J. Physiol.* for October 1929. In which Hsien Wu first suggested that the denatured protein is still the same molecule, only structure has been changed.) Anfinsen in various experiments showed that after denaturation by changed environment, if removing the denature agent, certain globular proteins can spontaneously refold to its native structure, Anfinsen (1973). The spontaneous renaturation suggests that protein folding does not need outside help, at least to the class of proteins in this chapter. Therefore, the fundamental law of thermodynamics asserts that in the environments in which a protein can fold, the native structure must have the minimum Gibbs free energy. The same is true for denaturation, under the denatured environment, the native structure no longer has the minimum Gibbs free energy, some other structure(s), will have the minimum Gibbs free energy. Thus let En present environment, any formula of Gibbs free energy should be stated as $G(\mathbf{X}, En)$ instead of just $G(\mathbf{X})$, unless the environment is specified like in this chapter. Let En_N be the physiological environment and En_U be some denatured environment, $\mathbf{X}_N$ be the native structure and $\mathbf{X}_U$ be one of the denatured stable structure in En_U, then the thermodynamic principle for both of protein folding and unfolding should be that

$$G(\mathbf{X}_N, En_N) < G(\mathbf{X}_U, En_N), \quad G(\mathbf{X}_N, En_U) > G(\mathbf{X}_U, En_U). \tag{55}$$

To check this, an experiment should be designed that can suddenly put proteins in a different environment. Formulas (1) and (2) should be written as $G(\mathbf{X}, En_N)$. Indeed, the chemical potentials μ_e and μ_i's are Gibbs free energies per corresponding particles, $\mu = u + Pv - Ts$. Two environment parameters, temperature T and pressure P, explicitly appear in μ, the inner energy u and entropy s may also implicitly depend on the environment. According to formulas (1) and (2), if $\mu_i < 0$, then make more H_i atoms to expose to water (make larger $A(M_{\mathbf{X}i})$) will reduce the Gibbs free energy. If $\mu_i > 0$, then the reverse will happen. Increase or reduce the H_i atoms' exposure to water $(A(M_{\mathbf{X}i}))$, the conformation has to change. The conformation changes to adjust until a conformation $\mathbf{X}_N$ is obtained, such that the net effect of any change of the conformation will either increase some H_i atoms' exposure to water while $\mu_i > 0$ or reduce H_i atoms' exposure to water while $\mu_i < 0$. In other words, the $G(\mathbf{X}, En_N)$ achieves its minimum at $G(\mathbf{X}_N, En_N)$. Protein folding, at least for the proteins considered in the assumptions, is explained very well by formulas (1) and (2).

In changed environment, the chemical potentials μ_e and μ_i's in formulas (1) and (2) changed their values. With the changed chemical potentials, $G(\mathbf{X}, En_U)$ has the same form as $G(\mathbf{X}, En_N)$ but different chemical potentials. Therefore, the structure $\mathbf{X}_U$ will be stable, according to the second inequality in (55), the process is exactly the same as described for the protein folding if the changing environment method does not include introducing new kinds (non-water) of particles, for example, if only temperature or pressure is changed.

Even in the new environment including new kinds of particles, formulas (1) and (2) can still partially explain the denaturation, only that more obstructs prevent the protein to denature to $\mathbf{X}_U$, but any way it will end in some structure other than the $\mathbf{X}_N$, the protein is denatured. Actually, this is a hint of how to modify the current formulas to extend to general proteins.

9.5. Explain hydrophobic effect and the role played by hydrogen bonding

In 1959, by reviewing the literature Kauzmann concluded that the hydrophobic effect is the main driving force in protein folding, Kauzmann (1959). Empirical correlation between hydrophobic free energy and aqueous cavity surface area was noted as early as by Reynolds *et.al.* (1974), giving justification of the hydrophobic effect. Various justifications of hydrophobic effect were published, based on empirical models of protein folding, for example, Dill (1990). But the debate continues to present, some still insist that it is the hydrogen bond instead of hydrophobic effect plays the main role of driving force in protein folding, for example, Rose *et. al* (2006). The theoretically derived formulas (1) and (2) can explain why the hydrophobic effect is indeed the driving force. A simulation of reducing hydrophobic area alone by Fang and Jing (2010) shows that the result is the appearance of regularly patterned intramolecular hydrogen bonds associated to the secondary structures.

In fact, according to formulas (1) and (2), if $\mu_i < 0$, then make more H_i atoms to appear in the boundary of $P_{\mathbf{X}}$ will reduce the Gibbs free energy. If $\mu_i > 0$, then the reverse will happen, reducing the exposure of H_i atoms to water will reduce the Gibbs free energy. This gives a theoretical explanation of the hydrophobic effect. The kinetic formulas $\mathbf{F}_i = -\nabla_{\mathbf{x}_i} G(\mathbf{X})$ (will be discussed later) is the force that push the conformation to change to the native structure.

The mechanics stated above works through the chemical potentials μ_i for various levels of hydrophobicity. In physiological environment, all hydrophobic H_i's will have positive μ_i, all hydrophilic H_i's will have negative μ_i. Thus changing conformation P_X such that the most hydrophilic H_i ($\mu_i = \min(\mu_1, \cdots, \mu_H)$) gets the first priority to appear on the boundary, and the most hydrophobic H_i ($\mu_i = \max(\mu_1, \cdots, \mu_H)$) gets the first priority to hide in the hydrophobic core to avoid contacting with water molecules, etc. One should keep in mind that all the time, the steric conditions (35) have to be obeyed.

But the hydrophobic effect is actually partially working through hydrogen bond formation. This is well presented in the chemical potentials in formulas (1) and (2). In fact, the values of the chemical potentials reflect the ability of the atoms or atom groups to form hydrogen bond, either with another atom group in the protein or with water molecules. This gives a way to theoretically or experimentally determine the values of hydrophilic chemical potentials: checking the actual energy value of the hydrogen bond.

According to Fikelstein and Ptitsyn (2002), energies of hydrogen bonds appearing in protein (intermolecular or intramolecular) are (the positive sign means that to break it energy is needed) and their energies are:

O–H : : : O (21 kJ mol^{-1} or 5.0 kcal mol^{-1}); O–H : : : N (29 kJ mol^{-1} or 6.9 kcal mol^{-1});
N–H : : : N (13 kJ mol^{-1} or 3.1 kcal mol^{-1}); N–H : : : O (8 kJ mol^{-1} or 1.9 kcal mol^{-1}).

For hydrophobic ones, it will be more complicated, common sense is that it reduces the entropy that certainly comes from the inability of forming hydrogen bonds with water molecules. Hence although hydrophobic effect is the driving force of protein folding, it works through the atom's ability or inability to form hydrogen bonds with water molecules.

How to explain the intramolecular hydrogen bonds? It seems that formulas (1) and (2) do not address this issue. The possible theory is that the amino acid sequence of a protein is highly selectable in evolution, in fact only a tiny number of amino acid sequences can really become a protein.

Indeed, suppose in average each species (or "kind" of prokaryote) has 10^5 proteins (*Homo sapiens* has around 3×10^5), and assume that per protein has 100 variants (versions with tiny difference in the peptide sequence of the protein), then there are at most 10^{47} peptide sequences that can really produce a natural protein. Now further suppose that only one in 10^{13} theoretically protein producing peptide sequences on the earth get a chance to be realized, then there will be at most 10^{60} possible protein producing peptide sequence. A huge number! The number of peptide sequences of length less than or equal to n is

$$N(n) = \sum_{i=1}^{n} 20^i = \frac{20^{n+1} - 20}{19} = \frac{20}{19}(20^n - 1) \cong 20^{n+0.0171} \cong 10^{1.301(n+0.0171)}. \tag{56}$$

The longest amino acid sequence in the record of ExPASy Proteomics Server has 35,213 residues. Then $N(35,213) > 10^{1.3\times 35,213} > 10^{45060}$ and the ratio of the number of potentially protein producing peptide sequences to the number of all possible sequence of length up to 35,213 is less than $10^{60}/10^{45060} = 10^{-45000}$, so tiny a number that it is undistinguishable from zero. Even assuming that the longest peptide sequence is only 400, the ratio is still less than 10^{-460}. How small a chance that a random peptide sequence happens to be a protein's peptide sequence!

With these highly specially selected peptide sequences, one can assume that while shrinking the various hydrophobic surfaces to form a hydrophobic core, residues are put in positions to form secondary structures and their associated hydrogen bonds. This sounds a little bit too arbitrary. But the huge number of candidate peptide sequences makes the evolutional selection not only possible but also probable. Moreover, a simulation of shrinking hydrophobic surface area alone indeed produced secondary structures and hydrogen bonds. The simulation was reported by Fang and Jing (2010). Without calculating any dihedral angles or electronic charges, without any arbitrary parameter, paying no attention to any particular atom's position, by just reducing hydrophobic surface area (there it was assumed that there are only two kinds of atoms, hydrophobic and hydrophilic), secondary structures and hydrogen bonds duly appeared. The proteins used in the simulation are 2i9c, 2hng, and 2ib0, with 123, 127, and 162 residues. No simulation of any kind of empirical or theoretical models had achieved such a success. More than anything, this simulation should prove that hydrophobic effect alone will give more chance of forming intramolecular hydrogen bonds. Indeed, pushing hydrophilic atoms to make hydrogen bonds with water molecules will give other non-boundary hydrophilic groups more chance to form intramolecular hydrogen bonds.

Again formula (2) can partly explain the success of this simulation, when there are only two hydrophobic classes in formula (2), the hydrophobic area presents the main positive part of the Gibbs free energy, reducing it is reducing the Gibbs free energy, no matter what is the chemical potential's real value.

9.6. Explanation of the successes of surface area models

In 1995, Wang *et al* (1995) compared 8 empirical energy models by testing their ability to distinguish native structures and their close neighboring compact non-native structures. Their models WZS are accessible surface area models with 14 hydrophobicity classes of atoms, $\sum_{i=1}^{14} \sigma_i A_i$. Each two combination of three targeting proteins were used to train WZS to get σ_i, hence there are three models WZS1, WZS2, and WZS3. Among the 8 models, all WZS's performed the best, distinguishing all 6 targeting proteins. The worst performer is the force field AMBER 4.0, it failed in distinguishing any of the 6 targets.

These testing and the successes of various surface area models such as Eisenberg and MacLachlan (1986), showed that instead of watching numerous pairwise atomic interactions, the surface area models, though looking too simple, have surprising powers. Now the formula (2) gives them a theoretic justification. On the other hand, the successes of these models also reenforce the theoretical results.

There is a gap between the accessible surface area model in Eisenberg and Maclanchlan (1986) and the experiment results (surface tension), as pointed out in Tuñón *et. al.* (1992). The gap disappeared when one uses the molecular surface area to replace the accessible surface area, in Tuñón *et. al.* (1992) it was shown that molecular surface area assigned of 72-73 cal/mol/Å^2 perfectly fits with the macroscopic experiment data. Later it was asserted that the molecular surface is the real boundary of protein in its native structure by Jackson and Sternberg (1993).

By the definition of Ω'_X, as shown in FIGURE 3 and FIGURE 4, water molecules contact to P_X must be outside the molecular surface M_X. Since the assessable surface is in the middle of the first hydration shell, it is better to use the molecular surface M_X as the boundary of the conformation $P_{\mathbf{X}}$. Moreover, the conversion of the mean numbers $N_i(\mathbf{X})$ to surface area, $N_i(\mathbf{X}) = \nu_i A(M_{\mathbf{X}.i})$, only works for the molecular surface, not for the accessible surface. This can explain the conclusions that molecular surface is a much better boundary than accessible surface as stated in Tuñón *et. al.* (1992) and Jackson and Sternberg (1993).

In fact, the advantage of the solvent accessible surface is that by definition of it one knows exactly each atom occupies which part of the surface, therefore, one can calculate its share in surface area. This fact may partly account why there are so many models based on the solvent accessible surface, even people knew the afore mentioned gap. For other surfaces, one has to define the part of surface that belongs to a specific hydrophobicity class. This was resolved in Fang (2005) via the distance function definition as is used here.

All surface area models neglected one element, the volume of the structure. As early as in the 1970's, Richards and his colleagues already pointed out that the native structure of globular proteins is very dense, or compact, (density $=$ 0.75, Richards (1977)). To make a conformation denser, obviously we should shrink the volume $V(\Omega_{\mathbf{X}})$. The model in Fang (2005) introduced volume term but kept the oversimplification of all atoms are either hydrophobic or hydrophilic. The derivation of formulas (1) and (2) shows that volume term should be counted, but it may be that $\nu_e \mu_e$ is very small, in that case, volume maybe really is irrelevant.

9.7. Coincidence with phenomenological mathematical model

If a theoretical result can be derived from two different disciplines, its possibility of correctness will be dramatically increased. Indeed, from a pure geometric consideration, a phenomenological mathematical model, $G(\mathbf{X}) = aV(\Omega_{\mathbf{X}}) + bA(M_{\mathbf{X}}) + cA(M_{\mathbf{X}1}), a, b, c > 0$ (it was assumed that there are only two hydrophobicity levels, hydrophobic and hydrophilic, the hydrophilic surface area $A(M_{\mathbf{X}2})$ is absorbed in $A(M_{\mathbf{X}})$ by $A(M_{\mathbf{X}2}) = A(M_{\mathbf{X}}) - A(M_{\mathbf{X}1})$), was created in Fang (2005). It was based on the well-known global geometric characteristics of the native structure of globular proteins: 1. high density; 2. smaller surface area; 3. hydrophobic core, as demonstrated and summarized in Richards (1977) and Novotny *et.al* (1984). So that to obtain the native structure, one should shrink the volume (increasing the density) and surface area, and form better hydrophobic core (reducing the hydrophobic surface area $A(M_{\mathbf{X}1})$) simultaneously and cohesively.

The coincidence of formula (2) and the phenomenological mathematical model of Fang (2005) cannot be just a coincidence. Most likely, it is the same natural law reflected in different disciplines. The advantage of formula (2) is that everything there has its physical meaning.

10. Applications

After the derivation it is suitable to point out some immediate applications of the formula $G(\mathbf{X})$.

10.1. Energy surface or landscape

An obvious application is the construction of Gibbs free energy surface or landscape. Empirical estimate is no longer needed, the Gibbs free energy formula $G : \mathfrak{X} \to \mathbb{R}$ gives a graph $(\mathbf{X}, G(\mathbf{X}))$ over the space $\mathfrak{X}$ (all eligible conformations for a given protein), and this is nothing but the Gibbs free energy surface. Mathematically it is a $3M$ dimensional hyper-surface. Its characteristics concerned by students of energy surface theory, such as how rugged it is? how many local minimums are there? is there a funnel? etc., can be answered by simple calculations of the formula.

Since the function G is actually defined on the whole $\mathbb{R}^{3M}$ (on an domain of $\mathbb{R}^{3M}$ containing all $\mathfrak{X}$ is enough), mathematical tools can be explored to study its graph, and compare the results with the restricted conformations. One important question is: Does the absolute minimum structure belongs to $\mathfrak{X}$?

10.2. Structure prediction

Prediction of protein structures is the most important method to reveal proteins' functions and working mechanics, it becomes a bottle neck in the rapidly developing life science. With more and more powerful computers, this problem is attacked in full front. Various models, homologous or *ab initio*, full atom model or coarse grained, with numerous parameters of which many are quite arbitrary, are used to achieve the goal. Although our computer power growths exponentially, prediction power does not follow that way. At this moment, one should take a deep breath and remind what the great physicist Fermi said: "There are two ways of doing calculations in theoretical physics. One way, and this is the way I prefer, is to have a clear physical picture of the process that you are calculating. The other way is to have a precise and self consistent mathematical formalism." And "I remember my friend Johnny von Neumann used to say, with four parameters I can fit an elephant, and with five I can make him wiggle his trunk." Quoted from Dyson (2004).

These remarks should also apply to any scientific calculation, not just theoretical physics. Look at the current situation, all *ab initio* prediction models are actually just empirical with many parameters to ensure some success. Fermi's comments remind us that a theory should be based on fundamental physical laws, and contain no arbitrary parameters. Look at formulas (1) and (2), one sees immediately that they are neat, precise and self consistent mathematical formulas. Furthermore, they including no arbitrary parameter, all terms in them have clear physical meanings. Chemical potentials μ_e and μ_i's, geometric constants ν_e and ν_i's, can be evalued by theory or experiments, they are not arbitrary at all.

But a theory has to be developed, tested, until justified or falsified. For interested researchers, the tasks are to determine the correct values of the chemical potentials in formula (1) and the geometric ratios ν_e and ν_i in formula (2). There are many estimates to them, but they are either for the solvent accessible surface area such as in Eisenberg and MaLachlan (1986) hence not suit to the experiment data as pointed out in Tuñón *et. al.*, or do not distinguish different hydrophobicity levels as in Tuñón *et. al.* (1992). To get the correct values of the chemical potentials and geometric constants, commonly used method of training with data can be employed, in which one can also test the formulas' ability of discriminating native and nearby compact non-native structures. After that, a direct test is to predict the native structure from the amino acid sequence of a protein by minimizing the following:

$$G(\mathbf{X}_N) = \inf_{\mathbf{X}\in\mathfrak{X}} G(\mathbf{X}). \tag{57}$$

and the gradient of G can be written as

$$\nabla G(\Theta) = \left(\frac{\partial G}{\partial \theta_1}, \cdots, \frac{\partial G}{\partial \theta_i}, \cdots, \frac{\partial G}{\partial \theta_L}\right)(\Theta) = (G_{\vec{L}_{\mathbf{b}_1}}, \cdots, G_{\vec{L}_{\mathbf{b}_i}}, \cdots, G_{\vec{L}_{\mathbf{b}_L}})(\mathbf{X}). \quad (67)$$

If the rotation around $\mathbf{b}_i$ with rotating angle $-sG_{\vec{L}_{b_i}}(\mathbf{X})$ on R_{b_i} and fix atoms in F_{b_i} be denoted as M_i, new conformation $\mathbf{Y}_s = M_L \circ M_{L-1} \circ \ldots \circ M_1$ will be obtained, where $s > 0$ is a suitable step length. That is to say, the dihedral angles of $\mathbf{Y}_s$ are

$$[\theta_1 - sG_{\vec{L}_{b_1}}(\mathbf{X}), \cdots, \theta_i - sG_{\vec{L}_{b_i}}(\mathbf{X}), \cdots, \theta_L - sG_{\vec{L}_{b_L}}(\mathbf{X})].$$

The order of rotations in fact is irrelevant, i.e., by any order, the same conformation $\mathbf{Y}_s$ will always be obtained, as proved in Fang and Jing (2008) and Appendix A. This way one can fast change the structure by simultaneous rotate around all rotatable bonds.

This actually is the Newton's fastest descending method, it reduces the Gibbs free energy $G(\mathbf{X})$ most efficiently. Afore mentioned simulations of Fang and Jing (2010) used this method.

10.4. Kinetics

There are evidence that some protein's native structure is not the global minimum of the Gibbs free energy, but only a local minimum. If the native structure of a protein achieves the global minimal value of the Gibbs free energy, the folding process is **thermodynamic**; if it is only a local minimum, the folding process is **kinetic**, Lazaridis and Karplus (2003).

With the formula (2) and the gradient just obtained, one actually has the kinetic in hand. In fact, for any atomic position $\mathbf{x}_i$, the kinetic force is $\mathbf{F}_i(\mathbf{X}) = -\nabla_{\mathbf{x}_i} G(\mathbf{X})$, Dai (2007). With formula (2) these quantities are readily calculable as mentioned above. The resulting Newton's fastest descending method will help us find the native structure, either in the thermodynamic case or in the kinetic case, here the thermodynamic and kinetic cases are combined by the Gibbs free energy formula (2) and its derivatives.

The moving along $-\nabla G$ method was used in the simulation in Fang and Jing (2010).

11. Conclusion

A quantum statistical theory of protein folding for monomeric, single domain, self folding globular proteins is suggested. The assumptions of the theory fit all observed realities of protein folding. The resulting formulas (1) and (2) do not have any arbitrary parameters and all terms in them have clear physical meaning. Potential energies involving pairwise interactions between atoms do not appear in them.

Formulas (1) and (2) have explanation powers. They give unified explanation to folding and denaturation, to the hydrophobic effect in protein folding and its relation with the hydrogen bonding. The formulas also explain the relative successes of surface area protein folding models. Relation between kinetic and thermodynamic of protein folding is discussed, driving force formula comes from the Gibbs free energy formula (2) are also given. Energy surface theory will be much easier to handle. The concept of $\triangle G$ is clarified.

Appendix

A. Calculations on the molecular surface

A.1. Rotation order

Let $P_{\mathbf{X}} = \cup_{i=1}^{N} B(\mathbf{x}_i, r_i)$ and $\mathbf{x}_{\alpha 0}$ and $\mathbf{x}_{\alpha 1}$ be bonded by b_α, the rotation line of b_α is $\mathbf{x}_{\alpha 0} + t\frac{\mathbf{x}_{\alpha 1}-\mathbf{x}_{\alpha 0}}{|\mathbf{x}_{\alpha 1}-\mathbf{x}_{\alpha 0}|} = \mathbf{x}_{\alpha 0} + t\mathbf{b}_\alpha$. Each b_α divides $\{\mathbf{x}_1, \cdots, \mathbf{x}_M\}$ into two groups F_α and R_α, balls in R_α will be rotated while balls in F_α will be fixed. Note that these partitions are independent of $P_{\mathbf{X}}$, they only depend on the molecular graph of the protein molecule. Let M_α be this rotation-fixation, it will be shown that

$$M_\alpha \circ M_\beta(\mathbf{X}) = M_\beta \circ M_\alpha(\mathbf{X}), \mathbf{X} \in (\mathbf{x}_1, \mathbf{x}_2, \cdots, \mathbf{x}_M), 1 \leq \alpha, \beta \leq L. \tag{1}$$

The formula of rotating a point X around a line $L : \mathbf{y} = \mathbf{x} + t\mathbf{b}$ ($|\mathbf{b}| = 1$) by an angle ω is $R(X) = \mathbf{x} + \mathbf{A}(\omega)(X - \mathbf{x})$. Let I be the identity matrix, $B = \mathbf{b}\mathbf{b}^T$ and Z_b the matrix such that the outer product $\mathbf{b} \wedge X = Z_b X$, then the orthonormal matrix $\mathbf{A}(\omega) = (1 - \cos\omega)B + \cos\omega I + \sin\omega Z_b$.

The topology of a protein molecule guarantees that if two bonds b_α and b_β such that $R_\alpha \subset R_\beta$, then $\{\mathbf{x}_{\alpha 0}, \mathbf{x}_{\alpha 1}\} \subset R_\beta$. Let b_1 and b_2 be two bonds and $L_1 : \mathbf{x} = \mathbf{x}_1 + t\mathbf{b}_1$ and $L_2 : \mathbf{x} = \mathbf{x}_2 + t\mathbf{b}_2$ be the two rotating lines and $X \in (\mathbf{x}_1, \mathbf{x}_2, \cdots, \mathbf{x}_N)$. To prove equation (1), there are only two cases to consider: $R_1 \subset R_2$ and $R_1 \cap R_2 = \emptyset$. In any case, if $X \in F_1 \cap F_2$, then $M_1 \circ M_2(X) = M_2 \circ M_1(X) = X$. If $X \in R_1 \subset R_2$, then

$$M_2 \circ M_1(X) = \mathbf{x}_2 + \mathbf{A}_2(\omega_2)(\mathbf{x}_1 - \mathbf{x}_2) + \mathbf{A}_2(\omega_2)\mathbf{A}_1(\omega_1)(X - \mathbf{x}_1). \tag{2}$$

On the other hand b_1 and hence L_1 itself will be rotated by M_2, $L_3 = M_2(L_1) = \mathbf{x}_3 + t\mathbf{b}_3$, where $\mathbf{x}_3 = \mathbf{x}_2 + \mathbf{A}_2(\omega_2)(\mathbf{x}_1 - \mathbf{x}_2)$, $\mathbf{b}_3 = \mathbf{A}_2(\omega_2)\mathbf{b}_1$. Since $X \in R_1 \subset R_2$ and $M_2(X) \in R_1$ (in the new conformation $M_2(P)$ where rotation around b_1 is rotation around L_3), $M_1 \circ M_2(X)$ will be the rotation R_3 around L_3 of $M_2(X)$ by angle ω_1, thus

$$M_1 \circ M_2(X) = \mathbf{x}_2 + \mathbf{A}_2(\omega_2)(\mathbf{x}_1 - \mathbf{x}_2) + \mathbf{A}_3(\omega_1)\mathbf{A}_2(\omega_2)(X - \mathbf{x}_1). \tag{3}$$

Let $\mathbf{v} \in \mathbb{R}^3$ be an arbitrary vector, writing $\mathbf{A}_1(\omega_1) = \mathbf{A}_1$, $\mathbf{A}_2(\omega_2) = \mathbf{A}_2$, and $\mathbf{A}_3(\omega_1) = \mathbf{A}_3$, then

$$\mathbf{A}_2\mathbf{A}_1\mathbf{v} = (1 - \cos\omega_1)(\mathbf{b}_1 \bullet \mathbf{v})\mathbf{A}_2\mathbf{b}_1 + \cos\omega_1 \mathbf{A}_2\mathbf{v} + \sin\omega_1 \mathbf{A}_2(\mathbf{b}_1 \wedge \mathbf{v}). \tag{4}$$

For any orthonormal matrix O, $(O\mathbf{b}_1) \bullet (O\mathbf{v}) = \mathbf{b}_1 \bullet \mathbf{v}$, $O(\mathbf{b}_1 \wedge \mathbf{v}) = (O\mathbf{b}_1) \wedge (O\mathbf{v})$. Then by $\mathbf{b}_3 = \mathbf{A}_2(\omega_2)\mathbf{b}_1$,

$$\begin{aligned}
\mathbf{A}_3\mathbf{A}_2\mathbf{v} &= (1 - \cos\omega_1)[\mathbf{b}_3 \bullet (\mathbf{A}_2\mathbf{v})]\mathbf{b}_3 + \cos\omega_1(\mathbf{A}_2\mathbf{v}) + \sin\omega_1 \mathbf{b}_3 \wedge (\mathbf{A}_2\mathbf{v}) \\
&= (1 - \cos\omega_1)(\mathbf{b}_1 \bullet \mathbf{v})\mathbf{A}_2\mathbf{b}_1 + \cos\omega_1 \mathbf{A}_2\mathbf{v} + \sin\omega_1 \mathbf{A}_2(\mathbf{b}_1 \wedge \mathbf{v}).
\end{aligned} \tag{5}$$

Since $\mathbf{v}$ was arbitrary, equations (2) to (5) show equation (1) is true.

If $R_1 \cap R_2 = \emptyset$ and $X \in R_2$, then X and $M_2(X) \in F_1$ hence $M_1 \circ M_2(X) = M_2(X) = M_2 \circ M_1(X)$.

The molecular surface is consisted of faces. Thus all integrals can be integrated piecewise on faces. There are three kinds of faces, convex, concave, and saddle, Connolly (1983). The formulas on each kind of face are given below. The notation $\dot{\mathbf{x}}$ means $L(\mathbf{x})$ with L the corresponding Lie vector field. All van der Waals radii r_i, as well as the probe radius r_p, are constants.

A.2. Convex face

A convex face is a piece of spherical surface lying on some $S_i = \partial B(\mathbf{x}_i, r_i)$ and bounded by circular arcs $\gamma_\nu, \nu = 1, \cdots, n_F$, let $\mathbf{v}_\nu^0$ and $\mathbf{v}_\nu^1$ be γ_ν's vertices and $\mathbf{c}_\nu$ and r_ν the center and radius of γ_ν's circle, $r_\nu\phi_\nu$ the arc length of γ_ν, $\mathbf{e}_3^\nu = (z_{\nu 1}, z_{\nu 2}, z_{\nu 3})$ be the unit vector in the direction of $(\mathbf{v}_\nu^0 - \mathbf{c}_\nu) \wedge (\mathbf{v}_\nu^1 - \mathbf{c}_\nu)$, $d_\nu = \mathbf{e}_3^\nu \bullet (\mathbf{c}_\nu - \mathbf{x}_i)$, $\mathbf{e}_1^\nu = \frac{\mathbf{v}_\nu^0 - \mathbf{c}_{ij\nu}}{r_\nu} = (x_{\nu 1}, x_{\nu 2}, x_{\nu 3})$, $\mathbf{e}_2^\nu = \mathbf{e}_3^\nu \wedge \mathbf{e}_1^\nu = (y_{\nu 1}, y_{\nu 2}, y_{\nu 3})$, $1 \le \nu \le n_F$. A point $\mathbf{x}$ on F has the form $\mathbf{x} = \mathbf{x}_i - r_i N$ and $X_\alpha(\mathbf{x}) = \dot{\mathbf{x}} - r_i \dot{N}$, by $N \bullet \dot{N} \equiv 0$ and the general divergence formula on sphere,

$$r_i \int_F (X_\alpha \bullet N) H \, d\mathcal{H}^2 = \int_F X_\alpha \bullet N \, d\mathcal{H}^2 = \frac{-1}{r_i} \dot{\mathbf{x}}_i \bullet \sum_{\nu=1}^{M} (X_\nu, Y_\nu, Z_\nu), \tag{6}$$

where

$$\begin{aligned} X_\nu = & \frac{r_\nu^2}{2} \{\phi_\nu z_{\nu 1} + \sin\phi_\nu [\cos\phi_\nu (x_{\nu 2} y_{\nu 3} + x_{\nu 3} y_{\nu 2}) + \sin\phi_\nu (y_{\nu 2} y_{\nu 3} - x_{\nu 2} x_{\nu 3})\} \\ & + r_\nu d_\nu z_{\nu 2} \left[y_{\nu 3} \sin\phi_\nu - x_{\nu 3} (1 - \cos\phi_\nu) \right], \end{aligned} \tag{7}$$

$$\begin{aligned} Y_\nu = & \frac{r_\nu^2}{2} \{\phi_\nu z_{\nu 2} + \sin\phi_\nu [\cos\phi_\nu (x_{\nu 3} y_{\nu 1} + x_{\nu 1} y_{\nu 3}) + \sin\phi_\nu (y_{\nu 1} y_{\nu 3} - x_{\nu 1} x_{\nu 3})]\} \\ & + r_\nu d_\nu z_{\nu 3} \left[y_{\nu 1} \sin\phi_\nu - x_{\nu 1} (1 - \cos\phi_\nu) \right], \end{aligned} \tag{8}$$

$$\begin{aligned} Z_\nu = & \frac{r_\nu^2}{2} \{\phi_\nu z_{\nu 3} + \sin\phi_\nu [\cos\phi_\nu (x_{\nu 1} y_{\nu 2} + x_{\nu 2} y_{\nu 1}) + \sin\phi_\nu (y_{\nu 1} y_{\nu 2} - x_{\nu 1} x_{\nu 2})]\} \\ & + r_\nu d_\nu z_{\nu 1} \left[y_{\nu 2} \sin\phi_\nu - x_{\nu 2} (1 - \cos\phi_\nu) \right]. \end{aligned} \tag{9}$$

A.3. Concave face

A concave face F is a spherical polygon on the probe sphere S when S is simultaneously tangent to 3 balls $B(\mathbf{x}_i, r_i), 1 \le i \le 3$. F is expressed by parameters $t_i \ge 0, i = 1, 2, 3$,

$$\mathbf{x} = \mathbf{p} + rN = \mathbf{p} + r \frac{t_1 \mathbf{x}_1 + t_2 \mathbf{x}_2 + t_3 \mathbf{x}_3 - \mathbf{p}}{|t_1 \mathbf{x}_1 + t_2 \mathbf{x}_2 + t_3 \mathbf{x}_3 - \mathbf{p}|}, \quad t_1 + t_2 + t_3 = 1. \tag{10}$$

$$\phi_t(\mathbf{x}) = \mathbf{p}(t) + rN(t) = \mathbf{p}(t) + r \frac{t_1 \mathbf{x}_1(t) + t_2 \mathbf{x}_2(t) + t_3 \mathbf{x}_3(t) - \mathbf{p}(t)}{|t_1 \mathbf{x}_1(t) + t_2 \mathbf{x}_2(t) + t_3 \mathbf{x}_3(t) - \mathbf{p}(t)|}, \tag{11}$$

$X_\alpha(\mathbf{x}) = \frac{d\phi_t(\mathbf{x})}{dt}\Big|_{t=0} = \dot{\mathbf{p}} + r\dot{N}$. Using $|\mathbf{p}(t) - \mathbf{x}_i(t)| = r_i + r =$ constant, let $b_i = (\mathbf{x}_i - \mathbf{p})\bullet\dot{\mathbf{x}}_i$, $\mathbf{b} = (b_1, b_2, b_3)^T$, $A = \begin{pmatrix} \mathbf{x}_1 - \mathbf{p} \\ \mathbf{x}_2 - \mathbf{p} \\ \mathbf{x}_3 - \mathbf{p} \end{pmatrix}$, then $\det A \neq 0$, $\dot{\mathbf{p}} = A^{-1}\mathbf{b}$. By $X_\alpha \bullet N = \dot{\mathbf{p}}\bullet N$,

$$r\int_F (X_\alpha \bullet N) H \, d\mathcal{H}^2 = -\int_F X_\alpha \bullet N \, d\mathcal{H}^2 = \frac{1}{r}\dot{\mathbf{p}}\bullet\sum_{i=1}^{3}(X_i, Y_i, Z_i). \tag{12}$$

Here the X_i, Y_i, and Z_i are the same as in equations (7) to (9).

Assume that $\mathbf{x}_1$ has different water association with $\mathbf{x}_2$ and $\mathbf{x}_3$, let $R_i = r_i + r$, $d_{ij} = |\mathbf{x}_i - \mathbf{x}_j|$, $y_{ij} = (R_i^2 - R_j^2)/2d_{ij}$. Then $f_P(\mathbf{x}) = (\mathbf{x} - \mathbf{p}) \bullet \mathbf{n}_k$, where $\mathbf{n}_k = (\mathbf{x}_k - \mathbf{x}_1)/d_{1k}$ is the directed unit normal of the dividing plane P_k (passing through $\mathbf{p}$ and $\mathbf{t}_{1k} = \frac{1}{2}(\mathbf{x}_1 + \mathbf{x}_k) + y_{1k}\mathbf{n}_k$ and perpendicular to it), $k = 2, 3$. The projection of $\partial W \cap F$ on the $\mathbf{x}_1\mathbf{x}_2\mathbf{x}_3$ plane is in the form of one or two curves γ_k, $(\{j,k\} = \{2,3\})$

$$t_k = A_k t_j + B_k,\ 0 \le t_j \le z_j,\ A_k = \frac{d_{1j}\cos\omega}{-d_{1k}},\ B_k = \frac{d_{1k} + 2y_{1k}}{2d_{1k}}, \tag{13}$$

where $\cos\omega = \frac{(\mathbf{x}_2 - \mathbf{x}_1)\bullet(\mathbf{x}_3 - \mathbf{x}_1)}{d_{12}d_{13}}$. $F \cap W$ is a spherical polygon with arcs γ_ν, $1 \le \nu \le n$, including some γ_k as above, so $\int_{W\cap F}(X_\alpha \bullet N) H d\mathcal{H}^2$ has the similar form as that in equation (12).

Let $\mathbf{A}_k = \mathbf{x}_j - \mathbf{x}_1 + A_k(\mathbf{x}_k - \mathbf{x}_1)$, $\mathbf{B}_k = B_k(\mathbf{x}_k - \mathbf{x}_1) + (\mathbf{x}_1 - \mathbf{p})$, $\mathbf{C}_k = \mathbf{B}_k \wedge \mathbf{A}_k$. Treating A_k and B_k as constants and setting $H_k = \dot{\mathbf{p}} \bullet \mathbf{C}_k$, $J_k = \dot{\mathbf{A}}_k \bullet \mathbf{C}_k$, and $K_k = \dot{\mathbf{B}}_k \bullet \mathbf{C}_k$. Let $a_k t_j^2 + b_k t_j + c_k = |\mathbf{A}_k t_j + \mathbf{B}_k|^2 > 0$, then $\Delta_k = 4a_k c_k - b_k^2 > 0$. By $\eta = N'_{t_j} \wedge N / |N'_{t_j}|$ and $d\mathcal{H}^1 = r|N'_{t_j}|dt_j$,

$$\begin{aligned} \int_{(\partial W\cap F)\cap\gamma_k} X_\alpha \bullet \eta d\mathcal{H}^1 = & \frac{2rH_k}{\sqrt{\Delta_k}}\left(\arctan\frac{2a_k z_j + b_k}{\sqrt{\Delta_k}} - \arctan\frac{b_k}{\sqrt{\Delta_k}}\right) \\ & + \frac{2r^2 J_k}{\Delta_k}\left(2\sqrt{c_k} - \frac{b_k z_j + 2c_k}{\sqrt{a_k z_j^2 + b_k z_j + c_k}}\right) \\ & + \frac{2r^2 K_k}{\Delta_k}\left(\frac{2a_k z_j + b_k}{\sqrt{a_k z_j^2 + b_k z_j + c_k}} - \frac{b_k}{\sqrt{c_k}}\right). \end{aligned} \tag{14}$$

Let $U_k = (\mathbf{A}_k \dot{\bullet} \mathbf{n}_k)$, $V_k = (\mathbf{B}_k \dot{\bullet} \mathbf{n}_k)$, $W_k = |\mathbf{C}_k \bullet \mathbf{n}_k| > 0$, then

$$\begin{aligned} \int_{\gamma_k} \frac{\frac{df_P}{dt}}{|\nabla_{M_P} f_P|} d\mathcal{H}^1 = & \frac{\pm 2r^2}{W_k}\left(\frac{(2a_k z_j + b_k)V_k}{\sqrt{a_k z_j^2 + b_k z_j + c_k}} - \frac{b_k V_k}{\sqrt{c_k}}\right. \\ & \left. + \ 2\sqrt{c_k}U_k - \frac{(b_k z_j + 2c_k)U_k}{\sqrt{a_k z_j^2 + b_k z_j + c_k}}\right), \end{aligned} \tag{15}$$

where the sign is determined by orientation.

A.4. Saddle face

A saddle face F is generated when the probe S simultaneously tangents to two balls $B(\mathbf{x}_1, r_1)$ and $B(\mathbf{x}_2, r_2)$, and rolls around the axis $\mathbf{e}_2 = \frac{\mathbf{x}_2 - \mathbf{x}_1}{d_{12}}$. The starting and stopping positions of the probe center is $\mathbf{p}$ and $\mathbf{q}$. Let $y = [(r_1+r)^2 - (r_2+r)^2]/2d_{12}$ and $\mathbf{t} = \frac{1}{2}(\mathbf{x}_1 + \mathbf{x}_2) + y\mathbf{e}_2$, $R = |\mathbf{p} - \mathbf{t}|$, $\mathbf{e}_1 = (\mathbf{p} - \mathbf{t})/R$, $\mathbf{e}_3 = \mathbf{e}_1 \wedge \mathbf{e}_2$, then F is parametrized by $0 \le \psi \le \psi_s$, $\theta_1 \le \theta \le \theta_2$,

$$\mathbf{x}(\psi, \theta) = \mathbf{t} + (R - r\cos\theta)(\cos\psi \mathbf{e}_1 + \sin\psi \mathbf{e}_3) + r\sin\theta \mathbf{e}_2, \tag{16}$$

where let $\omega_s = \arccos[(\mathbf{p} - \mathbf{t}) \bullet (\mathbf{q} - \mathbf{t})/R^2]$, then $\psi_s = \omega_s$ or $2\pi - \omega_s$. $\theta_1 = \arctan[-(d_{12} + 2y)/2R]$, $\theta_2 = \arctan[(d_{12} - 2y)/2R]$. These data are uniquely determined by the conformation P, see Connolly (1983). Let $\theta_k(t)$ and $\phi_s(t)$ be similarly defined for the conformation P_t, one can define $\phi_t(\psi) = \frac{\psi\psi_s(t)}{\psi_s}$, $\phi_t(\theta) = \frac{\theta_1(t)(\theta_2 - \theta) + \theta_2(t)(\theta - \theta_2)}{\theta_2 - \theta_1}$, and $U(\psi) = \cos\psi \mathbf{e}_1 + \sin\psi \mathbf{e}_3$, then for the same $0 \le \phi \le \phi_s$ and $\theta_1 \le \theta \le \theta_2$,

$$\phi_t(\mathbf{x}) = \mathbf{t}(t) + [R - r\cos\phi_t(\theta)]U(\phi_t(\psi)) + r\sin\phi_t(\theta)\mathbf{e}_2(t). \tag{17}$$

Let $\dot{U} = \cos\psi \dot{\mathbf{e}}_1 + \sin\psi \dot{\mathbf{e}}_3$, $U' = -\sin\psi \mathbf{e}_1 + \cos\psi \mathbf{e}_3$, then

$$\begin{aligned} X_\alpha(\mathbf{x}) &= \dot{\mathbf{t}} + (\dot{R} + r\dot{\theta}\sin\theta)U + (R - r\cos\theta)(\dot{U} + \dot{\psi}U') \\ &+ r\dot{\theta}\cos\theta \mathbf{e}_2 + r\sin\theta \dot{\mathbf{e}}_2. \end{aligned} \tag{18}$$

On F, $N = -\cos\theta U(\psi) + \sin\theta \mathbf{e}_2$, $\mathrm{d}\mathcal{H}^2 = r(R - r\cos\theta)\mathrm{d}\theta\mathrm{d}\psi$, $2H = \frac{R - 2r\cos\theta}{r(R - r\cos\theta)}$. Let $J = J(\psi_s) = \int_0^{\psi_s} U(\psi)\mathrm{d}\psi$, then

$$\begin{aligned} 4\int_F X_\alpha \bullet N \mathrm{d}\mathcal{H}^2 &= 4rR(\phi_s \dot{\mathbf{t}} \bullet \mathbf{e}_2 - RJ \bullet \dot{\mathbf{e}}_2)(\cos\theta_1 - \cos\theta_2) \\ &+ 4rR(\phi_s \dot{R} + \dot{\mathbf{t}} \bullet J)(\sin\theta_1 - \sin\theta_2) \\ &- r^2(\phi_s \dot{\mathbf{t}} \bullet \mathbf{e}_2 - RJ \bullet \dot{\mathbf{e}}_2)(\cos 2\theta_1 - \cos 2\theta_2) \\ &+ r^2(\phi_s \dot{R} + \dot{\mathbf{t}} \bullet J)[2(\theta_1 - \theta_2) + \sin 2\theta_1 - \sin 2\theta_2], \end{aligned} \tag{19}$$

$$\begin{aligned} 2\int_F (X_\alpha \bullet N) H \mathrm{d}\mathcal{H}^2 &= 2R(\phi_s \dot{\mathbf{t}} \bullet \mathbf{e}_2 - R \bullet \dot{\mathbf{e}}_2)(\cos\theta_1 - \cos\theta_2) \\ &+ 2R(\phi_s \dot{R} + \dot{\mathbf{t}} \bullet J)(\sin\theta_1 - \sin\theta_2) \\ &- r(\phi_s \dot{\mathbf{t}} \bullet \mathbf{e}_2 - RJ \bullet \dot{\mathbf{e}}_2)(\cos 2\theta_1 - \cos 2\theta_2) \\ &+ r(\phi_s \dot{R} + \dot{\mathbf{t}} \bullet J)[2(\theta_1 - \theta_2) + \sin 2\theta_1 - \sin 2\theta_2]. \end{aligned} \tag{20}$$

Assume that $\mathbf{x}_1$ is hydrophobic and $\mathbf{x}_2$ is not, then the dividing plane P passing through $\mathbf{p}$ and $\mathbf{t}$ and is perpendicular to $\mathbf{e}_2$. The curve $\partial W \cap F$ is given by $\mathbf{x}(\psi) = \mathbf{t} + (R - r)U(\psi)$, $0 \le \phi \le \phi_s$, on which $\mathrm{d}\mathcal{H}^1 = (R - r)\mathrm{d}\phi$. The hydrophobic surface integral on F then is the same as in equation (20), except $\theta_1 = 0$. Since on $\partial W \cap F$, $\eta = N' \wedge N = \mathbf{e}_2$, $\frac{\mathrm{d}\theta(t)}{\mathrm{d}t}|_{t=0} = \dot{\theta}_0 = \frac{\dot{\theta}_1\theta_2 - \dot{\theta}_2\theta_1}{\theta_2 - \theta_1}$, by equation (18),

$$\int_{\partial W \cap F} X_\alpha \bullet \eta \, \mathrm{d}\mathcal{H}^1 = (R - r)\phi_s(r\dot{\theta}_0 + \dot{\mathbf{t}} \bullet \mathbf{e}_2) - (R - r)^2 \dot{\mathbf{e}}_2 \bullet J, \tag{21}$$

Let $\mathbf{n}_j = \mathbf{e}_2$, then $f_{P_t}(\phi_t(\mathbf{x})) = [\phi_t(\mathbf{x}) - \mathbf{t}(t)] \bullet \mathbf{n}_j(t)$, $|\nabla_{M_P} f_P| = \mathbf{n}_j \bullet \eta = 1$, and $\frac{\mathrm{d}f_{P_t}(\phi_t(X))}{\mathrm{d}t}|_{t=0} = [\phi_t(X)^{\cdot} - \mathbf{t}(t)] \bullet \mathbf{e}_2 + [(R-r)U] \bullet \dot{\mathbf{e}}_2 = r\dot{\theta}_0$.

$$\int_{\partial W_P \cap F} \frac{\frac{\mathrm{d}f_P}{\mathrm{d}t}}{|\nabla_{M_P} f_P|} \mathrm{d}\mathcal{H}^1 = (R-r)\phi_s r\dot{\theta}_0. \tag{22}$$

Author details

Yi Fang

Department of Mathematics, Nanchang University, 999 Xuefu Road, Honggutan New District, Nanchang, 330031, China

12. References

[1] Anfinsen, C. B. (1973) Principles that govern the folding of protein chains. *Science* 181, 223-230.

[2] Bader, R. F. W. (1990) *Atoms in Molecules: A Quantum Theory.* (Clarendon Press · Oxford).

[3] Bailyn, M. (1994) *A Survey of Thermodynamics.* American Institute of Physics New York.

[4] Ben-Naim, A. (2012) Levinthal's question revisited, and answered. *Journal of Biomolecular Structure and Dynamics* 30(1), 113-124 (2012).

[5] Branden, C. and J. Tooze, J. (1999) *Introduction to Protein Structure.* (Second Edition, Garland).

[6] Connolly, M. L. (1983) Analytical molecular surface calculation. *J. Appl. Cryst.,* 16:548-558.

[7] Dai, X. (2007) *Advanced Statistical Physics.* (Fudan University Press, Shanghai).

[8] Dill, K. A. (1990) Dominant forces in protein folding. *Biochemistry,* 29 7133-7155.

[9] Dyson, F. (2004) A meeting with Enrico Fermi: How an intuitive physicist rescued a team from fruitless research. *Nature* 427, 297.

[10] Eisenberg, D and McLachlan, A. D. Solvation energy in protein folding and binding. *Nature* 319, 199-203.

[11] ExPASy Proteomics Server. http://au.expasy.org/sprot/relnotes/relstat.html

[12] Fang, Y. (2005) Mathematical protein folding problem. In: D. Hoffman, Ed, *Global Theory of Minimal Surfaces.* (*Proceedings of the Clay Mathematical Proceedings,* 2 2005) pp. 611-622.

[13] Fang, Y. and Jing, J. (2008) Implementation of a mathematical protein folding model. *International Journal of Pure and Applied Mathematics,* 42(4), 481-488.

[14] Fang, Y. and Jing, J. (2010) Geometry, thermodynamics, and protein. *Journal of Theoretical Biology,* 262, 382-390.

[15] Finkelstein, A. V. and Ptitsyn, O. B. (2002) *Protein Physics: A Course of Lectures.* Academic Press, An imprint of Elsevier Science, Amsterdam ... .

[16] Greiner, W., Neise, L., and Stöker, H. (1994) *Thermodynamics and Statistical Mechanics.* (Spriger-Verlag, New York, Berlin, ...).

[17] Jackson, R. M., and Sternberg, M. J. E. (1993) Protein surface area defined. *Nature* 366, 638.

[18] Kang, H. J. and Baker, E. N. (2011) Intramolecular isopeptide bonds: protein crosslinks built for stress? *TIBS,* 36(4), 229-237.

[19] Kauzmann, W. (1959) Some factors in the interpretation of protein denaturation. *Adv. Protein Chem.* 14, 1-63 (1959).

[20] Lazaridis, T., and Karplus, M. (2003) Thermodynamics of protein folding: a microscopic view. *Biophysical Chemistry,* 100, 367-395.

[21] Lee, B., and Richards, F. M. (1971) The interpretation of protein structures: estimation of static accessibility. *J. Mol. Biol.* 55:379-400.

[22] Novotny, J., Bruccoleri, R., and Karplus, M. (1984) An analysis of incorrectly folded protein models. Implications for structure predictions. *J. Mol. Biol.* 177, 787-818.

[23] Popelier, P. (2000) *Atoms in Molecules: An Introduction.* (Prentice Hall).

[24] Reynolds, J. A., Gilbert, D. B., and Tanford, C. (1974) Empirical correlation between hydrophobic free energy and aqueous cavity surface area. *Proc. Natl. Acad. Sci. USA* 71(8), 2925-2927.

[25] Richards, F. M. (1977) Areas, volumes, packing, and protein structure. *Ann. Rev. Biophys. Bioeng.* 6, 151-176.

[26] Rose, G. D., Fleming, P. J., Banavar, J. R., and Maritan, A. (2006) A backbone based theory of protein folding.
PNAS 103(45), 16623-16633.

[27] Tuñón, I., Silla, E., and Pascual-Ahuir, J. L. (1992) Molecular surface area and hydrophobic effect. Protien Engineering 5(8), 715-716.

[28] Wang, Y., Zhang, H., Li, W., and Scott, R. A. (1995) Discriminating compact nonnative structures from the native structure of globular proteins. *PNAS,* 92, 709-713.

[29] Wu, H. (1931) Studies on denaturation of proteins XIII. A theory of denaturation. *Chinese Journal nof Physiology* 4, 321-344 (1931). A preliminary report was read before the XIIIth International Congress of Physiology at Boston, Aug. 19-24, 1929 and published in the *Am. J Physiol.* for Oct. 1929. Reprinted in *Advances in Protein Chemicsty* 46, 6-26 (1995).

Chapter 4

Information Capacity of Quantum Transfer Channels and Thermodynamic Analogies

Bohdan Hejna

Additional information is available at the end of the chapter

http://dx.doi.org/10.5772/50466

1. Introduction

We will begin with a simple type of *stationary stochastic*[1] systems of quantum physics using them within a frame of the **Shannon** Information Theory and Thermodynamics but starting with their *algebraic representation.* Based on this algebraic description a model of information transmission in those systems by defining the Shannon information will be stated in terms of variable about the system state. Measuring on these system is then defined as a spectral decomposition of measured quantities - *operators*. The information capacity formulas, now of the *narrow-band* nature, are derived consequently, for the simple system governed by the **Bose–Einstein** (B–E) Law [bosonic (photonic) channel] and that one governed by the **Fermi-Dirac** (F–D) Law [fermionic (electron) channel]. *The not-zero value for the average input energy needed for information transmission existence in F–D systems* is stated [11, 12].

Further the *wide–band* information capacity formulas for B–E and F–D case are stated. Also the original *thermodynamic* capacity derivation for the wide–band photonic channel as it was stated by **Lebedev–Levitin** in 1966 is revised. This revision is motivated by apparent relationship between the B–E (photonic) wide–band information capacity and the *heat efficiency* for a certain *heat cycle,* being further considered as the demonstrating model for processes of information transfer in the original wide–band photonic channel. The information characteristics of a *model reverse* heat cycle and, by this model are analyzed, the information arrangement of which is set up to be most analogous to the structure of the photonic channel considered, we see the necessity of returning the transfer medium (the channel itself) to its initial state as a condition for a *sustain, repeatable* transfer. It is not regarded in [12, 30] where a single information transfer act only is considered. Or the return is

[1] We deal with such a system which is taking on at time $t = 0, 1, \ldots$ states θ_t from a state space Θ. If for any t_0 the relative frequencies I_B of events $B \subset \Theta$ is valid that $\frac{1}{T} \sum_{t=t_0+1}^{t_0+T} I_B(\theta_t)$ tends for $T \to \infty$ to probabilities $p_{t_0}(B)$ we speak about a *stochastic system*. If these probabilities do not depend on the beginning t_0, a *stationary stochastic system* is spoken about.

regarded, but by opening the whole transfer chain for covering these *return energy needs from its environment* not counting them in - *not within the transfer chain only* as we do now. *The result is the corrected capacity formula for wide–band photonic (B–E) channels being used for an information transfer organized cyclicaly.*

2. Information transfer channel

An *information, transfer* channel $\mathcal{K}$ is defined as an arranged *tri–partite* structure [5]

$$\mathcal{K} \overset{\text{Def}}{=} [X,\ \varepsilon,\ Y] \ \text{ where } X \overset{\text{Def}}{=} [A,\ p_X(\cdot)],\ \ Y \overset{\text{Def}}{=} [B,\ p_Y(\cdot)] \ \text{ and} \tag{1}$$

- X is an *input* stochastic quantity, a *source* of *input* messages $a \in A^+ \triangleq \mathcal{X}$, a *transceiver*,
- Y is an *output* stochastic quantity, a *source* of *output* messages $b \in B^+ \triangleq \mathcal{Y}$, a *receiver*,
- output messages $b \in \mathcal{Y}$ are *stochastic* dependent on input messages $a \in \mathcal{X}$ and they are received by the receiver of messages, Y,
- ε is the *maximal* probability of an error in the transfer of any symbol $x \in A$ in an input message $a \in \mathcal{X}$
[the maximal probability of erroneously receiving $y \in B$ (inappropriate for x) in an output message $b \in \mathcal{Y}$],
- A denotes a finite *alphabet* of elements x of the source of input messages,
- B denotes a finite alphabet of elements y of the source of output messages,
- $p_X(\cdot)$ is the *probability distribution* of evidence of any symbol $x \in A$ in an input message,
- $p_Y(\cdot)$ is the probability distribution of evidence of any symbol $y \in B$ in an output message.

The structure $(X,\ \mathcal{K},\ Y)$ or $(\mathcal{X},\ \mathcal{K},\ \mathcal{Y})$ is termed a *transfer (Shannon) chain*. The symbols $H(X)$ and $H(Y)$ respectively denote the *input information (Shannon) entropy* and the *output information (Shannon) entropy* of channel $\mathcal{K}$, *discrete* for this while,

$$H(X) \overset{\text{Def}}{=} -\sum_{x\in A} p_X(x)\,\ln p_X(x),\quad H(Y) \overset{\text{Def}}{=} -\sum_{y\in B} p_Y(y)\,\ln p_Y(y) \tag{2}$$

The symbol $H(X|Y)$ denotes the *loss entropy* and the symbol $H(Y|X)$ denotes the *noise entropy* of channel $\mathcal{K}$. These entropies are defined as follows,

$$H(X|Y) \overset{\text{Def}}{=} -\sum_A\sum_B p_{X,Y}(x,y)\,\ln p_{X|Y}(x|y),\quad H(Y|X) \overset{\text{Def}}{=} -\sum_A\sum_B p_{X,Y}(x,y)\ln p_{Y|X}(y|x) \tag{3}$$

where the symbol $p_{\cdot|\cdot}(\cdot|\cdot)$ denotes the *condition* and the symbol $p_{\cdot,\cdot}(\cdot,\cdot)$ denotes the *simultaneous* probabilities. For *mutual (transferred) usable* information, *transinformation* $T(X;Y)$ or $T(Y;X)$ is valid that

$$T(X;Y) = H(X) - H(X|Y) \ \text{ and } \ T(Y;X) = H(Y) - H(Y|X) \tag{4}$$

From (2) and (3), together with the definitions of $p_{\cdot|\cdot}(\cdot|\cdot)$ and $p_{\cdot,\cdot}(\cdot,\cdot)$, is provable prove that the transinformation is symmetric. Then the *equation of entropy (information) conservation* is valid

$$H(X) - H(X|Y) = H(Y) - H(Y|X) \tag{5}$$

The *information capacity* of the channel $\mathcal{K}$ (both discrete an continuous) is defined by the equation

$$C \stackrel{\text{Def}}{=} \sup I(X;Y) \tag{6}$$

over all possible probability distributions $q(\cdot),\ \ p(\cdot)$.It is the maximum (supremum) of the medium value of the usable amount of information about the input message x within the *output* message y.

Remark: For continuous distributions (densities) $p_{[\cdot]}(\cdot),\ p_{[\cdot|\cdot]}(\cdot|\cdot)$ on intervals $\mathcal{X},\ \mathcal{Y} \in \mathbb{R}$, $x \in \mathcal{X}, y \in \mathcal{Y}$ is

$$H(X) = -\int_{\mathcal{X}} p_X(x) \ln p_X \,\mathrm{d}x, \tag{7}$$

$$H(Y) = -\int_{\mathcal{Y}} p_Y(y) \ln p_Y(y) \,\mathrm{d}y \tag{8}$$

$$H(X|Y) = -\int_{\mathcal{X}}\int_{\mathcal{Y}} p_{X,Y}(x,y) \ln p_{X|Y}(x|y) \,\mathrm{d}x\mathrm{d}y,$$

$$H(Y|X) = -\int_{\mathcal{X}}\int_{\mathcal{Y}} p_{X,Y}(x,y) \ln p(y|x) \,\mathrm{d}x\mathrm{d}y$$

Equations (4), (5 are valid for both the quantities $H(\cdot)$ and $H(\cdot|\cdot)$, as well as for their respective changes $\Delta H(\cdot)[= H(\cdot)]$ and $\Delta H(\cdot|\cdot)[= H(\cdot|\cdot)]$.

3. Representation of physical transfer channels

The most simple way of description of *stationary* physical systems is an *eucleidian* space Ψ of their states expressed as *linear operators*.[2] This way enables the *mathematical* formulation of the term *(physical) state* and, generally, the term *(physical) quantity*.

Physical quantities $\boldsymbol{\alpha}$, associated with a physical system $\boldsymbol{\Psi}$ represented by the Eucleidian space Ψ are expressed by *symmetric* operators from the linear space $L(\Psi)$ of operators on Ψ, $\boldsymbol{\alpha} \in \mathbf{A} \subset L(\Psi)$ [7]. The supposition is that any physical quantity can achieve only those *real* values α which are the *eigenvalues* of the associated *symmetric operator* $\boldsymbol{\alpha}$ (symmetric *matrix* $[\alpha_{i,j}]_{n,n}$, $n = \dim \Psi$). They are elements of the *spectrum* $\mathbf{S}(\boldsymbol{\alpha})$ of the *operator* $\boldsymbol{\alpha}$. The eigenvalues $\alpha \in \mathbf{S}(\boldsymbol{\alpha}) \subset \mathbb{R}$ of the quantity $\boldsymbol{\alpha}$ being measured on the system $\boldsymbol{\Psi} \cong \Psi$ depend on the (inner) states $\boldsymbol{\theta}$ of this system $\boldsymbol{\Psi}$.

- The *pure states* of the system $\boldsymbol{\Psi} \cong \Psi$ are represented by *eigenvectors* $\psi \in \Psi$. It is valid that the *scalar project* $(\psi, \psi) = 1$; in *quantum physics* they are called *normalized wave functions*.
- The *mixed states* are nonnegative quantities $\boldsymbol{\theta} \in \mathbf{A}$; their *trace* [of an *square* matrix (operator) $\boldsymbol{\alpha}$] is defined

$$\mathrm{Tr}(\boldsymbol{\alpha}) \stackrel{\text{Def}}{=} \sum_{i=1}^{n} \alpha_{i,i} \quad \text{and, for } \boldsymbol{\alpha} \equiv \boldsymbol{\theta} \quad \text{is valid that} \quad \mathrm{Tr}(\boldsymbol{\theta}) = 1. \tag{9}$$

The *symmetric projector* $\pi\{\psi\} = \pi[\Psi(\{\psi\})]$ (orthogonal) on the one–dimensional *subspace* $\Psi(\{\psi\})$ of the space Ψ is nonnegative quantity for which $\mathrm{Tr}(\pi\{\psi\}) = 1$ is valid. The projector $\pi\{\psi\}$ *represents* [on the set of quantities $\mathbf{A} \subset L(\Psi)$] the pure state ψ of the system $\boldsymbol{\Psi}$. Thus, an arbitrary state of the system $\boldsymbol{\Psi}$ can be defined as a nonnegative quantity $\boldsymbol{\theta} \in \boldsymbol{\Theta} \subset \mathbf{A}$ for which $\mathrm{Tr}(\boldsymbol{\theta}) = 1$ is valid. For the *pure state* $\boldsymbol{\theta}$ is then valid that $\boldsymbol{\theta}^2 = \boldsymbol{\theta}$ and the *state space* $\boldsymbol{\Theta}$ of the system $\boldsymbol{\Psi}$ is defined as the set of all states $\boldsymbol{\theta}$ of the system $\boldsymbol{\Psi}$.

[2] The motivation is the *axiomatic theory of algebraic representation of physical systems* [7].

3.1. Probabilities and information on physical systems

3.1.0.1. Theorem:

For any state $\boldsymbol{\theta} \in \boldsymbol{\Theta}$ the pure states $\boldsymbol{\theta}_i = \pi\{\psi_i\}$ and numbers $q(i|\boldsymbol{\theta}) \geq 0$ exists, that

$$q(i|\boldsymbol{\theta}) \geq 0,\ \boldsymbol{\theta} = \sum_{i=1}^{n} q(i|\boldsymbol{\theta})\,\boldsymbol{\theta}_i \quad \text{where} \quad \sum_{i=1}^{n} q(i|\boldsymbol{\theta}) = 1 \tag{10}$$

3.1.0.2. Proof:

Let $D(\boldsymbol{\theta}) = \{D_\theta : \theta \in \mathbf{S}(\boldsymbol{\theta})\}$ is a *disjoint decomposition* of the set $\{1, 2, ..., n\}$ of indexes of the base $\{\psi_1, \psi_2, ..., \psi_n\}$ of Ψ. The set $\{\psi_i : i \in D_\theta\}$ is an orthogonal basis of the *eigenspace* $\Psi(\theta|\boldsymbol{\theta}) \subset \Psi$ of the operator $\boldsymbol{\theta}$ [for its eigenvalue $\theta \in \mathbf{S}(\boldsymbol{\theta})$]. Then

$$\text{card}\, D_\theta = \dim \Psi(\theta|\boldsymbol{\theta}) \geq 1 \ \text{ and } \ \pi[\Psi(\theta|\boldsymbol{\theta})] = \sum_{i \in D_\theta} \pi\{\psi_i\},\ \forall \theta \in \mathbf{S}(\boldsymbol{\theta}) \tag{11}$$

Let $q(i|\boldsymbol{\theta}) = \theta$ is taken for all $i \in D_\theta$, $\theta \in \mathbf{S}(\boldsymbol{\theta})$. By the *spectral decomposition theorem* [9],

$$\boldsymbol{\theta} = \sum_{\theta \in \mathbf{S}(\boldsymbol{\theta})} \theta\,\pi_\theta = \sum_{\theta \in \mathbf{S}(\boldsymbol{\theta})} \theta\pi[\Psi(\theta|\boldsymbol{\theta})] = \sum_{\theta \in \mathbf{S}(\boldsymbol{\theta})} \theta \sum_{i \in D_\theta} \pi\{\psi_i\} = \sum_{i=1}^{n} q(i|\boldsymbol{\theta})\,\pi\{\psi_i\} \tag{12}$$

$$\text{Tr}(\boldsymbol{\theta}) = \sum_{i=1}^{n} q(i|\boldsymbol{\theta}) \cdot \text{Tr}(\pi\{\psi_i\}) = \sum_{i=1}^{n} q(i|\boldsymbol{\theta}) = 1$$

The symbol $q(\cdot|\boldsymbol{\theta})$ denotes the *probability distribution* into *pure, canonic components* $\boldsymbol{\theta}_i$ of $\boldsymbol{\theta}$ is called:

- *canonic distribution (q-distribution)* of the state $\boldsymbol{\theta} \in \boldsymbol{\Theta}$.[3]

Further the two distribution defined on spectras of $\boldsymbol{\alpha} \in \mathbf{A}$ and $\boldsymbol{\theta} \in \boldsymbol{\Theta} \subset \mathbf{A}$ will be dealt:

- *dimensional distribution (d-distribution)* of the state $\boldsymbol{\theta} \in \boldsymbol{\Theta}$[4]

$$d(\theta|\boldsymbol{\theta}) \stackrel{\text{Def}}{=} \frac{\dim \Psi(\theta|\boldsymbol{\theta})}{\dim \Psi} = \frac{\dim \Psi(\theta|\boldsymbol{\theta})}{n},\quad \theta \in \mathbf{S}(\boldsymbol{\theta}) \tag{13}$$

- *distribution of measuring (p-distribution)* of the quantity $\boldsymbol{\alpha} \in \mathbf{A}$ in the state $\boldsymbol{\theta} \in \boldsymbol{\Theta}$,

$$p(\alpha|\boldsymbol{\alpha}|\boldsymbol{\theta}) \stackrel{\text{Def}}{=} \text{Tr}(\boldsymbol{\theta}\pi_\alpha),\quad \alpha \in \mathbf{S}(\boldsymbol{\alpha}) \tag{14}$$

where $\{\pi_\alpha : \alpha \in \mathbf{S}(\boldsymbol{\alpha})\}$ is the *spectral decomposition* of the *unit operator* $\mathbf{1}$. Due the nonnegativity of $\boldsymbol{\theta}\pi_\alpha$ is $p(\alpha|\boldsymbol{\alpha}|\boldsymbol{\theta}) = \text{Tr}(\boldsymbol{\alpha}\pi_\alpha) \geq 0$. By spectral decomposition of $\mathbf{1}$ and by definition of the *trace* $\text{Tr}(\cdot)$ is

$$\sum_{\alpha \in \mathbf{S}(\boldsymbol{\alpha})} p(\alpha|\boldsymbol{\alpha}|\boldsymbol{\theta}) = \text{Tr}\left(\boldsymbol{\theta} \sum_{\alpha \in \mathbf{S}(\boldsymbol{\alpha})} \pi_\alpha\right) = \text{Tr}(\boldsymbol{\theta}\mathbf{1}) = 1 \tag{15}$$

Thus the relation (14) defines the probability distribution on the spectrum of the operator $\mathbf{S}(\boldsymbol{\alpha})$.

---

[3] Or, the *system spectral distribution* SSD [11].

[4] Or, the *system distribution* of the system *spectral dimension* SDSD [11].

- The special case of the *p-distribution* is that *for measuring values of* $\boldsymbol{\alpha} \equiv \boldsymbol{\theta}$, $\boldsymbol{\theta} \in \boldsymbol{\Theta}$,

$$p(\theta|\boldsymbol{\theta}|\boldsymbol{\theta}) = \mathrm{Tr}(\boldsymbol{\theta}\pi_{\theta}) = \theta \dim \Psi(\theta|\boldsymbol{\theta}), \quad \theta \in \mathbf{S}(\boldsymbol{\theta}) \tag{16}$$

It is the case of *measuring the* $\boldsymbol{\theta}$ *itself*. Thus, by the equation (14), the term ***measuring*** **means just the spectral decomposition of the unit operator 1 within the spectral relation with** $\boldsymbol{\alpha}$ [4, 20]. The act of measuring of the quantity $\boldsymbol{\alpha} \in \mathbf{A}$ in (the system) state $\boldsymbol{\theta}$ gives the value α from the spectrum $\mathbf{S}(\boldsymbol{\alpha})$ with the probability $p(\alpha|\boldsymbol{\alpha}|\boldsymbol{\theta}) = \mathrm{Tr}(\boldsymbol{\theta}\pi_{\alpha})$ given by the p-distribution,

$$\sum_{\mathcal{S}(\alpha)} p(\alpha|\boldsymbol{\alpha}|\boldsymbol{\theta}) = \sum_{\mathcal{S}(\alpha)} \mathrm{Tr}(\boldsymbol{\theta}\pi_{\alpha}) = 1 \tag{17}$$

The *measured* **quantity** $\boldsymbol{\alpha}$ **in the state** $\boldsymbol{\theta}$ **is a** ***stochastic*** **quantity with its values [occuring with probabilities** $p(\cdot|\boldsymbol{\alpha}|\boldsymbol{\theta})$**] from its spectrum** $\mathbf{S}(\boldsymbol{\alpha})$. For its *mathematical expectation, medium value* is valid

$$E(\boldsymbol{\alpha}) = \sum_{\alpha \in \mathbf{S}(\boldsymbol{\alpha})} \alpha \,\mathrm{Tr}(\boldsymbol{\theta}\pi_{\alpha}) = \mathrm{Tr}\left(\boldsymbol{\theta} \sum_{\alpha \in \mathbf{S}(\boldsymbol{\alpha})} \alpha \pi_{\alpha}\right) = \mathrm{Tr}(\boldsymbol{\theta}\boldsymbol{\alpha}) = (\boldsymbol{\alpha}\psi, \psi) \tag{18}$$

Nevertheless, in the pure state $\boldsymbol{\theta} = \pi\{\psi\}$ the values $i = \alpha \in \mathbf{S}(\boldsymbol{\alpha})$ are measured, $\mathrm{Tr}(\boldsymbol{\theta}_i\boldsymbol{\alpha}) = (\boldsymbol{\alpha}\psi_i, \psi_i) = \alpha_{ii}$.

Let $\boldsymbol{\Psi}$ is an *arbitrary* stationary physical system and $\boldsymbol{\theta} \in \boldsymbol{\Theta} \subset \mathbf{A}$ is its arbitrary state. The *physical entropy* $\mathcal{H}(\boldsymbol{\theta})$ of the system $\boldsymbol{\Psi}$ *in the state* $\boldsymbol{\theta}$ is defined by the equality

$$\mathcal{H}(\boldsymbol{\theta}) \stackrel{\text{Def}}{=} -\mathrm{Tr}(\boldsymbol{\theta} \ln \boldsymbol{\theta}) \tag{19}$$

When $\{\pi_{\theta} : \theta \in \mathbf{S}(\boldsymbol{\theta})\}$ is the decomposition of **1** spectral equivalent with $\boldsymbol{\theta}$, then it is valid that

$$\boldsymbol{\theta} \ln \boldsymbol{\theta} = \sum_{\theta \in \mathbf{S}(\boldsymbol{\theta})} \theta \ln \theta \, \pi_{\theta} \text{ and } \mathcal{H}(\boldsymbol{\theta}) = -\sum_{\theta \in \mathbf{S}(\boldsymbol{\theta})} \theta \ln \theta \cdot \dim \Psi(\theta|\boldsymbol{\theta}) \tag{20}$$

3.1.0.3. Theorem:

For a physical system $\boldsymbol{\Psi}$ in any state $\boldsymbol{\theta} \in \boldsymbol{\Theta}$ is valid that

$$\begin{aligned} \mathcal{H}(\boldsymbol{\theta}) &= -\sum_{i=1}^{n} q(i|\boldsymbol{\theta}) \cdot \ln q(i|\boldsymbol{\theta}) = H[q(\cdot|\boldsymbol{\theta})] = \ln n - \sum_{\theta \in \mathbf{S}(\boldsymbol{\theta})} p(\theta|\boldsymbol{\theta}|\boldsymbol{\theta}) \cdot \ln \frac{p(\theta|\boldsymbol{\theta}|\boldsymbol{\theta})}{d(\theta|\boldsymbol{\theta})} \\ &= \ln n - I[p(\cdot|\boldsymbol{\theta}|\boldsymbol{\theta}) \,\|\, d(\cdot|\boldsymbol{\theta})] \end{aligned} \tag{21}$$

where $H(\cdot)$ is the *Shannon entropy*, $I(\cdot\|\cdot)$ is the *information divergence*, $p(\cdot|\boldsymbol{\theta}|\boldsymbol{\theta})$ is the p-distribution for the state $\boldsymbol{\theta}$, $q(\cdot|\boldsymbol{\theta})$ is the q-distribution for the state $\boldsymbol{\theta}$ and $d(\cdot|\boldsymbol{\theta})$ is the d-distribution for the state $\boldsymbol{\theta}$, $n = \dim \Psi$.

3.1.0.4. Proof:

The relations in (21) follows from (20) and from definition (10) of the distribution $q(\cdot|\boldsymbol{\theta})$. From definitions (13) and (16) of the other two distributions follows that [12, 38]

$$\begin{aligned} I[p(\cdot|\boldsymbol{\theta}|\boldsymbol{\theta}) \,\|\, d(\cdot|\boldsymbol{\theta})] &= \sum_{\theta \in \mathbf{S}(\boldsymbol{\theta})} \theta \dim \Psi(\theta|\boldsymbol{\theta}) \cdot \ln \frac{n\theta \cdot \dim \Psi(\theta|\boldsymbol{\theta})}{\dim \Psi(\theta|\boldsymbol{\theta})} \\ &= \sum_{\theta \in \mathbf{S}(\boldsymbol{\theta})} \theta \dim \Psi(\theta|\boldsymbol{\theta}) \cdot \ln n - \mathcal{H}(\boldsymbol{\theta}) = \ln n - \mathcal{H}(\boldsymbol{\theta}) \end{aligned} \tag{22}$$

Due to the quantities X, Y, $X|Y$, $Y|X$ describing the information transfer are in our algebraic description denoted as follows, $X \triangleq \boldsymbol{\theta}$, $Y \triangleq \boldsymbol{\alpha}$ or $Y \triangleq (\boldsymbol{\alpha}\|\boldsymbol{\theta})$, $(X|Y) \triangleq (\boldsymbol{\theta}|\boldsymbol{\alpha})$, $(Y|X) \triangleq (\boldsymbol{\alpha}|\boldsymbol{\theta})$. The laws of information transfer are writable in this way too:

$$C = \sup_{\boldsymbol{\alpha},\boldsymbol{\theta}} I(\boldsymbol{\alpha};\boldsymbol{\theta}), \quad I(\cdot\|\cdot) \equiv T(\cdot;\cdot) \tag{23}$$

$$\begin{aligned} I(\boldsymbol{\theta};\boldsymbol{\alpha}) &= I(\boldsymbol{\alpha};\boldsymbol{\theta}) = \mathcal{H}(\boldsymbol{\theta}) - H(\boldsymbol{\theta}|\boldsymbol{\alpha}) = H(\boldsymbol{\alpha}\|\boldsymbol{\theta}) - H(\boldsymbol{\alpha}|\boldsymbol{\theta}) = \mathcal{H}(\boldsymbol{\theta}) + H(\boldsymbol{\alpha}\|\boldsymbol{\theta}) - H(\boldsymbol{\theta},\boldsymbol{\alpha}) \\ H(\boldsymbol{\theta},\boldsymbol{\alpha}) &= H(\boldsymbol{\alpha},\boldsymbol{\theta}) = \mathcal{H}(\boldsymbol{\theta}) + H(\boldsymbol{\alpha}|\boldsymbol{\theta}) = H(\boldsymbol{\alpha}\|\boldsymbol{\theta}) + H(\boldsymbol{\theta}|\boldsymbol{\alpha}) \end{aligned} \tag{24}$$

4. Narrow-band quantum transfer channels

Let the symmetric operator $\boldsymbol{\varepsilon}$ of energy of quantum particle is considered, the spectrum of which eigenvalues ε_i is $\mathbf{S}(\boldsymbol{\varepsilon})$. Now the *equidistant* energy levels are supposed. In a pure state $\boldsymbol{\theta}_i$ of the measured (observed) system $\boldsymbol{\Psi}$ the eigenvalue $\varepsilon_i = i{\cdot}\varepsilon$, $\varepsilon > 0$. Further, the *output* quantity $\boldsymbol{\alpha}$ of the *observed* system $\boldsymbol{\Psi}$ is supposed (the system is *cell of the phase space B–E or F–D*) with the spectrum of eigenvalues $\mathbf{S}(\boldsymbol{\alpha}) = \{\alpha_0, \alpha_1, ..., \}$ being measured with probability distribution $\Pr(\cdot) = \{p(0), p(1), p(2), ...\}$

$$p(\alpha_k|\boldsymbol{\alpha}|\boldsymbol{\theta}_i) = \begin{cases} p(k-i) \text{ pro } k \geq i \\ 0 \qquad\qquad k < i \end{cases} \tag{25}$$

Such a situation arises when a particle with energy ε_i is *excited by an impact* from the output environment. The jump of energy level of the impacted particle is from ε_i up to ε_{i+j}, $i + j = k$. The output ε_{i+j} for the excited particle is measured (it is the value on the output of the channel $\mathcal{K} \cong \boldsymbol{\Psi}$. This transition j occurs with the probability distribution

$$\Pr(j),\ j \in \{0,\ 1,\ 2,\ ...\} \tag{26}$$

Let be considered the *narrow–band* systems (with one *constant* level of a particle energy) $\boldsymbol{\Psi}$ of B–E or F–D type [27] (denoted further by $\boldsymbol{\Psi}_{\text{B-E},\varepsilon}$, $\boldsymbol{\Psi}_{\text{F-D},\varepsilon}$).

- In the B–E system, *bosonic*, e.g. the *photonic gas* the B–E distribution is valid

$$\Pr(j) = (1-p)\cdot p^j, \quad j \in \{0,\ 1,\ ...\},\ p \in (0,1),\ \ p^{-\frac{\varepsilon}{k\Theta}} \tag{27}$$

- In the F–D system, *fermionic*, e.g. *electron gas* the F–D distribution is valid

$$\Pr(j) = \frac{p^j}{1+p}, \quad j \in \{0,1\},\ p \in (0,1),\ \ p^{-\frac{\varepsilon}{k\Theta}} \tag{28}$$

where parameter p is variable with *absolute temperature* $\Theta > 0$; k is the **Boltzman** constant.

Also a collision with a bundle of j particles with constant energies ε of each and absorbing the energy $j \cdot \varepsilon$ of the bundle is considerable. E.g., by $\boldsymbol{\Psi}$ (e.g. $\boldsymbol{\Psi}_{\text{B-E},\varepsilon}$ is the photonic gas) the monochromatic impulses with amplitudes $i \in \mathcal{S}$ are transferred, nevertheless generated from the environment of the same type but at the temperature T_W, $T_W > T_0$ where T_0 is the temperature of the transfer system $\boldsymbol{\Psi} \cong \mathcal{K}$ (the *noise* temperature).

It is supposed that both pure states $\boldsymbol{\theta}_i$ in the place where the input message is being *coded* - on the *input of the channel* $\boldsymbol{\Psi} \cong \mathcal{K}$ and, also, the measurable values of the quantity $\boldsymbol{\alpha}$ being observed on the place where the *output* message is *decoded* - on the *output of the channel* $\Psi \cong \mathcal{K}$ are arrangable in such a way that in a given i-th pure state θ_i of the system $\Psi \cong \mathcal{K}$ only the values $\alpha_k \in \mathcal{S}(\alpha)$, $k = i + j$ are measurable and, that the probability of measuring the k-th value is $\Pr(j) = Pr(k-i)$. This probability distribution describes the *additive noise* in the given channel $\Psi \cong \mathcal{K}$. Just it is this noise which creates observed values from the output spectrum $\mathcal{S}(\alpha) = \{\alpha_i, \alpha_{i+1}, ...\}$ being the selecting space of the stochastic quantity α.

The pure states with energy level $\varepsilon_i = i \cdot \varepsilon$ are achievable by sending i particles with energy ε of each. When the environment, through which these particles are going, generates a bundle of j particles with probability $\Pr(j)$ then, with the same probability the energy $\varepsilon_{i+j} = k{\cdot}\varepsilon$ is decoded on the output.

It is supposed, also, the infinite number of states θ_i, infinite spectrum $\mathbf{S}(\alpha)$ of the measured quantity α, then $\mathbf{S}(\alpha) = \{0, 1, 2, ...\}$, $\mathbf{S}(\alpha) = D(\theta)$; $[\mathbf{S}(\alpha) \stackrel{\triangle}{=} \mathbf{S},\ \alpha_k \stackrel{\triangle}{=} \alpha]$.

The *narrow–band, memory-less (quantum) channel, additive* (with additive noise) operating on the energy level $\varepsilon \in \mathcal{S}(\varepsilon)$ is defined by the tri–partite structure (1),

$$\mathcal{K}_\varepsilon = \{[\mathcal{S}, q(i|\theta)],\ p(\alpha|\alpha|\theta_i),\ [\mathcal{S},\ p(\alpha|\alpha|\theta)]\}. \tag{29}$$

4.1. Capacity of Bose–Einstein narrow–band channel

Let now $\theta_i \equiv i$, $i = 0, 1, ...$ are pure states of a system $\Psi_{\text{B-E},\varepsilon} \cong \mathcal{K}$ and let α is output quantity taking on in the state θ_i values $\alpha \in \mathbf{S}$ with probabilities

$$p(\alpha|\alpha|\theta_i) = (1-p)\, p^{\alpha - i} \tag{30}$$

Thus the distribution $p(\cdot|\alpha|\theta_i)$ is determined by the *forced* (inner–input) state $\theta_i = \pi\{\psi_i\}$ representing the coded input energy at the value $\varepsilon_i = i \cdot \varepsilon \in \mathbf{S}(\varepsilon)$, $\varepsilon = \text{const}$. For the *medium value* $\mathcal{W}$ of the input i is valid

$$\mathcal{W} = \sum_{i=0}^{\infty} i \cdot q(i|\theta) = E(\theta), \quad W = \varepsilon \cdot \mathcal{W} \tag{31}$$

The quantity W is the medium value of the energy coding the input signal i.

For the medium value of the number of particles $j = \alpha - i \geq 0$ with B–E statistics is valid

$$\sum_{j=0}^{\infty} j \cdot (1-p)\, p^j = (1-p) \cdot p \cdot \sum_{j=0}^{\infty} j\, p^{j-1} = (1-p) \cdot p \cdot \frac{\mathrm{d}}{\mathrm{d}p}\left[\frac{1}{1-p}\right] = \frac{p}{1-p}$$

The quantity $E(\alpha)$ is the medium value of the output quantity α and

$$E(\alpha) = \sum_{\alpha \in \mathbf{S}(\alpha)} \alpha \cdot p(\alpha|\alpha|\theta) \tag{32}$$

where $p(\alpha|\alpha|\theta)$ is the probability of measuring the eigenvalue $\alpha = k$ of the output variable α. This probability is defined by the state $\theta = \sum_{i=0}^{n} q(i|\theta)\, \theta_i$ of the system $\Psi_{\text{B-E},\varepsilon} \cong \mathcal{K}$

$$p(\alpha|\alpha|\theta) = \sum_{i=0}^{n} q(i|\theta) \cdot p(\alpha|\alpha|\theta_i) = (1-p) \cdot \sum_{i=0}^{n} q(i|\theta) \cdot p^{\alpha - i} \quad [= \text{Tr}(\theta \pi_\alpha)] \tag{33}$$

From the differential equation with the *condition* for $\alpha = 0$

$$p(\alpha|\boldsymbol{\alpha}|\boldsymbol{\theta}) = p(\alpha - 1|\boldsymbol{\alpha}|\boldsymbol{\theta}) \cdot p + (1-p) \cdot q(\alpha|\boldsymbol{\theta}),\ \forall \alpha \geq 1; \quad p(0|\boldsymbol{\alpha}|\boldsymbol{\theta}) = (1-p) \cdot q(0|\boldsymbol{\theta}) \tag{34}$$

follows, for the medium value $E(\boldsymbol{\alpha})$ of the output stochastic variable $\boldsymbol{\alpha}$, that

$$\begin{aligned} E(\boldsymbol{\alpha}) &= \sum_{\alpha\geq 1} \alpha \cdot p(\alpha - 1|\boldsymbol{\alpha}|\boldsymbol{\theta}) \cdot p + \sum_{\alpha \geq 1} \alpha \cdot (1-p) \cdot q(\alpha|\boldsymbol{\theta}) \\ &= \sum_{\alpha\geq 1} \alpha \cdot p(\alpha - 1|\boldsymbol{\alpha}|\boldsymbol{\theta}) \cdot p + \mathcal{W} \cdot (1-p) = p \cdot E(\boldsymbol{\alpha}) + p \cdot \sum_{\alpha\geq 1} p(\alpha - 1|\boldsymbol{\alpha}|\boldsymbol{\theta}) + \mathcal{W} \cdot (1-p) \end{aligned} \tag{35}$$

$$E(\boldsymbol{\alpha}) \cdot (1-p) = p + \mathcal{W} \cdot (1-p) \longrightarrow E(\boldsymbol{\alpha}) = \frac{p}{1+p} + \mathcal{W}, \quad \mathcal{W} = E(\boldsymbol{\theta}) > 0, \quad \boldsymbol{\theta} \in \boldsymbol{\Theta}_0 \tag{36}$$

The quantity $H(\boldsymbol{\alpha}\|\boldsymbol{\theta}_i)$ is the p-entropy of measuring $\boldsymbol{\alpha}$ for the input $i \in \mathbf{S}$ being represented by the pure state $\boldsymbol{\theta}_i$ of the system $\boldsymbol{\Psi}_{\text{B}-\text{E},\varepsilon}$

$$\begin{aligned} -H(\boldsymbol{\alpha}\|\boldsymbol{\theta}_i) &= \sum_{j \in \mathbf{S}} (1-p)\, p^j \cdot \ln[(1-p)\, p^j] \\ &= (1-p) \cdot \ln(1-p) \cdot \sum_j p^j - (1-p) \cdot p \cdot \ln p \cdot \sum_j j\, p^{j-1} \\ &= \ln(1-p) - (1-p) \cdot p \cdot \ln p \cdot \frac{\mathrm{d}}{\mathrm{d}p} \left[\frac{1}{1-p} \right] = -\frac{h(p)}{1-p}, \ \forall i \in \mathbf{S} \end{aligned} \tag{37}$$

where $h(p) \stackrel{\triangle}{=} -(1-p) \cdot \ln(1-p) - p \cdot \ln p$ is the Shannon entropy of Bernoulli distribution $\{p,\, 1-p\}$.

The quantity $H(\boldsymbol{\alpha}|\boldsymbol{\theta})$ is the conditional Shannon entropy of the stochastic quantity $\boldsymbol{\alpha}$ in the state $\boldsymbol{\theta}$ of the system $\boldsymbol{\Psi}_{\text{B}-\text{E},\varepsilon}$, not depending on the $\boldsymbol{\theta}$ (the *noise* entropy)

$$H(\boldsymbol{\alpha}|\boldsymbol{\theta}) = \sum_{i=0}^{n} q(i|\boldsymbol{\theta}) \cdot H(\boldsymbol{\alpha}\|\boldsymbol{\theta}_i) = \sum_{i=0}^{n} q(i|\boldsymbol{\theta}) \cdot \frac{h(p)}{1-p} = \frac{h(p)}{1-p} \tag{38}$$

For capacity $C_{\text{B}-\text{E}''}$ of the channel $\mathcal{K} \cong \boldsymbol{\Psi}_{\text{B}-\text{E}}$ is, following the capacity definition, valid

$$C_{\text{B}-\text{E}''} = \sup_{\boldsymbol{\theta} \in \boldsymbol{\Theta}_0} H(\boldsymbol{\alpha}\|\boldsymbol{\theta}) - H(\boldsymbol{\alpha}|\boldsymbol{\theta}) = \sup_{\boldsymbol{\theta} \in \boldsymbol{\Theta}_0} H(\boldsymbol{\alpha}\|\boldsymbol{\theta}) - \frac{h(p)}{1-p} \tag{39}$$

where the set $\boldsymbol{\Theta}_0 = \{\boldsymbol{\theta} \in \boldsymbol{\Theta},\ E(\boldsymbol{\theta}) = \mathcal{W} \geq 0\}$ represents the *coding procedure* of the input $i \in \mathbf{S}$. The quantity $H(\boldsymbol{\alpha}\|\boldsymbol{\theta}) = H(p(\cdot|\boldsymbol{\alpha}|\boldsymbol{\theta}))$ is the p-entropy of the output quantity $\boldsymbol{\alpha}$. Its supremum is determined by the **Lagrange** multipliers method:

$$H(\boldsymbol{\alpha}\|\boldsymbol{\theta}) = - \sum_{\alpha \in \mathbf{S}(\boldsymbol{\alpha})} p(\alpha|\boldsymbol{\alpha}|\boldsymbol{\theta}) \cdot \ln p(\alpha|\boldsymbol{\alpha}|\boldsymbol{\theta}) = - \sum_{\alpha \in \mathbf{S}(\boldsymbol{\alpha})} p_\alpha \cdot \ln p_\alpha, \quad p_\alpha \stackrel{\triangle}{=} p(\alpha|\boldsymbol{\alpha}|\boldsymbol{\theta}) \tag{40}$$

The conditions for determinating of the *bound* extreme are

$$\sum_{\alpha \in \mathbf{S}(\boldsymbol{\alpha})} p_\alpha = 1, \quad \sum_{\alpha \in \mathbf{S}(\boldsymbol{\alpha})} \alpha \cdot p_\alpha = E(\boldsymbol{\alpha}) = \text{const.} \tag{41}$$

The Lagrange function

$$L = -\sum_{\alpha} p_\alpha \cdot \ln p_\alpha - \lambda_1 \cdot \sum_{\alpha} p_\alpha + \lambda_1 - \lambda_2 \cdot \sum_{\alpha} \alpha \cdot p_\alpha + \lambda_2 E(\boldsymbol{\alpha}) \tag{42}$$

which gives the condition for the extreme, $\dfrac{\partial L}{\partial p_\alpha} = -\ln p_\alpha - 1 - \lambda_1 - \lambda_2 \cdot \alpha = 0$, yielding in

$$p_\alpha = e^{-1-\lambda_1} \cdot e^{-\lambda_2 \alpha} = p(\alpha|\boldsymbol{\alpha}|\boldsymbol{\theta}) \text{ and further, in } \sum_{\alpha} p_\alpha = \sum_{\alpha} \frac{e^{-1-\lambda_1}}{e^{\lambda_2 \alpha}} = \frac{e^{-1-\lambda_1}}{1-e^{-\lambda_2}} = 1 \tag{43}$$

Then for the medium value $E(\boldsymbol{\alpha})$ the following result is obtained

$$\begin{aligned} E(\boldsymbol{\alpha}) &= \sum_{\alpha} \alpha \cdot p_\alpha = \sum_{\alpha} \alpha \cdot e^{-1-\lambda_1} \cdot e^{-\lambda_2 \alpha} = -e^{-1-\lambda_1} \cdot \frac{\partial}{\partial \lambda_2} \sum_{\alpha} e^{-\lambda_2 \alpha} \\ &= -e^{-1-\lambda_1} \cdot \frac{\partial}{\partial \lambda_2} \left[\frac{1}{1-e^{-\lambda_2}} \right] = \frac{e^{-1-\lambda_1}}{(1-e^{-\lambda_2})^2} \cdot e^{-\lambda_2} = \frac{e^{-\lambda_2}}{1-e^{-\lambda_2}} \end{aligned} \tag{44}$$

By (35), (36) and for the parametr $p{=}e^{-\frac{\varepsilon}{k\Theta}} = \text{const.}$ ($\varepsilon = \text{const.}$ $\Theta = \text{const.}$) $E(\boldsymbol{\alpha})$ is a function of $\mathcal{W}$ only. For $\lambda_2 = \dfrac{\varepsilon}{kT_W}$ is $e^{-\lambda_2} = p(\mathcal{W})$ and $E(\boldsymbol{\alpha})$ is the medium value for $\boldsymbol{\alpha}$ with the *geometric* probability distribution

$$p(\cdot) = p(\cdot|\boldsymbol{\alpha}|\boldsymbol{\theta}) = [1 - p(\mathcal{W})] \cdot p(\mathcal{W})^\alpha, \ \ \alpha \in \mathbf{S}(\boldsymbol{\alpha}) \tag{45}$$

depending only on $\dfrac{\varepsilon}{kT_W}$ or, on the absolute temperature T_W respectively. Thus for $E(\boldsymbol{\alpha})$ is valid that

$$E(\boldsymbol{\alpha}) = \frac{p(\mathcal{W})}{1-p(\mathcal{W})}, \ \ p(\mathcal{W}) = e^{-\frac{\varepsilon}{kT_W}} \tag{46}$$

From (35) and (46) is visible that $p(\mathcal{W})$ or T_W respectively is the *only one* root of the equation

$$\frac{p(\mathcal{W})}{1-p(\mathcal{W})} = \frac{p}{1-p} + \mathcal{W}, \text{ resp. } \frac{e^{-\frac{\varepsilon}{kT_W}}}{1-e^{-\frac{\varepsilon}{kT_W}}} = \frac{e^{-\frac{\varepsilon}{kT_0}}}{1-e^{-\frac{\varepsilon}{kT_0}}} + \mathcal{W} \tag{47}$$

From (34) and (45) follows that for state $\boldsymbol{\theta} \in \boldsymbol{\Theta}_0$ or, for the q-distribution $q(\cdot|\boldsymbol{\theta})$ respectively, in which the value $H(\boldsymbol{\alpha}\|\boldsymbol{\theta})$ is maximal [that state in which $\boldsymbol{\alpha}$ achieves the distribution $q(\cdot|\boldsymbol{\theta}) = p(\cdot)$], is valid that

$$q(\alpha|\boldsymbol{\theta}) = \frac{1-p(\mathcal{W})}{1-p}, \ \alpha = 0 \ \text{ and } \ q(\alpha|\boldsymbol{\theta}) = \frac{1-p(\mathcal{W})}{1-p} [p(\mathcal{W}) - p] \cdot p(\mathcal{W})^{\alpha-1}, \ \alpha > 0 \tag{48}$$

For the *effective temperatutre* T_W *of coding input messages* the distribution (45) supremizes (maximizes) the p-entropy $H(\boldsymbol{\alpha}\|\boldsymbol{\theta})$ of $\boldsymbol{\alpha}$ and, by using (37) with $p(\mathcal{W})$, is gained that

$$\sup_{\boldsymbol{\theta} \in \boldsymbol{\Theta}_0} H(\boldsymbol{\alpha}\|\boldsymbol{\theta}) = \frac{h[p(\mathcal{W})]}{1-p(\mathcal{W})} \tag{49}$$

From (39) and (49) follows [12, 37] the capacity $C_{\mathrm{B-E},\varepsilon}$ of the narrow–band channel $\mathcal{K} \cong \boldsymbol{\Psi}_{\mathrm{B-E}}$

$$C_{\mathrm{B-E},\varepsilon} = \frac{h[p(\mathcal{W})]}{1-p(\mathcal{W})} - \frac{h(p)}{1-p} \tag{50}$$

By (35), (36) and (46) the medium value $\mathcal{W}$ of the input message $i \in \mathbf{S}$ is derived,

$$\mathcal{W} = \frac{p(\mathcal{W})}{1 - p(\mathcal{W})} - \frac{p}{1-p} \tag{51}$$

By (31) the condition for the *minimal* average energy W_{Krit} needed for coding the input message is

$$\mathcal{W} \geq 0 \quad \text{resp.} \quad W = \varepsilon \cdot \mathcal{W} \geq 0, \quad \varepsilon \in \mathbf{S}(\boldsymbol{\varepsilon}) \tag{52}$$

$$W \geq W_{Krit}, \; W_{Krit} = 0 \tag{53}$$

The relations (35), (36), (46) and (52), (53) yield in

$$E(\boldsymbol{\alpha}) = \frac{p(\mathcal{W})}{1 - p(\mathcal{W})} \geq \frac{p}{1-p} \quad \text{and, then} \quad p(\mathcal{W}) \geq p, \; 1 - p(\mathcal{W}) \leq 1 - p \tag{54}$$

From (47) and (54) follows that for the defined direction of the signal (messaage) transmission at the temperature T_W of its sending and decoding is valid that

$$\frac{p(\mathcal{W})}{1 - p(\mathcal{W})} > \frac{p}{1-p}, \; p(\mathcal{W}) > p, \; \mathcal{W} > 0 \tag{55}$$

$$p(\mathcal{W}) = e^{-\frac{\varepsilon}{kT_W}} \geq e^{-\frac{\varepsilon}{kT_0}} = p \quad \text{and thus} \quad T_W \geq T_0 \tag{56}$$

4.2. Capacity of Fermi–Dirac narrow–band channel

Let is now considered, in the same way as it was in the B–E system, the pure states $\boldsymbol{\theta}_i \equiv i$ of the system $\boldsymbol{\Psi}$ which are coding the input messages $i = 0, 1, \ldots$ and the output stochastic quantity $\boldsymbol{\alpha}$ having its selecting space $\mathbf{S}$. On the spectrum $\mathbf{S}$ probabilities of realizations $\alpha \in \mathbf{S}$ are defined,

$$p(\alpha|\boldsymbol{\alpha}|\boldsymbol{\theta}_i) = \frac{p^{\alpha - i}}{1+p}, \quad p \in (0,1), \; i = 0, 1, \ldots \tag{57}$$

expressing the additive stochastic transformation of an input i into the output α for wich is valid $\alpha = i$ or $\alpha = i + 1$.[5] The uniform energy level $\varepsilon =$ const. of particles is considered.

The quantity W is the mathematical expectation of the energy coding the input signal

$$W = \varepsilon \cdot \mathcal{W}, \quad \mathcal{W} = \sum_{i \in \mathbf{S}} i \cdot q(i|\boldsymbol{\theta}) = E(\boldsymbol{\theta}) \tag{58}$$

The medium value of a stochastic quantity with the F–D statistic is given by

$$\sum_{j \in \{0,1\}} j \cdot \frac{p^j}{1+p} = \frac{p}{1+p} \tag{59}$$

The quantity $E(\boldsymbol{\alpha})$ is the medium value of the output quantity $\boldsymbol{\alpha}$,

$$E(\boldsymbol{\alpha}) = \sum_{\alpha \in \mathbf{S}(\boldsymbol{\alpha})} \alpha \cdot p(\alpha|\boldsymbol{\alpha}|\boldsymbol{\theta}) \tag{60}$$

[5] In accordance with ***Pauli*** *excluding principle* (valid for *fermions*) and a given energetic level $\varepsilon \subset \mathbf{S}(\varepsilon)$.

where $p(\alpha|\boldsymbol{\alpha}|\boldsymbol{\theta})$ is probability of realization of $\alpha \in \mathbf{S}$ in the state $\boldsymbol{\theta}$ of the system $\boldsymbol{\Psi} \equiv \boldsymbol{\Psi}_{\mathrm{F-D},\varepsilon} \cong \mathcal{K}$,

$$\boldsymbol{\theta} = \sum_{i \in \mathbf{S}} q(i|\boldsymbol{\theta})\, \boldsymbol{\theta}_i, \quad p(\alpha|\boldsymbol{\alpha}|\boldsymbol{\theta}) = \sum_{i=0}^{n} q(i|\boldsymbol{\theta}) \cdot p(\alpha|\boldsymbol{\alpha}|\boldsymbol{\theta}_i) = \frac{1}{1+p} \cdot \sum_{i=0}^{n} q(i|\boldsymbol{\theta}) \cdot p^{\alpha-i} \tag{61}$$

From the differential equation

$$p(\alpha|\boldsymbol{\alpha}|\boldsymbol{\theta}) = \frac{q(\alpha-1|\boldsymbol{\theta}) \cdot p + q(\alpha|\boldsymbol{\theta})}{1+p}, \quad \alpha \geq 1 \tag{62}$$

with the condition for

$$\alpha = 0, \quad p(0|\boldsymbol{\alpha}|\boldsymbol{\theta}) = \frac{1}{1+p} \cdot q(0|\boldsymbol{\theta})$$

follows that for the medium value $E(\boldsymbol{\alpha})$ of the output stochastic variable $\boldsymbol{\alpha}$ is valid that

$$\begin{aligned} E(\boldsymbol{\alpha}) &= \frac{p}{1+p} \cdot \sum_{\alpha \geq 1} \alpha \cdot q(\alpha-1|\boldsymbol{\theta}) + \frac{1}{1+p} \cdot \sum_{\alpha \geq 1} \alpha \cdot q(\alpha|\boldsymbol{\theta}) \\ &= \frac{p}{1+p} \cdot \sum_{\alpha \geq 1} (\alpha-1) \cdot q(\alpha-1|\boldsymbol{\theta}) + \frac{p}{1+p} \cdot \sum_{\alpha \geq 1} q(\alpha-1|\boldsymbol{\theta}) + \frac{1}{1+p} \cdot \mathcal{W} \\ &= \frac{p}{1+p} \cdot \mathcal{W} + \frac{p}{1+p} + \frac{1}{1+p} \cdot \mathcal{W} \quad \longrightarrow \quad E(\boldsymbol{\alpha}) = \frac{p}{1+p} + \mathcal{W}, \ \mathcal{W} = E(\boldsymbol{\theta}), \ \boldsymbol{\theta} \in \boldsymbol{\Theta}_0 \end{aligned} \tag{63}$$

The quantity $H(\boldsymbol{\alpha}\|\boldsymbol{\theta}_i)$ is the p-entropy of measuring $\boldsymbol{\alpha}$ for the input $i \in \mathbf{S}$ being represented by the pure state $\boldsymbol{\theta}_i$ of the system $\boldsymbol{\Psi}_{\mathrm{F-D},\varepsilon}$

$$\begin{aligned} H(\boldsymbol{\alpha}\|\boldsymbol{\theta}_i) &= -\sum_{j=0}^{1} \frac{p^j}{1+p} \cdot \ln \frac{p^j}{1+p} = -\frac{1}{1+p} \cdot \ln \frac{1}{1+p} - \frac{p}{1+p} \cdot \ln \frac{p}{1+p} \\ &= -\left(1 - \frac{p}{1+p}\right) \cdot \ln\left(1 - \frac{p}{1+p}\right) - \frac{p}{1+p} \cdot \ln \frac{p}{1+p} = h\left(\frac{p}{1+p}\right), \quad \forall i \in \mathbf{S} \end{aligned} \tag{64}$$

The quantity $H(\boldsymbol{\alpha}|\boldsymbol{\theta})$ is the conditional (the *noise*) Shannon entropy of the stochastic quantity $\boldsymbol{\alpha}$ in the system state $\boldsymbol{\theta}$, but, independent on this $\boldsymbol{\theta}$,

$$H(\boldsymbol{\alpha}|\boldsymbol{\theta}) = \sum_{i=0}^{n} q(i|\boldsymbol{\theta}) \cdot H(\boldsymbol{\alpha}\|\boldsymbol{\theta}_i) = \sum_{i} q(i|\boldsymbol{\theta}) \cdot h\left(\frac{p}{1+p}\right) = h\left(\frac{p}{1+p}\right) \tag{65}$$

For capacity $C_{\mathrm{F-D},\varepsilon}$ of the channel $\mathcal{K} \cong \boldsymbol{\Psi}_{\mathrm{F-D},\varepsilon}$ is, by the capacity definition in (23)-(24), valid that

$$C_{\mathrm{F-D}''} = \sup_{\boldsymbol{\theta} \in \boldsymbol{\Theta}_0} H(\boldsymbol{\alpha}\|\boldsymbol{\theta}) - H(\boldsymbol{\alpha}|\boldsymbol{\theta}) = \sup_{\boldsymbol{\theta} \in \boldsymbol{\Theta}_0} H(\boldsymbol{\alpha}\|\boldsymbol{\theta}) - h\left(\frac{p}{1+p}\right) \tag{66}$$

where the set $\boldsymbol{\Theta}_0 = \{\boldsymbol{\theta} \in \boldsymbol{\Theta} : E(\boldsymbol{\theta}) = \mathcal{W} > 0\}$ represents the coding procedure.

The quantity $H(\boldsymbol{\alpha}\|\boldsymbol{\theta}) = H(p(\cdot|\boldsymbol{\alpha}|\boldsymbol{\theta}))$ is the p-entropy of the stochastic quantity $\boldsymbol{\alpha}$ in the state $\boldsymbol{\theta}$ of the system $\boldsymbol{\Psi}_{\mathrm{F-D},\varepsilon}$. Its supremum is determined by the **Lagrange** multipliers method in the same way as in B–E case and with the same results for the probility distribution $p(\cdot|\boldsymbol{\alpha}|\boldsymbol{\theta})$ (geometric) and the medium value $E(\boldsymbol{\alpha})$

$$p(\cdot) = p(\cdot|\boldsymbol{\alpha}|\boldsymbol{\theta}) = [1 - p(\mathcal{W})] \cdot p(\mathcal{W})^{\alpha}, \ \alpha \in \mathbf{S}(\boldsymbol{\alpha}) \tag{67}$$
$$E(\boldsymbol{\alpha}) = \frac{p(\mathcal{W})}{1 - p(\mathcal{W})}, \ p(\mathcal{W}) = e^{-\frac{\varepsilon}{kT_W}}$$

Again , the value $E(\boldsymbol{\alpha})$ depends on $\frac{\varepsilon}{kT_W}$, or on absolute temperature T_W respectively, only. By using $E(\boldsymbol{\alpha})$ in (63) it is seen that $p(\mathcal{W})$ or T_W respectively is the only one root of the equation [12, 30]

$$\frac{p(\mathcal{W})}{1 - p(\mathcal{W})} = \frac{p}{1+p} + \mathcal{W}, \ \text{resp.} \ \frac{e^{-\frac{\varepsilon}{kT_W}}}{1 - e^{-\frac{\varepsilon}{kT_W}}} = \frac{e^{-\frac{\varepsilon}{kT_0}}}{1 + e^{-\frac{\varepsilon}{kT_0}}} + \mathcal{W} \tag{68}$$

For the q-distribution $q(\cdot|\boldsymbol{\theta}) = p(\cdot)$ of states $\boldsymbol{\theta} \in \boldsymbol{\Theta}_0$, for which the relation (62) and (67) is gained, follows that

$$\frac{q(\alpha - 1|\boldsymbol{\theta}) \cdot p + q(\alpha|\boldsymbol{\theta})}{1+p} = [1 - p(\mathcal{W})] \cdot p(\mathcal{W})^{\alpha}, \quad \alpha \in \mathbf{S} \ \text{with conditions} \tag{69}$$
$$q(0|\boldsymbol{\theta}) = (1+p) \cdot [1 - p(\mathcal{W})] \ \text{and} \ q(1|\boldsymbol{\theta}) = (1+p) \cdot [1 - p(\mathcal{W})] \cdot [p(\mathcal{W}) - p];$$
$$q(\alpha|\boldsymbol{\theta}) = (1+p) \cdot [1 - p(\mathcal{W})] \cdot \left[\left(\sum_{i=0}^{\alpha-1} (-1)^i \cdot p(\mathcal{W})^{\alpha-i} \cdot p^i \right) + (-1)^{\alpha} \cdot p^{\alpha} \right], \ \alpha > 1$$

For the *effective temperatutre* T_W *of coding the input messages* the distribution (67) supremises (maximizes) the p-entropy $H(\boldsymbol{\alpha}\|\boldsymbol{\theta})$ of $\boldsymbol{\alpha}$ is valid, in the same way as in (49), that

$$\sup_{\theta \in \boldsymbol{\Theta}_0} H(\boldsymbol{\alpha}\|\boldsymbol{\theta}) = \frac{h[p(\mathcal{W})]}{1 - p(\mathcal{W})} \tag{70}$$

By using (70) in (66) the formula for the $C_{F-D,\varepsilon}$ capacity [12, 37] is gained

$$C_{F-D,\varepsilon} = \frac{h[p(\mathcal{W})]}{1 - p(\mathcal{W})} - h\left(\frac{p}{1+p}\right) \tag{71}$$

The medium value $\mathcal{W}$ of the input $i = 0, 1, 2, \ldots$ is limited by a minimal not-zero and positive 'bottom' value $\mathcal{W}_{Krit}$. From (58), (63) and (68) follows

$$E(\boldsymbol{\alpha}) = \frac{p(\mathcal{W})}{1 - p(\mathcal{W})} \geq \frac{p}{1-p}, \ p(\mathcal{W}) = e^{-\frac{\varepsilon}{kT_W}} \geq e^{-\frac{\varepsilon}{kT_0}} = p \ \text{and thus} \ T_W \geq T_0 \tag{72}$$

$$\mathcal{W} = \frac{p(\mathcal{W})}{1 - p(\mathcal{W})} - \frac{p}{1+p} \geq 0, \ \mathcal{W} \geq \frac{2p^2}{1-p^2} = \mathcal{W}_{Krit}, \text{resp.} \ W = \varepsilon \cdot \mathcal{W} \geq \varepsilon \cdot \frac{2e^{-2\frac{\varepsilon}{kT_0}}}{1 - e^{-2\frac{\varepsilon}{kT_0}}} \tag{73}$$

For the average coding energy W, when the channel $C_{F-D,\varepsilon}$ acts on a uniform energetic level ε, is

$$W \geq W_{Krit} = \frac{2p^2}{1-p^2} \tag{74}$$

For the F–D channel is then possible speak about the ***effect of the not-zero capacity when the difference between the coding temperatures*** T_W ***and the noise temperature*** T_0 **is zero.**[6] This phennomenon is, by necessity, *given by properties of cells of the F–D phase space.*

[6] Not *not-zero capacity* for zero input power as was stated in [37]. The (74) also repares small missprint in [11].

5. Wide–band quantum transfer channels

Till now the narrow–band variant of an information transfer channel $\mathcal{K}_\varepsilon$, $\varepsilon \in \mathbf{S}(\varepsilon)$, card $\mathbf{S}(\varepsilon) = 1$ has been dealt. Let is now considered the symmetric operator of energy $\boldsymbol{\varepsilon}$ of a particle, having the spectrum of eigenavalues

$$\mathbf{S}(\varepsilon) = \left\{0,\ \frac{h}{\tau},\ \frac{2h}{\tau},\ \dots,\ \frac{nh}{\tau},\ \dots\right\} = \left\{\frac{rh}{\tau}\right\}_{r=0,\ 1,\ \dots,\ n},\ \text{card}\,\mathbf{S}(\varepsilon) = n+1 \qquad (75)$$

where $\tau > 0$ denotes the time length of the input signal and h denotes **Planck** constant. The *multi–band physical transfer channel* **K**, *memory-less, with additive noise* is defend by the (arranged) set of narrow–band, *independent* components $\mathcal{K}_\varepsilon$, $\varepsilon \in \mathbf{S}(\varepsilon)$,

$$\mathbf{K} = \underset{\varepsilon\in\mathbf{S}(\varepsilon)}{\times} \mathcal{K}_\varepsilon = \underset{\varepsilon\in\mathbf{S}(\varepsilon)}{\times} \{\boldsymbol{i}_\varepsilon,\ p(\alpha_\varepsilon|\boldsymbol{\alpha}_\varepsilon|\boldsymbol{\theta}_{\varepsilon,i_\varepsilon}),\ \boldsymbol{\alpha}_\varepsilon\} = \{\boldsymbol{i},\ p(\overline{\alpha}|\boldsymbol{\alpha}|\boldsymbol{\theta}_{\overline{i}}),\ \boldsymbol{\alpha}\} \qquad (76)$$

$$\boldsymbol{i} = \underset{\varepsilon\in\mathbf{S}(\varepsilon)}{\times} \boldsymbol{i}_\varepsilon = \underset{\varepsilon\in\mathbf{S}(\varepsilon)}{\times} [\mathbf{S},\ q_\varepsilon(i_\varepsilon|\boldsymbol{\theta}_\varepsilon)],\ i_\varepsilon \in \mathbf{S} = \{0,\ 1,\ 2,\ \dots\}$$

$$\boldsymbol{\alpha} = \times_{\varepsilon\in\mathbf{S}(\varepsilon)} \boldsymbol{\alpha}_\varepsilon = \underset{\varepsilon\in\mathbf{S}(\varepsilon)}{\times} [\mathbf{S},\ p(\alpha_\varepsilon|\boldsymbol{\alpha}_\varepsilon|\boldsymbol{\theta}_\varepsilon)],\ \alpha_\varepsilon \in \mathbf{S} = \{0,\ 1,2,\ \dots\}$$

Due the independency of narrow–band components $\mathcal{K}_\varepsilon$ the vector quantities $\boldsymbol{i}_\varepsilon$, $\boldsymbol{\alpha}_\varepsilon$, $\boldsymbol{\theta}_\varepsilon$, $\boldsymbol{j}_\varepsilon$ are independent stochastic quantities too.

The simultaneous *q-distribution* of the input vector of i_ε and the simultaneous *p-distribution* of measuring the output vector of values α_ε (of the individual narrow–band components $\mathcal{K}_\varepsilon$) are

$$\prod_{\varepsilon\in\mathbf{S}(\varepsilon)} q_\varepsilon(i_\varepsilon|\boldsymbol{\theta}_\varepsilon) = q(\overline{i}|\boldsymbol{\theta}), \qquad \boldsymbol{\theta} = \underset{\varepsilon\in\mathbf{S}(\varepsilon)}{\times} \boldsymbol{\theta}_\varepsilon,\ \ \overline{i} \in \mathbf{S}(\boldsymbol{i}) \qquad (77)$$

$$\prod_{\varepsilon\in\mathbf{S}(\varepsilon)} p(\alpha_\varepsilon|\boldsymbol{\alpha}_\varepsilon|\boldsymbol{\theta}_\varepsilon) = p(\overline{\alpha}|\boldsymbol{\alpha}|\boldsymbol{\theta}), \quad \boldsymbol{\theta} = \underset{\varepsilon\in\mathbf{S}(\varepsilon)}{\times} \boldsymbol{\theta}_\varepsilon,\ \ \overline{\alpha} \in \mathbf{S}(\boldsymbol{\alpha}) \qquad (78)$$

The system of quantities $\boldsymbol{\theta}_\varepsilon$ (the set of states of the narrow–band components $\mathcal{K}_\varepsilon$) is the state $\boldsymbol{\theta}$ of the multi–band channel **K** in which the (*canonic*) *q-distribution* of the system **K** is defined. Values i', j', α'

$$\alpha' = j' + i';\ \alpha' = \sum_{\varepsilon\in\mathbf{S}(\varepsilon)} \alpha_\varepsilon,\ j' = \sum_{\varepsilon\in\mathbf{S}(\varepsilon)} j_\varepsilon,\ i' = \sum_{\varepsilon\in\mathbf{S}(\varepsilon)} i_\varepsilon;\ j_\varepsilon = \alpha_\varepsilon - i_\varepsilon \geq 0,\ \forall\varepsilon \in \mathbf{S}(\varepsilon),\ \text{card}\mathbf{S}_\varepsilon > 1 \qquad (79)$$

are the numbers of the input, output and additive (noise) particles of the *multi–band* channel **K**. In this channel the stochastic transformation of the input i' into the output α' is performed, being determined by additive stochastic transformations of the input i_ε into the output α_ε in individual narrow–band components $\mathcal{K}_\varepsilon$.

Realizations of the stochastic systems $\boldsymbol{i}$, $\boldsymbol{\alpha}$, $\boldsymbol{\theta}$, $\boldsymbol{j}$ are the *vectors (sequences)* $\overline{i}$, $\overline{\alpha}$, $\overline{\theta}$, $\overline{j}$

$$\overline{i} = (i_\varepsilon)_{\varepsilon\in\mathbf{S}(\varepsilon)},\ \ \overline{\alpha} = (\alpha_\varepsilon)_{\varepsilon\in\mathbf{S}(\varepsilon)},\ \ \overline{\theta} = (\theta_\varepsilon)_{\varepsilon\in\mathbf{S}(\varepsilon)},\ \ \overline{j} = (j_\varepsilon)_{\varepsilon\in\mathbf{S}(\varepsilon)};\ \overline{i},\overline{\alpha},\overline{j} \in \underset{\varepsilon\in\mathbf{S}(\varepsilon)}{\times} \mathbf{S}, \qquad (80)$$

$$i_\varepsilon,\ \alpha_\varepsilon,\ j_\varepsilon \in \mathbf{S},\ \ \theta_\varepsilon \in \mathbf{S}(\boldsymbol{\theta}_\varepsilon),\ \ \boldsymbol{\theta}_\varepsilon = \sum_{i_\varepsilon\in\mathbf{S}} \theta_\varepsilon\,\boldsymbol{\theta}_{\varepsilon,i_\varepsilon},\qquad \overline{\theta} \in \mathbf{S}(\boldsymbol{\theta})$$

For the probability of the additive stochastic transformation (77), (78) of input $\bar{i}$ into the output $\bar{\alpha}$ is valid

$$\prod_{\varepsilon \in \mathbf{S}(\varepsilon)} p(\alpha_\varepsilon | \boldsymbol{\alpha}_\varepsilon | \boldsymbol{\theta}_{\varepsilon,i_\varepsilon}) = p(\bar{\alpha}|\boldsymbol{\alpha}|\boldsymbol{\theta}_{\bar{i}}), \ \bar{\alpha} = \bar{j} + \bar{i}, \ \bar{i} \in \mathbf{S}(\boldsymbol{i}), \ \bar{j} \in \mathbf{S}(\boldsymbol{j}), \ \bar{\alpha} \in \mathbf{S}(\boldsymbol{\alpha}) \tag{81}$$

The symbol $\boldsymbol{\theta}_{\varepsilon,i_\varepsilon}$ denotes the pure state coding the input $i_\varepsilon \in \mathbf{S}$ of a narrow–band component $\mathcal{K}_\varepsilon$ and the state $\boldsymbol{\theta}_{\bar{i}} = \times_{\varepsilon \in \mathbf{S}(\varepsilon)} \boldsymbol{\theta}_{\varepsilon,i_\varepsilon}$ codes the input $\bar{i}$ for which

$$q(\bar{i}|\boldsymbol{\theta}) = q_{\overline{\theta}} = \prod_{\varepsilon \in \mathbf{S}(\varepsilon)} q_\varepsilon(\boldsymbol{\theta}_\varepsilon) \tag{82}$$

For the multi–band channel $\mathbf{K}$ the following quantities are defined:[7]

- the *p-entropy of the output* $\boldsymbol{\alpha}$

$$\begin{aligned} H(\boldsymbol{\alpha}\|\boldsymbol{\theta}) &= \sum_{\varepsilon \in \mathbf{S}(\varepsilon)} H(\boldsymbol{\alpha}_\varepsilon \| \boldsymbol{\theta}_\varepsilon) = - \sum_{\varepsilon \in \mathbf{S}(\varepsilon)} \sum_{\alpha_\varepsilon \in \mathbf{S}} p(\alpha_\varepsilon|\boldsymbol{\alpha}_\varepsilon|\boldsymbol{\theta}_\varepsilon) \cdot \ln p(\alpha_\varepsilon|\boldsymbol{\alpha}_\varepsilon|\boldsymbol{\theta}_\varepsilon) \\ &\leq \sum_{\varepsilon \in \mathbf{S}(\varepsilon)} \sup_{\boldsymbol{\theta}_\varepsilon} H(\boldsymbol{\alpha}_\varepsilon \| \boldsymbol{\theta}_\varepsilon) \end{aligned} \tag{83}$$

for which, following the output narrow–band B–E and F–D components $\mathcal{K}_\varepsilon \in \mathbf{K}$, is valid that

$$\sup_{\boldsymbol{\theta} \in \overline{\boldsymbol{\Theta}}_0} H(\boldsymbol{\alpha}\|\boldsymbol{\theta}) = \sup_{\boldsymbol{\theta} \in \overline{\boldsymbol{\Theta}}_0} \sum_{\varepsilon \in \mathbf{S}(\varepsilon)} H(\boldsymbol{\alpha}_\varepsilon \| \boldsymbol{\theta}_\varepsilon) = \sum_{\varepsilon \in \mathbf{S}(\varepsilon)} \sup_{\boldsymbol{\theta}_\varepsilon} H(\boldsymbol{\alpha}_\varepsilon \| \boldsymbol{\theta}_\varepsilon) = \sum_{\varepsilon \in \mathbf{S}(\varepsilon)} \frac{h[p_\varepsilon(\mathcal{W})]}{1 - p_\varepsilon(\mathcal{W})} \tag{84}$$

- the conditional *noise entropy* (entropy of the *multi–band B–E | F–D noise*)

$$H(\boldsymbol{\alpha}|\boldsymbol{\theta}) = \sum_{\varepsilon \in \mathbf{S}(\varepsilon)} H(\boldsymbol{\alpha}_\varepsilon|\boldsymbol{\theta}_\varepsilon) = \sum_{\varepsilon \in \mathbf{S}(\varepsilon)} \sum_{i \in \mathbf{S}} q(i|\boldsymbol{\theta}_{\varepsilon,i}) \cdot H(\boldsymbol{\alpha}_\varepsilon \| \boldsymbol{\theta}_{\varepsilon,i}) = \sum_{\varepsilon \in \mathbf{S}(\varepsilon)} \left[\frac{h(p_\varepsilon)}{1 - p_\varepsilon} \middle| h\left(\frac{p_\varepsilon}{1 + p_\varepsilon}\right) \right] \tag{85}$$

where $p_\varepsilon(W) = e^{-\frac{\varepsilon}{kT_{0W}}}$, $p_\varepsilon = e^{-\frac{\varepsilon}{kT_0}}$, $T_W \geq T_0 > 0$ and $h(p) = -p \ln p - (1-p) \ln(1-p)$.

- the *transinformation* $T(\boldsymbol{\alpha};\boldsymbol{\theta})$ and the *information capacity* $C(\mathbf{K})$,

$$\begin{aligned} C(\mathbf{K}) &= \sup_{\boldsymbol{\theta} \in \overline{\boldsymbol{\Theta}}_0} T(\boldsymbol{\alpha};\boldsymbol{\theta}) = \sup_{\boldsymbol{\theta} \in \overline{\boldsymbol{\Theta}}_0} H(\boldsymbol{\alpha}\|\boldsymbol{\theta}) - H(\boldsymbol{\alpha}|\boldsymbol{\theta}) \\ &= \sum_{\varepsilon \in \mathbf{S}(\varepsilon)} \frac{h[p_\varepsilon(\mathcal{W})]}{1 - p_\varepsilon(\mathcal{W})} - \sum_{\varepsilon \in \mathbf{S}(\varepsilon)} \left[\frac{h(p_\varepsilon)}{1 - p_\varepsilon} \middle| h\left(\frac{p_\varepsilon}{1 + p_\varepsilon}\right) \right] \end{aligned} \tag{86}$$

The set $\overline{\boldsymbol{\Theta}}_0 = \times_{\varepsilon \in \mathbf{S}(\varepsilon)} \{\boldsymbol{\theta}_\varepsilon \in \Theta_\varepsilon;\ E(\boldsymbol{\theta}_\varepsilon) = W_\varepsilon \geq 0\}$ represents a coding procedure of the input $\bar{i}$ of the $\mathbf{K}$ into $\boldsymbol{\theta}_{\bar{i}}$, [by transforming each input i_ε into pure state $\boldsymbol{\theta}_{\varepsilon,i_\varepsilon}$, $\forall \varepsilon \in \mathbf{S}(\varepsilon)$].

[7] Using the *chain rule* for simultaneous probabilities it is found that for information entropy of an independent stochastic system $\vec{X} = (X_1, X_2, ..., X_n)$ is valid that $H(\vec{X}) = \sum_i H(X_i|X_1, ...X_{i-1}) = \sum_i H(X_i)$. Thus the physical entropy $\mathcal{H}(\boldsymbol{\theta})$ of independent stochastic system, $\boldsymbol{\theta} = \{\boldsymbol{\theta}_\varepsilon\}_\varepsilon$, is the sum of $H_\varepsilon[q(\cdot|\boldsymbol{\theta}_\varepsilon)]$ over $\varepsilon \in \mathbf{S}_\varepsilon$ too.

5.1. Transfer channels with continuous energy spectrum

Let a spectrum of energy with the finite cardinality $n+1$ and a finite time interval $\tau > 0$ are considered

$$\mathbf{S}(\varepsilon) = \{\varepsilon_r\}_{r=0,\,1,\,\dots,\,n} = \left\{\frac{rh}{\tau}\right\}_{r=0,\,1,\,\dots,\,n}, \quad \Delta\varepsilon = \frac{h}{\tau},\ \varepsilon_r = \frac{rh}{\tau} = r\cdot\Delta\varepsilon, \tag{87}$$

$$\text{card}\,\mathbf{S}(\varepsilon) = \frac{\varepsilon_n}{\Delta\varepsilon} = n+1,\ n\cdot\Delta\varepsilon = n\cdot\frac{h}{\tau} = \varepsilon_n,\ \frac{\tau}{n} = \frac{h}{\varepsilon_n} = \text{const.}$$

For a transfer channel with the continuous spectrum of energies of particles and with the *band–width* equal to card $\mathbf{S}(\varepsilon) = \frac{\varepsilon_n}{h}$, is valid that

$$\lim_{\tau\to\infty}\varepsilon_r = \lim_{\tau\to\infty}\frac{rh}{\tau} = \lim_{\tau\to\infty} r\cdot\Delta\varepsilon \stackrel{\triangle}{=} r\,\mathrm{d}\,\varepsilon \ \text{resp.}\ \lim_{\tau\to\infty}\frac{1}{\tau} = \frac{\mathrm{d}\varepsilon}{h}, \quad \mathbf{S}(\varepsilon) = \langle 0, \varepsilon_n) \tag{88}$$

But the infinite wide–band and infinite number of particles ($\tau \longrightarrow \infty,\ n \longrightarrow \infty$) will be dealt with. Then

$$\mathbf{S}(\varepsilon) = \{\varepsilon_r\}_{r=0,\,1,\,\dots} = \left\{\frac{rh}{\tau}\right\}_{r=0,\,1,\,\dots} = \lim_{\tau\to\infty}\varepsilon_r = \lim_{\tau\to\infty}\frac{rh}{\tau} = \lim_{\varepsilon\to 0} r\,\Delta\varepsilon = r\,\mathrm{d}\,\varepsilon \tag{89}$$

and thus the *wide–band spectrum* $\mathbf{S}(\varepsilon)$ *of energies* is

$$\lim_{\tau\to\infty}\frac{1}{\tau} = \frac{\mathrm{d}\varepsilon}{h}, \quad \mathbf{S}(\varepsilon) = \langle 0, \infty) \tag{90}$$

With the denotation $\alpha_\varepsilon \stackrel{\triangle}{=} \alpha,\ i \stackrel{\triangle}{=} i_\varepsilon,\ j \stackrel{\triangle}{=} j_\varepsilon,\ i_\varepsilon,\ j_\varepsilon \in \mathbf{S},\ \alpha_\varepsilon \in \mathbf{S}, \varepsilon \in \mathbf{S}(\varepsilon)$ For the p-entropy of the output $\boldsymbol{\alpha}$ of the wide–band transfer channel $\mathbf{K}_{\mathrm{B-E|F-D}}$ is valid that

$$H(\boldsymbol{\alpha}\|\boldsymbol{\theta}) = \lim_{\tau\to\infty}\frac{1}{\tau}\sum_{\varepsilon\in\mathbf{S}(\varepsilon)} H(\boldsymbol{\alpha}_\varepsilon\|\boldsymbol{\theta}_\varepsilon) = \lim_{\tau\to\infty} -\frac{1}{\tau}\sum_{\varepsilon\in\mathbf{S}(\varepsilon)}\sum_{\alpha_\varepsilon\in\mathbf{S}} p(\alpha_\varepsilon|\boldsymbol{\alpha}_\varepsilon|\boldsymbol{\theta}_\varepsilon)\cdot\ln p(\alpha_\varepsilon|\boldsymbol{\alpha}_\varepsilon|\boldsymbol{\theta}_\varepsilon) \tag{91}$$

$$= -\frac{1}{h}\int_0^\infty \left[\sum_{\alpha\in\mathbf{S}} p(\alpha|\boldsymbol{\alpha}_\varepsilon|\boldsymbol{\theta}_\varepsilon)\cdot\ln p(\alpha|\boldsymbol{\alpha}_\varepsilon|\boldsymbol{\theta}_\varepsilon)\right]\mathrm{d}\varepsilon$$

$$\sup_{\boldsymbol{\theta}} H(\boldsymbol{\alpha}\|\boldsymbol{\theta}) = \lim_{\tau\to\infty}\frac{1}{\tau}\sum_{\varepsilon\in\mathbf{S}(\varepsilon)}\sup_{\theta_\varepsilon} H(\boldsymbol{\alpha}_\varepsilon\|\boldsymbol{\theta}_\varepsilon) = \frac{1}{h}\int_0^\infty \frac{h[p_\varepsilon(\mathcal{W})]}{1-p_\varepsilon(\mathcal{W})}\,\mathrm{d}\varepsilon$$

For conditional entropy of the wide–band transfer channel $\mathbf{K}_{\mathrm{B-E|F-D}}$ [entropy of the wide–band noise independent on the system ($\mathbf{K}_{\mathrm{B-E|F-D}}$) state $\boldsymbol{\theta}$] and for its information capacity is valid, by (85) and (86)

$$H(\boldsymbol{\alpha}|\boldsymbol{\theta}) = \lim_{\tau\to\infty}\frac{1}{\tau}\sum_{\varepsilon\in\mathbf{S}(\varepsilon)} H(\boldsymbol{\alpha}_\varepsilon|\boldsymbol{\theta}_\varepsilon) = \lim_{\tau\to\infty}\frac{1}{\tau}\sum_{\varepsilon\in\mathbf{S}(\varepsilon)}\sum_{i\in\mathbf{S}} q(i|\boldsymbol{\theta}_{\varepsilon,i})\cdot H(\boldsymbol{\alpha}_\varepsilon\|\boldsymbol{\theta}_{\varepsilon,i}) \tag{92}$$

$$= \frac{1}{h}\int_0^\infty H(\boldsymbol{\alpha}_\varepsilon|\boldsymbol{\theta}_\varepsilon)\,\mathrm{d}\varepsilon = \frac{1}{h}\int_0^\infty \left[\frac{h(p_\varepsilon)}{1-p_\varepsilon}\middle| h\left(\frac{p_\varepsilon}{1+p_\varepsilon}\right)\right]\mathrm{d}\varepsilon$$

$$\mathrm{C}(\mathbf{K}_{\mathrm{B-E|F-D}}) = = \frac{1}{h}\int_0^\infty \frac{h[p_\varepsilon(\mathcal{W})]}{1-p_\varepsilon(\mathcal{W})}\mathrm{d}\varepsilon - \frac{1}{h}\int_0^\infty \left[\frac{h(p_\varepsilon)}{1-p_\varepsilon}\middle| h\left(\frac{p_\varepsilon}{1+p_\varepsilon}\right)\right]\mathrm{d}\varepsilon \tag{93}$$

By (52) and (74) the average number of particles on the input of a narrow–band component $\mathcal{K}_\varepsilon$ is

$$\mathcal{W}_\varepsilon = \sum_{i\in\mathbf{S}} i \cdot q(i|\boldsymbol{\theta}_\varepsilon) \geq \left[0 \,\middle|\, \frac{2p_\varepsilon^2}{1-p_\varepsilon^2}\right] \tag{94}$$

and then for the *whole average number of input particles* $\mathcal{W}'$ of the wide–band transfer channel **K** is obtained

$$\mathcal{W}' = \frac{1}{h}\int_0^\infty \mathcal{W}_\varepsilon \,\mathrm{d}\varepsilon, \;\; W = \frac{1}{h}\int_0^\infty \varepsilon\, \mathcal{W}_\varepsilon \,\mathrm{d}\varepsilon \tag{95}$$

where W is *whole input energy* and T_W is the *effective coding temperature* being supposingly at the value T_{W_ε}, $T_W = T_{W_\varepsilon}$, $\forall\varepsilon \in \mathbf{S}(\boldsymbol{\varepsilon})$.

5.2. Bose–Einstein wide–band channel capacity

By derivations (86) and (92), (93) is valid that [12]

$$\mathsf{C}(\mathbf{K}_{\mathrm{B-E}}) = \frac{1}{h}\int_0^\infty \frac{h[p_\varepsilon(\mathcal{W})]}{1-p_\varepsilon(\mathcal{W})}\,\mathrm{d}\varepsilon - \frac{1}{h}\int_0^\infty \frac{h(p_\varepsilon)}{1-p_\varepsilon}\,\mathrm{d}\varepsilon \tag{96}$$

For the first or, for the second integral respectively, obviously is valid

$$\frac{1}{h}\int_0^\infty \frac{h[p_\varepsilon(W)]}{1-p_\varepsilon(W)}\,\mathrm{d}\varepsilon = \frac{\pi^2 \mathrm{k} T_W}{3h} \quad \text{resp.} \quad \frac{1}{h}\int_0^\infty \frac{h(p_\varepsilon)}{1-p_\varepsilon}\,\mathrm{d}\varepsilon = \frac{\pi^2 \mathrm{k} T_0}{3h} \tag{97}$$

Then, for the capacity of the wide–band B–E transfer channel $\mathbf{K}_{\mathrm{B-E}}$ is valid

$$\mathsf{C}(\mathbf{K}_{\mathrm{B-E}}) = \frac{\pi^2 \mathrm{k}}{3h}(T_W - T_0) = \frac{\pi^2 \mathrm{k} T_W}{3h}\cdot\frac{T_W - T_0}{T_W} \triangleq \frac{\pi^2 \mathrm{k} T_W}{3h}\cdot \eta_{\max},\; T_W \geq T_0 \tag{98}$$

and for the *whole average output energy* is valid

$$\lim_{\tau\to\infty}\frac{1}{\tau}\sum_{\varepsilon\in\mathbf{S}(\boldsymbol{\varepsilon})} \varepsilon \frac{p_\varepsilon(\mathcal{W})}{1-p_\varepsilon(\mathcal{W})} = \frac{1}{h}\int_0^\infty \varepsilon \frac{p_\varepsilon(\mathcal{W})}{1-p_\varepsilon(\mathcal{W})}\,\mathrm{d}\varepsilon = -\frac{\mathrm{k}^2 T_W^2}{h}\int_0^1 \frac{\ln(1-t)}{t}\,dt = \frac{\pi^2 \mathrm{k}^2 T_W^2}{6h} \tag{99}$$

For the *whole average energy* of the B–E *noise* must be valid

$$\lim_{\tau\to\infty}\frac{1}{\tau}\sum_{\varepsilon\in\mathbf{S}(\boldsymbol{\varepsilon})} \varepsilon \frac{p_\varepsilon}{1-p_\varepsilon} = \frac{1}{h}\int_0^\infty \varepsilon \frac{p_\varepsilon}{1-p_\varepsilon}\,\mathrm{d}\varepsilon = \frac{\pi^2 \mathrm{k}^2 {T_0}^2}{6h} \tag{100}$$

From the relations (79) among the energies of the output α', of the noise j' and the input i',

$$\frac{\pi^2 \mathrm{k}^2 T_W^2}{6h} = \frac{\pi^2 \mathrm{k}^2 {T_0}^2}{6h} + W \tag{101}$$

the effective coding temperature T_W is derivable, $T_W = T_0 \cdot \sqrt{1 + \frac{6hW}{\pi^2 \mathrm{k}^2 {T_0}^2}}$. Using it in (98) gives

$$\mathsf{C}(\mathbf{K}_{\mathrm{B-E}}) = \frac{\pi^2 \mathrm{k} T_0}{3h}\left(\sqrt{1 + \frac{6hW}{\pi^2 \mathrm{k}^2 {T_0}^2}} - 1\right) \tag{102}$$

For $T_0 \to 0$ the *quantum aproximation* of $C(\mathbf{K}_{B-E})$, independent on the heat noise energy (deminishes whith temperture's aiming to absolute 0° K)

$$\lim_{T_0\to 0} C(\mathbf{K}_{B-E}) = \lim_{T_0\to 0}\left(\sqrt{\frac{\pi^4 k^2 T_0{}^2}{9h^2}+\pi^2\frac{2W}{3h}}-\frac{\pi^2 kT_0}{3h}\right) = \pi\sqrt{\frac{2W}{3h}} \tag{103}$$

The *classical approximation* of $C(\mathbf{K}_{B-E})$ is gaind for temperatures $T_0 \gg 0$ ($T_0 \to \infty$ respectively). It is near to value $\frac{W}{kT_0}$, the Shannon capacity of the wide–band **Gaussian** channel with the whole noise energy kT_0 and with the whole average input energy W. For T_0 from (101), great enough, is gained that[8]

$$C(\mathbf{K}_{B-E}) \doteq \frac{\pi^2 kT_0}{3h}\left(\frac{3hW}{\pi^2 k^2 T_0{}^2}\right) = \frac{W}{kT_0} \tag{104}$$

5.3. Fermi–Dirac wide–band channel capacity

By derivations (86) and (92), (93) is valid that [12]

$$C(\mathbf{K}_{F-D}) = \frac{1}{h}\int_0^\infty \frac{h[p_\varepsilon(W)]}{1-p_\varepsilon(W)}\,d\varepsilon - \frac{1}{h}\int_0^\infty h\left(\frac{p_\varepsilon}{1+p_\varepsilon}\right)d\varepsilon = \frac{\pi^2 kT_W}{3h} - \frac{1}{h}\int_0^\infty h\left(\frac{p_\varepsilon}{1+p_\varepsilon}\right)d\varepsilon \tag{105}$$

For the second integral obviously is valid

$$\frac{1}{h}\int_0^\infty h\left(\frac{p_\varepsilon}{1+p_\varepsilon}\right)d\varepsilon = \frac{\pi^2 kT_0}{6h} \tag{106}$$

By figuring (105) the capacity of the wide–band F–D channel $\mathbf{K}_{F-D}$ is gained,

$$C(\mathbf{K}_{F-D}) = \frac{\pi^2 k}{3h}\left(T_W - \frac{T_0}{2}\right) \tag{107}$$

and for $T_W > T_0$is writable

$$C(\mathbf{K}_{F-D}) = C(\mathbf{K}_{B-E})\cdot\frac{2T_W - T_0}{2T_W - 2T_0} \tag{108}$$

For the *whole average output energy* is valid the same as for the B–E case,

$$\frac{1}{h}\int_0^\infty \varepsilon\frac{p_\varepsilon(W)}{1-p_\varepsilon(W)}\,d\varepsilon = \frac{\pi^2 k^2 T_W^2}{6h} \tag{109}$$

For the *whole average F–D wide–band noise* energy is being derived

$$\lim_{\tau\to\infty}\frac{1}{\tau}\sum_{\varepsilon\in\mathbf{S}(\varepsilon)}\varepsilon\frac{p_\varepsilon}{1+p_\varepsilon} = \frac{1}{h}\int_0^\infty \varepsilon\frac{e^{-\frac{\varepsilon}{kT_0}}}{1+e^{-\frac{\varepsilon}{kT_0}}}\,d\varepsilon = \frac{k^2T_0{}^2}{h}\int_0^\infty x\,\frac{e^{-x}}{1+e^{-x}}\,dx \tag{110}$$
$$= -\frac{k^2T_0{}^2}{h}\int_0^1 \frac{\ln t}{t+1}\,dt = \frac{\pi^2 k^2 T_0{}^2}{12h}$$

---

[8] For $|x| < 1$, $\sqrt{1+x} = 1 + \frac{1}{2}x - \frac{1}{8}x^2 + \ldots \doteq 1 + \frac{1}{2}x$ where $x = \frac{6hW}{\pi^2 k^2 T_0{}^2} < 1$.

From the relation (79) among the whole output, input, and noise energy,

$$\frac{\pi^2 k^2 T_W^2}{6h} = \frac{\pi^2 k^2 {T_0}^2}{12h} + W \tag{111}$$

follows the effective coding temperature $T_W = T_0 \cdot \sqrt{\frac{1}{2} + \frac{6hW}{\pi^2 k^2 {T_0}^2}}$. Using it in (107) the result is [24]

$$C(\mathbf{K}_{\text{F-D}}) = \frac{\pi^2 k T_0}{3h} \left(\sqrt{\frac{1}{2} + \frac{6hW}{\pi^2 k^2 {T_0}^2}} - \frac{1}{2} \right) \tag{112}$$

For $T_0 \to 0$ the *quantum approximation* capacity $C(\mathbf{K}_{\text{F-D}})$, independent on heat noise energy kT_0 is gaind (the same as in the B–E case (103),

$$\lim_{T_0 \to 0} C(\mathbf{K}_{\text{F-D}}) = \lim_{T_0 \to 0} \left(\sqrt{\frac{\pi^4 k^2 {T_0}^2}{9h^2} \cdot \frac{1}{2} + \pi^2 \frac{2W}{3h}} - \frac{\pi^2 k T_0}{3h} \cdot \frac{1}{2} \right) = \pi \sqrt{\frac{2W}{3h}} \tag{113}$$

The *classical approximation* of the capacity $C(\mathbf{K}_{\text{F-D}})$ is gained for $T_0 \gg 0$[9]

$$\begin{aligned} C(\mathbf{K}_{\text{F-D}}) &= \frac{\pi^2 k T_0}{3h} \left[\frac{1}{\sqrt{2}} \sqrt{1 + \frac{12hW}{\pi^2 k^2 {T_0}^2}} - \frac{1}{2} \right] \doteq \frac{\pi^2 k T_0}{3h} \left[\frac{1}{\sqrt{2}} \left(1 + \frac{6hW}{\pi^2 k^2 {T_0}^2} \right) - \frac{1}{2} \right] \\ &= \frac{\pi^2 k T_0}{6h} \left(\sqrt{2} - 1 \right) + \sqrt{2} \frac{W}{kT_0} \quad \left[\underset{T_0 \to \infty}{\longrightarrow} \frac{\pi^2 k T_0}{6h} \left(\sqrt{2} - 1 \right), \; W = \text{const.} \geq W_{crit} \right] \end{aligned} \tag{114}$$

By (74) the condition for the medium value of the input particles of a narrow–band component $\mathcal{K}_\varepsilon$, $\varepsilon \in \mathbf{S}(\varepsilon)$, of the channel $\mathbf{K}_{\text{F-D}}$ is valid, $\mathcal{W}_\varepsilon \geq \frac{2p_\varepsilon^2}{1 - p_\varepsilon^2}$, from which the condition for the whole input energy of the wide–band channel $\mathbf{K}_{\text{F-D}}$ follows. By (95) it is gaind, for $T_W \geq T_0 > 0$, that[10]

$$W \geq \lim_{\tau \to \infty} \frac{1}{\tau} \sum_{\varepsilon \in \mathbf{S}(\varepsilon)} \varepsilon \mathcal{W}_\varepsilon \geq \lim_{\tau \to \infty} \frac{1}{\tau} \sum_{\varepsilon \in \mathbf{S}(\varepsilon)} \varepsilon \frac{2p_\varepsilon^2}{1 - p_\varepsilon^2} \doteq \frac{2}{h} \int_0^\infty \varepsilon \frac{e^{-2\frac{\varepsilon}{kT_0}}}{1 - e^{-2\frac{\varepsilon}{kT_0}}} \, d\varepsilon = \frac{\pi^2 k^2 {T_0}^2}{12h} = W_{crit} > 0 \tag{115}$$

6. Physical information transfer and thermodynamics

Whether the considered information transfers are narrow–band or wide–band, their algebraic-information description remains the same. So let be considered an arbitrary *stationary physical system* $\mathbf{\Psi}$ of these two band–types as usable for information transfer.

Let a system state $\theta' = \sum_{i=1}^{n} q(i|\theta') \, \pi\{\psi_i'\} \in \Theta$ of the system $\mathbf{\Psi}$ is the *successor (follower, equivocant)* of the system state $\theta = \sum_{i=1}^{n} q(i|\theta) \, \pi\{\psi_i\} \in \Theta$, $\theta \longrightarrow \theta'$ is written. The

[9] For $\sqrt{1+x} \doteq 1 + \frac{1}{2}x$ when $|x| < 1$; $x = \frac{12hW}{\pi^2 k^2 {T_0}^2}$.

[10] If, in the special case of F–D channel, it is considered that the value W given by the number of electrons as the average energy of the *modulating current* entering into a wire, over a *time unit*, then it is the average power on the electric resistor $R = 1\Omega$ too.

distribution $q(i|\theta') = \sum_{j=1}^{n} p(i|j)\, q(j|\theta),\ p(i|j) = (\psi'_i, \psi_j)^2 = u_{ji}^2$ and $\theta \longrightarrow \theta'$, ensures existence of the *transformation matrix* $[u_{i,j}]$ of a *base* of the space $\Psi = \{\Psi\}_{i=1}^{n}$ into the base $\{\Psi'\}_{i=1}^{n}$.[11]

From the relation $\theta \longrightarrow \theta'$ also is visible that it is *reflexive* and *transitive* relation between states and, thus, it defines (a partial) *arrangment* on the set space Θ. The *terminal, maximal* state for this arrangement is the equilibrial state θ^+ of the system Ψ: it is the successor of an arbitrary system state, including itself.

The *statistic, Shannon, information)* entropy $H(\cdot)$ is a generalization of the physical entropy$\mathcal{H}(\theta)$. The quantity I-divergence $I(\cdot\|\cdot)$ is, by (21), a generalization of the physical quantity $I(p\|d) = \mathcal{H}(\theta^+) - \mathcal{H}(\theta)$ where the state

$$\theta^+ = \frac{1}{n}\sum_{i=1}^{n} \theta_i \in \Theta \quad \text{for} \quad \theta_i = \pi\{\psi_i\}, \quad i = 1, 2, ..., n \tag{116}$$

is the *equlibrial state* of the system Ψ. The probability distribution into the *canonic components* θ_i of θ^+ is *uniform* and thus

$$\mathcal{H}(\theta^+) = \ln n = \ln \dim(\Psi) \tag{117}$$

Information divergence $I(p\|d) \geq 0$ expresses the *distance* of the two probability distributions $q(\cdot|\theta)$ and $q(\cdot|\theta^+)$ of states (stochastic quantities)

$$\theta = [\mathbf{S}, q(\cdot|\theta)] \quad \text{and} \quad \theta^+ = [\mathbf{S}, q(\cdot|\theta^+)] \tag{118}$$

In the physical sense the divergence $I(p\|d)$ is a *measure of a not-equilibriality* of the state θ of the physical (let say a thermodynamic) system Ψ. Is maximized in the initial (starting), not-equilibrium state of the (time) evolution of the Ψ. It is clear that $I(p\|d) \equiv T(\alpha; \theta)$

6.0.0.5. $\mathcal{H}$-Theorem, *II. Second Principle of Thermodynamics*:

Let for states $\theta, \theta' \in \Theta$ of the system Ψ is valid that $\theta \rightarrow \theta'$. Then

$$\mathcal{H}(\theta') \geq \mathcal{H}(\theta) \tag{119}$$

and the equality arises for $\theta = \theta'$ only [12, 38].

6.0.0.6. Proof:

(a) For a *strictly convex* function $f(u) = u \cdot \ln u$ the ***Jensen*** *inequality* is valid [23]

$$f\left[\sum_{j=1}^{n} p(i|j)\, q(j|\theta)\right] \leq \sum_{j=1}^{n} p(i|j)\, f[q(j|\theta)], \quad i = 1, 2, ..., n \tag{120}$$

$$\sum_{i=1}^{n} f\left[\sum_{j=1}^{n} p(i|j)\, q(j|\theta)\right] \leq \sum_{i=1}^{n} p(i|j) \sum_{j=1}^{n} f[q(j|\theta)]$$

$$= \sum_{j=1}^{n} f[q(j|\theta)] = \sum_{j=1}^{n} q(j|\theta) \ln q(j|\theta) = -H[q(\cdot|\theta)] = -\mathcal{H}(\theta) \quad \text{due to} \quad \sum_{i=1}^{n} p(i|j) = 1$$

---

[11] It is the matrix of the *unitary operator* $\mathbf{u}(t)$ expressing the time evolution of the system Ψ.

and for distributions $q(i|\boldsymbol{\theta}')$ is valid $\mathcal{H}(\boldsymbol{\theta}') \geq \mathcal{H}(\boldsymbol{\theta})$:

$$\sum_{i=1}^{n} f\left(\sum_{j=1}^{n} p(i|j)\, q(j|\boldsymbol{\theta})\right) = \sum_{i=1}^{n} q(i|\boldsymbol{\theta}') \ln q(i|\boldsymbol{\theta}') = -\mathcal{H}(\boldsymbol{\theta}') \;\; [\leq -\mathcal{H}(\boldsymbol{\theta})]$$

(b) The equality in (119) arises if and only if the index permutation $[i(1),\, i(2),\, ...,\, i(n)]$ exists that $p(i|j) = \delta[i|i(j)],\ j = 1,\, 2,\, ...,\, n;$ then $q[i(j)|\boldsymbol{\theta}'] = q(j|\boldsymbol{\theta}),\ j = 1,\, 2,\, ...$
Let a fixed j is given. Then, when $0 = p(i|j) = (\psi_i', \psi_j)^2,\ i \neq i(j)$, the orthogonality is valid

$$\Psi(\psi_j|\boldsymbol{\psi}_j) \perp \left[\bigoplus_{i \neq i(j)} \Psi(\psi_i'|\boldsymbol{\psi}_i')\right], \quad \boldsymbol{\psi}_j = \pi\{\psi_j\} = \boldsymbol{\theta}_j, \quad \boldsymbol{\psi}_i' = \pi\{\psi_i'\} = \boldsymbol{\theta}_i' \tag{121}$$

and, consequently, $\psi_j \in \Psi[\psi_{i(j)}'|\boldsymbol{\psi}_{i(j)}'],\ p[i(j)|j] = (\psi_{i(j)}', \psi_j)^2 = 1$. It results in $\psi_j = \psi_{i(j)}'$. This prooves that the equality $\mathcal{H}(\boldsymbol{\theta}') = \mathcal{H}(\boldsymbol{\theta})$ implies the equality $q(j|\boldsymbol{\theta})\, \pi\{\psi_j\} = q[i(j)|\boldsymbol{\theta}]\, \pi\{\psi_{i(j)}'\}$ and $\boldsymbol{\theta} = \boldsymbol{\theta}'$.

$\mathcal{H}$-theorem says, that a **reversible transition is not possible between any two different states** $\boldsymbol{\theta} \neq \boldsymbol{\theta}'$. From the inequality (119) also follows that any state $\boldsymbol{\theta} \in \boldsymbol{\Theta}$ of the system $\boldsymbol{\Psi}$ is the successor of itself, $\boldsymbol{\theta} \rightarrow \boldsymbol{\theta}$ and, that any **reversibility** of the **relation $\boldsymbol{\theta} \rightarrow \boldsymbol{\theta}'$ (the transition $\boldsymbol{\theta}' \rightarrow \boldsymbol{\theta}$) is not possible within the system only, it is not possible without openning this system $\boldsymbol{\Psi}$**. The difference

$$\mathcal{H}(\boldsymbol{\theta}^+) - \mathcal{H}(\boldsymbol{\theta}) = \max_{\boldsymbol{\theta}' \in \boldsymbol{\Theta}} \mathcal{H}(\boldsymbol{\theta}') - \mathcal{H}(\boldsymbol{\theta}) = H\left[q(\cdot|\boldsymbol{\theta}^+)\right] - H\left[q(\cdot|\boldsymbol{\theta})\right] \tag{122}$$

reppresents the information-theoretical expressing of the ***Brillouin** (maximal) entropy defect* ΔH (the Brillouin *negentropic information principle* [2, 30]). For the state $\boldsymbol{\theta}^+$ is valid that $\boldsymbol{\theta} \rightarrow \boldsymbol{\theta}^+, \forall \boldsymbol{\theta} \in \boldsymbol{\Theta}$. It is also called the *terminal* state or the (*atractor* of the time evolution) of the system $\boldsymbol{\Psi}$.[12]

6.0.0.7. Gibbs Theorem:

For all $\boldsymbol{\theta},\ \tilde{\boldsymbol{\theta}} \in \boldsymbol{\Theta}$ of the system $\boldsymbol{\Psi}$ is valid

$$\mathcal{H}(\boldsymbol{\theta}) \leq -\mathrm{Tr}(\boldsymbol{\theta} \ln \tilde{\boldsymbol{\theta}}) \tag{123}$$

and the equality arises only for $\boldsymbol{\theta} = \tilde{\boldsymbol{\theta}}$ [38].

6.0.0.8. Proof:

Let for $\boldsymbol{\theta},\ \tilde{\boldsymbol{\theta}} \in \boldsymbol{\Theta}$ is valid that $\boldsymbol{\theta} = \sum_{i=1}^{n} q(i|\boldsymbol{\theta})\, \pi\{\psi_i\},\ \tilde{\boldsymbol{\theta}} = \sum_{i=1}^{n} q(i|\tilde{\boldsymbol{\theta}})\, \pi\{\psi_i'\}$ and let the operators $\boldsymbol{\alpha},\ \boldsymbol{\theta}$ are *commuting* $\boldsymbol{\alpha\theta} = \boldsymbol{\theta\alpha},\ D(\boldsymbol{\alpha}) = \{D_\alpha : \alpha \in \mathbf{S}(\boldsymbol{\alpha})\},\ D(\boldsymbol{\theta}) = \{D_\theta : \theta \in \mathbf{S}(\boldsymbol{\theta})\}$, are their spectral decompositions. Let be the state $\boldsymbol{\theta}'$ the successor of $\boldsymbol{\theta},\ \boldsymbol{\theta} \rightarrow \boldsymbol{\theta}'$ and relations $p(\alpha|\boldsymbol{\alpha}|\boldsymbol{\theta}) = \sum_{i \in D_\alpha} q(i|\boldsymbol{\theta}') = p(\alpha|\boldsymbol{\alpha}|\boldsymbol{\theta}')$ and $\mathcal{H}(\boldsymbol{\theta}') \geq \mathcal{H}(\boldsymbol{\theta})$ are valid. For the matrix (θ_{ij}) of the

[12] In this sense, the physical entropy $\mathcal{H}(\boldsymbol{\theta})$ (19), (21) determines the direction of the *thermodynamic time arrow* [2], $\dfrac{\mathcal{H}(\boldsymbol{\theta}') - \mathcal{H}(\boldsymbol{\theta})}{\Delta t} = \dfrac{\partial \mathcal{H}}{\partial t} \geq 0, \quad \Delta t = t_{\theta'} - t_\theta > 0$. The equality occurs in the *equlibrial (stationary)* state $\boldsymbol{\theta}^+$ of the system $\boldsymbol{\Psi}$ and its environment.

operator $\boldsymbol{\theta}$ in the base $\{\psi_1', \psi_2', \ldots, \psi_n'\}$ is obtained that $\theta_{ij} = \sum_{k=1}^{n} u_{ki}\, u_{kj}\, q(k|\boldsymbol{\theta})$ and thus for operators $\ln\tilde{\boldsymbol{\theta}}$ and $\mathrm{Tr}(\boldsymbol{\theta}\ln\tilde{\boldsymbol{\theta}})$ is valid that

$$\ln\tilde{\boldsymbol{\theta}} = \sum_{i=1}^{n} \ln q(i|\tilde{\boldsymbol{\theta}})\, \pi\{\psi_i'\} \text{ and } -\mathrm{Tr}(\boldsymbol{\theta}\ln\tilde{\boldsymbol{\theta}}) = -\sum_{i=1}^{n}\left[\sum_{k=1}^{n} u_{ki}^2\, q(k|\boldsymbol{\theta})\right] \ln q(i|\tilde{\boldsymbol{\theta}}) \tag{124}$$
$$= -\sum_{i=1}^{n} q(i|\boldsymbol{\theta}') \ln q(i|\tilde{\boldsymbol{\theta}})$$

For the information divergence of the distributions $q(\cdot|\boldsymbol{\theta}')$, $q(\cdot|\tilde{\boldsymbol{\theta}})$ and the entropy $\mathcal{H}(\boldsymbol{\theta}')$ is valid that

$$I[q(\cdot|\boldsymbol{\theta}')\|q(\cdot|\tilde{\boldsymbol{\theta}})] = \sum_{i=1}^{n} q(i|\boldsymbol{\theta}') \ln \frac{q(i|\boldsymbol{\theta}')}{q(i|\tilde{\boldsymbol{\theta}})} \geq 0, \quad -\mathcal{H}(\boldsymbol{\theta}') \geq \sum_{i=1}^{n} q(i|\boldsymbol{\theta}') \ln q(i|\tilde{\boldsymbol{\theta}}). \tag{125}$$

By (119) for $\boldsymbol{\theta} \to \boldsymbol{\theta}'$ is writable that $\mathcal{H}(\boldsymbol{\theta}) \leq \mathcal{H}(\boldsymbol{\theta}') \leq -\mathrm{Tr}(\boldsymbol{\theta}\ln\tilde{\boldsymbol{\theta}})$. By (123) $-\mathrm{Tr}(\boldsymbol{\theta}'\ln\tilde{\boldsymbol{\theta}}) \geq \mathcal{H}(\boldsymbol{\theta}') \geq \mathcal{H}(\boldsymbol{\theta})$ are valid; the first equality is for

$$I[q(\cdot|\boldsymbol{\theta}')\|q(\cdot|\tilde{\boldsymbol{\theta}})] = 0, \;\; \mathcal{H}(\boldsymbol{\theta}') = \mathcal{H}(\tilde{\boldsymbol{\theta}}), \;\; q(i|\boldsymbol{\theta}') = q(i|\tilde{\boldsymbol{\theta}}), \quad i = 1, 2, \ldots, n, \;\; \boldsymbol{\theta}' = \tilde{\boldsymbol{\theta}} \tag{126}$$

the second equality is for $\boldsymbol{\theta}' = \boldsymbol{\theta}$. **The Gibbs theorem expresses, in the deductive (matematical-logical) way, the phenomenon of Gibbs paradox.**[13]

From formulas (47), (55), (56) and (68), (72), (73) for the narrow-band B–E and F–D capacities follows that

$$e^{-\frac{\varepsilon}{kT_W}} \cdot e^{\frac{\varepsilon}{kT_0}} \geq 1, \;\; e^{\frac{\varepsilon}{kT_0}\left(\frac{T_W - T_0}{T_W}\right)} \geq e^0; \; \varepsilon > 0, \; T_0 > 0 \longrightarrow T_W \geq T_0 \longrightarrow \frac{T_W - T_0}{T_W} \triangleq \eta_{\max} \geq 0 \tag{127}$$

and it is seen that the quantity temperature is decisive for studied information transfers. The last relation envokes, inevitably, such an opinion, that these transfers are able be modeled by a *direct* reversible **Carnot** cycle with efficiency $\eta_{max} \in (0, 1\rangle)$. Conditions leading to $C_{[\cdot|\cdot]} < 0$ mean, in such a *direct* thermodynamic model, that its efficiency should be $\eta_{\max} < 0$. This is the contradiction with the *Equivalence Principle of Thermodynamics* [19]; expresses only that the transfer is running in the opposite direction (as for temperatures).

As for B–E channel; for the supposition $W < 0$ the inequalities $T_W < T_0$ and $p(\mathcal{W}) < p$ would be gained which is the *contradiction* with (35), (36) and (47). It would be such a situation with the *information is transferred in a different direction and under a different operation mode.* Our sustaining on the meaning about the original organization of the transfer, for $T_W > T_0$, then leads to the contradiction mentioned above saying only that we are convinced mistakenly about the actual direction of the information transfer. In the case $T_W = T_0$ for the capacity $C_{B-E''}$ from (50) is valid that $C_{B-E''} = 0$. Then $W = W_{Krit}\, [= 0]$ for $p(\mathcal{W}) = p$.

As for F–D channel; for the supposition $\mathcal{W} < \frac{2p^2}{1-p^2}\, T_W < T_0$ and $p(\mathcal{W}) < p$ is gained which is the contradiction with (68). For $T_W = T_0$ is for C_{F-D} from (71) valid

[13] Derived by the information-thermodynamic way together with the *I.* and *II. Thermodynamic Principle* and with the *Equivalence Principle of Thermodynamics in [16, 17, 19].*

$C_{F-D} = \dfrac{h(p)}{1-p} + \dfrac{p}{1+p} \cdot \ln p - \ln(1+p)$.[14]

Let be noticed yet the relations between the wide–band B–E and F–D capacities and the model heat efficiency η_{max}. For the B–E capacity (98) is gained that

$$C(\mathbf{K}_{B-E}) = \frac{\pi^2 kT_W}{3h}\left(\frac{T_W - T_0}{T_W}\right) = \frac{\pi^2 kT_W}{3h}\eta_{\max} \underset{\eta_{\max}\to 1}{\longrightarrow} \frac{\pi^2 kT_W}{3h} = C^{\max}(\mathbf{K}_{B-E}) \quad (128)$$

$$C^{\max}(\mathbf{K}_{B-E}) = \sup_{\boldsymbol{\theta}} H(\boldsymbol{\alpha}\|\boldsymbol{\theta}) = H(\boldsymbol{i}) = \mathcal{H}(\boldsymbol{\theta})$$

$$C(\mathbf{K}_{B-E}) > 0,\ T_W > T_0, \quad C(\mathbf{K}_{B-E}) \underset{\substack{T_W\to T_0\\(\eta_{\max}\to 0)}}{\longrightarrow} 0,\ T_W \longrightarrow T_0$$

It is the information capacity for such a direct Carnot cycle where $H(X) = \dfrac{\pi^2 kT_W}{3h} = C^{\max}(\mathbf{K}_{B-E})$.

For the wide-band F–D capacity from (105) is valid

$$C(\mathbf{K}_{F-D}) = \frac{\pi^2 kT_W}{3h} - \frac{\pi^2 kT_0}{6h},\ T_W \geq T_0 \text{ and for } T_W > T_0, \quad (129)$$

$$C(\mathbf{K}_{F-D}) = \frac{\pi^2 kT_W}{3h}\cdot\frac{2T_W - T_0}{2T_W} = \frac{\pi^2 kT_W}{3h}\cdot\frac{2T_W - T_0}{2(T_W - T_0)}\cdot\eta_{\max}$$

$$= C(\mathbf{K}_{B-E})\cdot\frac{2T_W - T_0}{2(T_W - T_0)}$$

Due to $1 - \eta_{\max} = \dfrac{T_0}{T_W}$ is valid $T_0 = T_W(1-\eta_{\max})$ and also $C(\mathbf{K}_{F-D}) = \dfrac{\pi^2 kT_W}{6h}\cdot(1+\eta_{\max})$. Then,

$$C\ (\mathbf{K}_{F-D}) = \underset{\eta_{\max}\to 1}{\longrightarrow} \frac{\pi^2 kT_W}{3h}, \quad (130)$$

$$C\ (\mathbf{K}_{F-D}) \underset{\substack{T_W\to T_0\\(\eta_{\max}\to 0)}}{\longrightarrow} \frac{1}{2}H(\boldsymbol{i}) = \frac{1}{2}\mathcal{H}(\boldsymbol{\theta}) = \frac{\pi^2 kT_W}{6h}$$

$$C\ (\mathbf{K}_{F-D}) \in \left\langle \frac{\pi^2 kT_W}{6h}, \frac{\pi^2 kT_W}{3h}\right\rangle = \left\langle \frac{1}{2}\mathcal{H}(\boldsymbol{\theta}), \mathcal{H}(\boldsymbol{\theta})\right\rangle$$

Again the phenomenon of the not-zero capacity is seen here when the difference between the coding temperature T_W and the noise temperature T_0 is zero. Capacities $C(\mathbf{K}_{F-D}) \geq 0$ are, surely, considerable for $T_W \in \langle \frac{T_0}{2}, T_0\rangle$ and being given by the property of the F–D phase space cells. Capacities $C(\mathbf{K}_{B-E}) < 0$ and $C(\mathbf{K}_{F-D}) < 0$ are without sense for the given direction of information transfer.

[14] Nevertheless the capacity C_{F-D} for this case $W < W_{Krit}$ is set in [12, 13]. *Similar* results as this one and (74) are gained for the **Maxwell–Boltzman** (M–B) system in [13].

Nevertheless, it will be shown that all these processes themselves are not organized cyclically 'by themselves'.

Further the relation between the information transfer in a wide–band B–E (photonic) channel *organized in a cyclical way* and a relevant (reverse) heat cycle will be dealt with. But, firstly, the way in which the capacity formula for an information transfer system of photons was derived in [30] will be reviewed.

6.1. Thermodynamic derivation of wide–band photonic capacity

A transfer channel is now created by the electromagnetic radiation of a system $\mathcal{L} \cong \mathbf{K}_{\mathrm{L-L}}$ of photons being emitted from an *absolute black body* at temperature T_0 and within a frequency bandwidth of $\Delta\nu = R^+$, where ν is the frequency. Then the energy of such radiation is the *energy* of *noise*. A source of *input messages, signals* transmits monochromatic electromagnetic impulses (numbers a_i of photons) into this environment with an average *input energy W*. This source is defined by an alphabet of input messages, signals $\{a_i\}_{i=1}^{n}$, with a probability distribution $p_i = p(a_i), i = 1, 2, \ldots, n$.[15] The *output (whole, received)* signal is created by *additive* superposition of the input signal and the noise signal. The input signal a_i, within a frequency ν, is represented by the *occupation* number $m = m(\nu)$, which equates to the number of photons of an input field with an energy level $\varepsilon(\nu) = h\nu$. The output signal is represented by the occupation number $l = l(\nu)$. The noise signal, created by the number of photons emitted by absolute black body radiation at temperature T_0, is represented by the occupation number $n = n(\nu)$. The medium values of these quantities (*spectral* densities of the input, noise and output photonic stream) are denoted as $\overline{m}$, $\overline{n}$ and $\overline{l}$. In accordance with the ***Planck*** *radiation* law, the spectral density $\overline{r}$ of a photonic stream of absolute black body radiation at temperature Θ and within frequency ν, is given by the ***Planck*** *distribution*,

$$\overline{r}(\nu) = \frac{p(\nu,\Theta)}{1-p(\nu,\Theta)}, \quad \overline{n}(\nu) = \frac{p(\nu,T_0)}{1-p(\nu,T_0)}, \quad \overline{l}(\nu) = \frac{p(\nu,T_W)}{1-p(\nu,T_W)}, \quad p(\nu,\Theta) = e^{-\dfrac{h\nu}{k\Theta}} \tag{131}$$

Thus, for the average energy P of radiation at temperature Θ within the bandwidth $\Delta\nu = R^+$ is gained that

$$P(\Theta) = \int_0^\infty \overline{\varepsilon}(\nu,\Theta)\mathrm{d}\nu = \frac{\pi^2 \mathrm{k}^2 \Theta^2}{6\hbar} \quad \text{where } \overline{\varepsilon}(\nu,\Theta) = \overline{r}(\nu)h\nu \text{ and } \frac{\mathrm{d}P(\Theta)}{\mathrm{d}\Theta} = \frac{\pi^2\mathrm{k}^2\Theta}{3\hbar}. \tag{132}$$

Then, for the average noise energy P_1 at temperature T_0, and for the average output energy P_2 at temperature T_W, both of which occur within the bandwidth $\Delta\nu = R^+$ is valid that

$$P_1(T_0) = \frac{\pi^2\mathrm{k}^2{T_0}^2}{6\hbar}, \quad P_2(T_W) = \frac{\pi^2\mathrm{k}^2{T_W}^2}{6\hbar}. \tag{133}$$

The entropy H of radiation at temperature ϑ is derived from ***Clausius*** *definition* of *heat entropy* and thus

$$H = \int_0^\vartheta \frac{1}{\mathrm{k}\Theta}\frac{\mathrm{d}P(\Theta)}{\mathrm{d}\Theta}\mathrm{d}\Theta = \frac{\pi^2\mathrm{k}\vartheta}{3\hbar} = \frac{2P(\vartheta)}{\mathrm{k}\vartheta} \tag{134}$$

[15] To distinguish between two frequencies mutually deferring at an infinitesimally small $\mathrm{d}\nu$ is needed, in accordance with ***Heisenberg*** *uncertainty principle*, a time interval spanning the infinite length of time, $\Delta t \longrightarrow \infty$; analog of the *thermodynamic stationarity*.

7.1. Triangular heat cycle

Elementary change $d\Theta$ of temperature Θ of the *environment* of the general continuous cycle $\mathcal{O}$ and thus of its working *medium* $\mathcal{L}$ (both are in the *diathermic contact at* Θ) causes the elementary reversible change of the heat $Q*(\Theta)$ *delivered, (radiated)* into $\mathcal{L}$, just about the value $\delta Q*(\Theta)$,

$$\delta Q*(\Theta) = \frac{\partial Q*(\Theta)}{\partial \Theta} d\Theta, \quad Q*(\Theta_W) = \int_0^{\Theta_W} \frac{\partial Q*(\Theta)}{\partial \Theta} d\Theta \tag{147}$$

The heat $Q*(\Theta_W)$ is the whole heat delivered (reversibly) into $\mathcal{L}$ (at the *end* temperature Θ_W). For the infinitezimal heat $\delta Q*(\Theta)$ delivered (reversibly) into $\mathcal{L}$ at temperature Θ and in accordance with the *Clausius* definition of heat entropy $S*_\mathcal{L}$ [22] is valid that

$$\delta Q*(\Theta) = \Theta \cdot dS*_\mathcal{L}(\Theta), \quad dS*_\mathcal{L}(\Theta) = \frac{\delta Q*(\Theta)}{\Theta} \tag{148}$$

For the whole change of entropy $\Delta S*_\mathcal{L}(\Theta_W)$, or for the entropy $S*_\mathcal{L}(\Theta_W)$ respectively, delivered into the medium $\mathcal{L}$ by its heating within the temperature interval $(0, \Theta_W\rangle$, is valid that

$$\Delta S*_\mathcal{L}(\Theta_W) = \int_0^{\Theta_W} dS*_\mathcal{L}(\Theta) = \int_0^{\Theta_W} \frac{\delta Q*(\Theta)}{\Theta} = \int_0^{\Theta_W} \frac{\frac{\partial Q*(\Theta)}{\partial \Theta} d\Theta}{\Theta} = S*_\mathcal{L}(\Theta_W) \tag{149}$$

when $S*_\mathcal{L}(0) \stackrel{\text{Def}}{=} 0$ is set down. By (148) for the whole heat $Q*(\Theta_W)$ deliverd into $\mathcal{L}$ within the temperature interval $\Theta \in (0, \Theta_W\rangle$ also is valid that

$$Q*(\Theta_W) = \int_0^{\Theta_W} \Theta dS*_\mathcal{L}(\Theta) \tag{150}$$

Then, by *medium value* theorem[19] is valid that $\overline{\Theta_{(0,\Theta_W)}} = \frac{0 + \Theta_W}{2} = \frac{\Theta_W}{2}$ and

$$Q*(\Theta_W) = \int_{S*_\mathcal{L}(0)}^{S*_\mathcal{L}(\Theta_W)} \Theta dS*_\mathcal{L}(\Theta) = [S*_\mathcal{L}(\Theta_W) - S*_\mathcal{L}(0)] \cdot \overline{\Theta_{(0,\Theta_W)}} \tag{151}$$

For the extremal values T_0 a T_W of the cooler temperature Θ of $\mathcal{O}$ and by (151)

$$Q*_0 \stackrel{\triangle}{=} Q*(T_0) = \int_0^{T_0} \delta Q*_W(\Theta) \quad \text{and} \quad Q*_W \stackrel{\triangle}{=} Q*(T_W) = \int_0^{T_W} \delta Q*_W(\Theta) \tag{152}$$

$$Q*_0 = \int_{S*_\mathcal{L}(0)}^{S*_\mathcal{L}(T_0)} \Theta dS*_\mathcal{L}(\Theta) = [S*_\mathcal{L}(T_0) - S*_\mathcal{L}(0)] \cdot \overline{\Theta_{(0,T_0)}}, \quad \overline{\Theta_{(0,T_0)}} = \frac{T_0}{2}$$

$$Q*_W = \int_{S*_\mathcal{L}(0)}^{S*_\mathcal{L}(T_W)} \Theta dS*_\mathcal{L}(\Theta) = [S*_\mathcal{L}(T_W) - S*_\mathcal{L}(0)] \cdot \overline{\Theta_{(0,T_W)}}, \quad \overline{\Theta_{(0,T_W)}} = \frac{T_W}{2}$$

With $S*_\mathcal{L}(0) = 0$ for the (end) temperatures Θ, T_0, T_W of $\mathcal{L}$ and the relevant heats and their entropies is valid

$$Q*(\Theta) = S*_\mathcal{L}(\Theta) \cdot \frac{\Theta}{2} \quad \text{and then} \quad S*_\mathcal{L}(\Theta) = \frac{2Q*(\Theta)}{\Theta} \tag{153}$$

$$Q*_W = S*_\mathcal{L}(T_W) \cdot \frac{T_W}{2} \quad \text{and then} \quad S*_\mathcal{L}(T_W) = \frac{2Q*_W}{T_W}$$

$$Q*_0 = S*_\mathcal{L}(T_0) \cdot \frac{T_0}{2} \quad \text{and then} \quad S*_\mathcal{L}(T_0) = \frac{2Q*_0}{T_0}$$

[19] Of Integral Calculus.

For the change $\Delta S*_{\mathcal{L}}$ of the thermodynamic entropy $S*_{\mathcal{L}}$ of the system $\mathcal{L}$, at the tememperature Θ running through the interval $\langle T_0, T_W \rangle$, by gaining heat from its environment (the environment of the cycle $\mathcal{O}$), is valid

$$\Delta S*_{\mathcal{L}} = S*_{\mathcal{L}}(T_W) - S*_{\mathcal{L}}(T_0) = 2\left(\frac{Q*_W}{T_W} - \frac{Q*_0}{T_0}\right) = \int_{T_0}^{T_W} \frac{\delta Q*(\Theta)}{\Theta} \tag{154}$$

By (149) for the entropy $S*_{\mathcal{L}}(\Theta_W)$ of $\mathcal{L}$ at variable temperature $\Theta \in \langle 0, \Theta_W \rangle$, $\Theta_W \leq T_W$, is gained that

$$\begin{aligned} S*_{\mathcal{L}}(\Theta_W) &= \int_0^{\Theta_W} \left(\frac{\partial}{\partial\Theta}\left[\frac{S*_{\mathcal{L}}(\Theta)\cdot\Theta}{2}\right]\right)\frac{d\Theta}{\Theta} = 2\cdot\frac{1}{2}\int_0^{\Theta_W} dS*_L(\Theta) \\ &= \frac{1}{2}\int_0^{\Theta_W} S*'_{\mathcal{L}}(\Theta)d\Theta + \frac{1}{2}\int_0^{\Theta_W} S*_{\mathcal{L}}(\Theta)\frac{d\Theta}{\Theta} \end{aligned} \tag{155}$$

and then

$$S*_{\mathcal{L}}(\Theta_W) = \int_0^{\Theta_W} S*_{\mathcal{L}}(\Theta)\frac{d\Theta}{\Theta} = \int_0^{\Theta_W} dS*_{\mathcal{L}}(\Theta) \quad \left[= \frac{2Q*(\Theta_W)}{\Theta_W}\right]$$

and thus

$$S*_{\mathcal{L}}(\Theta)\frac{d\Theta}{\Theta} = dS*_{\mathcal{L}}(\Theta) \tag{156}$$

By the result of derivation (155)-(156) for an arbitrary temperature Θ of medium $\mathcal{L}$ is valid that[20]

$$S*_{\mathcal{L}}(\Theta) = l\cdot\Theta, \quad l = \frac{2Q*(\Theta)}{\Theta^2} \quad \longrightarrow \quad Q*(\Theta) = \lambda\cdot\Theta^2, \quad \lambda = \frac{l}{2} \tag{157}$$

Obviously, from (154) for $\Theta \in \langle T_0, T_W \rangle$ is derivable that

$$\Delta S*_{\mathcal{L}} = l\cdot(T_W - T_0) = l\cdot T_W\cdot(1-\beta), \quad \beta = \frac{T_0}{T_W} \tag{158}$$

Let such a reverse cycle is given that the medium $\mathcal{L}$ of which takes, through the elementary isothermal expansions at temperatures $\Theta \in \langle T_0, T_W \rangle$, the whole heat ΔQ_0

$$\Delta Q_0 = \int_{T_0}^{T_W} \delta Q*(\Theta) = \int_{T_0}^{T_W} l\Theta d\Theta = \frac{l}{2}\cdot\left(T_W{}^2 - T_0^2\right) = Q*_W - Q*_0 \tag{159}$$

or, with medium values

$$\begin{aligned} \Delta Q_0 &= \int_{T_0}^{T_W} \delta Q*(\Theta) = \int_{S*_{\mathcal{L}}(T_0)}^{S*_{\mathcal{L}}(T_W)} \Theta dS*_{\mathcal{L}}(\Theta) \\ &= \overline{\Theta_{W_{\langle T_0, T_W\rangle}}}\cdot[S*_{\mathcal{L}}(T_W) - S*_{\mathcal{L}}(T_0)] = \frac{T_W + T_0}{2}\cdot 2\left[\frac{Q*_W}{T_W} - \frac{Q*_0}{T_0}\right] \end{aligned}$$

and thus equivalently

$$\Delta Q_0 = (T_W + T_0)\cdot\lambda\cdot[T_W - T_0] = \lambda T_W{}^2\cdot(1-\beta^2), \quad \beta = \frac{T_0}{T_W}$$

For a reverse reversible Carnot cycle $\mathcal{O}'_{rrev}$, equivalent with the just considered general continuous heat cycle $\mathcal{O}$, drawing up the same heat ΔQ_0, consumpting the same mechanical

---

[20] If $\int \frac{f(x)}{x}dx = \int df(x)$, or $\frac{df(x)}{f(x)} = \frac{dx}{x}$, then $\ln|f(x)| = \ln|x| + \ln L$, $L > 0$, $\ln|f(x)| = \ln(L\cdot|x|)$, $f(x) = l\cdot x$, $l \in \mathbb{R}$.

work ΔA and giving, at its higher temperature t_W (the average temperature of the heater of our general cycle), the same heat ΔQ_W, is valid that $\Delta Q_0 = \Delta Q_W \cdot \gamma$ where $\gamma = \frac{\frac{T_0+T_W}{2}}{t_W}$ is the *transform ratio.*

Then

$$\Delta Q_W = \Delta Q_0 \cdot \frac{2t_W}{T_0 + T_W} = \frac{l}{2} \cdot \left(T_W{}^2 - T_0{}^2\right) \cdot \frac{2t_W}{T_0 + T_W} = l \cdot t_W \cdot (T_W - T_0) \tag{160}$$

$$\Delta A = \Delta Q_W \cdot (1 - \gamma) = l \cdot t_W \cdot (T_W - T_0) \cdot \left(1 - \frac{T_0 + T_W}{2t_W}\right) \tag{161}$$
$$= \frac{l}{2} \cdot (T_W - T_0) \cdot (2t_W - T_0 - T_W)$$

For the elementary work $\delta A(\cdot, \cdot)$ corresponding with the heat $Q*(\Theta)$ pumped out (reversibly) from $\mathcal{L}$ at the (end, output) temperature Θ of $\mathcal{L}$ and for the entropy $S*_{\mathcal{L}}(\Theta)$ of the whole environment of $\mathcal{O}$ (including $\mathcal{L}$ with $\mathcal{O}$) is valid

$$S*_{\mathcal{L}}(\Theta)\frac{\mathrm{d}\Theta}{\Theta} = \frac{Q*(\Theta)\frac{\mathrm{d}\Theta}{\Theta}}{\Theta} \stackrel{\triangle}{=} \frac{\delta A(\Theta, \mathrm{d}\Theta)}{\Theta} = \mathrm{d}S*_{\mathcal{L}}(\Theta) = l \cdot \mathrm{d}\Theta \tag{162}$$

$$\delta A(\Theta, \mathrm{d}\Theta) = S*_{\mathcal{L}}(\Theta)\mathrm{d}\Theta = l \cdot \Theta\mathrm{d}\Theta \text{ and } \delta A(\mathrm{d}\Theta, \mathrm{d}\Theta) = l \cdot \mathrm{d}\Theta\mathrm{d}\Theta = \mathrm{d}S*_{\mathcal{L}}(\Theta)\mathrm{d}(\Theta) \stackrel{\triangle}{=} \delta A$$
$$\left(\int_{T_0}^{\Theta} l\mathrm{d}\theta\right)\mathrm{d}\Theta = l \cdot (\Theta - T_0)\mathrm{d}\Theta \stackrel{\triangle}{=} \delta A(\Theta, \mathrm{d}\Theta; T_0) \tag{163}$$

For the whole work $\Delta A(\Theta_W; T_0)$ consumpted by the general reverse cycle $\mathcal{O}$ between temperatures T_0 and Θ_W, being coverd by elementary cycles (162), is valid that

$$\Delta A(\Theta_W; T_0) = \int_{T_0}^{\Theta_W}\left[\int_{T_0}^{\Theta} \mathrm{d}S*_{\mathcal{L}}(\theta)\right]\mathrm{d}\Theta = \int_{T_0}^{\Theta_W}\left[\int_{T_0}^{\Theta} l\mathrm{d}\theta\right]\mathrm{d}\Theta = l \cdot \int_{T_0}^{\Theta_W}(\Theta - T_0)\mathrm{d}\Theta \tag{164}$$
$$= \frac{l}{2} \cdot (\Theta_W{}^2 - T_0{}^2) - l \cdot T_0(\Theta_W - T_0) = \frac{l}{2} \cdot \Theta_W{}^2 + \frac{l}{2} \cdot T_0{}^2 - \frac{2l}{2} \cdot T_0\Theta_W = \lambda \cdot (\Theta_W - T_0)^2$$
$$\stackrel{\triangle}{=} \oint_{\mathcal{O}(\Theta_W, T_0)} \delta A = \lambda \cdot T_W{}^2 \cdot (1 - \beta)^2, \text{ when it is valid that } \Theta_W = T_W,\ \beta = \frac{T_0}{T_W}$$

But, then for the results for ΔA in (160), (161) and (164) follows that

$$t_W = \frac{T_W + \Theta_W}{2} = T_W \text{ and then } t_W = T_W = \text{const.} \tag{165}$$

Thus our general cycle $\mathcal{O}$ is of a triangle shape, $\mathcal{O} \stackrel{\triangle}{=} \mathcal{O}_{rrev\triangle}$ with the *apexes*

$$[lT_0, T_0],\ [lT_W, T_W],\ [lT_0, T_W] \tag{166}$$

and its efficiency is $1 - \gamma = \frac{1 - \beta}{2} = \frac{1}{2} \cdot \eta_{\max}$. Thus, the return to the *initial (starting) state* of the medium $\mathcal{L}$ is possible by using the oriented *abscissas* (in the $S - T$ diagram)

$$\overrightarrow{[lT_W, T_W], [lT_0, T_W]} \text{ and } \overrightarrow{[lT_0, T_W], [lT_0, T_0]} \tag{167}$$

For works $\delta A(\Theta, \mathrm{d}\Theta; T_0)$ of elementary Carnot cycles covering cycle $\mathcal{O}_{rrev\triangle}$ (166), the range of their working temperatures is $\mathrm{d}\Theta$ and, for the given heater temperature $\Theta \in \langle T_0, \Theta_W\rangle$ is, by (162)-(163) valid that

$$
\begin{aligned}
\delta A(\Theta, \mathrm{d}\Theta; T_0) &= \Delta Q_W(\Theta) \cdot \frac{\mathrm{d}\Theta}{\Theta} = l \cdot \Theta \cdot (\Theta - T_0)\,\frac{\mathrm{d}\Theta}{\Theta} = l \cdot (\Theta - T_0)\mathrm{d}\Theta \qquad (168)\\
&= [S*_{\mathcal{L}}(\Theta) - S*_{\mathcal{L}}(T_0)] \cdot \mathrm{d}\Theta\\
\Delta Q_W(\Theta) &= l\Theta\,(\Theta - T_0) \ \text{ and for } \gamma \text{ used in (160) is gained that}\\
\Delta Q_0(\Theta) &= \Delta Q_W(\Theta) \cdot \gamma(\Theta) = l \cdot \Theta \cdot (\Theta - T_0) \cdot \frac{\Theta + T_0}{2\Theta} = \lambda \cdot \left(\Theta^2 - {T_0}^2\right)
\end{aligned}
$$

For the whole heats ΔQ_0 a ΔQ_W being changed mutually between the working medium $\mathcal{L}$ of the whole triangular cycle $\mathcal{O}_{rrev\triangle}$ and its environment (166), and for the work ΔA, in its equivalent Carnot cycle $\mathcal{O}'_{rrev}$ with working temperatures $\dfrac{T_0 + T_W}{2}$ and T_W, will be valid that[21]

$$
\Delta Q_0 = \int_{T_0}^{T_W} \left[\int_0^{\Theta} l\mathrm{d}\theta\right] \mathrm{d}\Theta = \frac{l}{2} \cdot {T_W}^2 \cdot (1-\beta^2) \stackrel{\triangle}{=} W \left[= l \cdot (T_W - T_0) \cdot T_0 + \frac{l}{2} \cdot (T_W - T_0)^2\right] \qquad (169)
$$

$$
\begin{aligned}
\Delta Q_W &= \int_{T_0}^{T_W} \left[\int_0^{T_W} l\mathrm{d}\theta\right] \mathrm{d}\Theta = l \cdot (T_W - T_0) \cdot T_W = l \cdot {T_W}^2 \cdot (1 - \beta)\\
\Delta A &= \frac{1}{2} \int_{T_0}^{T_W} \left[\int_{T_0}^{T_W} l\mathrm{d}\theta\right] \mathrm{d}\Theta = \cdot l{T_W}^2 \cdot (1 - \beta) - l \cdot {T_W}^2 \cdot (1 - \beta) \qquad (170)\\
&= l \cdot {T_W}^2 \cdot (1 - \beta) \cdot [1 - \frac{l}{2}(1 + \beta)] = \frac{l}{2} \cdot {T_W}^2 \cdot (1 - \beta)^2 = \oint_{\mathcal{O}_{rrev\triangle}} \delta A
\end{aligned}
$$

7.2. Capacity corections for wide–band photonic transfer channel

The average *output* energy $P_2(\Theta_W)$ of the message being received within interval $(0, T_W\rangle$ of the temperature Θ of the medium $\mathcal{L} \cong \mathbf{K}_{\mathrm{L-L}}$ from [30], when $0 < \Theta_0 \leq \Theta \leq \Theta_W$ and $\Theta_0 \leq T_0$ and $\Theta_W \leq T_W$ are valid, is given by the sum of the *input* average energy $W(\Theta_W, \Theta_0)$ and the average energy $P_1(\Theta_0)$ of the *additive noise*

$$
P_2(\Theta_W) = P_1(\Theta_0) + W(\Theta_W, \Theta_0) \qquad (171)
$$

The output message bears the *whole average output information* $H_2(\Theta_W)$. By the medium value theorem is possible, for a certain maximal temperature $\Theta_W \leq T_W$ of the temperature $\Theta \in (0, \Theta_W\rangle$, consider that the receiving of the output message is performed at the *average (constant) temperature* $\overline{\Theta_W} = \dfrac{\Theta_W}{2}$. Then for the whole change of the output information entropy $\Delta H_2 \stackrel{\triangle}{=} H_2(\Theta_W)$ [the thermodynamic entropy $S*_{\mathcal{L}}(\Theta_W)$ in information units] is valid

$$
H_2(\Theta_W) = \frac{P_2(\Theta_W)}{\mathrm{k}\overline{\Theta_W}} = \frac{P_1(\Theta_0) + W(\Theta_W, \Theta_0)}{\mathrm{k}\overline{\Theta_W}} \stackrel{\triangle}{=} H_1(\Theta_W, \Theta_0) + H[W(\Theta_W, \Theta_0)] \qquad (172)
$$

---

[21] Further it willbe layed down $\lambda = \dfrac{\pi^2 k^2}{6\hbar}$, $l = \dfrac{\pi^2 k^2}{3\hbar} = 2 \cdot \lambda$.

By (153) is valid that $Q*(\Theta) = \lambda\Theta^2$ and $\delta Q*(\Theta) = l\Theta d\Theta$. Thus for $\Theta \in (0, T_0\rangle$ a $\Theta \in (0, T_W\rangle$ is valid

$$dH_{\Theta_W}(Y) = \frac{\delta Q*(\Theta_W)}{k\Theta_W} = \frac{l}{k}d\Theta_W, \; dH_{\Theta_W}(Y|X) = \frac{2l\Theta d\Theta}{k\Theta_W} \tag{173}$$

With $\Theta_W = T_W$ and $\Theta_0 = T_0$ and with the reducing temperature $\frac{T_W}{2}$ is possible to write

$$P_2 = W + P_1 \triangleq Q*_W = \lambda T_W{}^2 \triangleq Y \tag{174}$$

$$H_2 \triangleq H(Y) = \int_0^{T_W} \frac{l}{k} d\Theta_W = \frac{l}{k}T_W = \frac{P_2}{k\frac{T_W}{2}} = \frac{2(W+P_1)}{kT_W} = \frac{2\lambda T_W{}^2}{kT_W} = \frac{lT_W}{k}$$

$$P_1 \triangleq Q*_0 = \lambda T_0{}^2 \triangleq Y|X$$

$$H_1 \triangleq H(Y|X) = \int_0^{T_0} \frac{2l\Theta d\Theta}{kT_W} = \frac{l}{kT_W}\cdot T_0{}^2 = \frac{P_1}{k\frac{T_W}{2}} = \frac{2\lambda T_0{}^2}{kT_W} = \frac{lT_0{}^2}{kT_W}$$

$$W = Q*_W - Q*_0 \triangleq X$$

$$H[W(T_W, T_0)] \triangleq H(X) = \frac{W}{k\frac{T_W}{2}} = \frac{2\lambda}{kT_W}\cdot(T_W{}^2 - T_0{}^2) \tag{175}$$

By the channel equation (4), (5) and by equations (23)-(24) and also by definitions (174)-(175) and with the loss entropy $H(X|Y) = 0$ it must be valid for the transinformation $T(\cdot;\cdot)$ that

$$T(Y;X) = H(Y) - H(Y|X) = \frac{l}{k}\cdot(T_W - T_0)\cdot(1+\beta) = H(X) \tag{176}$$

$$T(X;Y) = H(X) - H(X|Y) = H(X) = T(X,Y) \text{ and by using } l = \frac{\pi^2 k^2}{3\hbar}, \; \beta = \frac{T_0}{T_W},$$

$$T(Y;X) = \frac{\pi^2 k}{3h}\cdot T_W\cdot(1-\beta^2) = \frac{\pi^2 k}{3h}\cdot T_W\cdot(1-\beta)\cdot(1+\beta) = C_{T_0,W}(\mathbf{K}_{L-L})\cdot(1+\beta)$$

For the given extremal temperatures T_0, T_W the value $T(X;Y)$ stated this way is the only one, and thus also, it is the information capacity $C'_{T_0,T_W}(W)$ of the channel $\mathbf{K}_{L-L}$ (the *first* correction)

$$C'_{T_0,T_W} = (W) = T(X;Y) = \frac{\pi^2 kT_0}{3h}\cdot\left(\sqrt{1+\frac{6h\cdot W}{\pi^2 k^2 T_0{}^2}} - 1\right)\cdot(1+\beta) \tag{177}$$

The information capacity correction (177) of the wide–band photonic channel $\mathbf{K}_{L-L}$ [30], stated this way, is $(1+\beta)$-times *higher* than the formulas (102) and (138) say. The reason is in using two different information descriptions of the oriented abscissa $\overrightarrow{[l0,0],[lT_W,T_W]}$ in derivation (138) and (177) which abscissa $\overrightarrow{[l0,0],[lT_W,T_W]}$ is on one line in $S-T$ diagram and is composed from two oriented abscissas,

$$\overrightarrow{[l0,0],[lT_0,T_0]} \text{ and } \overrightarrow{[lT_0,T_0],[lT_W,T_W]} \tag{178}$$

The first abscissa represents the *phase of noise generation* and the second one the *phase of input signal generation*. The whole composed abscissa represents the *phase of whole output signal generation*.

For the **sustaining, in the sense repeatable, cyclical information transfer, the renewal of the initial or starting state of the transfer channel $K_{L-L} \cong \mathcal{L}$, after any *individual information transfer act* - the *sending input and receiving output message* has been accomplished, is needed.**
Nevertheless, in derivations of the formulas (102), (177) and (138) **this return of the physical medium $\mathcal{L}$**, after accomplishing any individual information transfer act, into the starting state **is either not considered, or, on the contrary, is considered, but by that the whole transfer chain is opened *to cover the energetic needs for this return transition from another, outer resources* than from those ones within the transfer chain itself**. In both these two cases the channel equation is fulfilled. This enables any individual act of information transfer be realized by *external and forced out, repeated starting* of each this individual transfer act.[22] [23]

If for the creation of a cycle the resources of the transfer chain are used only, the need for *another correction*, this time in (177) arises. To express it it will be used the *full cyclical thermodynamic analogy* **K** of $\mathbf{K}_{L-L}$ used *cyclically*, $\mathbf{K}'_{L-L}$. The information transfer will be modeled by the cyclical thermodynamic process $\mathcal{O}_{rrev\triangle}$ of reversible changes in the channel $\mathbf{K} \cong \mathbf{K}'_{L-L} \cong \mathcal{L}$ and without opening the transfer chain. (Also $\mathbf{K} \cong \mathbf{K}'_{B-E}$).

7.2.1. Return of transfer medium into initial state, second correction

Now the further correction for capacity formulas (102), (138) and (177) will be dealt with for that case that the *return* of the medium $\mathcal{L}$ into its initial, starting state is performed *within the transfer chain* only. It will be envisiged by a triangular reverse heat cycle $\mathcal{O}_{rrev\triangle}$ created by the oriented abscissas within the apexes in the $S-T$ diagram (166), $[lT_0, T_0]$, $[lT_W, T_W]$, $[lT_0, T_W]$. The abstract experiment from [30] will be now, formally and as an analogy, realized by this *reverse* and reversible heat cycle $\mathcal{O}_{rrev\triangle} \equiv \mathcal{O}'_{rrev}$, described informationaly, and thought as modeling information transfer process in a channel $\mathbf{K} \cong \mathbf{K}'_{L-L|B-E} \cong \mathcal{L}$. Thus the denotation $\mathbf{K} \equiv \mathbf{K}_\triangle$ is usable. By (153)-(157) it will be

$$Q*(\Theta) = \frac{\pi^2 k^2 \Theta^2}{6h} = \frac{l}{2} \cdot \Theta^2, \;\; Q*_W = \frac{l}{2} \cdot T_W{}^2, \;\; Q*_0 = \frac{l}{2} \cdot T_0{}^2 \tag{179}$$

The working temperature Θ_0 of cooling and Θ_W of heating are changing by (157),

$$\Theta_0 = \frac{1}{l} \cdot S*_{\mathcal{L}}(\Theta_0) \in \langle T_0, T_W \rangle \;\text{ and }\; \Theta_W = T_W = \text{const.} \tag{180}$$

and the heat entropy $S*_{\mathcal{L}}(\Theta)$ of the medium $\mathcal{L}$ is changing by (155)-(156),

$$\mathrm{d}S*_{\mathcal{L}} = \frac{\partial Q*(\Theta)}{\partial \Theta} \mathrm{d}\Theta \cdot \frac{1}{\Theta} \;\text{ and then }\; S*_{\mathcal{L}}(\Theta) = l \cdot \Theta = \frac{2Q*(\Theta)}{\Theta} \tag{181}$$

Using integral (149) it is possible to write that

$$Q*(\Theta) = \int_0^\Theta \delta Q*(\theta) = \int_0^\Theta \frac{\partial Q*(\theta)}{\partial \theta} \mathrm{d}\theta = \int_0^\Theta l\theta \mathrm{d}\theta \tag{182}$$

---

[22] For these both cases is not possible to construct a construction-relevant heat cycles *described in a proper information way*.

[23] But the modeling by the direct cycle such as in (128) is possible for the *II. Principle of Thermodynamics* is valid in any case and giving the possibility of the cycle description.

For the whole heats ΔQ_0 and ΔQ_W being changed mutually between $\mathcal{L}$ with the cycle $\mathcal{O}_{rrev\triangle}$ and its environment and, for the whole work ΔA for the equivalent Carnot cycle $\mathcal{O}'_{rrev}$ with working temperatures $\frac{T_0 + T_W}{2}$ and T_W is valid, by (169)-(170), that

$$W = \Delta Q_0 = \frac{l}{2} \cdot T_W{}^2 \cdot (1-\beta^2) \stackrel{\triangle}{=} X, \quad \Delta Q_W = l \cdot T_W{}^2 \cdot (1-\beta) \stackrel{\triangle}{=} Y \tag{183}$$
$$\Delta Q_W - \Delta Q_0 = \Delta A = \frac{l}{2} \cdot T_W{}^2 \cdot (1-\beta)^2 \stackrel{\triangle}{=} Y|X$$

For the whole work ΔA delivered into the cycle $\mathcal{O}_{rrev\triangle}$, at the temperature T_W, and the entropy $S*_{\mathcal{L}}$ of its working medium $\mathcal{L}$ is valid

$$\frac{\Delta A}{T_W} = \oint_{\mathcal{O}_{rrev\triangle}} \frac{\delta A}{T_W} = \int_{T_0}^{T_W} l(\Theta - T_0) \cdot \frac{d\Theta}{T_W} \tag{184}$$
$$\oint_{\mathcal{O}_{rrev\triangle}} \frac{\delta A}{T_W} = \frac{1}{2} \int_{T_0}^{T_W} \left[\int_{T_0}^{T_W} dS*_{\mathcal{L}}(\theta) \right] \frac{d\Theta}{T_W} = \int_{T_0}^{T_W} [S*_{\mathcal{L}}(T_W) - S*_{\mathcal{L}}(T_0)] \frac{d\Theta}{2T_W}$$
$$= \frac{l}{2T_W} \cdot (T_W - T_0)^2 = \frac{l}{2} \cdot T_W \cdot (1-\beta)^2 = \frac{\Delta A}{T_W}$$

Following (4), (5) and (23) and the triangular shape of the cycle $\mathcal{O}_{rrev\triangle}$, the changes of information entropies by expressions (142), (169)-(170) are defined, valid for the equivalent $\mathcal{O}'_{rrev}$[24], see (142),

$$H(X) \stackrel{\text{Def}}{=} \frac{\Delta Q_0}{kT_W} = \int_{T_0}^{T_W} \left[\int_0^{\Theta} \frac{\delta Q*(\theta)}{\theta} \right] \frac{d\Theta}{kT_W} = \int_{T_0}^{T_W} \left[\int_0^{\Theta} l d` \right] \frac{1}{kT_W} d\Theta \tag{185}$$
$$H(Y) \stackrel{\text{Def}}{=} \frac{\Delta Q_W}{kT_W} = \int_{T_0}^{T_W} \left[\int_0^{T_W} \frac{\delta Q*(\theta)}{\theta} \right] \frac{d\Theta}{kT_W} = \int_{T_0}^{T_W} \left[\int_0^{T_W} l d` \right] \frac{1}{kT_W} d\Theta$$
$$H(Y|X) \stackrel{\text{Def}}{=} \frac{\Delta A}{kT_W} = \frac{1}{2} \int_{T_0}^{T_W} \left[\int_{T_0}^{T_W} \frac{\delta Q*(\theta)}{\theta} \right] \frac{d\Theta}{kT_W} = \frac{1}{2} \int_{T_0}^{T_W} \left[\int_{T_0}^{T_W} l d\theta \right] \frac{1}{kT_W} d\Theta$$
$$T(Y;X) = H(Y) - H(Y|X) = \frac{l}{2kT_W} \int_{T_0}^{T_W} \left[2 \int_0^{T_W} d\theta - \int_{T_0}^{T_W} d\theta \right] d\Theta$$

and by figguring these formulas with $l = \frac{\pi^2 k}{3h}$ is gained that

$$H(X) = \frac{l}{2k} \cdot T_W \cdot (1-\beta^2) = \frac{\pi^2 k T_W}{6h} \cdot (1-\beta^2)$$
$$H(Y) = \frac{l}{k} \cdot T_W \cdot (1-\beta) = \frac{\pi^2 k T_W}{3h} \cdot (1-\beta)$$
$$H(Y|X) = \frac{l}{2k} \cdot T_W \cdot (1-\beta)^2 = \frac{\pi^2 k T_W}{6h} \cdot (1-\beta)^2 = \oint_{\mathcal{O}_{rrev}} \frac{\delta A}{kT_W}$$
$$T(Y;X) = \frac{l}{k} \cdot T_W \cdot (1-\beta) - \frac{l}{2k} \cdot T_W \cdot (1-\beta)^2 = \frac{l}{2k} \cdot T_W \cdot (1-\beta) \cdot (2-1+\beta)$$
$$= \frac{l}{2k} \cdot T_W \cdot (1-\beta^2) = \frac{\pi^2 k T_W}{6h} \cdot (1-\beta^2) = H(X) = T(X;Y)$$
$$H(X|Y) = 0$$

[24] In accordance with the input energy delivered and the extremal temperatures used in [30].

It is visible that the quantity $H(Y)$ $[= H(X) + H(Y|X)]$ is introduced correctly, for by (185) is valid that

$$H(Y) = \frac{l}{2\mathrm{k}} \cdot T_W \cdot (1-\beta)^2 + \frac{l}{2\mathrm{k}} \cdot T_W \cdot (1-\beta^2) = \frac{l}{2\mathrm{k}} \cdot T_W \cdot (1-\beta) \cdot (1-\beta+1+\beta) \quad (186)$$
$$= \frac{l}{\mathrm{k}} \cdot T_W \cdot (1-\beta)$$

For the transinformation and the information capacity of the transfer organized this way is valid (177),

$$T(X;Y) = \frac{1}{2} C'_{T_0,T_W}(W) \quad (187)$$

With the extremal temperatures T_0 and T_W the information capacity $C*_{T_0,T_W}(W)$ is given by

$$T(X;Y) = C*_{T_0,T_W}(W) \text{ and then } C^{\max} = \lim_{T_0 \longrightarrow T_W} C*_{T_0,T_W}(W) = H(Y) \quad (188)$$

From the difference $\Delta Q_0 \stackrel{\triangle}{=} W = Q*_0 - Q*_W$ (in $\mathcal{L}$) follows that the temperature $T_W = T_0 \cdot \sqrt{1 + \frac{6h \cdot W}{\pi^2 \mathrm{k}^2 T_0{}^2}}$.

Then, for the transinformation, in the same way as in (187), is now valid

$$T(X;Y) = \frac{\pi^2 k}{6h} \cdot \left(1-\beta^2\right) = \frac{\pi^2 k}{3h} \cdot (T_W - T_0) \cdot \frac{1+\beta}{2} = \frac{\pi^2 \mathrm{k} T_0}{3h} \cdot \left(\sqrt{1 + \frac{6h \cdot W}{\pi^2 \mathrm{k}^2 T_0{}^2}} - 1\right) \cdot \frac{1+\beta}{2} \quad (189)$$

The transformation $T(X;Y)$ is the capacity $C(\mathbf{K}_\triangle)$ and it is possible to write

$$T(X;Y) = C(\mathbf{K}_\triangle) = C*_{T_0,T_W}(W) = \frac{\pi^2 \mathrm{k} T_0}{6hT_W} \cdot \left(\sqrt{1 + \frac{6h \cdot W}{\pi^2 \mathrm{k}^2 T_0{}^2}} - 1\right) \cdot (T_0 + T_W) \quad (190)$$
$$= C*_{T_W}(W) = \frac{W}{\mathrm{k} T_W} = C*_{T_0}(W) = \frac{W}{\mathrm{k} T_0 \cdot \sqrt{1 + \frac{6h \cdot W}{\pi^2 \mathrm{k}^2 T_0{}^2}}} \stackrel{\triangle}{=} C(\mathbf{K}'_{\mathrm{L-L|B-E}})$$

which value is $2\times$ *less* than (177) and $\frac{2}{1+\beta}\times$ *less* than (138).

For $T_0 \longrightarrow 0$ the *quantum approximation* $C(W)$ of the capacity $C*_{T_0,T_W}(W)$ is obtained, independent on the noise energy (the noise power deminishes near the abslute $0°\,K$)

$$C(W) = \lim_{T_0 \to 0} \left(\sqrt{\frac{\pi^4 \mathrm{k}^2 T_0{}^2}{6^2 \hbar^2} + \pi^2 \frac{W}{6h}} - \frac{\pi^2 \mathrm{k} T_0}{6h}\right) \cdot (1+\beta) = \pi \cdot \sqrt{\frac{W}{6h}} \quad (191)$$

The *classical aproximation* $C_{T_0}(W)$ of $C*_{T_0,T_W}(W)$ is gained for $T_0 \gg 0$. This value is near Shannon capacity of the wide–band *Gaussian* channel with noise energy $\mathrm{k}T_0$ and with the whole average input energy (energy) W; in the same way as in (104) is now gained

$$C_{T_0}(W) \doteq \frac{\pi^2 \mathrm{k} T_0}{6h} \left(\frac{3h \cdot W}{\pi^2 \mathrm{k}^2 T_0{}^2}\right) \cdot (1+\beta) = \frac{W}{2\mathrm{k} T_0} \cdot (1+\beta) \longrightarrow \frac{W}{\mathrm{k} T_0} \quad (192)$$

The mutual difference of results (189) and and (102), (138) [12, 30] is given by **the necessity of the returning the transfer medium, the channel $\mathbf{K}_\triangle \cong \mathbf{K}'_{\mathrm{L-L}|B-E} \cong \mathcal{L}$ into its initial state after each individual information transfer act has been accomplished and, by the relevant temperatutre reducing of the heat ΔQ_0** $\left[\text{by } T_W \text{ in (183)-(189)}\right]$. Thus, our thermodynamic cyclical model $\mathbf{K}_\triangle \cong \mathcal{O}_{rrev\triangle}$ for the repeatible information transfer through the channel $\mathbf{K}'_{\mathrm{L-L|B-E}}$ is of the information capacity (189), while in [12, 30] the information capacity of the *one-act* information transfer is stated.[25] By (189) the *whole energy costs* for the cyclical information transfer considered is countable.[26]

8. Conclusion

After each completed 'transmission of an input message and receipt of an output message' ('one-act' transfer) the transferring system must be reverted to its starting state, otherwise the constant (in the sense repeatable) flow of information could not exist. The author believes that either the opening of the chain was presupposed in the original derivation in [30], or that the return of transferring system to its starting state was not considered at all, it was not counted-in. In our derivations this needed state transition is considered be powered within the transfer chain itself, without its openning. Although our derivation of the information capacity for a cyclical case (using the cyclic thermodynamic model) results in a lower value than the original one it seems to be more exact and its result as more precise from the *theoretic point of view*, extending and not ceasing the previous, *original* result [12, 30] which *remains* of its *technology-drawing value*. Also it forces us in being aware and respecting of the *global costs* for (any) communication and its evaluation and, as such, it is of a *gnoseologic character*.

Author details

Bohdan Hejna

Institute of Chemical Technology Prague, Department of Mathematics, Studentská 6, 166 28 Prague 6, Czech Republic

9. References

[1] Bell, D. A. *Teorie informace*; SNTL: Praha, 1961.

[2] Brillouin, L. *Science and Information Theory*; Academia Press: New York, 1963.

[3] Cholevo, A. S. On the Capacity of Quantum Communication Channel. *Problems of Information Transmission* 1979, 15 (4), 3–11.

[4] Cholevo, A. S. *Verojatnostnyje i statističeskije aspekty kvantovoj teorii*; Nauka: Moskva, 1980.

[5] Cover, T. M.; Thomas, J. B. *Elements of Information Theory*; Wiley: New York, 1991.

[6] Davydov, A. S. *Kvantová mechanika*; SPN: Praha, 1978.

[7] Emch, G. G. *Algebraičeskije metody v statističeskoj mechanike i kvantovoj těorii polja*; Mir: Moskva, 1976.

---

[25] For one-act information transfer the choose between two information descriptions is possible which result in capacity (138) or (177).

[26] For the energy $T(X;Y) \cdot T_W$ on the output the energy $\frac{2}{1+\beta} \cdot T_W \times$ *greater* is needed on the input of the transfer channel which is in accordance with the *II. Principle of thermodynamics* for $\frac{2}{1+\beta} > 1$ is valid.

[8] Gershenfeld, N. Signal entropy and the thermodynamics of computation. *IBM Systems Journal* 1996, *35* (3/4), 577–586. DOI 10.1147/sj.353.0577.
[9] Halmos, P. R. *Konečnomernyje vektornyje prostranstva*; Nauka: Moskva, 1963.
[10] Hašek, O.; Nožička, J. *Technická mechanika pro elektrotechnické obory II.*; SNTL: Praha, 1968.
[11] Hejna, B.; Vajda, I. Information transmission in stationary stochastic systems. *AIP Conf. Proc.* 1999, *465* (1), 405–418. DOI: 10.1063/1.58272.
[12] Hejna, B. Informační kapacita stacionárních fyzikálních systému. Ph.D. Dissertation, ÚTIA AV ČR, Praha, FJFI ČVUT, Praha, 2000.
[13] Hejna, B. Generalized Formula of Physical Channel Capacities. *International Journal of Control, Automation, and Systems* 2003, 15.
[14] Hejna, B. Thermodynamic Model of Noise Information Transfer. In *AIP Conference Proceedings*, Computing Anticipatory Systems: CASYS'07 - Eighth International Conference; Dubois, D., Ed.; American Institute of Physics: Melville, New York, 2008; pp 67–75. ISBN 978-0-7354-0579-0. ISSN 0094-243X.
[15] Hejna, B. Proposed Correction to Capacity Formula for a Wide-Band Photonic Transfer Channel. In *Proceedings of International Conference Automatics and Informatics'08*; Atanasoff, J., Ed.; Society of Automatics and Informatics: Sofia, Bulgaria, 2008; pp VII-1–VIII-4.
[16] Hejna, B. Gibbs Paradox as Property of Observation, Proof of II. Principle of Thermodynamics. In *AIP Conf. Proc.*, Computing Anticipatory Systems: CASYS'09: Ninth International Conference on Computing, Anticipatory Systems, 3–8 August 2009; Dubois, D., Ed.; American Institute of Physics: Melville, New York, 2010; pp 131–140. ISBN 978-0-7354-0858-6. ISSN 0094-243X.
[17] Hejna, B. *Informační termodynamika I.: Rovnovážná termodynamika přenosu informace*; VŠCHT Praha: Praha, 2010. ISBN 978-80-7080-747-7.
[18] Hejna, B. *Informační termodynamika II.: Fyzikální systémy přenosu informace*; VŠCHT Praha: Praha, 2011. ISBN 978-80-7080-774-3.
[19] Hejna, B. Information Thermodynamics, *Thermodynamics - Physical Chemistry of Aqueous Systems*, Juan Carlos Moreno-Pirajaán (Ed.), ISBN: 978-953-307-979-0, InTech, 2011 Available from:
http://www.intechopen.com/articles/show/title/information-thermodynamics
[20] Helstroem, C. W. *Quantum Detection and Estimation Theory*; Pergamon Press: London, 1976.
[21] Horák, Z.; Krupka, F. *Technická fyzika*; SNTL/ALFA: Praha, 1976.
[22] Horák, Z.; Krupka, F. *Technická fysika*; SNTL: Praha, 1961.
[23] Jaglom, A. M.; Jaglom, I. M. *Pravděpodobnost a teorie informace*; Academia: Praha, 1964.
[24] Chodasevič, M. A.; Sinitsin, G. V.; Jasjukevič, A. S. *Ideal fermionic communication channel in the number-state model*;Division for Optical Problems in Information technologies, Belarus Academy of Sciences, 2000.
[25] Kalčík, J.; Sýkora, K. *Technická termomechanika*; Academia: Praha, 1973.
[26] Karpuško, F. V.; Chodasevič, M. A. *Sravnitel'noje issledovanije skorostnych charakteristik bozonnych i fermionnych kommunikacionnych kanalov*; Vesšč NAN B, The National Academy of Sciences of Belarus, 1998.
[27] Landau, L. D.; Lifschitz, E. M. *Statistical Physics*, 2nd ed.; Pergamon Press: Oxford, 1969.
[28] Landauer, M. Irreversibility and Heat Generation in the Computing Process. *IBM J. Res. Dev.* 2000, *44* (1/2), 261.
[29] Lavenda, B. H. *Statistical Physics*; Wiley: New York, 1991.

[30] Lebedev, D.; Levitin, L. B. Information Transmission by Electromagnetic Field. *Information and Control* 1966, *9*, 1–22.
[31] Maršák, Z. *Termodynamika a statistická fyzika*; ČVUT: Praha, 1995.
[32] Marx, G. *Úvod do kvantové mechaniky*; SNTL: Praha, 1965.
[33] Moore, W. J. *Fyzikální chemie*; SNTL: Praha, 1981.
[34] Prchal, J. *Signály a soustavy*; SNTL / ALFA: Praha, 1987.
[35] Shannon, C. E. A Mathematical Theory of Communication. *The Bell Systems Technical Journal* 1948, *27*, 379–423, 623–656.
[36] Sinitsin, G. V.; Chodasevič, M. A.; Jasjukevič, A. S. *Elektronnyj kommunikacionnyj kanal v modeli svobodnogo vyroždennogo fermi-gaza*; Belarus Academy of Sciences, 1999.
[37] Vajda, I. *Teória informácie a štatistického rozhodovania*; Alfa: Bratislava, 1982.
[38] Watanabe, S. *Knowing and Guessing*; Wiley: New York, 1969.

Chapter 5

Thermodynamics' Microscopic Connotations

A. Plastino, Evaldo M. F. Curado and M. Casas

Additional information is available at the end of the chapter

http://dx.doi.org/10.5772/51370

1. Introduction

Thermodynamics is the science of energy conversion. It involves heat and other forms of energy, mechanical one being the foremost one. Potential energy is the capacity of doing work because of the position of something. Kinetic energy is due to movement, depending upon mass and speed. Since all objects have structure, they possess some internal energy that holds such structure together, a kind of strain energy. As for work, there are to kinds of it: internal and external. The later is work done on "something". The former is work effected within something, being a capacity. Heat is another king of energy, the leit-motif of thermodynamics. Thermodynamics studies and interrelates the macroscopic variables, such as temperature, volume, and pressure that are employed to describe thermal systems and concerns itself with phenomena that can be experimentally reproducible.

In thermodynamics one is usually interested in special system's states called equilibrium ones. Such states are steady ones reached after a system has stabilized itself to such an extent that it no longer keeps changing with the passage of time, as far as its macroscopic variables are concerned. From a thermodynamics point of view a system is defined by its being prepared in a certain, specific way. The system will always reach, eventually, a unique state of thermodynamic equilibrium, univocally determined by the preparation-manner. Empiric reproducibility is a fundamental requirement for physics in general an thermodynamics in particular. The main source of the strength, or robustness, of thermodynamics, lies on the fact does it deals only with phenomena that are experimentally reproducible.

Historically, thermodynamics developed out the need for increasing the efficiency of early steam engines, particularly through the work of the French physicist Nicolas Sadi-Carnot (1824) who believed that a heat engine's efficiency was to play an important role in helping France win the Napoleonic Wars. Scottish physicist Lord Kelvin was the first to formulate a succint definition of thermodynamics in 1854: "Thermodynamics is the subject of the relation of heat to forces acting between contiguous parts of bodies, and the relation of heat to electrical agency". Chemical thermodynamics studies the role of entropy in the process of chemical reactions and provides the main body of knowledge of the field. Since Boltzmann in the 1870's, statistical thermodynamics, or statistical mechanics, that are microscopic theories, began to explain macroscopic thermodynamics via statistical predictions on the collective motion of atoms.

1.1. Thermodynamics' laws

The laws of physics are established scientific regularities regarded as universal and invariable facts of the universe. A "law" differs from hypotheses, theories, postulates, principles, etc., in that it constitutes an analytic statement. A theory starts from a set of axioms from which all laws and phenomena should arise via adequate mathematical treatment. The principles of thermodynamics, often called "its laws", count themselves amongst the most fundamental regularities of Nature [1]. These laws define fundamental physical quantities, such as temperature, energy, and entropy, to describe thermodynamic systems and they account for the transfer of energy as heat and work in thermodynamic processes. An empirically reproducible distinction between heat and work constitutes the "hard-core" of thermodynamics. For processes in which this distinction cannot be made, thermodynamics remains silent. One speaks of four thermodynamics' laws:

- The zeroth law of thermodynamics allows for the assignment of a unique temperature to systems that are in thermal equilibrium with each other.
- The first law postulates the existence of a quantity called the internal energy of a system and shows how it is related to the distinction between energy transfer as work and energy transfer as heat. The internal energy is conserved but work and heat are not defined as separately conserved quantities. Alternatively, one can reformulate the first law as stating that perpetual motion machines of the first kind can not exist.
- The second law of thermodynamics expresses the existence of a quantity called the entropy S and states that for an isolated macroscopic system S never decreases, or, alternatively, that perpetual motion machines of the second kind are impossible.
- The third law of thermodynamics refers to the entropy of a system at absolute zero temperature ($T = 0$) and states that it is impossible to lower T in such a manner that reaches the limit $T = 0$.

Classical thermodynamics accounts for the exchange of work and heat between systems with emphasis in states of thermodynamic equilibrium. Thermal equilibrium is a condition *sine qua non* for *macroscopically specified systems* only. It shoul be noted that, at the microscopic (atomic) level all physical systems undergo random fluctuations. Every finite system will exhibit statistical fluctuations in its thermodynamic variables of state (entropy, temperature, pressure, etc.), but these are negligible for macroscopically specified systems. Fluctuations become important for microscopically specified systems. Exceptionally, for macroscopically specified systems found at critical states, fluctuations are of the essence.

1.2. The Legendre transform

The Legendre transform is an operation that transforms one real-valued function of f a real variable x into another f_T, of a different variable y, maintaining constant its information content. The derivative of the function f becomes the argument to the function f_T.

$$f_T(y) = xy - f(x);\; y = f'(x) \Rightarrow \text{ reciprocity.} \tag{1}$$

The Legendre transform its own inverse. It is used to get from Lagrangians the Hamiltonian formulation of classical mechanics.

Legendre' reciprocity relations constitute thermodynamics' essential formal ingredient [2]. In general, for two functions I and α one has

$$I(A_1,\ldots,A_M) = \alpha + \sum_{k=1}^{M} \lambda_k A_k, \tag{2}$$

with the A_i extensive variables and the λ_i intensive ones. Obviously, the Legendre transform main goal is that of changing the identity of our relevant independent variables. For α we have

$$\alpha(\lambda_1,\ldots,\lambda_M) = I - \sum_{k=1}^{M} \lambda_k \langle A_k \rangle . \tag{3}$$

The three operative reciprocity relations become [2]

$$\frac{\partial \alpha}{\partial \lambda_k} = -\langle A_k \rangle\,; \qquad \frac{\partial I}{\partial \langle A_k \rangle} = \lambda_k\,; \qquad \frac{\partial I}{\partial \lambda_i} = \sum_k^M \lambda_k \frac{\partial \langle A_k \rangle}{\partial \lambda_i}, \tag{4}$$

the last one being the so-called Euler theorem.

1.3. The axioms of thermodynamics

Thermodynamics can be regarded as a formal logical structure whose *axioms* are empirical facts [2], which gives it a unique status among the scientific disciplines [1]. The four axioms given below are equivalent to the celebrated laws of thermodynamics of the prevous Subsection [2].

- For every system there exists a quantity E, the internal energy, such that a unique E_s−value is associated to each and every state s. The difference $E_{s1} - E_{s2}$ for two different states s_1 and s_2 in a closed system is equal to the work required to bring the system, while adiabatically enclosed, from one state to the other.
- There exist particular states of a system, the equilibrium ones, that are uniquely determined by E and by a set of extensive macroscopic parameters A_ξ, $\xi = 1,\ldots,M$. The number and characteristics of the A_ξ depends on the nature of the system.
- For every system there exists a state function $S(E, \forall A_\xi)$ that (i) always grows if internal constraints are removed and (ii) is a monotonously (growing) function of E. S remains constant in quasi-static adiabatic changes.

- S and the temperature $T = [\frac{\partial E}{\partial S}]_{A_1,\ldots,A_M}$ vanish for the state of minimum energy and are non-negative for all other states.

From the second and 3rd. Postulates one extracts the following two essential assertions

1. **Statement 3a)** for every system there exists a state function S, a function of E and the A_ξ

$$S = S(E, A_1, \ldots, A_M). \tag{5}$$

2. **Statement 3b)** S is a monotonous (growing) function of E, so that one can interchange the roles of E and S in (5) and write

$$E = E(S, A_1, \ldots, A_M), \tag{6}$$

Eq. (6) clearly indicates that

$$dE = \frac{\partial E}{\partial S} dS + \sum_{\xi} \frac{\partial E}{\partial A_\xi} dA_\xi \Rightarrow dE = TdS + \sum_{\xi} P_\xi dA_\xi, \tag{7}$$

with P_ξ generalized pressures and the temperature T defined as [2]

$$T = \left(\frac{\partial E}{\partial S}\right)_{[\forall A_\xi]}. \tag{8}$$

Eq. (7) will play a key-role in our future considerations. If we know $S(E, A_1, \ldots, A_n)$ or, equivalently because of monotonicity, $E(S, A_1, \ldots, A_n))$ we have a *complete* thermodynamic description of a system [2]. For experimentalists, it is often more convenient to work with *intensive* variables defined as follows [2].

Let $S \equiv A_0$. The intensive variable associated to the extensive A_i, to be called P_i are the derivatives

$$P_0 \equiv T = [\frac{\partial E}{\partial S}]_{A_1,\ldots,A_n}, \ 1/T = \beta. \tag{9}$$

$$P_j \equiv \lambda_j / T = [\frac{\partial E}{\partial A_j}]_{S,A_1,\ldots,A_{j-1},A_{j+1},\ldots,A_n}. \tag{10}$$

Any one of the Legendre transforms that replaces any s extensive variables by their associated intensive ones (β, λ's will be Lagrange multipliers in SM)

$$L_{r_1,\ldots,r_s} = E - \sum_j P_j A_j, \ (j = r_1, \ldots, r_s)$$

contains the same information as either S or E. The transform $L_{r_1,\ldots,r_s}$ is a function of $n - s$ extensive and s intensive variables. This is called the *Legendre invariant structure of thermodynamics*. As we saw above, this implies certain relationships amongst the relevant system's variables, called the *reciprocity relations* (RR), that are crucial for the microscopic discussion of Thermodynamics.

2. Classical statistical mechanics

In 1903 Gibbs formulated the first axiomatic theory for statistical mechanics [1, 3], revolving around the concept of phase space. The phase space (PS) precise location is given by generalized coordinates and momenta. Gibbs' postulates properties of an imaginary (Platonic) ad-hoc notion: the "ensemble" (a mental picture). The ensemble consists of extremely many (N) independent systems, all identical in nature with the one of actual physical interest, but differing in PS-location. That is, the original system is to be mentally repeated many times, each with a different arrangement of generalized coordinates and momenta. Here Liouville's theorem of volume conservation in phase space for Hamiltonian motion plays a crucial role. The ensemble amounts to a distribution of N PS-points, representative of the actual system. N is large enough that one can properly speak of a density D at any PS-point $\phi = q_1, \ldots, q_N; p_1, \ldots, p_N$, with $D = D(q_1, \ldots, q_N; p_1, \ldots, p_N, t) \equiv D(\phi)$, with t the time, and, if we call $d\phi$ the volume element,

$$N = \int d\phi \, D; \quad \forall t. \tag{11}$$

Randomly extracting a system from the ensemble, the probability of selecting it being located in a neighborhood of ϕ would yield

$$P(\phi) = D(\phi)/N. \tag{12}$$

Consequently,

$$\int P \, d\phi = 1. \tag{13}$$

Liouville's theorem follows from the fact that, since phase-space points can not be "destroyed", if

$$N_{12} = \int_{\phi_1}^{\phi_2} D \, d\phi, \tag{14}$$

then

$$\frac{dN_{12}}{dt} = 0. \tag{15}$$

An appropriate analytical manipulation involving Hamilton's canonical equations of motion then yields the theorem in the form [1]

$$\dot{D} + \sum_i^N \frac{\partial D}{\partial p_i} \dot{p}_i + \sum_i^N \frac{\partial D}{\partial q_i} \dot{q}_i = 0, \tag{16}$$

entailing the PS-conservation of density.

Equilibrium means simply $\dot{D} = 0$, i. e.,

$$\sum_i^N \frac{\partial D}{\partial p_i} \dot{p}_i + \sum_i^N \frac{\partial D}{\partial q_i} \dot{q}_i = 0. \tag{17}$$

2.1. The classical axioms

Gibbs refers to PS-location as the "phase" of the system [1, 3]. The following statements completely explain in microscopic fashion the corpus of classical equilibrium thermodynamics [1].

- The probability that at time t the system will be found in the dynamical state characterized by ϕ equals the probability $P(\phi)$ that a system randomly selected from the ensemble shall possess the phase ϕ will be given by Eq. (12) above.
- All phase-space neighborhoods (cells) have the same a priori probability.
- D depends only upon the system's Hamiltonian.
- The time-average of a dynamical quantity F equals its average over the ensemble, evaluated using D.

3. Information

Information theory (IT) treats information as data communication, with the primary goal of concocting efficient manners of encoding and transferring data. IT is a branch of applied mathematics and electrical engineering, involving the quantification of information, developed by Claude E. Shannon [4] in order to i) find fundamental limits on signal processing operations such as compressing data and ii) finding ways of reliably storing and communicating data. Since its 1948-inception it has considerably enlarged its scope and found applications in many areas that include statistical inference, natural language processing, cryptography, and networks other than communication networks. A key information-measure (IM) was originally called (by Shannon) entropy, in principle unrelated to thermodynamic entropy. It is usually expressed by the average number of bits needed to store or communicate one symbol in a message and quantifies the uncertainty involved in predicting the value of a random variable.Thus, a degree of knowledge (or ignorance) is associated to any normalized probability distribution $p(i)$, $(i = 1, \ldots, N)$, determined by a functional $I[\{p_i\}]$ of the $\{p_i\}$ [4–7] which is precisely Shannon's entropy. IT was la axiomatized in 1950 by Kinchin [8], on the basis of four axioms, namely,

- I is a function ONLY of the $p(i)$,
- I is an absolute maximum for the uniform probability distribution,
- I is not modified if an $N+1$ event of probability zero is added,
- Composition law.

As for the last axiom, consider two sub-systems $[\Sigma^1, \{p^1(i)\}]$ and $[\Sigma^2, \{p^2(j)\}]$ of a composite system $[\Sigma, \{p(i,j)\}]$ with $p(i,j) = p^1(i)\, p^2(j)$. Assume further that the conditional probability distribution (PD) $Q(j|i)$ of realizing the event j in system 2 for a fixed $i-$event in system 1. To this PD one associates the information measure $I[Q]$. Clearly,

$$p(i,j) = p^1(i)\, Q(j|i). \tag{18}$$

Then Kinchin's fourth axiom states that

$$I(p) = I(p^1) + \sum_i p^1(i)\, I(Q(j|i)). \tag{19}$$

An important consequence is that, out of the four Kinchin axioms one finds that Shannons's measure

$$S = -\sum_{i=1}^{N} p(i) \ln [p(i)], \tag{20}$$

gives us the only way of complying with Kinchin's axioms.

4. Statistical mechanics and information theory

It has been argued [9] that the statistical mechanics (SM) of Gibbs is a juxtaposition of subjective, probabilistic ideas on the one hand and objective, mechanical ideas on the other. From the mechanical viewpoint, the vocables "statistical mechanics" suggest that for solving physical problems we ought to acknowledge a degree of uncertainty as to the experimental conditions. Turning this problem around, it also appears that the purely statistical arguments are incapable of yielding any physical insight unless some mechanical information is a priori assumed [9]. This is the conceptual origin of the link SM-IT pioneered by Jaynes in 1957 via his Maximum Entropy Principle (MaxEnt) [5, 6, 10] which allowed for reformulating SM in information terms. Since IT's central concept is that of information measure (IM)

Descartes' scientific methodology considers that truth is established via the agreement between two *independent* instances that can neither suborn nor bribe each other: analysis (purely mental) and experiment [11]. The analytic part invokes mathematical tools and concepts: Mathematics' world $\Leftrightarrow$ Laboratory. The mathematical realm is called Plato's Topos Uranus (TP). Science in general, and physics in particular, may thus be seen as a [TP $\Leftrightarrow$ "Experiment"] two-way bridge. TP concepts are related to each other in the form of "laws" that adequately describe the relationships obtaining among suitable chosen variables that describe the phenomenon at hand. In many cases these laws are integrated into a comprehensive theory (e.g., classical electromagnetism, based upon Maxwell's equations) [1, 12–15].

Jaynes' MaxEnt ideas describe thermodynamics via the link [IT as a part of TP]$\Leftrightarrow$ [Thermal experiment], or in a more general scenario: [IT] $\Leftrightarrow$ [Phenomenon at hand]. It is clear that the relation between an information measure and entropy is [IM] $\Leftrightarrow$ [Entropy S]. One can then assert that an IM is not necessarily an entropy, since the first belongs to the Topos Uranus and the later to the laboratory. Of course, in some special cases an association $IM \Leftrightarrow$ entropy S can be established. Such association is both useful and proper in very many situations [5].

If, in a given scenario, N distinct outcomes ($i = 1, \ldots, N$) are possible, three alternatives are to be considered [6]:

1. Zero ignorance: predict with certainty the actual outcome.
2. Maximum ignorance: Nothing can be said in advance. The N outcomes are equally likely.

3. Partial ignorance: we are given the probability distribution $\{P_i\}$; $i = 1, \ldots, N$.

If our state of knowledge is appropriately represented by a set of, say, M expectation values, then the "best", least unbiased probability distribution is the one that [6]

- reflects just what we know, without "inventing" unavailable pieces of knowledge [5, 6] and, additionally,
- maximizes ignorance: the truth, all the truth, *nothing but* the truth [6].

Such is the MaxEnt rationale. In using MaxEnt, one is not maximizing a physical entropy, but only maximizing ignorance in order to obtain the least biased distribution compatible with the a priori knowledge.

Statistical mechanics and thereby thermodynamics can be formulated on an information theory basis if the density operator $\hat{\rho}$ is obtained by appealing to Jaynes' maximum entropy principle (MaxEnt), that can be stated as follows:

Assume that your prior knowledge about the system is given by the values of M expectation values $< A_1 >, \ldots, < A_M >$. In such circumstances $\hat{\rho}$ is uniquely determined by extremizing $I(\hat{\rho})$ subject to M constraints given, namely, the M conditions $< A_j >= Tr[\hat{\rho}\,\hat{A}_j]$, a procedure that entails introducing M Lagrange multipliers λ_i. Additionally, since normalization of $\hat{\rho}$ is necessary, a normalization Lagrange multiplier ξ should be invoked. The procedure immediately leads one [6] to realizing that $I \equiv S$, the equilibrium Boltzmann's entropy, if the a priori knowledge $< A_1 >, \ldots, < A_M >$ refers only to extensive quantities. Of course, I, once determined, *affords for complete thermodynamical information for the system of interest* [6].

5. A new micro-macroscopic way of accounting for thermodynamics

Gibbs' and MaxEnt approaches satisfactorily describe equilibrium thermodynamics. We will here search for a new, different alternative able to account for thermodynamics from first principles. Our idea is to give axiom-status to Eq. (7), *which is an empirical statement*. Why? Because neither in Gibbs' nor in MaxEnt's axioms we encounter a direct connection with actual thermal data. By appealing to Eq. (7) we would instead be actually employing empirical information. This is our rationale.

Consequently, we will concoct a new SM-axiomatics by giving postulate status to the following macroscopic statement:

Axiom (1)

$$dE = TdS + \sum_{\nu} P_{\nu} dA_{\nu}. \tag{21}$$

This is a macroscopic postulate to be inserted into a microscopic axiomatics' corpus.

We still need *some* amount of microscopic information, since we are building up a microscopic theory. We wish to add as little as possible, of course (Ockham's razor). At this point it is useful to remind the reader of Kinchin's postulates, recounted above. We will content ourselves with borrowing for our theoretical concerns just his his first axiom. Thus, we conjecture at this point, and will prove below, that the following assertion suffices for our theoretical purposes:

Axiom (2) If there are W microscopic accessible states labelled by i, whose microscopic probability we call p_i, then

$$S = S(p_1, p_2, \ldots, p_W). \tag{22}$$

Thus, we are actually taking as a postulate something that is actually known from both quantum and classical mechanics.

Axiom (3) The internal energy E and the external parameters A_ν are to be considered as the expectation values of suitable operators, that is, the hamiltonian H and the hermitian operators $\mathcal{R}_\nu$ (i.e., $A_\nu \equiv <\mathcal{R}_\nu>$). Thus, the A_ν (and also E) will depend on the eigenvalues of these operators *and* on the probability set. (Note that energy eigenvalues depend of course upon the $\mathcal{R}_\nu$.

The reader will immediately realize that Axiom (2) is just a way of re-expressing Boltzmann's "atomic" conjecture. Thus, macroscopic quantities become statistical averages evaluated using a microscopic probability distribution [16]. Our present three new axioms are statements of fact. What do we mean? That they are borrowed from either experiment or pre-existent theories. Somewhat surprisingly, our three axioms do not actually incorporate any knew knowledge at all. The merely re-express known previous notions. Ockham's razor at its best! Our theory could no be more economical.

We need now to prove that the above three postulates allow one to reconstruct the imposing edifice of statistical mechanics. We will tackle this issue by showing below that they our axioms are equivalent to those of Jaynes' [17]. At this point we need to recall the main goal of statistical mechanics, namely, finding the probability distribution (or the density operator) that best describes our physical system. In order to do so Jaynes appealed to his MaxEnt postulate, that we restate below for the sake of fixing notation.

MaxEnt axiom: assume your prior knowledge about the system is given by the values of M expectation values

$$A_1 \equiv <\mathcal{R}_1>, \ldots, A_R \equiv <\mathcal{R}_M>. \tag{23}$$

Then, ρ is uniquely fixed by extremizing the information measure $I(\rho)$ subject to $\rho-$normalization plus the constraints given by the M conditions constituting our assumed foreknowledge

$$A_\nu = <\mathcal{R}_\nu> = Tr[\rho\,\mathcal{R}_\nu]. \tag{24}$$

This leads, after a Lagrange-constrained extremizing process, to the introduction of M Lagrange multipliers λ_ν, that one assimilates to the generalized pressures P_ν. The truth, the whole truth, nothing but the truth [6]. Jaynes rationale asserts that if the entropic measure that reflects our ignorance were not of maximal character, we would actually be *inventing* information not at hand.

While working through his variational process, Jaynes discovers that, after multiplying by Boltzmann's constant k_B the right-hand-side of his expression for the information measure, it converts itself into an entropy, $I \equiv S$, the equilibrium thermodynamic one, with the caveat that $A_1 = <\mathcal{R}_1>, \ldots, A_M = <\mathcal{R}_M>$ refer to extensive quantities. Having ρ, his universal form $I(\rho)$ *yields complete microscopic information with respect to the system of interest.* To achieve our ends one needs now just to prove that the new axiomatics, with (21) and (22), is equivalent to MaxEnt.

6. New connection between macroscopic and microscopic approaches

In establishing our new connections between the micro- and macro-scenarios we shall work with the classical instance only, since the corresponding quantum treatment constitute in this sense just a straightforward extension.

Our main idea is to pay attention to the generic change $p_i \to p_i + dp_i$ as constrained by Eq. (21). In other word, we insist on studying the change dp_i that takes place in such a manner that (21) holds. Our main macroscopic quantities S, A_j, and E will vary with dp_i. These changes are not arbitrary but are constrained by (21). Note here an important advantage to be of our approach. We need *not* specify beforehand the information measure employed.

Since several possibilities exist (see for instance Gell-Mann and Tsallis [18]), this entails that the choice of information nature is not predetermined by macroscopic thermodynamics. For a detailed discussion of this issue see Ferri, Martinez, and Plastino [19].

The pertinent ingredients at hand are

- an arbitrary, smooth function $f(p)$ permitting one expressing the information measure via

$$I \equiv S(\{p_i\}) = \sum_i p_i f(p_i), \tag{25}$$

  such that $S(\{p_i\})$ is a concave function,
- M quantities A_ν representing values of extensive quantities $\langle \mathcal{R}_\nu \rangle$, that adopt, for a micro-state i, the value a_i^ν with probability p_i,
- still another arbitrary smooth, monotonic function $g(p_i)$ ($g(0) = 0$; $g(1) = 1$). With the express purpose of employing generalized, non-Shannonian entropies, we slightly generalize here the expectation-value definitions by recourse to g via (26):

$$A_\nu \equiv \langle \mathcal{R}_\nu \rangle = \sum_i^W a_i^\nu\, g(p_i); \;\; \nu = 2, \ldots, M, \tag{26}$$

$$E = \sum_i^W \epsilon_i\, g(p_i), \tag{27}$$

  where ϵ_i is the energy associated to the microstate i.

We take $A_1 \equiv E$ and pass to a consideration of the probability variations dp_i that should generate accompanying changes dS, dA_ν, and dE in, respectively, S, the A_ν, and E.

The essential issue at hand is that of enforcing compliance with

$$dE - TdS + \sum_{\nu=1}^W dA_\nu \lambda_\nu = 0, \tag{28}$$

with T the temperature and λ_ν generalized pressures. By recourse to (25), (26), and (27) we i) recast now (28) for

$$p_i \to p_i + dp_i, \tag{29}$$

and ii) expand the resulting equation up to first order in the dp_i.

Remembering that the Lagrange multipliers λ_ν are identical to the generalized pressures P_ν of Eq. (7), one thus encounter, after a little algebra [20–26],

$$C_i^{(1)} = [\sum_{\nu=1}^{M} \lambda_\nu \, a_i^\nu + \epsilon_i]$$

$$C_i^{(2)} = -T \frac{\partial S}{\partial p_i}$$

$$\sum_i [C_i^{(1)} + C_i^{(2)}] dp_i \equiv \sum_i K_i dp_i = 0, \tag{30}$$

so that, appropriately rearranging things

$$T_i^{(1)} = f(p_i) + p_i f'(p_i)$$
$$T_i^{(2)} = -\beta[(\sum_{\nu=1}^{M} \lambda_\nu \, a_i^\nu + \epsilon_i) \, g'(p_i) - K],$$
$$(\beta \equiv 1/kT), \tag{31}$$

and we are in a position to recast (30) in the fashion

$$T_i^{(1)} + T_i^{(2)} = 0; \;\; (for \;\; any \;\; i), \tag{32}$$

an expression whose importance will become manifest later on.

Eqs. (30) or (32) yield one and just one p_i-expression, as demonstrated in Refs. [20–26]. However, it will be realized below that, at this stage, an explicit expression for this probability distribution is not required.

We pass now to traversing the opposite road that leads from Jaynes' MaxEnt procedure and ends up with *our* present equations. This entails extremization of S subject to constraints in E, A_ν, and normalization. For details see [20–26].

Setting $\lambda_1 \equiv \beta = 1/T$ one has

$$\delta_{p_i}[S - \beta \langle H \rangle - \sum_{\nu=2}^{M} \lambda_\nu \langle \mathcal{R}_\nu \rangle - \xi \sum_i p_i] = 0, \tag{33}$$

(normalization Lagrange multiplier ξ) is easily seen in the above cited references to yield as a solution the very set of Eqs. (30). The detailed proof is given in the forthcoming Section. Eqs. (30) arise then from two different approaches:

- our methodology, based on Eqs. (21) and (22), and
- following the well known MaxEnt route.

Accordingly, we see that both MaxEnt and our axiomatics co-imply one another. They are indeed equivalent ways of constructing equilibrium statistical mechanics. As a really relevant fact

One does not need to know the analytic form of $S[p_i]$ neither in Eqs. (30) nor in (33).

7. Proof

Here we prove that Eqs. (30) can be derived from the MaxEnt approach (33). One wishes to extremize S subject to the constraints of fixed valued for i) U, ii) the M values A_ν (entailing Lagrange multipliers (1) β and (2) M γ_ν), and iii) normalization (Lagrange multiplier ξ). One has also

$$A_\nu = \langle \mathcal{R}_\nu \rangle = \sum_i p_i\, a_i^\nu, \tag{34}$$

with $a_i^\nu = \langle i|\mathcal{R}_\nu|i\rangle$ the matrix elements in the basis $\langle i\rangle$ of $\mathcal{R}_\nu$. The ensuing variational problem one faces, with $U = \sum_i p_i \epsilon_i$, is

$$\delta_{\{p_i\}} \left[S - \beta U - \sum_{\nu=1}^{M} \gamma_\nu A_\nu - \xi \sum_i p_i \right] = 0, \tag{35}$$

that immediately leads, for $\gamma_\nu = \beta\lambda_\nu$, to

$$\delta_{p_m} \sum_i \left(p_i f(p_i) - [\beta p_i (\sum_{\nu=1}^{M} \lambda_\nu\, a_i^\nu + \epsilon_i) + \xi p_i] \right) = 0, \tag{36}$$

so that the the following two quantities vanish

$$\begin{aligned} & f(p_i) + p_i f'(p_i) - [\beta(\textstyle\sum_{\nu=1}^{M} \lambda_\nu\, a_i^\nu + \epsilon_i) + \xi] \\ & \Rightarrow \text{ if } \xi \equiv \beta K, \\ & f(p_i) + p_i f'(p_i) - \beta(\textstyle\sum_{\nu=1}^{M} \lambda_\nu a_i^\nu + \epsilon_i) + K] \\ & \Rightarrow 0 = T_i^{(1)} + T_i^{(2)}. \end{aligned} \tag{37}$$

We realize now that (32) and the last equality of (37) are one and the same equation. MaxEnt does lead to (32).

8. Conclusions

We have formally proved above that our axiomatics allows one to derive MaxEnt equations and viceversa. Thus, our treatment provides an alternative foundation for equilibrium statistical mechanics. We emphasized that, opposite to what happens with both Gibbs' and Jaynes' axioms, our postulates have zero new informational content. Why? Because they are borrowed either from experiment or from pre-existing theories, namely, information theory and quantum mechanics.

The first and second laws of thermodynamics are two of physics' most important empirical facts, constituting pillars to our present view of Nature. Statistical mechanics (SM) adds an underlying microscopic substratum able to explain not only these two laws but the whole of thermodynamics itself [2, 6, 27–30]. Basic SM-ingredient is a microscopic probability distribution (PD) that controls microstates-population [27]. Our present ideas yield a detailed picture, from a new perspective [20–26], of how changes in the independent external thermodynamic parameters affect the micro-state population and, consequently, the entropy and the internal energy.

Acknowledgement

This work was partially supported by the MEC Grant FIS2005-02796 (Spain).

Author details

A. Plastino
Universidad Nacional de La Plata, Instituto de Física (IFLP-CCT-CONICET), C.C. 727, 1900 La Plata, Argentina
Physics Departament and IFISC-CSIC, University of Balearic Islands, 07122 Palma de Mallorca, Spain

Evaldo M. F. Curado
Centro Brasileiro de Pesquisas Fisicas, Rio de Janeiro, Brazil

M. Casas
Physics Departament and IFISC-CSIC, University of Balearic Islands, 07122 Palma de Mallorca, Spain

9. References

[1] R. B. Lindsay and H. Margenau, *Foundations of physics*, NY, Dover, 1957.
[2] E. A. Desloge, *Thermal physics* NY, Holt, Rhinehart and Winston, 1968.
[3] J. Willard Gibbs, *Elementary Principles in Statistical Mechanics*, New Haven, Yale University Press, 1902.
[4] C. E. Shannon, Bell System Technol. J. 27 (1948) 379-390.
[5] E. T. Jaynes *Papers on probability, statistics and statistical physics*, edited by R. D. Rosenkrantz, Dordrecht, Reidel, 1987.
[6] A. Katz, *Principles of Statistical Mechanics, The information Theory Approach*, San Francisco, Freeman and Co., 1967.
[7] T. M. Cover and J. A. Thomas, *Elements of information theory*, NY, J. Wiley, 1991.
[8] A. Plastino and A. R. Plastino in *Condensed Matter Theories*, Volume 11, E. Ludeña (Ed.), Nova Science Publishers, p. 341 (1996).
[9] D. M. Rogers, T. L. Beck, S. B. Rempe, *Information Theory and Statistical Mechanics Revisited*, ArXiv 1105.5662v1.
[10] D. J. Scalapino in *Physics and probability. Essays in honor of Edwin T. Jaynes* edited by W. T. Grandy, Jr. and P. W. Milonni (Cambridge University Press, NY, 1993), and references therein.
[11] B. Russell, *A history of western philosophy* (Simon & Schuster, NY, 1945).
[12] P. W. Bridgman *The nature of physical theory* (Dover, NY, 1936).
[13] P. Duhem *The aim and structure of physical theory* (Princeton University Press, Princeton, New Jersey, 1954).
[14] R. B. Lindsay *Concepts and methods of theoretical physics* (Van Nostrand, NY, 1951).

[15] H. Weyl *Philosophy of mathematics and natural science* (Princeton University Press, Princeton, New Jersey, 1949).
[16] D. Lindley, *Boltzmann's atom*, NY, The free press, 2001.
[17] W.T. Grandy Jr. and P. W. Milonni (Editors), *Physics and Probability. Essays in Honor of Edwin T. Jaynes*, NY, Cambridge University Press, 1993.
[18] M. Gell-Mann and C. Tsallis, Eds. *Nonextensive Entropy: Interdisciplinary applications*, Oxford, Oxford University Press, 2004.
[19] G. L. Ferri, S. Martinez, A. Plastino, Journal of Statistical Mechanics, P04009 (2005).
[20] E. Curado, A. Plastino, Phys. Rev. E 72 (2005) 047103.
[21] A. Plastino, E. Curado, Physica A 365 (2006) 24
[22] A. Plastino, E. Curado, International Journal of Modern Physics B 21 (2007) 2557
[23] A. Plastino, E. Curado, Physica A 386 (2007) 155
[24] A. Plastino, E. Curado, M. Casas, Entropy A 10 (2008) 124
[25] International Journal of Modern Physics B 22, (2008) 4589
[26] E. Curado, F. Nobre, A. Plastino, Physica A 389 (2010) 970.
[27] R.K. Pathria, *Statistical Mechanics* (Pergamon Press, Exeter, 1993).
[28] F. Reif, *Statistical and thermal physics* (McGraw-Hill, NY, 1965).
[29] J. J.Sakurai, *Modern quantum mechanics* (Benjamin, Menlo Park, Ca., 1985).
[30] B. H. Lavenda, *Statistical Physics* (J. Wiley, New York, 1991); B. H. Lavenda, *Thermodynamics of Extremes* (Albion, West Sussex, 1995).

Section 3

Property Prediction and Thermodynamics

Chapter 6

Group Contribution Methods for Estimation of Selected Physico-Chemical Properties of Organic Compounds

Zdeňka Kolská, Milan Zábranský and Alena Randová

Additional information is available at the end of the chapter

http://dx.doi.org/10.5772/49998

1. Introduction

Thermodynamic data play an important role in the understanding and design of chemical processes. To determine values of physico-chemical properties of compounds we can apply experimental or non-experimental techniques. Experimental techniques belong to the most correct, accurate and reliable. All experimental methods require relevant technical equipment, time necessary for experiment, sufficient amount of measured compounds of satisfactory purity. Compound must not affect technical apparatus and should not be decomposed during experiment. Other aspect is a valid legislation, which limits a usage of dangerous compounds by any users.

If due to any of these conditions mentioned above causes the experiment cannot be realized, some non-experimental approaches can be applied.

2. Non-experimental approaches to determine physico-chemical properties of compounds

If due to any of conditions results in that experimental determination cannot be realized and data on physico-chemical property are necessary, we have to employ some non-experimental approaches, either calculation methods or estimation ones. Due to the lack of experimental data for several industrially important compounds, different estimation methods have been developed to provide missing data. Estimation methods include those based on theory (e.g. statistical thermodynamics or quantum mechanics), various empirical relationships (correlations of required property with variable, experimentally determined compound characteristics, e.g. number of carbon atoms in their molecule, molecular weight, normal boiling temperature, etc.), and several classes of "additivity-principle" methods

(Baum, 1989; Pauling et al., 2001). Estimation methods can be divided into several groups from many aspects, e.g. into methods based on theoretical, semi-theoretical relations and the empirical ones. Books and papers of last decades divide estimation methods depending on the required input data into QPPR or QSPR approaches (Baum, 1989). QPPR methods (Quantity-Property-Property-Relationship) are input data-intensive. They require for calculation of searched value property knowledge of other experimental data. We can use them successfully only when we have input data. On the other hand QSPR (Baum, 1989) methods (Quantity-Structure-Property-Relationship) need only knowledge of the chemical structure of a compound to predict the estimated property. QSPR methods use some structural characteristics, such as number of fragments (atoms, bonds or group of atoms in a molecule), topological indices or other structural information, molecular descriptors, to express the relation between the property and molecular structure of compound (Baum, 1989; Pauling et al., 2001; Gonzáles et al., 2007a). Empirical and group contribution methods seem to be the most suitable (Pauling et al., 2001; Majer et al., 1989) due to their simplicity, universality and fast usage.

2.1. Group contribution methods

Group contribution methods are presented as empirical QSPR approaches. The easiest models were based on study of property on number of carbon atoms n_C or methylen groups n_{CH2} in molecules of homological series. In Fig. 1 is presented dependence of normal boiling temperature T_b on number of carbon atoms n_C (bottom axis) or methylen groups n_{CH2} (top axis) in molecules of homological series *n*-alkanols C_1-C_{12} (Majer & Svoboda, 1985; NIST database). As we can see, this dependence is clearly linear in some range of n_C=C_2-C_{10}. But increasing discrepancy is evident either for low number of carbon atoms C_1 or for higher one n_C>C_{10}. Due to these departures from linear behaviour some parameters covering structural effects on property were inclusive to these easy models (e.g. Chickos et al., 1996). From these approaches structural fragments and subsequently group contribution methods have been established.

Group contribution methods are based on the so called "additive principle". That means any compound can be divided into fragments, usually atoms, bonds or group of atoms, etc. All fragments have a partial value called a contribution. These contributions are calculated from known experimental data. Property of a compound is obtained by a summing up the values of all contributions presented in the molecule. Example of division of molecule of ethanol into atomic, bond and group fragments is presented in Fig. 2. When we divide this molecule into atomic fragments, the total value of property X of ethanol is given by summing up the values for two carbon atom contributions $X_{(C)}$, six hydrogen atom contributions $X_{(H)}$ and one oxygen atom contribution $X_{(O)}$. The second way is the division of ethanol molecules into the following bond fragments with their contribution: $X_{(C-C)}$, $X_{(C-O)}$, $X_{(C-H)}$ and $X_{(O-H)}$. Due to increasing quality and possibility of computer technique a fragmentation into more complex group structural fragments is applied in present papers (Baum, 1989; Pauling et al., 2001). Some of ways to divide the molecule ethanol into group structural fragments are presented in Fig. 2. Ethanol molecule can be divided either into: (i)

CH_3-, -CH_2- and -OH, or: (ii) CH_3- and –CH_2OH. More complex compounds are described by more complex structural fragments.

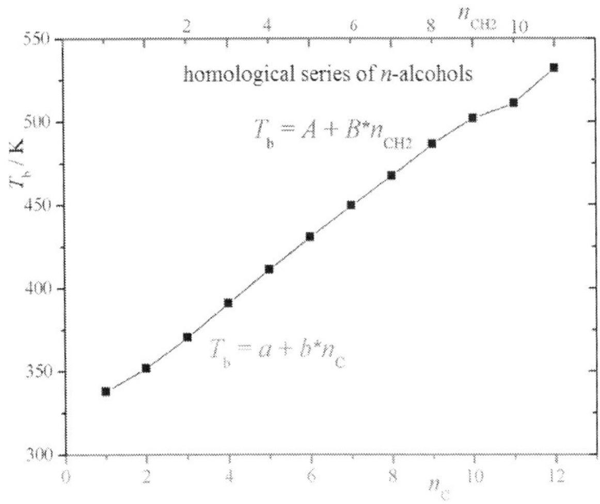

Figure 1. Dependence of normal boiling temperature of *n*-alkanols in homological series C_1-C_{12}

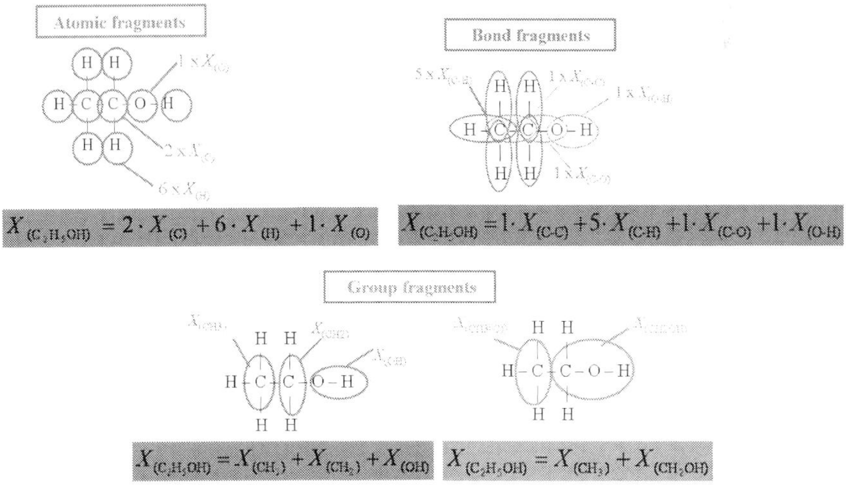

Figure 2. Example of division of ethanol molecule into atomic, bond and group structural fragments

Group contribution methods are essentially empirical estimation methods. A large variety of these models have been designed during last centuries, differing in a field of their applicability and in the set of experimental data. They were developed to estimate, e.g. critical properties (Lydersen, 1955; Ambrose, 1978; Ambrose, 1979; Joback & Reid, 1987;

Gani & Constantinou, 1996; Poling et al., 2001; Marrero & Gani, 2001; Brown et al., 2010; Monago & Otobrise, 2010; Sales-Cruz et al., 2010; Manohar & Udaya Sankar, 2011; Garcia et al., 2012), parameters of state equations (Pereda et al., 2010; Schmid & Gmehling, 2012), acentric factor (Constantinou & Gani, 1994; Brown et al., 2010; Monago & Otobrise, 2010), activity coefficients (Tochigi et al., 2005; Tochigi & Gmehling, 2011), vapour pressure (Poling et al., 2001; Miller, 1964), liquid viscosity (Joback & Reid, 1987; Conte et al., 2008; Sales-Cruz et al., 2010), gas viscosity (Reichenberg, 1975), heat capacity (Joback & Reid, 1987; Ruzicka & Domalski, 1993a; Ruzicka & Domalski, 1993b; Kolská et al., 2008), enthalpy of vaporization (e.g. Chickos et al., 1995; Chickos & Wilson, 1997; Marrero & Gani, 2001; Kolská et al. 2005, etc.), entropy of vaporization (Chiskos et al., 1998; Kolská et al. 2005), normal boiling temperature (Joback & Reid, 1987; Gani & Constantinou, 1996; Marrero & Gani, 2001), liquid thermal conductivity (Nagvekar & Daubert, 1987), gas thermal conductivity (Chung et al., 1984), gas permeability and diffusion coefficients (Yampolskii et al., 1998), liquid density (Campbell & Thodos, 1985; Sales-Cruz et al., 2010; Shahbaz et al., 2012), surface tension (Brock, 1955; Conte et al., 2008; Awasthi et al., 2010), solubility parameters of fatty acid methyl esters (Lu et al., 2011), flash temperatures (Liaw & Chiu, 2006; Liaw et al., 2011). Large surveys of group contribution methods for enthalpy of vaporization and liquid heat capacity have been presented in references (Zábranský et al., 2003; Kolská, 2004; Kolská et al., 2005; Kolská et. al, 2008; Zábranský et al, 2010a). Group-contribution-based property estimation methods ca be also used to predict the missing UNIFAC group-interaction parameters for the calculation of vapor-liquid equilibrium (Gonzáles at al., 2007b).

Group contribution methods can be used for pure compounds, even inorganic compounds (e.g. Williams, 1997; Briard et al., 2003), organometallic compounds (e.g. Nikitin et al., 2010) and also for mixtures (e.g. Awasthi et al., 2010; Papaioannou et al., 2010; Teixeira et al., 2011; Garcia et al., 2012). Also e.g. estimation of thermodynamic properties of polysacharides was presented (Lobanova et al., 2011). Discussion about determination of properties of polymers has been also published (Satyanarayana et al., 2007; Bogdanic, 2009; Oh & Bae, 2009). Property models based on the group contribution approach for lipid technology have been also presented (Díaz-Tovar et al., 2007).

During last years also models for ionic liquids and their variable properties were developed, e.g. for density, thermal expansion and viscosity of cholinium-derived ionic liquids (Costa et al., 2011; Costa et al., 2012), viscosity (Adamová et al., 2011), the glass-transition temperature and fragility (Gacino et al., 2011), experimental data of mixture with ionic liquid were compared with group contribution methods (Cehreli & Gmehling, 2010) or thermophysical properties were studied (Gardas et al., 2010).

Some of these group contribution methods were developed for only limited number of compounds, for some family of compounds, e.g. for fluorinated olefins (Brown et al., 2010), hydrocarbons (Chickos et al., 1995), fatty acid methyl esters (Lu et al., 2011), etc., most of approaches were established for a wide range of organic compounds.

2.1.1. Group contribution methods by Marrero-Gani

In this chapter for most of estimations the modified group contribution method by Marrero and Gani (Marrero & Gani, 2001; Kolská et al., 2005; Kolská et. al. 2008) was applied, which has been originally developed for estimation of different thermodynamic properties at one temperature only (Constantinou & Gani, 1996; Marrero & Gani, 2001). Determination of group contribution parameters is performed in three levels, primary, secondary and third. At first, all compounds are divided into the primary (first) order group contributions. This primary level uses contributions from simple groups that allow description of a wide variety of organic compounds. Criteria for their creation and calculation have been described (Marrero & Gani, 2001; Kolská et al., 2005; Kolská et al., 2008). The primary level groups, however, are insufficient to capture a proximity effect (they do not implicate an influence of their surroundings) and differences between isomers. Using primary level groups enables to estimate correctly properties of only simple and monofunctional compounds, but the estimation errors for more complex substances are higher. The primary level contributions provide an initial approximation that is improved at the second level and further refined at the third level, if that is possible and necessary. The higher levels (second and third) involve polyfunctional and structural groups that provide more information about a molecular structure of more complex compounds. These higher levels are able to describe more correctly polyfunctional compounds with at least one ring in a molecule, or non-ring chains including more than four carbon atoms in a molecule, and multi-ring compounds with a fused or non-fused aromatic or non-aromatic rings. The differences between some isomers are also able to distinguish by these higher levels. Complex polycyclic compounds or systems of fused aromatic or nonaromatic rings are described by the third order contributions. They are still bigger and more complex than the first, even the second order ones. The multilevel scheme enhances the accuracy, reliability and the range of application of group contribution method for an almost all classes of organic compounds.

After these all three levels the total value of predicted property X is obtained by the summing up of all group contributions, which occur in the molecule. First order groups, second and third order ones, if they are in.

$$X = x_0 + \sum_{i=1}^{n} N_i C_{xi} + \varpi \sum_{j=1}^{m} M_j D_{xj} + z \sum_{k=1}^{o} O_k E_{xk} \quad (1)$$

where X stands the estimated property, x_0 is an adjustable parameter for the relevant property, C_{xi} is the first-order group contribution of type i, D_{xj} is the second-order group contribution of type j, E_{xk} is the third-order group contributions of the type k and N_i, M_j, O_k denote the number of occurrences of individual group contributions. The more detail description of parameters calculation is mentioned in original papers (Marrero & Gani, 2001; Kolská et al., 2005; Kolská et al., 2008)

To develop reliable and accurate group contribution model three important steps should be realized: (i) to collect input database, rather of critically assessed experimental data, from which parameters, group contributions, would be calculated; (ii) to design structural fragments for description of all chemical structures for compounds of input database; (iii) to divide all chemical structures into defined structural fragments correctly. It can be realized either manually, when databases of chemical structures and structural fragments inclusive several members only, either via computer program, when databases contain hundreds of compounds and structural fragments are more complex. To calculate group contribution parameters for thermophysical properties the ProPred program has been used (Marrero, 2002). Description and division chemical structures for other estimations have been made handy. Molecular structures for electronical splitting of all compounds from the basic data set were input in the Simplified Molecular Input Line Entry Specification, so-called the SMILES format (Weininger et al., 1986; Weininger, 1988; Weininger et al., 1989; Weininger, 1990).

For more universal usage of computer fragmentation a suitable computer program has been developed (Kolská & Petrus, 2010). The main goal of the newly developed program is to provide a powerful tool for authors using group contribution methods for automatic fragmentation of chemical structures.

2.2. Estimation of selected physico-chemical properties of compounds

The models for estimation of several physical or physico-chemical properties of pure organic compounds, such as enthalpy of vaporization, entropy of vaporization (Kolská et al., 2005), liquid heat capacity (Kolská et al., 2008) and a Nafion swelling (Randová et al, 2009) is presented below. Most of them are developed to estimate property at constant temperature 298.15 K and at normal boiling temperature (Kolská et al., 2005; Kolská et al., 2008; Randová et al., 2009), liquid heat capacity as a temperature dependent (Kolská et al., 2008). Hitherto unpublished results for estimation of a flash temperature or organic compounds and for determination of reactivation abbility of reactivators of acetylcholinesterase inhibited by inhibitor are presented in this chapter.

2.2.1. Enthalpy of vaporization and entropy of vaporization

Enthalpy of vaporization, ΔH_V, entropy of vaporization, ΔS_V are important thermodynamic quantities of a pure compound, necessary for chemical engineers for modelling of many technological processes with evaporation, for extrapolation and prediction of vapour pressure data, or for estimation of the other thermodynamic properties, e.g. solubility parameters. It can be also used for extrapolation and prediction of vapour pressure data.

There are several methods to determine these properties, experiment-based and model-based. Experiment-based methods, such as calorimetry or gas chromatography, provide generally reliable data of good accuracy. In the case of model-based methods, we can distinguish several groups of methods on the basis of the input information they require. Methods based on the Clausius-Clapeyron equation and vapour pressure data, variable empirical correlations, methods based on the tools of statistical thermodynamics or quantum mechanics. During last decades the group contribution methods are widely used for their universality and simplicity. More rich survey of estimation methods for enthalpy of vaporization is presented in papers (Kolská, 2004; Kolská et al., 2005).

Large databases of critically assessed data have been used for group contribution calculations: data for 831 compounds have been used for estimations at 298.15 K, and data for 589 compounds have been used for estimations at the normal boiling temperature. Organic compounds were divided into several classes (aliphatic and acyclic saturated and unsaturated hydrocarbons, aromatic hydrocarbons, halogenated hydrocarbons, compounds containing oxygen, nitrogen or sulphur atoms and miscellaneous compounds). Especially calorimetrically measured experimental data from the compilation (Majer & Svoboda, 1985) and data from some other sources mentioned in original paper (Kolská et al., 2005) were employed.

Results for estimations of these three properties are presented in the following Tables, Table 1 for enthalpy of vaporization at 298.15 K, Table 2 for enthalpy of vaporization at normal boiling temperature and Table 3 for entropy of vaporization at normal boiling temperature, *NC* means a number of compounds used for development of model and contributions calculation, *NG* is number of applied structural fragments (groups), *AAE* is absolute average error and *ARE* is average relative error (Kolská et al., 2005).

Table 1 shows that values for 831 compounds were used for estimation of enthalpy of vaporization at 298.15 K. When only first level groups were used, the prediction was performed with the *AAE* and the *ARE* of 1.3 kJ/mol and 2.8%, resp. Values of 116 group contributions were calculated at this step. Then, 486 compounds were described by the second order groups. Prediction of these compounds improved after the use of these contributions from the value of 1.3 kJ/mol to 0.8 kJ/mol (from 2.8% to 1.8%) in comparison when using only the first level groups. At the end only 55 compounds were suitable for refining by the third order groups. The results were refined from the values of 1.4 kJ/mol to 1.1 kJ/mol (from 2.5% to 2.1%). The total prediction error was cut down from the value of 1.3 kJ/mol to 1.0 kJ/mol for *AAE* and from 2.8% to 2.2 % for *ARE* after usage of all three-level groups, as it is obvious from this table. A similar pattern of results for other predicted properties are presented in Tables 2 and 3.

Estimation level	*NC*	*NG*	*AAE* / kJ/mol	*ARE* / %
FIRST	**831**	**116**	**1.3**	**2.8**
SECOND	**486**	**91**	**0.8**	**1.8**
486 compounds	**after only the FIRST**		**(1.3)**	**(2.8)**
THIRD	**55**	**15**	**1.1**	**2.1**
55 compounds	**After FIRST + SECOND**		**(1.4)**	**(2.5)**
ALL LEVELS	**831**	**222**	**1.0**	**2.2**

Table 1. Results for estimation of enthalpy of vaporization at 298.15 K (Kolská et al., 2005)

Estimation level	*NC*	*NG*	*AAE* / kJ/mol	*ARE* / %
FIRST	589	111	1.2	3.2
SECOND	377	100	0.9	2.5
377 compounds	after only the FIRST		(1.2)	(3.4)
THIRD	23	14	1.1	2.1
23 compounds	After FIRST + SECOND		(1.3)	(2.7)
ALL LEVELS	589	225	0.9	2.6

Table 2. Results for estimation of enthalpy of vaporization at normal boiling temperature (Kolská et al., 2005)

Estimation level	*NC*	*NG*	*AAE* / J/(K·mol)	*ARE* / %
FIRST	589	111	2.1	2.2
SECOND	377	100	1.8	1.9
377 compounds	after only the FIRST		(2.3)	(2.4)
THIRD	23	14	1.9	1.9
23 compounds	After FIRST + SECOND		(2.5)	(2.5)
ALL LEVELS	589	225	1.7	1.8

Table 3. Results for estimation of entropy of vaporization at normal boiling temperature (Kolská et al., 2005)

As an example of the use of all three levels we have chosen the molecule of 1,1,4,7-tetramethylindane. Its chemical structure is shown in Fig. 3 and its division into individual first, second and third order groups with the result for vaporization enthalpy at 298.15 K is presented in Table 4. When we sum up all group contribution of the first level, we have got value of 64.48 kJ/mol. The first level provides an initial approximation with the relative error of estimated value exceeding 5 % in comparison with experimental value 61.37 kJ/mol. Estimated value of vaporization enthalpy at 298.15 K is then improved at the second level and further refined at the third level, after those the relative error reduced to 1.2 %.

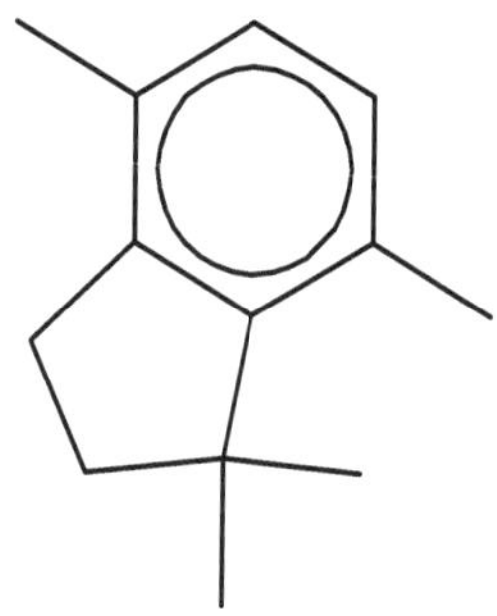

Figure 3. Chemical structure of 1,1,4,7-tetramethylindane

Estimation level	Group fragment no.	Group fragment definition	Its frequency	Group contribution value for ΔHv at 298.15 K / kJ/mol
FIRST	x_0	**Adjustable parameter**	1	**9.672**
	1	CH3	2	2.266
	13	aCH	2	4.297
	15	aC fused with nonaromatic subring	2	6.190
	18	aC-CH3	2	8.121
	107	CH2 (cyclic)	2	4.013
	109	C (cyclic)	1	3.667
Estimated value				64.48
SECOND	55	Ccyc-CH3	2	-1.355
Estimated value				63.59
THIRD	6	aC-CHncyc (fused rings) (n in 0..1)	2	0.279
	19	AROM.FUSED[2]s1s4	1	-0.615
Estimated value				62.13

Table 4. Results for estimation of enthalpy of vaporization at 298.15 K for 1,1,4,7-tetramethylindane, aC means carbon atom in aromatic ring, abbreviation cyc is used for cycle (Kolská et al., 2005)

Group contribution methods by Ducros (Ducros et al., 1980; Ducros et al., 1981; Ducros & Sannier, 1982; Ducros & Sannier, 1984), by Chickos (Chickos et al., 1996), the empirical method, equations nos. 6 and 7 by Vetere (Vetere, 1995) and method by Ma and Zhao (Ma & Zhao, 1993) were used for comparison of results obtained in this work for estimation at 298.15 K and at normal boiling temperature, resp. While the new approach (Kolská et al., 2005) was applied for enthalpy of vaporization at 298.15 K for 831 organic compounds with the *ARE* of 2.2 %, the Ducros´s method could be applied to only 526 substances with the *ARE* of 3.1 % and the Chickos´s one for 800 compounds with the *ARE* of 4.7 %. For comparison of the results of estimation at the normal boiling temperature the new model provided for 589 compounds, the *ARE* was 2.6 % for enthalpy of vaporization and 1.8 % for entropy of vaporization (Kolská et al., 2005), the Vetere´s method was capable of estimating the values of for the same number of compounds with the following results: 4.6 % (Eq. 6, Vetere, 1995) and 3.4 % (Eq. 7, Vetere, 1995), model by Ma and Zhao (Ma & Zhao, 1993) for 549 compounds with the *ARE* of 2.5 %. The error for the enthalpy of vaporization, based on an independent set of various 74 compounds not used for correlation, has been determined to be 2.5%. Group contribution description and values for next usage of readers are presented in original paper (Kolská et al., 2005).

2.2.2. *Liquid heat capacity*

Isobaric heat capacity of liquid C^l_p is an important thermodynamic quantity of a pure compound. Its value must be known for the calculation of an enthalpy difference required for the evaluation of heating and cooling duties. Liquid heat capacity also serves as an input parameter for example in the calculation of temperature dependence of enthalpy of vaporization, for extrapolation of vapour pressure and the related thermal data by their simultaneous correlation, etc.

In work (Kolská et al., 2008) the three-level group contribution method by Marrero and Gani (Marrero & Gani, 2001) mentioned above, which is able to calculate liquid heat capacity at only one temperature 298.15 K, was applied, and this approach has been extended to estimate heat capacity of liquids as a function of temperature. Authors have employed the combination of equation for the temperature dependence of heat capacity and the model by Marrero and Gani to develop new model (Kolská et al., 2008).

For parameter calculation 549 organic compounds of variable families of compounds were taken. In Table 5 are presented results of this estimation. *NG* means number of applied structural groups and *ARE* is the average relative error. More detailed results are presented in original paper (Kolská et al., 2008).

Estimation level	*NG*	*ARE* / %
First	111	1.9
Second	88	1.6
Third	25	1.5

Table 5. Results for estimation of liquid heat capacity in temperature range of pure organic compounds (Kolská et al., 2008)

Also these estimated values were compared with results obtained by other estimation methods (Zábranský & Růžička, 2004; Chickos et al., 1993) for the basic dataset (compounds applied for parameter calculation) and also for 149 additional compounds not used in the parameter calculation (independent set). The first method (Zábranský & Růžička, 2004) was applied for all temperature range, the method proposed by Chickos (Chickos et al., 1993) was only used for temperature 298.15 K with the following results: new model was applied for 404 compounds with *ARE* of 1.5 %, the older method by Zábranský (Zábranský & Růžička, 2004) for the same number of compounds with the *ARE* of 1.8 % and the Chickos's one for 399 compounds with the *ARE* of 3.9 %.

For the heat capacity of liquids authors used recommended data from the compilations by (Zábranský et al., 1996; Zábranský et al., 2001). Because the experimental data are presented permanently, it is necessary to update database of critically assessed and recommended data. Therefore authors's work has been also aimed at updating and extending two publications prepared earlier within the framework of the IUPAC projects (Zábranský et al., 1996; Zábranský et al., 2001). These publications contain recommended data on liquid heat capacities for almost 2000 mostly organic compounds expressed in terms of parameters of correlating equations for temperature dependence of heat capacity. In new work (Zábranský et al., 2010b) authors collected experimental data on heat capacities of pure liquid organic and inorganic compounds that have melting temperature below 573 K published in the primary literature between 1999 and 2006. Data from more than 200 articles are included into the database. Compounds were divided into several families, such as hydrocarbons (saturated, cyclic, unsaturated, aromatic), halogenated hydrocarbons containing atoms of fluorine, chlorine, iodine, bromine, compounds containing oxygen (alcohols, phenols, ethers, ketones, aldehydes, acids, esters, heterocycles, other miscellaneous compounds), compounds containing nitrogen (amines, nitriles, heterocycles, other miscellaneous compounds), compounds containing sulphur (thioles, sulphides, heterocycles) and compounds containing silicon. Also data of organometallic compounds, compounds containing atoms of phosphorus and boron as well as some inorganic compounds were included. Also the list of families of compounds has been extended by a new group denoted as ionic liquids due to an increased interest in physical-chemical properties of these compounds in recent years. Data for approximately 40 ionic liquids were included. Altogether new data for almost 500 compounds, out of them about 250 compounds were not covered the in previous works (Zábranský et al., 1996; Zábranský et al., 2001), were compiled and critically evaluated.

2.2.3. *Nafion swelling*

Prediction of the physical and chemical properties of pure substances and mixtures is a serious problem in the chemical process industries. One of the possibilities for prediction of the properties is the group contribution method. The anisotropic swelling of Nafion 112 membrane in pure organic liquids (solvents) was monitored by an optical method. Nafion is a poly(tetrafluoroethylene) (PTFE) polymer with perfluorovinyl pendant side chains ended by sulfonic acid groups. The PTFE backbone guarantees a great chemical stability in both reducing and oxidizing environments. Nafion membrane is important in chemical industry. It is used in fuel cells, membrane reactors, gas dryers, production of NaOH, etc. (Randová et al., 2009). In many applications Nafion is immersed in liquid, which significantly affects the membrane properties, namely swelling and transport properties of permeates (Randová et al., 2009). The change in the size of the membrane sample is taken as a measure of swelling. All experimental data were presented (Randová et al., 2009) and these results were used as a basis for application of the group contribution method to the relative expansion in equilibrium. From a total of 38 organic liquids under study, 26 were selected as an evaluational set from which the group and structural group contributions were assigned. The remaining 12 compounds were used as the testing set.

Due to limited number of compounds the more complex and known group contribution methods could not been taken. Authors have to develop new group structural fragments. The proposed method utilizes the four kinds of the structural units: constants, C-backbone, functional groups, and molecular geometry (Randová et al., 2009). Constants were presented as alcohols, ketones, ethers, esters, carboxylic acids. As C-backbone were taken groups CH3, -CH2- and >CH-. Functional groups as hydroxyl OH-, carbonyl -C=O and ether –O- and fragments for molecular geometry for cycles and branched chains were taken. The relative expansions A_{exp} (for the drawing direction) and/or B_{exp} (for the perpendicular direction) were calculated from the side lengths of the dry membrane sample (a_{10}, a_{20}, b_{10}, b_{20}) and the side lengths of the swelled membrane sample in equilibrium (a_1, a_2, b_1, b_2) according to the eq. (3). Description of mentioned sizes is presented in Fig. 4.

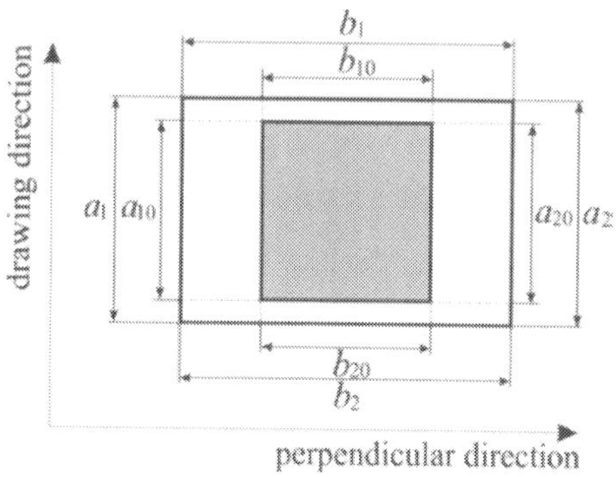

Figure 4. Description of membrane dimensions a_{10}, a_{20}, b_{10}, b_{20} side lengths of the dry membrane and a_1, a_2, b_1, b_2 side lengths of the swelled membrane in equilibrium (Randová et al., 2009)

$$u_{a1} = \frac{a_1 - a_{10}}{a_{10}} \quad , \quad u_{a2} = \frac{a_2 - a_{20}}{a_{20}} \quad , \quad u_{b1} = \frac{b_1 - b_{10}}{b_{10}} \quad , \quad u_{b2} = \frac{b_2 - b_{20}}{b_{20}} \tag{3}$$

Calculation approach is presented in original paper (Randová et al., 2009). Value of ±1.5% in relative expansions was determined to be the experimental error. Maximum differences between the experimental and calculated relative expansions in both sets did not exceed the value of ±3% (Randová et al., 2009).

The values of 13 contributions for individual membrane relative expansions were determined on the basis of experimental data on relative expansion of Nafion membrane. Obtained results are in good agreement with experimental data. Maximum differences between experimental and calculated values are nearly the same, only twice greater than the experimental error.

2.2.4. *Flash temperature of organic compounds*

The flash temperature T_f and lower flammability limit (LFL) are one of the most important variables to consider when designing chemical processes involving flammable substances. These characteristics are not fundamental physical points. Flash temperature is one of the most important variables used to characterize fire and explosion hazard of liquids. The flash temperature is defined as the lowest temperature at which vapour above liquid forms flammable mixture with air at a pressure 101 325 Pa. Usual approach for flash temperature estimation is linear relationship between flash temperature T_f and normal boiling temperature T_b (Dvořák, 1993). Some models for flash temperature were presented earlier (Liaw & Chiu, 2006; Liaw et al., 2011).

In this work to estimate flash temperature of organic compound authors applied the modified group contribution method (Kolská et al., 2005) and calculate group contribution values data for 186 compounds (Steinleitner, 1980) were used. The database for calculation of parameters contains data for aliphatic and acyclic saturated and unsaturated hydrocarbons, aromatic hydrocarbons, alcohols, halogenated hydrocarbons, compounds containing oxygen, nitrogen or sulphur atoms and miscellaneous compounds. To collect more data for development of reliable method was not able due to that all databases collect some values obtained via closed cup type measuring method and others measured by open cup one and data both of methods vary.

Flash temperature was calculated by relationship (4) similar to eq. (1):

$$T_f = T_f^\circ + \sum_{i=1}^{n} N_i C_i + \varpi \sum_{j=1}^{m} M_j D_j + z \sum_{k=1}^{o} O_k E_k \tag{4}$$

where T_f^o is an adjustable parameter, C_i is the first-order group contribution of type i, D_j is the second-order group contribution of type j, E_k is the third-order group contribution of the type k and N_i, M_j, O_k denote the number of occurrences of individual group contributions. Determination of contributions and of adjustable parameters was performed by a three-step regression procedure (Marrero & Gani, 2001). To evaluate the method error the following statistical quantities for each compound, absolute error AE (eq. 5) and relative error ARE (eq. 6) were used:

$$AE[T_f] = \left| (T_f)_{\text{exp}} - (T_f)_{est} \right| \tag{5}$$

$$RE[T_f] = \left(\frac{\left| (T_f)_{\text{exp}} - (T_f)_{est} \right|}{(T_f)_{\text{exp}}} \right) \cdot 100 \tag{6}$$

where subscripts "exp" and "est" mean experimental and estimated value of the flash temperature. 186 compounds from the basic data set were described by the first level group contributions (Kolská et al., 2005). From this large database only 114 compounds could be selected to be described by the original second level groups as defined earlier (Kolská et al., 2005). The total absolute and the relative average errors for all 186 compounds were equal to 6.3 K and 2.0 %. Results for individual estimation levels are presented in Table 6.

Estimation level	NC	AAE / K	ARE / %
FIRST	186	7.9	2.4
SECOND	105	5.7	1.8
THIRD	11	2.9	0.8
ALL LEVELS	186	6.5	2.0

Table 6. Results for Estimation of flash temperature, NC is number of compounds

Individual calculated structural fragments of the first, second and third estimation levels are presented in Tables 7-9, resp.

Structural fragment	Contribution / K	Structural fragment	Contribution / K	Structural fragment	Contribution / K
T_f^o	194.35	aCH	12.39	aC-OH	85.26
CH3	5.38	aC	21.15	CH2Cl	45.56
CH2	13.28	aC	26.84	CHCl	42.83
CH	15.77	aC	31.53	CCl	37.51
C	13.59	aN	25.51	CHCl2	67.42
CH2=CH	11.51	aC-CH3	28.32	CCl3	100.38
CH=CH	34.57	aC-CH2	37.11	aC-Cl	50.70
CH2=C	19.63	aC-CH	37.38	aC-F	53.47
CH=C	27.52	aC-C	19.88	aC-Br	64.97
C=C	29.27	aC-CH=CH2	50.66	-I	78.85
CH#C	14.94	OH	64.22	-Br	59.25
C#C	15.04	-SH	55.62	CH=CH	18.21
-F	2.99	CH2	10.89	CH=C	37.59
-Cl	29.32	CH	22.89	N	52.99
CH2SH	56.33	C	-9.50	O	-3.39

Table 7. Group contribution of the first level for estimation of flash temperature

Structural fragment	Contribution / K	Structural fragment	Contribution / K	Structural fragment	Contribution / K
(CH3)2CH	-1.45	CHm=CHn-Cl (m,n in 0..2)	-0.50	CHcyc-OH	-1.31
(CH3)3C	-3.98	aC-CHn-X (n in 1..2) X: Halogen	1.04	Ccyc-CH3	-0.23
CH(CH3)CH(CH3)	7.68	aC-CHn-OH (n in 1..2)	5.51	>Ncyc-CH3	-1.11E-17
CH(CH3)C(CH3)2	22.69	aC-CH(CH3)2	1.80	AROMRINGs1s2	-2.24
CHn=CHm-CHp=CHk (k,m,n,p in 0..2)	0.53	aC-CF3	0.13	AROMRINGs1s3	1.72
CH3-CHm=CHn (m,n in 0..2)	-2.01	(CHn=C)cyc-CH3 (n in 0..2)	0.46	AROMRINGs1s4	-0.84
CH2-CHm=CHn (m,n in 0..2)	2.23	CHcyc-CH3	-5.70	AROMRINGs1s2s4	-2.84
CHp-CHm=CHn (m,n in 0..2; p in 0..1)	3.78	CHcyc-CH2	17.80	AROMRINGs1s2s4s5	6.22
CHOH	-3.92	CHcyc-CH=CHn (n in 1..2)	8.52	PYRIDINEs3s5	9.98E-18
COH	-4.98	CHcyc-C=CHn (n in 1..2)	-1.20	(CH=CHOCH=CH)cyc	-4.24
CHm(OH)CHn(OH) (m,n in 0..2)	13.41	CHcyc-Cl	1.59	(3 F)	-0.13
				(perFlouro)	2.66E-17

Table 8. Group contribution of the second level for estimation of flash temperature

Structural fragment	Contribution / K	Structural fragment	Contribution / K	Structural fragment	Contribution / K
OH-(CHn)m-OH (m>2, n in 0..2)	-33.36	CHcyc-CHcyc (different rings)	-1.87	AROM.FUSED[2]	8.69
aC-aC (different rings)	-6.04	CH multiring	0.98	AROM.FUSED[4a]	-26.07
aC-CHncyc (fused rings) (n in 0..1)	-4.34	aC-CHm-aC (different rings) (m in 0..2)	12.92		

Table 9. Group contribution of the third level for estimation of flash temperature

2.2.5. Reactivation ability of some reactivators of acetylcholinesterase

In the last years regarding to valid legislation on dangerous compounds it is necessary to know many of important characteristics of chemical compounds. Due to this new models for their estimation were developed. New models for estimation of reactivation ability of reactivators for acetylcholinesterase inhibited by (i) chloropyrifos (*O*,*O*-diethyl *O*-3,5,6-trichloropyridin-2-yl phosphorothioate) as a representative of organophosphate insecticide and by (ii) sarin ((*RS*)-propan-2-yl methylphosphonofluoridate) as a representative of nerve agent is now presented. Both of these family compounds, organophosphate pesticide and nerve agent, are highly toxic and have the same effect to living organisms, which is based on an inhibition of acetylcholinesterase (AChE). New compounds able to reactivate the inhibited AChE, so-called reactivators of AChE, are synthesized. Reactivation ability of these reactivators is studied using standard reactivation *in vitro* test (Kuča & Kassa, 2003). Reactivation ability of reactivators means the percentage of original activity of AChE (Kuča & Patočka, 2004). New models for determination values of reactivation ability of reactivators AChE inhibited by (i) chloropyrifos and (ii) sarin have been developed. Concentration of reactivators was $c=1\cdot10^{-3}$ mol·dm^{-3}. In comparison with previous cases (estimations of thermophysical properties) authors have only less experimental data for development of model (about 20 for each of cases). Due to their long names and complex chemical structures these compounds in this chapter only are presented as their codes taken from original papers (Kuča & Kassa, 2003; Kuča et al., 2003a; Kuča et al., 2003b; Kuča et al., 2003c; Kuča & Patočka, 2004; Kuča & Cabal, 2004a; Kuča & Cabal, 2004b; Kuča et. al., 2006. Data of reactivation ability for these reactivators were given by the mentioned author team (Kuča et al.). Classical group contribution method includes groups describing some central atom, central atom with its bonds, or central atom with its nearest surrounding. However these models commonly used experimental data of hundreds or thousands compounds for parameters calculation. Due to for much small database in these cases it was necessary to design new fragments depending on the molecular structures available compounds. Structural fragments in this work cover larger and more complex part of molecules in comparison with other papers focused to group contribution methods. Reactivation potency is given in the group contribution method by the following relation, eq. 7:

$$R_{\mathrm{p}} = \sum_{i=1}^{n} x_i \cdot R_{\mathrm{p}i} \tag{7}$$

where R_{pi} is value of individual fragment i presented in molecule by which it contributes to total value of R_p, x is number of frequency of this fragment i in molecule. Parameters R_{pi} were obtained by minimization function S_{Rp}, eq. 8:

$$S_{Rp} = \sum_{i=1}^{m} \left(R_{pi,\text{calc}} - R_{pi,\text{exp}} \right)^2 \tag{8}$$

where suffix exp presents experimental data and suffix calc the calculated values of R_p, m is number of compounds in dataset. The results obtained by this new approach were compared with experimental data using the following statistical quantities - an absolute error of individual compounds AE (eq. 9) and the average absolute error of dataset AAE (eq. 10):

$$AE_i = R_{pi,\text{calc}} - R_{pi,\text{exp}} \tag{9}$$

$$AAE = \sum_{i=1}^{m} \left(\frac{\left| R_{pi,\text{calc}} - R_{pi,\text{exp}} \right|}{m} \right) \tag{10}$$

Parameters of new model were calculated from the experimental data of the basic dataset. For model for reactivators AChE inhibited by chloropyrifos the input database included data of reactivation ability R_p for 24 reactivators (K 135, K 078, TO 096, TO 100, K 076, TO 094, TO 063, TO 097, TO 098, K 347, TO 231, K 117, K 074, K 033, K 106, K 107, K 110, K 114, HI-6, K 282, K 283, K 285, K 129, K 099) of concentration $c = 1 \cdot 10^{-3}$ mol·dm^{-3}. Values for 17 groups with the AAE of 1.85 % of R_p were calculated. Designed groups with their calculated values of R_{pi} are presented in Table 10. These calculated parameters were tested on the test set of 5 independent compounds (TO 238, K 111, K 113, Methoxime, K 280) of which experimental data were not applied to group contributions determination. The AAE of R_p prediction for this test-set was 1.45 %. Table 11 presents experimental data and predicted values for these 5 independent compounds. Also illustration of usage of this method for two compounds from this test set is added below.

As it is clear from Table 10 the highest values of contributions are given for fragments P_3, P_7 for monoaromatic reactivators and P_{11}, P_{12} and P_{14} for two aromatic rings in reactivator molecule. On the other hand the smallest contribution (the negative ones) to total value of reactivation ability yields fragments P_5 a P_6 for monoaromatic compounds and P_{16} and P_{17} for two aromatic ring reactivators. These values resulted in fact that reactivation ability of new reactivators for reactivation AChE inhibited by chloropyrifos should be increased by presence of the following functional groups in molecules: another quarternary nitrogen atom in aliphatic ring bonded to aromatic quarternary nitrogen atom, the oxime groups in *para-* or *meta-* positions and presence of other aliphatic rings bonded to aromatic ring in other position than quarternary nitrogen and oxime groups. In all cases it is clear that reactivation ability decreases with presence of cycle ring, double bond and also in a less range with the presence of oxygen atoms presented in molecules. Also *ortho-* position of oxime group does not contribute positively.

no.	Fragment description	R_{pi} / %	no.	Fragment description	R_{pi} / %
P_1	Oxime group (=NOH) in position *o*- due to a quarternary nitrogen N^+ atom in aromatic ring	26.365	P_9	two oxime groups in positions *o*- due to a quarternary nitrogen atom in aromatic ring	26.580
P_2	Oxime group (=NOH) in position *p*- due to a quarternary nitrogen N^+ atom in aromatic ring	15.365	P_{10}	two oxime groups in positions *m*- due to a quarternary nitrogen atom N^+ in aromatic ring	15.737
P_3	Other quarternary nitrogen atom N^+ with 4 CH_x- groups in molecule, in aliphatic ring bonded to nitrogen atom N in aromatic ring	46.792	P_{11}	two oxime groups in positions *p*- due to a quarternary nitrogen atom N^+ in aromatic ring	47.105
P_4	Number of members bonded in aliphatic ring after the group P_3	-1.047	P_{12}	two oxime groups, one in position *o*-, other in position *m*- due to a quarternary nitrogen atom N^+ in aromatic ring	52.105
P_5	Cycle ring bonded to nitrogen atom N in aromatic ring	-10.365	P_{13}	two oxime groups, one in position *o*-, other in position *p*- due to a quarternary nitrogen atom N^+ in aromatic ring	25.842
P_6	Oxygen atom O bonded in aliphatic ring bonded to one aromatic ring	-32.437	P_{14}	two oxime groups, one in position *m*-, other in position *p*- due to a quarternary nitrogen atom N^+ in aromatic ring	56.105
P_7	Presence of other aliphatic ring bonded to aromatic one	88.073	P_{15}	Oxygen atom O bonded in aliphatic ring between two aromatic rings	-5.842
P_8	Number of members bonded in aliphatic ring following group N-CH_x- (nitrogen atom N is a part of aromatic ring), (which are not included in other groups)	-2.344	P_{16}	Cycle between two aromatic rings	-10.474
			P_{17}	Double bond between two aromatic rings	-22.105

Table 10. List of structural fragments and their values for estimation of reactivation ability of reactivators for acetylcholinesterase inhibited by chloropyrifos

Reactivator	$R_{p,exp}$ / %	$R_{p,calc}$ / %	Deviation / %
TO 238	48.00	49.55	1.55
K 111	8.00	5.26	-2.74
K 113	37.00	36.63	-0.37
Methoxime	45.00	47.11	2.11
K 280	4.00	4.48	0.48

Table 11. Results for estimation of reactivation ability of the test dataset of 5 reactivators of acetylcholinesterase inhibited by chloropyrifos

Illustration of new method for reactivation ability prediction of two reactivators (TO 238 and K 280) of which experimental data were not used for parameters calculation follows.

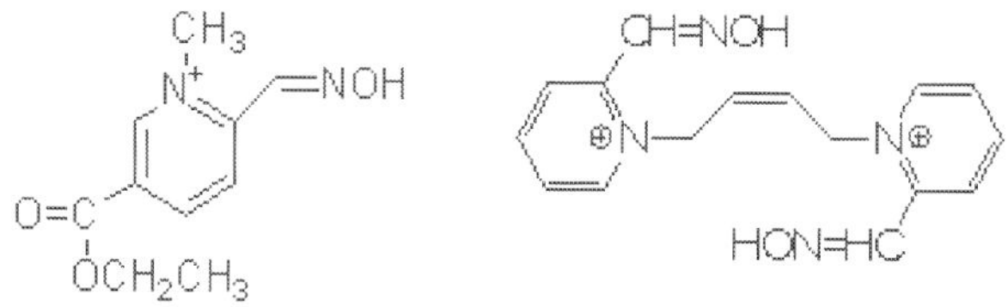

TO 238 K 280

Figure 5. Chemical structure of two reactivators of acetylcholinesterase signed as TO 238 and K 280

Example of usage of the new model for reactivation ability prediction for TO 280 reactivator:

$R_{p,calc}$(TO 238) = $P_1 + 2{\cdot}P_6 + P_7$ = 26.365 + 2·(-32.437) + 88.073 = 49.546 %
$R_{p,exp}$(TO 238) = 48.00 %
$AE = R_{p,calc}$(TO 238) - $R_{p,exp}$(TO 238) = 1.55 %.

Example of usage of the new model for reactivation ability prediction for K 280 reactivator:

$R_{p,calc}$(K 280) = $P_9 + P_{17}$ = 26.580 + (-22.105) = 4.475 %
$R_{p,exp}$(K 280) = 4.00 %
$AE = R_{p,calc}$(K 280) - $R_{p,exp}$(K 280) = 0.48 %.

For model development for reactivators AChE inhibited by sarin the input database included data of reactivation ability R_p for 18 reactivators (K 127, K 128, K 141, K 276, K 311, K 277, K 077, K 142, K 131, K 100, K 233, K 194, K 191, K 067, K 119, K 053, Pralidoxime, HI-6) of concentration $c=1{\cdot}10^{-3}$ mol·dm^{-3} were taken. Due to the smaller database in comparison with the chloropyrifos-inhibited case it was not possible to apply the same structural fragments. Values for 11 new structural different groups with the *AAE* of 3.39 % of R_p have been calculated. Designed groups with their calculated values of R_{pi} are presented in Table 12. These calculated parameters were tested on the test set of 4 independent compounds (TO 055, TO 058, K 197, Obidoxime) of which experimental data were not applied to group contributions determination. The *AAE* of R_p prediction for this test-set was 2.18 %. Table 13 presents experimental data and predicted values for 4 independent compounds.

no.	Fragment description	R_{pi} / %	no.	Fragment description	R_{pi} / %
P_1	Quarternary nitrogen atom N inclusive in aromatic ring	22.50	P_7	Other member of ring between two quarternary nitrogen atoms N^+ or/and bonded at the last quarternary nitrogen atom N^+ of molecule	-6.41
P_2	Presence of oxime group	-31.21	P_8	Presence of oxygen atom O in molecule other than mentioned in the following group	2.16
P_3	*ortho-* position of substituent on aromatic ring	46.03	P_9	Presence of group >C=O in molecule	7.88
P_4	*meta-* position of substituent on aromatic ring	14.49	P_{10}	Presence of group $-NH_x (x = 0, .., 2)$ in molecule	-12.20
P_5	*para-* position of substituent on aromatic ring	40.01	P_{11}	Presence of a double bond between two carbon atoms in a ring between two quarternary nitrogen atoms N^+ in molecule	1.66
P_6	Presence of cycle in a molecule	-10.03			

Table 12. List of structural fragments and their values for estimation of reactivation ability of reactivators for acetylcholinesterase inhibited by sarin

As it is shown in Table 12, the highest and the positive values of group contributions are given for fragments P_1, P_3-P_5, P_8 and P_9. On the other hand the smallest contribution (the negative ones) to the total value of reactivation ability yield fragments P_6, P_7 and P_{10}. Also the value of fragment P_2 for oxime group seems to have a negative effect to the total value but it should be said, that the oxime group has to be summed up with some group for its position on aromatic ring. It results in a fact that the oxime group in *meta-* position has the negative influence to the total value of reactivation ability, on the other hand the total value of R_p increases with oxime group in positions of *ortho-* or *para-*. These values resulted in fact that reactivation ability of new reactivators for reactivation AChE inhibited by sarin should be increased by the presence of the following function groups in molecules: another quarternary nitrogen atom in aromatic ring, the oxime groups in *ortho-* or *para–* positions, presence of oxygen atom or group >C=O in molecule. It is clear that reactivation ability decreases with presence of cycle ring and also with presence of the group NH_x $(x = 0, .., 2)$ in molecules. Also *meta-* position of oxime group, as same as the longer ring $(CH_x)_n$ $(x = 0, .., 2)$ bonded at quarternary nitrogen atoms, that means group P_7, do not contribute positively.

Reactivator	$R_{p,exp}$ / %	$R_{p,calc}$ / %	Deviation / %
TO 055	30.00	32.38	2.38
TO 058	25.00	27.63	2.63
K 197	4.00	4.08	0.08
Obidoxime	41.00	44.70	3.70

Table 13. Results for estimation of reactivation ability of the test dataset of 4 reactivators of acetylcholinesterase inhibited by sarin

Illustration of new method for reactivation ability prediction of two reactivators (TO 055 and TO 058) of which experimental data were not used for parameters calculation follows.

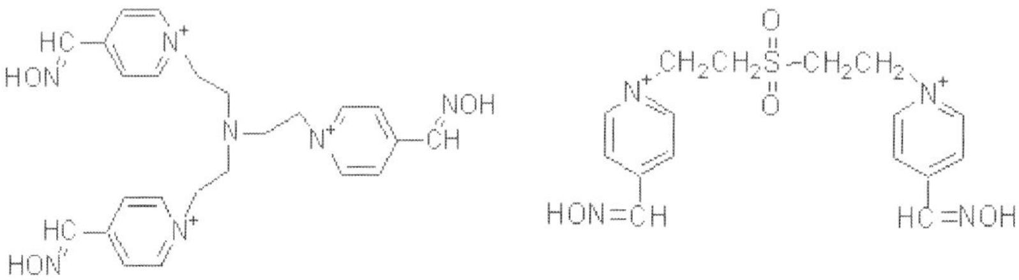

TO 055 TO 058

Figure 6. Chemical structure of two reactivators of acetylcholinesterase signed as TO 055 and TO 058

Example of usage of the new model for reactivation ability prediction for TO 055 reactivator: $R_{p,calc}(\text{TO 055}) = 3 \cdot P_1 + 3 \cdot P_2 + 3 \cdot P_5 + 3 \cdot P_6 + 3 \cdot P_7 + P_{10} = 3 \cdot (22.50) + 3 \cdot (-31.21) + 3 \cdot (40.01) + 3 \cdot (-10.03) + 3 \cdot (-6.41) + (-12.20) = 32.38\ \%$; $R_{p,exp}(\text{TO 055}) = 30.00\ \%$

$AE = R_{p,calc}(\text{TO 055}) - R_{p,exp}(\text{TO 055}) = 2.38\ \%$.

Example of usage of the new model for reactivation ability prediction for TO 055 reactivator: $R_{p,calc}(\text{TO 058}) = 2 \cdot P_1 + 2 \cdot P_2 + 2 \cdot P_5 + 2 \cdot P_6 + 3 \cdot P_7 + 2 \cdot P_8 = 2 \cdot (22.50) + 2 \cdot (-31.21) + 2 \cdot (40.01) + 2 \cdot (-10.03) + 3 \cdot (-6.41) + 2 \cdot (2.16) = 27.63\ \%$; $R_{p,exp}(\text{TO 058}) = 25.00\ \%$

$AE = R_{p,calc}(\text{TO 058}) - R_{p,exp}(\text{TO 058}) = 2.63\ \%$.

As it is clear, in comparison with the previous cases, these models are applicable only for the same inhibitors but for new reactivators of ACHE inhibited by the same inhibitors (the first for chloropyrifos, the second one for sarin). But on the other hand, it can be also used as a tool for easy prediction of reactivation potency of some newly synthesized reactivators without any other *in vitro* standard tests.

3. Conclusion

Most of the industrial applications and products contain a mixture of many components and for the production it is important to know the properties of individual substance and the properties of aggregates. The accomplishments of all of these experiments are too expensive and time-consuming, so the calculation or estimation methods are good way to solve this problem. The group contribution methods are the important and favourible estimation method, because they permit to determine value of property of extant or hypothetic compound. Group contribution methods are the suitable tool for estimation of many physico-chemical quantities of pure compounds and mixtures too as it was showed and confirmed above for some cases. It can be used for estimation of pure compounds, as well as mixtures, for one temperature estimation, as well as for temperature range, etc. The biggest advantage of these methods is they need knowledge only chemical structure of compounds without any other input information. The presented models have been developed for estimation of many variable properties, enthalpy of vaporization, entropy of vaporization, liquid heat capacity, swelling of Nafion, flash temperature and reactivation ability of reactivators of acetylcholinesterase inhibited by organophosphate compounds. Proposed models and their structural fragments, accuracy and reliability depend mainly on frequency of input data and their accuracy, correctness and reliability. The most of presented models of group contribution methods, not only in the cases presented in this chapter, can be applied for the wide variety of organic compounds, when groups describing these molecules are presented. Some of models can be applied from only limited families of compounds due to their parameters were calculated only for limited database of compounds. Group contribution methods can be applied either for estimation or prediction of properties at one temperature or as a temperature function depending on their development. The accuracy of developed models is the higher, the input database is more reliable.

Author details

Zdeňka Kolská
Faculty of Science, J. E. Purkyně University, Usti nad Labem, Czech Republic

Milan Zábranský and Alena Randová
Institute of Chemical Technology, Prague, Czech Republic

Acknowledgement

This work was supported by the GA CR under the project P108/12/G108. Authors also thank to Ing. Michal Karlík from ICT Prague, Czech Republic, for data for flash temperature estimation and Prof. Kamil Kuča from Department of Toxicology, Faculty of Millitary Health Science Hradec Kralove, Czech Republic, for data on reactivation ability.

4. References

Baum, E. J. (1997). *Chemical Property Estimation: Theory and Application*, CRC Press LLC, ISBN 978-0873719384, Boca Raton, USA

Poling, B. E.; Prausnitz, J. M. & O´Connell, J. P. (2001). *The Properties of Gases and Liquids*, fifth edition. McGraw-Hill, ISBN 0-07-011682-2, New York, USA

González, H. E.; Abildskov, J.; & Gani, R. (2007). Computer-aided framework for pure component properties and phase equilibria prediction for organic systems. *Fluid Phase Equilibria*, Vol.261, No.1-2, pp. 199–204, ISSN 0378-3812

Majer, V.; Svoboda, V. & Pick, J. (1989). *Heats of Vaporization of Fluids*. Elsevier, ISBN 0-444-98920-X. Amsterdam, Netherlands

Majer, V. & Svoboda V. (1985). *Enthalpies of Vaporization of Organic Compounds, Critical Review and Data Compilation*. IUPAC. Chemical Data Series No. 32, Blackwell, Oxford, ISBN 0-632-01529-2

NIST database. http://webbook.nist.gov/chemistry/

Chickos, J. S.; Acree, W. E. Jr. & Liebman J. F. (1998). Phase Change Enthalpies and Entropies. In: *Computational Thermochemistry: Prediction and Estimation of Molecular* Thermodynamics. D. Frurip and K. Irikura, (Eds.), ACS Symp. Ser. 677, pp. 63-91, Washington, D. C.

Lydersen, A. L. (1955). *Estimation of Critical Properties of Organic Compounds*. Eng. Exp. Stn. rept. 3; University of Wisconsin College of Engineering: Madison, WI

Ambrose, D. (1978). *Correlation and Estimation of Vapour-Liquid Critical Properties. I. Critical Temperatures of Organic Compounds*. National Physical Laboratory, Teddington: NPL Rep. Chem. 92

Ambrose, D. (1979). *Correlation and Estimation of Vapour-Liquid Critical Properties. II. Critical Pressures and Volumes of Organic Compounds*. National Physical Laboratory, Teddington: NPL Rep. Chem. 98

Joback, K. G. & Reid, R. C. (1987). Estimation of pure-component properties from group-contributions. *Chemical Engineering Communications*, Vol.57, No.1-6, pp. 233-243, ISSN 0098-6445

Gani, R. & Constantinou, L. (1996). Molecular structure based estimation of properties for process design. *Fluid Phase Equilibria*, Vol.116, No.1-2, pp. 75-86, ISSN 0378-3812

Marrero J. & Gani R. (2001). Group-contribution based estimation of pure component properties. *Fluid Phase Equilibria*, Vol.183, Special Issue, pp. 183-208, ISSN 0378-3812

Brown, J. S.; Zilio, C. & Cavallini, A. (2010). Thermodynamic properties of eight fluorinated olefins. *International Journal of Refrigeration*, Vol.33, No.2, pp. 235-241, ISSN 0140-7007

Monago, K. O. & Otobrise, C. (2010). Estimation of pure-component properties of fatty acid sand esters from group contributions. *Journal of Chemical Society of Nigeria*, Vol.35, No.2, pp. 142-148

Sales-Cruz, M.; Aca-Aca, G.; Sanchez-Daza, O. & Lopez-Arenas, T. (2010). Predicting critical properties, density and viscosity of fatty acids, triacylglycerols and methyl esters by group contribution methods. *Computer-Aided Chemical Engineering*, Vol.28, pp. 1763-1768

Manohar, B. & Udaya Sankar, K. (2011). Prediction of solubility of Psoralea corylifolia L. Seed extract in supercritical carbon dioxide by equation of state models. *Theoretical Foundations of Chemical Engineering*, Vol.45, No.4, pp. 409-419, ISSN 0040-5795

Garcia, M.; Alba, J.; Gonzalo, A.; Sanchez, J. L. & Arauzo, J. (2012). Comparison of Methods for Estimating Critical Properties of Alkyl Esters and Its Mixtures. *Journal of Chemical & Engineering Data*, Vol.57, No.1, pp. 208-218, ISSN 0021-9568

Pereda, S.; Brignole, E. & Bottini, S. (2010). Equations of state in chemical reacting systems. In: *Applied Thermodynamics of Fluids*, Goodwin, A. R. H.; Sengers, J. V. & Peters, C. J. (Eds.), 433-459, Royal Society of Chemistry; 1st Ed., ISBN 978-1847558060, Cambridge, UK

Schmid, B. & Gmehling, J. (2012). Revised parameters and typical results of the VTPR group contribution equation of state. *Fluid Phase Equilibria*, Vol.317, pp. 110-126, ISSN 0378-3812

Constantinou, L. & Gani, R. (1994). New Group Contribution Method for Estimating Properties of Pure Compounds. *AIChE Journal*, Vol.40, No.10, pp. 1697-1710, ISSN ISSN 0001-1541

Tochigi, K.; Kurita, S.; Okitsu, Y.; Kurihara, K. & Ochi, K. (2005). Measurement and Prediction of Activity Coefficients of Solvents in Polymer Solutions Using Gas Chromatography and a Cubic-Perturbed Equation of State with Group Contribution. *Fluid Phase Equilibria*, Vol.228, No. Special Issue, pp. 527-533, ISSN 0378-3812

Tochigi, K. & Gmehling, J. (2011). Determination of ASOG Parameters-Extension and Revision. *Journal of Chemical Engineering of Japan*, Vol.44, No.4, pp. 304-306, ISSN 0021-9592

Miller, D. G. (1964). Estimating Vapor Pressures-Comparison of Equations. *Industrial and Engineering Chemistry*, Vol.56, No.3, pp. 46-&, ISSN 0019-7866

Conte, E.; Martinho, A.; Matos, H. A.; & Gani, R. (2008). Combined Group-Contribution and Atom Connectivity Index-Based Methods for Estimation of Surface Tension and Viscosity. *Industrial & Engineering Chemistry Research*, Vol.47, No.20, pp. 7940–7954, ISSN 0888-5885

Reichenberg, D. (1975). New Methods for Estimation of Viscosity Coefficients of Pure Gases at Moderate Pressures (With Particular Reference To Organic Vapors). *AIChE Journal*, Vol.21, No.1, 181-183, ISSN 0001-1541

Ruzicka, V. & Domalski, E. S. (1993a). Estimation of the Heat-Capacities of Organic Liquids as a Function of Temperature Using Group Additivity. 1. Hydrocarbon Compounds. *Journal of Physical and Chemical Reference Data*, Vol.22, No.3, pp. 597-618, ISSN 0047-2689.

Ruzicka, V. & Domalski, E. S. (1993b). Estimation of the Heat-Capacities of Organic Liquids as a Function of Temperature Using Group Additivity. 2. Compounds of Carbon, Hydrogen, Halogens, Nitrogen, Oxygen, and Sulfur. *Journal of Physical and Chemical Reference Data*, Vol.22, No.3, pp. 619-657, ISSN 0047-2689

Kolská, Z.; Kukal, J.; Zábranský, M. & Růžička, V. (2008). Estimation of the Heat Capacity of Organic Liquids as a Function of Temperature by a Three-Level Group Contribution Method. *Industrial & Engineering Chemistry Research*, Vol.47, No.6, pp. 2075-2085, ISSN 0888-5885

Chickos, J. S.; Hesse, D. G.; Hosseini, S.; Liebman, J. F.; Mendenhall, G. D.; Verevkin, S. P.; Rakus, K.; Beckhaus, H.-D. & Ruechardt, C. (1995). Enthalpies of vaporization of some highly branched hydrocarbons. *Journal of Chemical Thermodynamics*, Vol.27, No.6, pp. 693-705, ISSN 0021-9614

Chickos, J. S. & Wilson, J. A. (1997). Vaporization Enthalpies at 298.15K of the n-Alkanes from C21-C28 and C30. *Journal of Chemical and Engineering Data*, Vol.42, No.1, pp. 190-197, ISSN 0021-9568

Kolská, Z.; Růžička, V. & Gani, R. (2005). Estimation of the Enthalpy of Vaporization and the Entropy of Vaporization for Pure Organic Compounds at 298.15 K and at Normal Boiling Temperature by a Group Contribution Method. *Industrial & Engineering Chemistry Research*, Vol.44, No.22, pp. 8436-8454, ISSN 0888-5885

Nagvekar, M. & Daubert, T. E. (1987). A Group Contribution Method for Liquid Thermal-Conductivity. *Industrial & Engineering Chemistry Research*, Vol.26, No.7, pp. 1362-1365, ISSN 0888-5885

Chung, T. H.; Lee, L. L. & Starling, K. E. (1984). Applications of Kinetic Gas Theories and Multiparameter Correlation for Prediction of Dilute Gas Viscosity and Thermal-Conductivity. *Industrial & Engineering Chemistry Fundamentals*, Vol.23, No.1, pp. 8-13, ISSN 0196-4313

Yampolskii, Y.; Shishatskii, S.; Alentiev, A. & Loza, K. (1998). Group Contribution Method for Transport Property Predictions of Glassy Polymers: Focus on Polyimides and Polynorbornenes. *Journal of Membrane Science*, Vol.149, No.2, pp. 203-220, ISSN 0376-7388

Campbell, S. W. & Thodos, G. (1985). Prediction of Saturated-Liquid Densities and Critical Volumes for Polar and Nonpolar Substances. Journal of Chemical and Engineering Data, Vol.30, No.1, pp. 102-111, ISSN 0021-9568

Shahbaz, K.; Baroutian, S.; Mjalli, F. S.; Hashim, M. A. & AlNashef, I. M. (2012). Densities of ammonium and phosphonium based deep eutectic solvents: Prediction using artificial intelligence and group contribution techniques. *Thermochimica Acta*, Vol.527, pp. 59-66, ISSN 0040-6031

Brock, J. R. & Bird, R. B. (1955). Surface Tension and the Principle of Corresponding States. *AIChE Journal*, Vol.1, No.2, pp. 174-177, ISSN 0001-1541

Awasthi, A.; Tripathi, B. S. & Awasthi, A. (2010). Applicability of corresponding-states group-contribution methods for the estimation of surface tension of multicomponent liquid mixtures at 298.15 K. *Fluid Phase Equilibria*, Vol.287, No.2, pp. 151-154, ISSN 0378-3812

Lu, X.; Yang, Y. & Ji, J. (2011). Application of solubility parameter in solubility study of fatty acid methyl esters . *Zhongguo Liangyou Xuebao*, Vol.26, No.6, pp. 60-65

Liaw, H. & Chiu, Y. (2006). A general model for predicting the flash point of miscible mixtures. *Journal of Hazardous Materials*, Vol.137, No.1, pp. 38-46, ISSN 0304-3894

Liaw, H.; Gerbaud, V. & Li, Y. (2011). Prediction of miscible mixtures flash-point from UNIFAC group contribution methods. *Fluid Phase Equilibria*, Vol.300, No.1-2, pp. 70-82, ISSN 0378-3812

Zábranský, M.; Růžička, V. & Malijevský, A. (2003). Odhadové metody tepelných kapacit čistých kapalin. *Chemické Listy*, Vol.97, No.1, pp. 3-8, ISSN 0009-2770

Kolská, Z. (2004). Odhadové metody pro výparnou entalpii. *Chemické Listy*, Vol.98, No.6, pp. 328-334, ISSN 0009-2770

Zábranský, M.; Kolská, Z.; Růžička, V. & Malijevský, A. (2010a). The Estimation of Heat Capacities of Pure Liquids. In *Heat capacities: liquids, solutions and vapours*. T.M. Letcher & E. Wilhelm (Eds.), 421-435, The Royal Society of Chemistry, ISBN 978-0-85404-176-3, London

Gonzalez, H. E.; Abildskov, J.; Gani, R.; Rousseaux, P. & Le Bert, B. (2007). A Method for Prediction of UNIFAC Group Interaction Parameters. *AIChE Journal*, Vol. 53, No.6, pp. 1620-32, ISSN 0001-1541

Williams, J. D. (1997). *Prediction of melting and heat capacity of inorganic liquids by the method of group contributions*. Thesis, New Mexico State Univ., Las Cruces, NM, USA

Briard, A. J.; Bouroukba, M.; Petitjean, D. & Dirand, M. (2003). Models for estimation of pure n-alkanes' thermodynamic properties as a function of carbon chain length. *Journal of Chemical and Engineering Data*, Vol.48, No.6, pp. 1508-1516, ISSN 0021-9568

Nikitin, E. D.; Popov, A. P.; Yatluk, Y. G. & Simakina, V. A. (2010). Critical Temperatures and Pressures of Some Tetraalkoxytitaniums. *Journal of Chemical and Engineering Data*, Vol.55, No.1, pp. 178-183, ISSN 0021-9568

Papaioannou, V.; Adjiman, C. S.; Jackson, G. & Galindo, A. (2010). Group Contribution Methodologies for the Prediction of Thermodynamic Properties and Phase Behavior in Mixtures. In: *Molecular Systems Engineering*. Pistikopoulos E. N.; Georgiadis, M. C.; Due, V.; Adjiman, C. S.; Galindo, A. (Eds.), 135-172, Wiley-VCH, ISBN 978-3-527-31695-3, Weinheim

Teixeira, M. A.; Rodriguez, O.; Mota, F. L.; Macedo, E. A. & Rodrigues, A. E. (2010). Evaluation of Group-Contribution Methods To Predict VLE and Odor Intensity of Fragrances. *Industrial & Engineering Chemistry Research*, Vol.50, No.15, pp. 9390-9402, ISSN 0888-5885

Lobanova, O.; Mueller, K.; Mokrushina, L. & Arlt, W. (2011). Estimation of Thermodynamic Properties of Polysaccharides. *Chemical Engineering & Technology*, Vol.34, No.6, pp. 867-876, ISSN 0930-7516

Satyanarayana, K. C.; Gani, R.; & Abildskov, J. (2007). Polymer property modeling using grid technology for design of structured products. *Fluid Phase Equilibria*, Vol.261, No.1-2, pp. 58–63, ISSN 0378-3812

Bogdanic, G. (2009). Additive Group Contribution Methods for Predicting the Properties of Polymer Systems. In: *Polymeric Materials*. Nastasovic, A. B. & Jovanovic, S. M. (Eds.), 155-197, Transworld Research Network, ISSN 978-81-7895-398-4, Kerala

Oh, S. Y. & Bae, Y. C. (2009). Group contribution method for group contribution method for estimation of vapor liquid equilibria in polymer solutions. *Macromolecular Research*, Vol.17, No.11, pp. 829-841, ISSN 1598-5032

Díaz-Tovar, C.; Gani, R. & Sarup, B. (2011). Lipid technology: Property prediction and process design/analysis in the edible oil and biodiesel industries. *Fluid Phase Equilibria*, Vol.302, No.1-2, pp. 284–293, ISSN 0378-3812

Costa, A. J. L.; Esperanca, J. M. S. S.; Marrucho, I. M. & Rebelo, L. P. N. (2011). Densities and Viscosities of 1-Ethyl-3-methylimidazolium n-Alkyl Sulfates. *Journal of Chemical & Engineering Data*, Vol.56, No.8, pp. 3433-3441, ISSN 0021-9568

Costa, A. J. L.; Soromenho, M. R. C.; Shimizu, K.; Marrucho, I. M.; Esperanca, J. M. S. S.; Lopes, J. N. C. & Rebelo, L. P. N. (2012). Density, Thermal Expansion and Viscosity of Cholinium-Derived Ionic Liquids. *Chemphyschem: a European journal of chemical physics and physical chemistry*, Ahead of Print, ISSN 1439-7641

Adamova, G.; Gardas, Ramesh L.; Rebelo, L. P. N.; Robertson, A. J. & Seddon, K. R. (2011). Alkyltrioctylphosphonium Chloride Ionic Liquids: Synthesis and Physicochemical Properties. *Dalton Transactions*, Vol.40, No.47, pp. 12750-12764, ISSN 1477-9226

Gacino, F. M.; Regueira, T.; Lugo, L.; Comunas, M. J. P. & Fernandez, J. (2011). Influence of Molecular Structure on Densities and Viscosities of Several Ionic Liquids. *Journal of Chemical & Engineering Data*, Vol.56, No.12, pp. 4984-4999, ISSN 0021-9568

Cehreli, S. & Gmehling, J. (2010). Phase Equilibria for Benzene-Cyclohexene and Activity Coefficients at Infinite Dilution for the Ternary Systems with Ionic Liquids. *Fluid Phase Equilibria*, Vol.295, No.1, pp. 125-129, ISSN 0378-3812

Gardas, R. L.; Ge, R.; Goodrich, P.; Hardacre, C.; Hussain, A. & Rooney, D. W. (2010). Thermophysical Properties of Amino Acid-Based Ionic Liquids. *Journal of Chemical & Engineering Data*, Vol.55, No.4, pp. 1505-1515, ISSN 0021-9568

Marrero J. (2002). Programm *ProPred*, Version 3.5, Jorge Marrero, Department of Chemical Engineering, DTU Denmark, released date: May 15, 2002

Weininger, D.; Weininger, A. & Weininger, J. (1986). Smiles. A Modern Chemical Language and Information System. *Chemical Design Automation News*, Vol.1, No.8, pp. 2-15, ISSN 0886-6716

Weininger, D. (1988). Smiles. A Chemical Language and Information-System. 1. Introduction to Metodology and Encoding Rules. *Journal of Chemical Information and Computer Sciences*, Vol.28, No.1, pp. 31-36, ISSN 0095-2338

Weininger, D.; Weininger, A. & Weininger, J. (1989). Smiles. 2. Algorithm for Generation of Unique Smiles Notation. *Journal of Chemical Information and Computer Sciences*, Vol.29, No.2, pp. 97-101, ISSN 0095-2338

Weininger, D. (1990). Smiles. 3. Depict – Graphical Depiction of Chemical Structures. *Journal of Chemical Information and Computer Sciences*, Vol.30, No.3, pp. 237-243, ISSN 0095-2338

Kolská, Z. & Petrus P. (2010). Tool for group contribution methods – computational Fragmentation *Collection of Czechoslovak Chemical Communications*, Vol.75, No.4, pp. 393–404, ISSN 0010-0765

Randová, A.; Bartovská, L.; Hovorka, Š.; Poloncarzová, M.; Kolská, Z. & Izák, P. (2009). Application of the Group Contribution Approach to Nafion Swelling. *Journal of Applied Polymer Science*, Vol.111, No.4, pp. 1745-1750, ISSN 0021-8995

Ducros, M.; Gruson, J. F. & Sannier H. (1980). Estimation of enthalpies of vaporization for liquid organic compounds. Part 1. Applications to alkanes, cycloalkanes, alkenes, benzene hydrocarbons, alkanethiols, chloro- and bromoalkanes, nitriles, esters, acids, and aldehydes. *Termochimica Acta*, Vol.36, No.1, pp. 39-65, ISSN 0040-6031

Ducros, M.; Gruson, J. F. & Sannier H. (1981). Estimation of the enthalpies of vaporization of liquid organic compounds. Part 2. Ethers, thioalkanes, ketones and amines. *Termochimica Acta*, Vol.44, No.2, pp. 131-140, ISSN 0040-6031

Ducros, M. & Sannier, H. (1982). Estimation of the enthalpies of vaporization of liquid compounds. Part 3. Application to unsaturated hydrocarbons. *Termochimica Acta*, Vol.54, No.1-2, pp. 153-157, ISSN 0040-6031

Ducros, M. & Sannier, H. (1984). Determination of vaporization enthalpies of liquid organic compounds. Part 4. Application to organometallic compounds. *Termochimica Acta*, Vol.75, No.3, pp. 329-340, ISSN 0040-6031

Vetere A. (1995). Methods to predict the vaporization enthalpies at the normal boiling temperature of pure compounds revisited. *Fluid Phase Equilibria*, Vol.106, No.1-2, pp. 1-10 , ISSN 0378-3812

Ma, P. & Zhao, X. (1993). Modified Group Contribution Method for Predicting the Entropy of Vaporization at the Normal Boiling Point. *Industrial & Engineering Chemistry Research*, Vol.32, No.12, pp. 3180-3183, ISSN 0888-5885

Zábranský, M. & Růžička, V. (2004). Estimation of the Heat Capacities of Organic Liquids as a Function of Temperature Using Group Additivity. An Amendment. *J. Phys. Chem. Ref. Data*, Vol.33, No.4, pp. 1071-1081, ISSN 0047-2689

Chickos, J. S.; Hesse, D. G. & Liebman, J. F. (1993). A Group Additivity Approach for the Estimation of Heat-Capacities of Organic Liquids and Solids at 298 K. *Structural Chemistry*, Vol.4, No.4, pp. 261-269, ISSN 1040-0400

Zábranský, M.; Růžička, V.; Majer, V. & Domalski E. S. (1996). Heat capacity of liquids: Volume II. Critical review and recommended values. *Journal of Physical and Chemical Reference Data*, Monograph, 815-1596, American Chemical Society: Washington, D.C., ISSN 1063-0651

Zábranský, M.; Růžička, V. & Domalski E. S. (2001). Heat Capacity of Liquids: Critical Review and Recommended Values. Supplement I. *Journal of Physical and Chemical Reference Data*, Vol.30, No.5, pp. 1199-1689

Zábranský, M.; Kolská, Z.; Růžička, V. & Domalski, E. S. (2010b). Heat Capacity of Liquids: Critical Review and Recommended Values. Supplement II. *Journal of Physical and Chemical Reference Data*, Vol.39, No.1, pp. 013103/1-013103/404, ISSN 0047-2689

Dvořák, O. (1993). Estimation of the flash point of flammable liquids. *Chemický průmysl*, Vol.43, No.5, pp. 157-158., ISSN 0009-2789

Steinleitner, H. G. (1980). *Tabulky hořlavých a nebezpečných látek*, Transl. Novotný, V.; Benda, E. Svaz požární ochrany ČSSR , 1st ed. Praha

Kuča, K. & Kassa, J. (2003). A Comparison of the Ability of a New Bisperidinium Oxime-1,4-(hydroxyiminomethylpyridinium)-4-(4-carbamoylpyridinium)butane Dibromide and Currently used Oximes to Reactivate Nerve Agent-Inhibited Rat Brain Acetylcholinesterase by In Vitro Methods. *Journal of Enzyme Inhibition*, Vol.18, No.6, pp. 529-533, ISSN 8755-5093

Kuča, K.; Bielavský, J.; Cabal, J. & Kassa, J. (2003a). Synthesis of a New Reactivator of Tabun-Inhibited Acetylcholnesterase. *Bioorganic & Medicinal Chemistry Letters*, Vol.13, No.20, pp. 3545-3547, ISSN 0960-894X

Kuča, K.; Bielavský, J.; Cabal, J. & Bielavská, M. (2003b). Synthesis of a Potential Reactivator of Acetylcholinesterase-(1-(4-hydroxyiminomethylpyridinium)-3-(carbamoylpyridinium)-propane dibromide. Tetrahedron Letters, Vol.44, No.15, pp. 3123-3125, ISSN 0040-4039

Kuča, K.; Patočka, J. & Cabal, J. (2003c). Reactivation of Organophosphate Inhibited Acetyylcholinesterase Activity by α,ω-bis-(4-hydroxyiminomethylpyridinium)alkanes in vitro. *Journal of Applied Biomedicine*, Vol.1, No.4, pp. 207-211. ISSN 1214-0287

Kuča K. & Patočka J. (2004). Reactivation of Cyclosarine-Inhibited Rad Brain Acetylcholinesterase by Pyridinium-Oximes. *Journal of Enzyme Inhibition and Medicinal Chemistry*. Vol.19, No.1, pp. 39-43, ISSN 1475-6366

Kuča K. & Cabal, J. (2004a). In Vitro Reaktivace Acetylcholinesterázy Inhibované O-Isopropylmethylfluorofofonátem užitím biskvarterního oximu HS-6. *Česká a Slovenská Farmacie*, Vol.53, No.2, pp. 93-95, ISSN 1210-7816

Kuča K. & Cabal, J. (2004b). In Vitro Reactivation of Tabun-Inhibited Acetylcholinesterase Using New Oximes – K027, K005, K033 and K048. *Central European Journal of Public Health*, Vol.12, No.Suppl, pp. S59-S61, ISSN 1210-7778

Kuča K.; Jun, D. & Musílek, K. (2006). Structural Requirements of Acetylcholinesterase Reactivators. *Mini-Reviews in Medicinal Chemistry*, Vol.6, No.3, pp. 269-277, ISSN 1389-5575/06

Chapter 7

Thermodynamic Properties and Applications of Modified van-der-Waals Equations of State

Ronald J. Bakker

Additional information is available at the end of the chapter

http://dx.doi.org/10.5772/50315

1. Introduction

Physical and chemical properties of natural fluids are used to understand geological processes in crustal and mantel rock. The fluid phase plays an important role in processes in diagenesis, metamorphism, deformation, magmatism, and ore formation. The environment of these processes reaches depths of maximally 5 km in oceanic crusts, and 65 km in continental crusts, e.g. [1, 2], which corresponds to pressures and temperatures up to 2 GPa and 1000 °C, respectively. Although in deep environments the low porosity in solid rock does not allow the presence of large amounts of fluid phases, fluids may be entrapped in crystals as fluid inclusions, i.e. nm to µm sized cavities, e.g. [3], and fluid components may be present within the crystal lattice, e.g. [4]. The properties of the fluid phase can be approximated with equations of state (Eq. 1), which are mathematical formula that describe the relation between intensive properties of the fluid phase, such as pressure (p), temperature (T), composition (x), and molar volume (V_m).

$$p\left(T, V_m, x\right) \tag{1}$$

This pressure equation can be transformed according to thermodynamic principles [5], to calculate a variety of extensive properties, such as entropy, internal energy, enthalpy, Helmholtz energy, Gibbs energy, et al., as well as liquid-vapour equilibria and homogenization conditions of fluid inclusions, i.e. dew point curve, bubble point curve, and critical points, e.g. [6]. The partial derivative of Eq. 1 with respect to temperature is used to calculate total entropy change (dS in Eq. 2) and total internal energy change (dU in Eq. 3), according to the Maxwell's relations [5].

$$dS = \left(\frac{\partial p}{\partial T}\right)_{V, n_T} dV \tag{2}$$

INTECH
open science | open minds

$$dU = \left[T \cdot \left(\frac{\partial p}{\partial T} \right)_{V,n_T} - p \right] dV \tag{3}$$

where n_T is the total amount of substance in the system. The enthalpy (H) can be directly obtained from the internal energy and the product of pressure and volume according to Eq. 4.

$$H = U + p \cdot V \tag{4}$$

The Helmholtz energy (A) can be calculated by combining the internal energy and entropy (Eq. 5), or by a direct integration of pressure (Eq. 1) in terms of total volume (Eq. 6).

$$A = U - TS \tag{5}$$

$$dA = -pdV \tag{6}$$

The Gibbs energy (G) is calculated in a similar procedure according to its definition in Eq. 7.

$$G = U + p \cdot V - T \cdot S \tag{7}$$

The chemical potential (μ_i) of a specific fluid component (i) in a gas mixture or pure gas (Eq. 8) is obtained from the partial derivative of the Helmholtz energy (Eq. 5) with respect to the amount of substance of this component (n_i).

$$\mu_i = \left(\frac{\partial A}{\partial n_i} \right)_{T,V,n_j} \tag{8}$$

The fugacity (f) can be directly obtained from chemical potentials (Eq. 9) and from the definition of the fugacity coefficient (φ_i) with independent variables V and T (Eq. 10).

$$RT \ln \left(\frac{f_i}{f_i^0} \right) = \mu_i - \mu_i^0 \tag{9}$$

where μ_i^0 and f_i^0 are the chemical potential and fugacity, respectively, of component i at standard conditions (0.1 MPa).

$$RT \ln \varphi_i = \int_V^\infty \left[\left(\frac{\partial p}{\partial n_i} \right)_{T,V,n_j} - \frac{RT}{V} \right] dV - RT \ln z . \tag{10}$$

where φ_i and z (compressibility factor) are defined according to Eq. 11 and 12, respectively.

$$\varphi_i = \frac{f_i}{x_i \cdot p} \tag{11}$$

$$z = \frac{pV}{n_T RT} \tag{12}$$

2. Two-constant cubic equation of state

The general formulation that summarizes two-constant cubic equations of state according to van der Waals [7], Redlich and Kwong [8], Soave [9], and Peng and Robinson [10] is illustrated in Eq. 13 and 14, see also [11]. In the following paragraphs, these equations are abbreviated with *Weos, RKeos, Seos,* and *PReos*.

$$p = \frac{RT}{V_m - \zeta_1} - \frac{\zeta_2}{V_m \cdot (V_m + \zeta_3) + \zeta_4 \cdot (V_m - \zeta_4)} \tag{13}$$

$$p = \frac{n_T RT}{V - n_T \zeta_1} - \frac{n_T^2 \zeta_2}{V^2 + n_T \zeta_3 V + n_T \zeta_4 V - n_T^2 \zeta_4^2} \tag{14}$$

where p is pressure (in MPa), T is temperature (in Kelvin), R is the gas constant (8.3144621 J·mol^{-1}K^{-1}), V is volume (in cm^3), V_m is molar volume (in cm^3·mol^{-1}), n_T is the total amount of substance (in mol). The parameters ζ_1, ζ_2, ζ_3, and ζ_4 are defined according to the specific equations of state (Table 1), and are assigned specific values of the two constants a and b, as originally designed by Waals [7]. The a parameter reflects attractive forces between molecules, whereas the b parameter reflects the volume of molecules.

	W	RK	S	PR
ζ_1	b	b	b	b
ζ_2	a	$a \cdot T^{-0.5}$	a	a
ζ_3	-	b	b	b
ζ_4	-	-	-	b

Table 1. Definitions of ζ_1, ζ_2, ζ_3, and ζ_4 according to van der Waals (W), Redlich and Kwong (RK), Soave (S) and Peng and Robinson (PR).

This type of equation of state can be transformed in the form of a cubic equation to define volume (Eq. 15) and compressibility factor (Eq. 16).

$$a_0 V^3 + a_1 V^2 + a_2 V + a_3 = 0 \tag{15}$$

$$b_0 z^3 + b_1 z^2 + b_2 z + b_3 = 0 \tag{16}$$

where a_0, a_1, a_2, and a_3 are defined in Eq. 17, 18, 19, and 20, respectively; b_0, b_1, b_2, and b_3 are defined in Eq. 21, 22, 23, and 24, respectively.

$$a_0 = p \tag{17}$$

$$a_1 = n_T p \cdot (\zeta_3 + \zeta_4 - \zeta_1) - n_T RT \tag{18}$$

$$a_2 = -n_T^2 p \cdot \left(\zeta_4^2 + \zeta_1\zeta_3 + \zeta_1\zeta_4\right) - n_T^2 RT \cdot \left(\zeta_3 + \zeta_4\right) + n_T^2\zeta_2 \tag{19}$$

$$a_3 = n_T^3 p \cdot \zeta_1\zeta_4^2 + n_T^3 RT \cdot \zeta_4^2 - n_T^3 \cdot \zeta_1\zeta_2 \tag{20}$$

$$b_0 = \left(\frac{RT}{p}\right)^3 \tag{21}$$

$$b_1 = \left(\frac{RT}{p}\right)^2 \cdot \left(\zeta_3 + \zeta_4 - \zeta_1 - \frac{RT}{p}\right) \tag{22}$$

$$b_2 = \left(\frac{RT}{p}\right) \cdot \left(-\zeta_4^2 - \left(\zeta_3 + \zeta_4\right) \cdot \left(\zeta_1 + \frac{RT}{p}\right) + \frac{\zeta_2}{p}\right) \tag{23}$$

$$b_3 = \left(\zeta_1 + \frac{RT}{p}\right) \cdot \zeta_4^2 - \frac{\zeta_1\zeta_2}{p} \tag{24}$$

The advantage of a cubic equation is the possibility to have multiple solutions (maximally three) for volume at specific temperature and pressure conditions, which may reflect coexisting liquid and vapour phases. Liquid-vapour equilibria can only be calculated from the same equation of state if multiple solution of volume can be calculated at the same temperature and pressure. The calculation of thermodynamic properties with this type of equation of state is based on splitting Eq. 14 in two parts (Eq. 25), i.e. an ideal pressure (from the ideal gas law) and a departure (or residual) pressure, see also [6].

$$p = p_{ideal} + p_{residual} \tag{25}$$

where

$$p_{ideal} = \frac{n_T RT}{V} \tag{26}$$

The residual pressure ($p_{residual}$) can be defined as the difference (Δp, Eq. 27) between ideal pressure and reel pressure as expressed in Eq. 14 .

$$\Delta p = p_{residual} = -\frac{n_T RT}{V} + \frac{n_T RT}{V - n_T\zeta_1} - \frac{n_T^2\zeta_2}{V^2 + n_T\zeta_3 V + n_T\zeta_4 V - n_T^2\zeta_4^2} \tag{27}$$

The partial derivative of pressure with respect to temperature (Eq. 28) is the main equation to estimate the thermodynamic properties of fluids (see Eqs. 2 and 3).

$$\frac{\partial p}{\partial T} = \frac{\partial p_{ideal}}{\partial T} + \frac{\partial \Delta p}{\partial T} \tag{28}$$

where

$$\frac{\partial \Delta p}{\partial T} = -\frac{n_T R}{V} + \frac{n_T R}{V - n_T\zeta_1} + \frac{n_T RT}{\left(V - n_T\zeta_1\right)^2}\cdot\frac{\partial(n_T\zeta_1)}{\partial T} - \frac{1}{V^2 + n_T\zeta_3 V + n_T\zeta_4 V - n_T{}^2\zeta_4{}^2}\cdot\frac{\partial(n_T{}^2\zeta_2)}{\partial T} + \frac{n_T{}^2\zeta_2}{\left(V^2 + n_T\zeta_3 V + n_T\zeta_4 V - n_T{}^2\zeta_4{}^2\right)^2}\cdot\frac{\partial\left(n_T\zeta_3 V + n_T\zeta_4 V - n_T{}^2\zeta_4{}^2\right)}{\partial T} \quad (29)$$

The parameters ζ_1, ζ_3, and ζ_4 are usually independent of temperature, compare with the b parameter (Table 1). This reduces Eq. 29 to Eq. 30.

$$\frac{\partial \Delta p}{\partial T} = -\frac{n_T R}{V} + \frac{n_T R}{V - n_T\zeta_1} - \frac{1}{V^2 + n_T\zeta_3 V + n_T\zeta_4 V - n_T{}^2\zeta_4{}^2}\cdot\frac{\partial(n_T{}^2\zeta_2)}{\partial T} \quad (30)$$

Other important equations to calculate thermodynamic properties of fluids are partial derivatives of pressure with respect to volume (Eq. 31 and 32).

$$\frac{\partial p}{\partial V} = -\frac{n_T RT}{\left(V - n_T\zeta_1\right)^2} + \frac{n_T{}^2\zeta_2}{\left(V^2 + n_T\zeta_3 V + n_T\zeta_4 V - n_T{}^2\zeta_4{}^2\right)^2}\cdot\left(2V + n_T\zeta_3 + n_T\zeta_4\right) \quad (31)$$

$$\frac{\partial^2 p}{\partial V^2} = \frac{2n_T RT}{\left(V - n_T\zeta_1\right)^3} - \frac{2n_T{}^2\zeta_2}{\left(V^2 + n_T\zeta_3 V + n_T\zeta_4 V - n_T{}^2\zeta_4{}^2\right)^3}\cdot\left(2V + n_T\zeta_3 + n_T\zeta_4\right)^2 + \frac{2n_T{}^2\zeta_2}{\left(V^2 + n_T\zeta_3 V + n_T\zeta_4 V - n_T{}^2\zeta_4{}^2\right)^2} \quad (32)$$

Eqs. 31 and 32 already include the assumption that the parameters ζ_1, ζ_2, ζ_3, and ζ_4 are independent of volume. Finally, the partial derivative of pressure in respect to the amount of substance of a specific component in the fluid mixture (n_i) is also used to characterize thermodynamic properties of fluid mixtures (Eq. 33).

$$\frac{\partial p}{\partial n_i} = \frac{RT}{V - n_T\zeta_1} + \frac{n_T RT}{\left(V - n_T\zeta_1\right)^2}\cdot\frac{\partial(n_T\zeta_1)}{\partial n_i} - \frac{1}{V^2 + n_T\zeta_3 V + n_T\zeta_4 V - n_T{}^2\zeta_4{}^2}\cdot\frac{\partial(n_T{}^2\zeta_2)}{\partial n_i} + \frac{n_T{}^2\zeta_2}{\left(V^2 + n_T\zeta_3 V + n_T\zeta_4 V - n_T{}^2\zeta_4{}^2\right)^2}\cdot\left[\left(\frac{\partial(n_T\zeta_3)}{\partial n_i} + \frac{\partial(n_T\zeta_4)}{\partial n_i}\right)\cdot V - 2n_T\zeta_4\frac{\partial(n_T\zeta_4)}{\partial n_i}\right] \quad (33)$$

3. Thermodynamic parameters

The entropy (S) is obtained from the integration defined in Eq. 2 at constant temperature (Eqs. 34 and 35).

$$\int_{S_0}^{S_1} dS = \int_{V_0}^{V_1} \left(\frac{\partial p}{\partial T}\right)_{V,n_T} dV \tag{34}$$

$$S_1 = S_0 + \int_{V_0}^{V_1} \left(\frac{\partial p_{ideal}}{\partial T} + \frac{\partial \Delta p}{\partial T}\right) dV \tag{35}$$

The limits of integration are defined as a reference ideal gas at S_0 and V_0, and a real gas at S_1 and V_1. This integration can be split into two parts, according to the ideal pressure and residual pressure definition (Eqs. 25, 26, and 27). The integral has different solutions dependent on the values of ζ_3 and ζ_4: Eq. 36 for $\zeta_3 = 0$ and $\zeta_4 = 0$, and Eqs. 37 and 38 for $\zeta_3 > 0$.

$$S_1 = S_0 + n_T R \ln\left(\frac{V_1}{V_0}\right) + n_T R \ln\left(\frac{V_1 - n_T\zeta_1}{V_0 - n_T\zeta_1} \cdot \frac{V_0}{V_1}\right) + \left(\frac{1}{V_1} - \frac{1}{V_0}\right) \cdot \frac{\partial (n_T{}^2 \zeta_2)}{\partial T} \tag{36}$$

$$\begin{aligned} S_1 = {} & S_0 + n_T R \ln\left(\frac{V_1}{V_0}\right) + n_T R \ln\left(\frac{V_1 - n_T\zeta_1}{V_0 - n_T\zeta_1} \cdot \frac{V_0}{V_1}\right) \\ & - \frac{1}{q} \cdot \frac{\partial (n_T{}^2 \zeta_2)}{\partial T} \cdot \ln\left(\frac{2V_1 + n_T(\zeta_3 + \zeta_4) - q}{2V_1 + n_T(\zeta_3 + \zeta_4) + q}\right) + \frac{1}{q} \cdot \frac{\partial (n_T{}^2 \zeta_2)}{\partial T} \cdot \ln\left(\frac{2V_0 + n_T(\zeta_3 + \zeta_4) - q}{2V_0 + n_T(\zeta_3 + \zeta_4) + q}\right) \end{aligned} \tag{37}$$

where

$$q = n_T \sqrt{4\zeta_4{}^2 + (\zeta_3 + \zeta_4)^2} \tag{38}$$

The *RKeos* and *Seos* define q as $n_T b$, whereas in the *PReos* q is equal to $n_T b\sqrt{8}$, according to the values for ζ_3 and ζ_4 listed in Table 1. Eqs. 36 and 37 can be simplified by assuming that the lower limit of the integration corresponds to a large number of V_0. As a consequence, part of the natural logarithms in Eqs. 36 and 37 can be replaced by the unit value 1 or 0 (Eqs. 39, 40, and 41).

$$\lim_{V_0 \to \infty} \left(\frac{V_0}{V_0 - n_T\zeta_1}\right) = 1 \tag{39}$$

$$\lim_{V_0 \to \infty} \left(\frac{1}{V_0}\right) = 0 \tag{40}$$

$$\lim_{V_0 \to \infty} \left(\frac{2V_0 + n_T(\zeta_3 + \zeta_4) - q}{2V_0 + n_T(\zeta_3 + \zeta_4) + q}\right) = 1 \tag{41}$$

The entropy change that is caused by a volume change of ideal gases corresponds to the second term on the right-hand side of Eqs. 36 and 37. This term can be used to express the behaviour of an ideal mixture of perfected gases. Each individual gas in a mixture expands from their partial volume (v_i) to the total volume at a pressure of 0.1 MPa, which results in a new expression for this term (Eq. 42)

$$n_T R \ln\left(\frac{V_1}{V_0}\right)_{ideal.mix} = \sum_i \left[n_i R \ln\left(\frac{V_1}{v_i}\right)\right] \tag{42}$$

where n_i is the amount of substance of component i in the fluid mixture. In addition, the partial volume of an ideal gas is related to the standard pressure p_0 (0.1 MPa) according to the ideal gas law (Eq. 43, compare with Eq. 26).

$$v_i = \frac{n_i RT}{p_0} \tag{43}$$

Finally, the entropy of fluid phases containing gas mixtures at any temperature and total volume according to the two-constant cubic equation of state is given by Eq. 44 for $\zeta_3 = 0$ and $\zeta_4 = 0$, and Eq. 45 for $\zeta_3 > 0$.

$$S = S_0 + \sum_i \left[n_i R \ln\left(\frac{p_0 V}{n_i RT}\right)\right] + n_T R \ln\left(\frac{V - n_T\zeta_1}{V}\right) + \frac{1}{V}\cdot\frac{\partial(n_T{}^2\zeta_2)}{\partial T} \tag{44}$$

$$S = S_0 + \sum_i \left[n_i R \ln\left(\frac{p_0 V}{n_i RT}\right)\right] + n_T R \ln\left(\frac{V - n_T\zeta_1}{V}\right) - \frac{1}{q}\cdot\frac{\partial(n_T{}^2\zeta_2)}{\partial T}\cdot\ln\left(\frac{2V + n_T(\zeta_3+\zeta_4) - q}{2V + n_T(\zeta_3+\zeta_4) + q}\right) \tag{45}$$

The subscripts "1" for the upper limit of integration is eliminated to present a pronounced equation. The standard state entropy (S_0) of a mixture of ideal gases is defined according to the arithmetic average principle (Eq. 46).

$$S_0 = \sum_i n_i \cdot s_i^0 \tag{46}$$

where s_i^0 is the molar entropy of a pure component i in an ideal gas mixture at temperature T.

The internal energy (U, see Eq. 3) is obtained from the pressure equation (Eq. 14) and its partial derivative with respect to temperature (Eqs. 28 and 30):

$$\int_{U_0}^{U_1} dU = \int_{V_0}^{V_1} \left(T\cdot\frac{\partial p}{\partial T} - p\right) dV \tag{47}$$

$$U_1 = U_0 + \int_{V_0}^{V_1} \left(\frac{1}{V^2 + n_T\zeta_3 V + n_T\zeta_4 V - n_T{}^2\zeta_4{}^2} \cdot \left(n_T{}^2\zeta_2 - T\frac{\partial(n_T{}^2\zeta_2)}{\partial T}\right)\right) dV \tag{48}$$

Similar to the integral in the entropy definition (see Eqs. 44 and 45), Eq. 48 has different solutions dependent on the values of ζ_3 and ζ_4: Eq. 49 for $\zeta_3 = 0$ and $\zeta_4 = 0$, and Eq. 50 for $\zeta_3 > 0$.

$$U = U_0 - \frac{1}{V}\cdot\left(n_T^2\zeta_2 - T\frac{\partial(n_T^2\zeta_2)}{\partial T}\right) \tag{49}$$

$$U = U_0 + \frac{1}{q}\cdot\left(n_T^2\zeta_2 - T\frac{\partial(n_T^2\zeta_2)}{\partial T}\right)\cdot\ln\left(\frac{2V + n_T(\zeta_3+\zeta_4) - q}{2V + n_T(\zeta_3+\zeta_4) + q}\right) \tag{50}$$

The definition of q is given in Eq. 38. The standard state internal energy (U_0) of a mixture of ideal gases is defined according to the arithmetic average principle (Eq. 51).

$$U_0 = \sum_i n_i \cdot u_i^0 \tag{51}$$

where u_i^0 is the molar internal energy of a pure component i in an ideal gas mixture at temperature T.

Enthalpy (Eq. 52 for $\zeta_3 = 0$ and $\zeta_4 = 0$, and Eq. 53 for $\zeta_3 > 0$), Helmholtz energy (Eq. 55 for $\zeta_3 = 0$ and $\zeta_4 = 0$, and Eq. 56 for $\zeta_3 > 0$), and Gibbs energy (Eq. 58 for $\zeta_3 = 0$ and $\zeta_4 = 0$, and Eq. 59 for $\zeta_3 > 0$) can be obtained from the definitions of pressure, entropy and internal energy according to standard thermodynamic relations, as illustrated in Eq. 4, 5, and 7. Standard state enthalpy (H_0), standard state Helmholtz energy (A_0), and standard state Gibbs energy (G_0) of an ideal gas mixture at 0.1 MPa and temperature T are defined in Eqs. 54, 57, and 60, respectively.

$$H = U_0 + \frac{n_T RTV}{V - n_T\zeta_1} - \frac{1}{V}\cdot\left(2n_T^2\zeta_2 - T\frac{\partial(n_T^2\zeta_2)}{\partial T}\right) \tag{52}$$

$$\begin{aligned} H = U_0 &+ \frac{n_T RTV}{V - n_T\zeta_1} - \frac{n_T^2\zeta_2 V}{V^2 + n_T\zeta_3 V + n_T\zeta_4 V - n_T^2\zeta_4^2} \\ &+ \frac{1}{q}\cdot\left(n_T^2\zeta_2 - T\frac{\partial(n_T^2\zeta_2)}{\partial T}\right)\cdot\ln\left(\frac{2V + n_T(\zeta_3+\zeta_4) - q}{2V + n_T(\zeta_3+\zeta_4) + q}\right) \end{aligned} \tag{53}$$

$$H_0 = U_0 + n_T RT \tag{54}$$

$$A = U_0 - TS_0 - \sum_i\left[n_i RT\ln\left(\frac{p_0 V}{n_i RT}\right)\right] - n_T RT\ln\left(\frac{V - n_T\zeta_1}{V}\right) - \frac{n_T^2\zeta_2}{V} \tag{55}$$

$$\begin{aligned} A = U_0 &- TS_0 - \sum_i\left[n_i RT\ln\left(\frac{p_0 V}{n_i RT}\right)\right] - n_T RT\ln\left(\frac{V - n_T\zeta_1}{V}\right) \\ &+ \frac{n_T^2\zeta_2}{q}\cdot\ln\left(\frac{2V + n_T(\zeta_3+\zeta_4) - q}{2V + n_T(\zeta_3+\zeta_4) + q}\right) \end{aligned} \tag{56}$$

$$A_0 = U_0 - TS_0 \tag{57}$$

$$G = U_0 - TS_0 - RT\sum_i \left[n_i \ln\left(\frac{p_0 V}{n_i RT}\right)\right] - n_T RT \ln\left(\frac{V - n_T\zeta_1}{V}\right) + \frac{n_T RTV}{V - n_T\zeta_1} - \frac{2n_T^2\zeta_2}{V} \tag{58}$$

$$\begin{aligned} G = {} & U_0 - TS_0 - RT\sum_i \left[n_i \ln\left(\frac{p_0 V}{n_i RT}\right)\right] - n_T RT \ln\left(\frac{V - n_T\zeta_1}{V}\right) + \frac{n_T RTV}{V - n_T\zeta_1} \\ & + \frac{n_T^2\zeta_2}{q}\cdot \ln\left(\frac{2V + n_T(\zeta_3+\zeta_4) - q}{2V + n_T(\zeta_3+\zeta_4) + q}\right) - \frac{n_T^2\zeta_2 V}{V^2 + n_T\zeta_3 V + n_T\zeta_4 V - n_T^2\zeta_4^2} \end{aligned} \tag{59}$$

$$G_0 = U_0 - TS_0 + n_T RT \tag{60}$$

The Helmholtz energy equation (Eqs. 55, 56, and 57) is used for the definition of chemical potential (μ_i) of a component in either vapour or liquid phase gas mixtures (compare with Eq. 8), Eq. 61 for $\zeta_3 = 0$ and $\zeta_4 = 0$, and Eq. 62 for $\zeta_3 > 0$, calculated with two-constant cubic equations of state.

$$\begin{aligned} \mu_i = {} & \frac{\partial U_0}{\partial n_i} - T\frac{\partial S_0}{\partial n_i} - RT\ln\left(\frac{p_0 V}{n_i RT}\right) + RT - RT\ln\left(\frac{V - n_T\zeta_1}{V}\right) \\ & + \frac{n_T RT}{V - n_T\zeta_1}\cdot\frac{\partial(n_T\zeta_1)}{\partial n_i} - \frac{1}{V}\cdot\frac{\partial n_T^2\zeta_2}{\partial n_i} \end{aligned} \tag{61}$$

$$\begin{aligned} \mu_i = {} & \frac{\partial U_0}{\partial n_i} - T\frac{\partial S_0}{\partial n_i} - RT\ln\left(\frac{p_0 V}{n_i RT}\right) + RT - RT\ln\left(\frac{V - n_T\zeta_1}{V}\right) + \frac{n_T RT}{V - n_T\zeta_1}\cdot\frac{\partial(n_T\zeta_1)}{\partial n_i} \\ & + \left(\frac{\partial n_T^2\zeta_2}{\partial n_i}\cdot\frac{1}{q} - \frac{n_T^2\zeta_2}{q^2}\cdot\frac{\partial q}{\partial n_i}\right)\cdot\ln\left(\frac{2V + n_T(\zeta_3+\zeta_4) - q}{2V + n_T(\zeta_3+\zeta_4) + q}\right) \\ & + \frac{n_T^2\zeta_2}{q}\cdot\frac{1}{2V + n_T(\zeta_3+\zeta_4) - q}\cdot\left(\frac{\partial n_T\zeta_3}{\partial n_i} + \frac{\partial n_T\zeta_4}{\partial n_i} - \frac{\partial q}{\partial n_i}\right) \\ & - \frac{n_T^2\zeta_2}{q}\cdot\frac{1}{2V + n_T(\zeta_3+\zeta_4) + q}\cdot\left(\frac{\partial n_T\zeta_3}{\partial n_i} + \frac{\partial n_T\zeta_4}{\partial n_i} + \frac{\partial q}{\partial n_i}\right) \end{aligned} \tag{62}$$

where

$$\frac{\partial q}{\partial n_i} = \frac{1}{q}\cdot\left[4n_T\zeta_4\frac{\partial n_T\zeta_4}{\partial n_i} + n_T(\zeta_3+\zeta_4)\cdot\left(\frac{\partial n_T\zeta_3}{\partial n_i} + \frac{\partial n_T\zeta_4}{\partial n_i}\right)\right] \tag{63}$$

The definitions of the partial derivative of q in respect to amount of substance (Eq. 63) according to $\zeta_3 = \mathrm{b}$ and $\zeta_4 = 0$ [8, 9] is illustrated in Eq. 64, and $\zeta_3 = \mathrm{b}$ and $\zeta_4 = \mathrm{b}$ [10] in Eq. 65.

$$\frac{\partial q}{\partial n_i} = \frac{\partial (n_T b)}{\partial n_i} \quad (64)$$

$$\frac{\partial q}{\partial n_i} = \sqrt{8} \cdot \frac{\partial (n_T b)}{\partial n_i} \quad (65)$$

The fugacity coefficient (φ_i) is defined according to Eqs. 9 and 10 from the difference between the chemical potential of a real gas mixture and an ideal gas mixture at standard conditions (0.1 MPa), see Eq. 66 for $\zeta_3 = 0$ and $\zeta_4 = 0$, and Eq. 67 for $\zeta_3 > 0$. Fugacity coefficient defined in Eq. 66 is applied to *Weos* and Eq. 67 is applied to *RKeos*, *Seos*, and *PReos*.

$$RT\ln(\varphi_i) = -RT\ln\left(\frac{pV}{n_T RT}\right) - RT\ln\left(\frac{V - n_T\zeta_1}{V}\right) + \frac{n_T RT}{V - n_T\zeta_1} \cdot \frac{\partial(n_T\zeta_1)}{\partial n_i} - \frac{1}{V} \cdot \frac{\partial n_T^2\zeta_2}{\partial n_i} \quad (66)$$

$$\begin{aligned} RT\ln(\varphi_i) = & -RT\ln\left(\frac{pV}{n_T RT}\right) - RT\ln\left(\frac{V - n_T\zeta_1}{V}\right) + \frac{n_T RT}{V - n_T\zeta_1} \cdot \frac{\partial(n_T\zeta_1)}{\partial n_i} \\ & + \left(\frac{\partial n_T^2\zeta_2}{\partial n_i} \cdot \frac{1}{q} - \frac{n_T^2\zeta_2}{q^2} \cdot \frac{\partial q}{\partial n_i}\right) \cdot \ln\left(\frac{2V + n_T(\zeta_3 + \zeta_4) - q}{2V + n_T(\zeta_3 + \zeta_4) + q}\right) \\ & + \frac{n_T^2\zeta_2}{q} \cdot \frac{1}{2V + n_T(\zeta_3 + \zeta_4) - q} \cdot \left(\frac{\partial n_T\zeta_3}{\partial n_i} + \frac{\partial n_T\zeta_4}{\partial n_i} - \frac{\partial q}{\partial n_i}\right) \\ & - \frac{n_T^2\zeta_2}{q} \cdot \frac{1}{2V + n_T(\zeta_3 + \zeta_4) + q} \cdot \left(\frac{\partial n_T\zeta_3}{\partial n_i} + \frac{\partial n_T\zeta_4}{\partial n_i} + \frac{\partial q}{\partial n_i}\right) \end{aligned} \quad (67)$$

4. Spinodal

The stability limit of a fluid mixture can be calculated with two-constant cubic equations of state, e.g. see [6]. This limit is defined by the spinodal line, i.e. the locus of points on the surface of the Helmholtz energy or Gibbs energy functions that are inflection points, e.g. see [12] and references therein. The stability limit occurs at conditions where phase separation into a liquid and vapour phase should take place, which is defined by the binodal. Metastability is directly related to spinodal conditions, for example, nucleation of a vapour bubble in a cooling liquid phase within small constant volume cavities, such as fluid inclusions in minerals (< 100 μm diameter) occurs at conditions well below homogenization conditions of these phases in a heating experiment. The maximum temperature difference of nucleation and homogenization is defined by the spinodal. In multi-component fluid systems, the partial derivatives of the Helmholtz energy with respect to volume and amount of substance of each component can be arranged in a matrix that has a determinant (D_{spin}) equal to zero (Eq. 68) at spinodal conditions.

$$D_{spin} = \begin{vmatrix} A_{VV} & A_{n_1 V} & A_{n_2 V} & \cdots \\ A_{V n_1} & A_{n_1 n_1} & A_{n_2 n_1} & \cdots \\ A_{V n_2} & A_{n_1 n_2} & A_{n_2 n_2} & \cdots \\ \vdots & \vdots & \vdots & \ddots \end{vmatrix} = 0 \tag{68}$$

This matrix is square and contains a specific number of columns that is defined by the number of differentiation variables, i.e. volume and number of components in the fluid mixture minus 1. The individual components of this matrix are defined according to Eqs. 69, 70, 71, 72, 73, and 74. The exact definition of these components according to two-constant cubic equations of state can be obtained from the web site http://fluids.unileoben.ac.at (see also [6]).

$$A_{VV} = \left(\frac{\partial^2 A}{\partial V^2} \right)_{n_1, n_2, \cdots} \tag{69}$$

$$A_{n_1 n_1} = \left(\frac{\partial^2 A}{\partial {n_1}^2} \right)_{n_2, V, \cdots} \tag{70}$$

$$A_{n_2 n_2} = \left(\frac{\partial^2 A}{\partial {n_2}^2} \right)_{n_1, V, \cdots} \tag{71}$$

$$A_{n_1 V} = \left(\frac{\partial^2 A}{\partial n_1 \partial V} \right)_{n_2, \cdots} = A_{V n_1} \tag{72}$$

$$A_{n_2 V} = \left(\frac{\partial^2 A}{\partial n_2 \partial V} \right)_{n_1, \cdots} = A_{V n_2} \tag{73}$$

$$A_{n_1 n_2} = \left(\frac{\partial^2 A}{\partial n_1 \partial n_2} \right)_{V, \cdots} = A_{n_2 n_1} \tag{74}$$

The determinant in Eq. 68 is calculated with the Laplacian expansion that contains "*minors*" and "*cofactors*", e.g. see [13]. The mathematical computation time increases exponential with increasing number of components. Therefore, the *LU* decomposition [14] can be applied in computer programming to reduce this time.

The spinodal curve, binodal curve and critical point of a binary CO_2-CH_4 mixture with $x(CO_2)$ = 0.9 are illustrated in Figure 1, which are calculated with the *PReos* [10]. The spinodal has a small loop near the critical point, and may reach negative pressures at lower temperatures. The binodal remains within the positive pressure part at all temperatures. The binodal is obtained from equality of fugacity (Eq. 66 and 67) of each component in both liquid and vapour phase, and marks the boundary between a homogeneous fluid mixture and fluid immiscibility [6, 15].

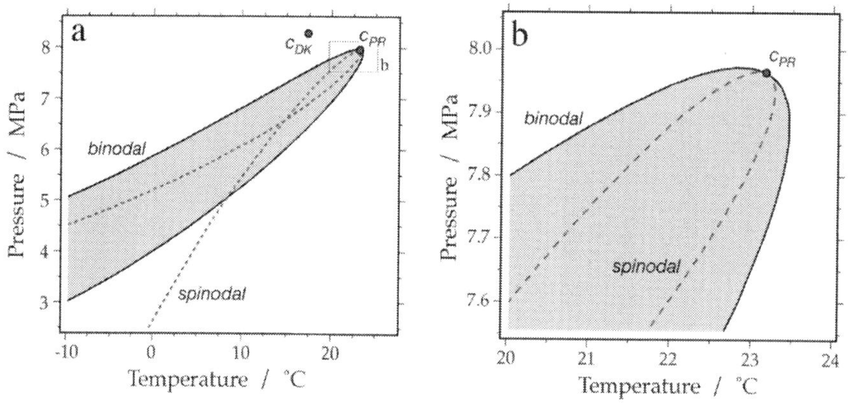

Figure 1. (**a**) Temperature-pressure diagram of a binary CO_2-CH_4 fluid mixture, with $x(CO_2) = 0.9$. The shaded area illustrates *T-p* condition of immiscibility of a CO_2-rich liquid phase and a CH_4-rich vapour phase (the binodal). The red dashed line is the spinodal. All lines are calculated with the equation of state according to *PReos* [10]. The calculated critical point is indicated with c_{PR}. c_{DK} is the interpolated critical point from experimental data [16]. (**b**) enlargement of (**a**) indicated with the square in thin lines.

5. Pseudo critical point

The pseudo critical point is defined according to the first and second partial derivatives of pressure with respect to volume (Eqs. 31 and 32). This point is defined in a *p-V* diagram where the inflection point and extremum coincide at a specific temperature, i.e. Eqs. 31 and 32 are equal to 0. The pseudo critical point is equal to the critical point for pure gas fluids, however, the critical point in mixtures cannot be obtained from Eqs. 31 and 32. The pseudo critical point estimation is used to define the two-constants (a and b) for pure gas fluids in cubic equations of state according to the following procedure. The molar volume of the pseudo critical point that is derived from Eqs 31 and 32 is presented in the form of a cubic equation (Eq. 75).

$$0 = V_m^3 - 3\zeta_1 \cdot V_m^2 + \left[3\zeta_4^2 - 3\zeta_1 \cdot (\zeta_3 + \zeta_4)\right] \cdot V_m - \zeta_1 \cdot \left[\zeta_4^2 + (\zeta_3 + \zeta_4)^2\right] + \zeta_4^2 \cdot (\zeta_3 + \zeta_4) \quad (75)$$

The solution of this cubic equation can be obtained from its reduced form, see page 9 in [15]:

$$x^3 + f \cdot x + g = 0 \quad (76)$$

where

$$f = 3 \cdot \left[\zeta_4^2 - \zeta_1 \cdot (\zeta_1 + \zeta_3 + \zeta_4)\right] \quad (77)$$

$$g = -2\zeta_1^3 + \zeta_1 \cdot \left[2\zeta_4^2 - (\zeta_3 + \zeta_4)^2\right] + (\zeta_3 + \zeta_4) \cdot \left[\zeta_4^2 - 3\zeta_1^2\right] \tag{78}$$

$$V_m = x + \zeta_1 \tag{79}$$

The values of f and g in terms of the b parameters for the individual two-constant cubic equations of state are given in Table 2. The molar volume at pseudo critical conditions is directly related to the b parameter in each equation of state: *Weos* Eq. 80; *RKeos* Eq. 81; *Seos* also Eq. 81; and *PReos* Eq. 82.

$$Weos: \quad V_m^{pc} = 3 \cdot b \tag{80}$$

$$RKeos: \quad V_m^{pc} = \frac{b}{\sqrt[3]{2} - 1} \approx 3.847322 \cdot b \tag{81}$$

$$PReos: \quad V_m^{pc} = \left[1 + Q\right] \cdot b \approx 3.951373 \cdot b \tag{82}$$

where Q is defined according to Eq. 83, the superscript "*pc*" is the abbreviation for "*pseudo critical*".

$$Q = \left(4 + \sqrt{8}\right)^{1/3} + \left(4 - \sqrt{8}\right)^{1/3} \tag{83}$$

Equation of state	*f*	*g*	*b in Eqs. 80-82*	*b in Eqs. 94-96*	*difference*
van der Waals [7]	$-3b^2$	$-2b^3$	31.3727	42.8453	37 %
Redlich and Kwong [8]	$-6b^2$	$-6b^3$	24.4633	29.6971	21 %
Soave [9]	$-6b^2$	$-6b^3$	24.4633	29.6971	21 %
Peng and Robinson [10]	$-6b^2$	$-8b^3$	23.8191	26.6656	12 %

Table 2. Definitions of f and g according to Eq. 77 and 78, respectively. The values of b are calculated for the critical conditions of pure CO_2: $V_{m,C}$ = 94.118 $cm^3 \cdot mol^{-1}$, T_C = 304.128 K and p_C = 7.3773 MPa [18]. The last column gives the percentage of difference between the values of b (Eqs. 80-82 and 94-96).

The temperature at pseudo critical conditions is obtained from the combination of Eqs. 80-82 and the first partial derivative of pressure with respect to volume (Eq. 31).

$$Weos: \quad T^{pc} = \frac{8}{27} \cdot \frac{\zeta_2}{bR} \approx 0.29629630 \cdot \frac{\zeta_2}{bR} \tag{84}$$

$$RKeos: \quad T^{pc} = \left(\sqrt[3]{2} - 1\right)^2 \cdot \left(\frac{3\zeta_2}{bR}\right) \approx 0.20267686 \cdot \frac{\zeta_2}{bR} \tag{85}$$

$$PReos: \quad T^{pc} = \frac{2Q + 4}{\left(Q + 4 + 2/Q\right)^2} \cdot \left(\frac{3\zeta_2}{bR}\right) \approx 0.17014442 \cdot \frac{\zeta_2}{bR} \tag{86}$$

where Q is defined according to Eq. 83. The order of equations (84, 85, 86) is according to the order of equations of state in Eq. 80, 81,and 82. The parameter ζ_2 is used in Eqs. 84, 85 and, 86 instead of the constant a (see Table 1). Eq. 87 illustrates the transformation of Eq. 85 for the *RKeos* [8] by substitution of ζ_2 according to its value given in Table 1.

$$T^{pc} = \left(\sqrt[3]{2} - 1\right)^{4/3} \cdot \left(\frac{3a}{bR}\right)^{2/3} \tag{87}$$

Any temperature dependency of the a constant has an effect on the definition of the pseudo critical temperature. The pressure at pseudo critical condition (Eqs. 88-90) is obtained from a combination of the pressure equation (Eq.14), pseudo critical temperature (Eqs. 84-87) and pseudo critical molar volume (Eqs. 80-82).

$$Weos: \quad p^{pc} = \frac{1}{27} \cdot \frac{\zeta_2}{b^2} \approx 0.03703704 \cdot \frac{\zeta_2}{b^2} \tag{88}$$

$$RKeos: \quad p^{pc} = \left(\sqrt[3]{2} - 1\right)^3 \cdot \frac{\zeta_2}{b^2} \approx 0.01755999 \cdot \frac{\zeta_2}{b^2} \tag{89}$$

$$PReos: \quad p^{pc} = \frac{Q^2 - 2}{\left(Q^2 + 4Q + 2\right)^2} \cdot \frac{\zeta_2}{b^2} \approx 0.01227198 \cdot \frac{\zeta_2}{b^2} \tag{90}$$

where Q is defined according to Eq. 83. The order of equations (88, 89, and 90) is according to the order of equations of state in Eqs. 80, 81, and 82. These equations define the relation between the a and b constant in two-constant cubic equations of state and critical conditions, i.e. temperature, pressure, and molar volume of pure gas fluids. Therefore, knowledge of these conditions from experimental data can be used to determine the values of a (or ζ_2) and b, which can be defined as a function of only temperature and pressure (Eqs. 91-93, and 94-96, respectively).

$$Weos: \quad \zeta_2 = \frac{27}{64} \cdot \frac{R^2 T_C^{\ 2}}{p_C} = 0.421875 \cdot \frac{R^2 T_C^{\ 2}}{p_C} \tag{91}$$

$$RKeos: \quad \zeta_2 = \frac{1}{9 \cdot \left(\sqrt[3]{2} - 1\right)} \cdot \frac{R^2 T_C^{\ 2}}{p_C} \approx 0.42748024 \cdot \frac{R^2 T_C^{\ 2}}{p_C} \tag{92}$$

$$PReos: \quad \zeta_2 = \frac{\left(Q^2 + 4Q + 2\right)^2 \cdot \left(Q^2 - 2\right)}{4Q^2 \cdot \left(Q + 2\right)^2} \cdot \frac{R^2 T_C^{\ 2}}{p_C} \approx 0.45723553 \cdot \frac{R^2 T_C^{\ 2}}{p_C} \tag{93}$$

$$Weos: \quad b = \frac{1}{8} \cdot \frac{RT_C}{p_C} = 0.125 \cdot \frac{RT_C}{p_C} \tag{94}$$

$$RKeos: \quad b = \frac{\left(\sqrt[3]{2}-1\right)}{3} \cdot \frac{RT_C}{p_C} \approx 0.08664035 \cdot \frac{RT_C}{p_C} \tag{95}$$

$$PReos: \quad b = \frac{\left(Q^2-2\right)}{2Q^2 \cdot \left(Q+2\right)} \cdot \frac{RT_C}{p_C} \approx 0.07779607 \cdot \frac{RT_C}{p_C} \tag{96}$$

where T_C and p_C are the critical temperature and critical pressure, and Q is defined according to Eq. 83. The order of equations (91-93, and 94-96) is according to the order of equations of state in Eqs. 80-82. Comparison of the value of b calculated with experimental critical volume (Eqs. 80, 81 and 82) and critical temperature and pressure (Eqs. 94, 95, and 96) is illustrated in Table 2. The difference indicates the ability of a specific equation of state to reproduce fluid properties of pure gases. A large difference indicates that the geometry or morphology of the selected equation of state in the *p-V-T-x* parameter space is not exactly reproducing fluid properties of pure gases. The empirical modifications of the van-der-Waals equation of state according to Peng and Robinson [10] result in the most accurate equation in Table 2 (11% for pure CO_2).

6. Critical point and curve

The critical point is the highest temperature and pressure in a pure gas system where boiling may occur, i.e. where a distinction can be made between a liquid and vapour phase at constant temperature and pressure. At temperatures and pressures higher than the critical point the pure fluid is in a homogeneous supercritical state. The critical point of pure gases and multi-component fluid mixtures can be calculated exactly with the Helmholtz energy equation (Eqs. 55-57) that is obtained from two-constant cubic equations of state, e.g. see [17, 18], and it marks that part of the surface described with a Helmholtz energy function where two inflection points of the spinodal coincide. Therefore, the conditions of the spinodal are also applied to the critical point. In addition, the critical curve is defined by the determinant (D_{crit}) of the matrix illustrated in Eq. 97, see also [6].

$$D_{crit} = \begin{vmatrix} A_{VV} & A_{n_1 V} & A_{n_2 V} & \cdots \\ A_{V n_1} & A_{n_1 n_1} & A_{n_2 n_1} & \cdots \\ \vdots & \vdots & \vdots & \vdots \\ D_V & D_{n_1} & D_{n_2} & \ddots \end{vmatrix} = 0 \tag{97}$$

The number of rows in Eq.97 is defined by the differentiation variables volume and number of components minus 2. The last row is reserved for the partial derivatives of the determinant D_{spin} from Eq. 68:

$$D_V = \frac{\partial D_{spin}}{\partial V} \tag{98}$$

$$D_{n_1} = \frac{\partial D_{spin}}{\partial n_1} \tag{99}$$

$$D_{n_2} = \frac{\partial D_{spin}}{\partial n_2} \tag{100}$$

The derivatives of the spinodal determinant (Eqs. 98-100) are calculated from the sum of the element-by-element products of the matrix of "cofactors" (or adjoint matrix) of the spinodal (Eq. 101) and the matrix of the third derivatives of the Helmholtz energy function (Eq. 102).

$$\begin{vmatrix} C_{VV} & C_{n_1 V} & C_{n_2 V} & \cdots \\ C_{V n_1} & C_{n_1 n_1} & C_{n_2 n_1} & \cdots \\ C_{V n_2} & C_{n_1 n_2} & C_{n_2 n_2} & \cdots \\ \vdots & \vdots & \vdots & \ddots \end{vmatrix} \tag{101}$$

$$\begin{vmatrix} A_{VVK} & A_{n_1 VK} & A_{n_2 VK} & \cdots \\ A_{V n_1 K} & A_{n_1 n_1 K} & A_{n_2 n_1 K} & \cdots \\ A_{V n_2 K} & A_{n_1 n_2 K} & A_{n_2 n_2 K} & \cdots \\ \vdots & \vdots & \vdots & \ddots \end{vmatrix} \tag{102}$$

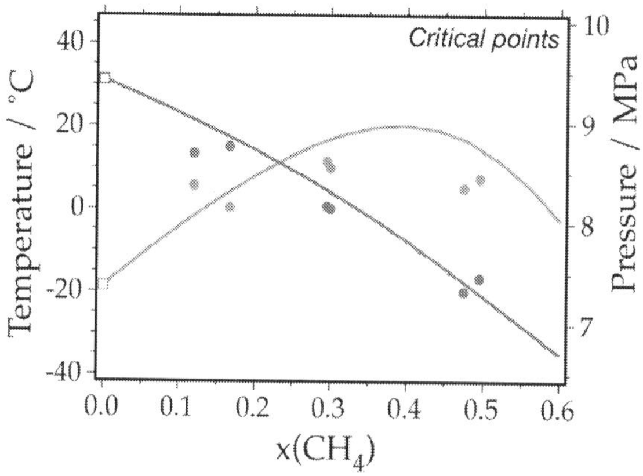

Figure 2. Calculated critical points of binary CO_2-CH_4 fluid mixtures in terms of temperature (red line) and pressure (green line), obtained from the *PReos* [10]. Solid circles are experimental data [16, 19]. The open squares are the critical point of pure CO_2 [20].

where C_{xy} are the individual elements in the matrix of "cofactors", as obtained from the Laplacian expansion. The subscript K refers to the variable that is used in the third differentiation (volume, amount of substance of the components 1 and 2. To reduce computation time in software that uses this calculation method, the *LU* decomposition has been used to calculate the determinant in Eq. 97. The determinants in Eqs. 68 and 97 are both used to calculate exactly the critical point of any fluid mixture and pure gases, based on two-constant cubic equations of state that define the Helmholtz energy function.

An example of a calculated critical curve, i.e. critical points for a variety of compositions in a binary fluid system, is illustrated in Figure 2. The prediction of critical temperatures of fluid mixtures corresponds to experimental data [16, 19], whereas calculated critical pressures are slightly overestimated at higher fraction of CH_4. This example illustrates that the *PReos* [10] is a favourable modification that can be used to calculate sub-critical conditions of CO_2-CH_4 fluid mixtures.

7. Mixing rules and definitions of ζ_1 and ζ_2

All modifications of the van-der-Waals two-constant cubic equation of state [7] have an empirical character. The main modifications are defined by Redlich and Kwong, Soave and Peng and Robinson (see Table 1), and all modification can by summarized by specific adaptations of the values of ζ_1, ζ_2, ζ_3, *and* ζ_4 to fit experimental data. The original definition [7] of ζ_1 (b) and ζ_2 (a) for pure gases is obtained from the pseudo critical conditions (Eqs. 91-93, and 94-96). This principle is adapted in most modifications of the van-der-Waals equation of state, e.g. *RKeos* [8]. Soave [9] and Peng and Robinson [10] adjusted the definition of ζ_2 with a temperature dependent correction parameter α (Eqs. 103-105).

$$\zeta_2 = a_C \cdot \alpha \tag{103}$$

$$\alpha = \left[1 + m \left(1 - \sqrt{\frac{T}{T_C}} \right) \right]^2 \tag{104}$$

$$m = \sum_{i=0,1,2} m_i \cdot \omega^i \tag{105}$$

where a_c is defined by the pseudo critical conditions (Eqs. 91-93), and ω is the acentric factor. The summation in Eq. 105 does not exceed $i = 2$ for Soave [9] and Peng and Robinson [10]. The definition of the acentric factor is arbitrary and chosen for convenience [5] and is a purely empirical modification. These two equations of state have different definitions of pseudo critical conditions (see Eqs. 91-93 and 94-96), therefore, the values of m_i must be different for each equation (Table 3).

	Soave [9]	*Peng and Robinson [10]*
m_0	0.480	0.37464
m_1	1.574	1.54266
m_2	-0.176	-0.26992

Table 3. Values of the constant m_i in Eq. 105.

The two-constant cubic equation of state can be applied to determine the properties of fluid mixtures by using "*mixing rules*" for the parameters ζ_1 and ζ_2 which are defined for individual pure gases according to pseudo critical conditions. These mixing rules are based on simplified molecular behaviour of each component (*i* and *j*) in mixtures [21, 22] that describe the interaction between two molecules:

$$\zeta_1^{mix} = \sum_i \sum_j x_i \cdot \zeta_1(i) \tag{106}$$

$$\zeta_2^{mix} = \sum_i \sum_j x_i x_j \cdot \zeta_2(i,j) \tag{107}$$

where

$$\zeta_2(i,j) = \sqrt{\zeta_2(i) \cdot \zeta_2(j)} \tag{108}$$

These mixing rules have been subject to a variety of modifications, in order to predict fluid properties of newly available experimental data of mixtures. Soave [9] and Peng and Robinson [10] modified Eq. 108 by adding an extra correction factor (Eq. 109).

$$\zeta_2(i,j) = \left(1 - \delta_{ij}\right) \cdot \sqrt{\zeta_2(i) \cdot \zeta_2(j)} \tag{109}$$

where δ_{ij} has a constant value dependent on the nature of component *i* and *j*.

8. Experimental data

As mentioned before, modifications of two-constant cubic equation of state was mainly performed to obtain a better fit with experimental data for a multitude of possible gas mixtures and pure gases. Two types of experimental data of fluid properties were used: 1. homogeneous fluid mixtures at supercritical conditions; and 2. immiscible two-fluid systems at subcritical conditions (mainly in petroleum fluid research). The experimental data consist mainly of pressure, temperature, density (or molar volume) and compositional data, but can also include less parameters. Figure 3 gives an example of the misfit between the first type of experimental data for binary CO_2-CH_4 mixtures [19] and calculated fluid properties with *RKeos* [8] at a constant temperature (15 °C). The *RKeos* uses the pseudo critical defined parameters ζ_1 and ζ_2 (Eqs. 92 and 95) and mixing rules according to Eqs. 106-108 and is only approximately reproducing the fluid properties of CO_2-CH_4 mixtures at subcritical conditions

Experimental data of homogeneous supercritical gas mixtures in the ternary CO_2-CH_4-N_2 system [23] are compared with the two-constant cubic equations of state in Table 4. The *Weos* [7] clearly overestimates (up to 14.1 %) experimentally determined molar volumes at 100 MPa and 200 °C. The *Seos* [9] is the most accurate model in Table 4, but still reach deviations of up to 2.3 % for CO_2-rich gas mixtures. The *PReos* [10] gives highly underestimated molar volumes at these conditions.

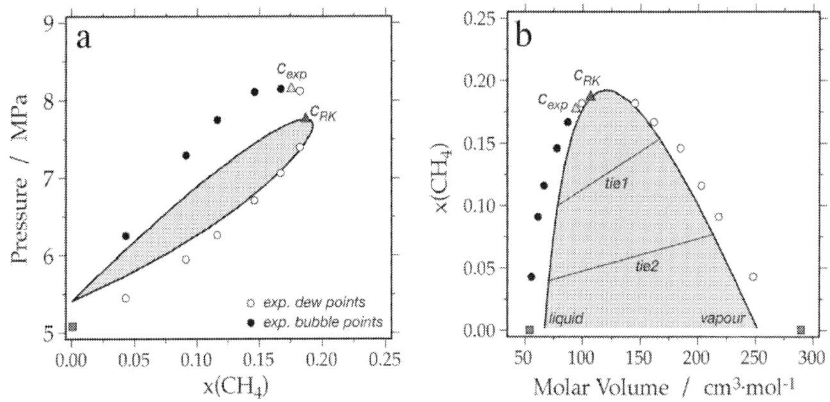

Figure 3. Modelled immiscibility of binary CO_2-CH_4 gas mixtures (shaded areas) in a pressure - amount CH_4 fraction diagram (**a**) and amount CH_4 fraction - molar volume diagram (**b**) at 15 °C. The solid and open circles are experimental data [16]. The red squares are the properties of pure CO_2 [20]. The yellow triangle (C_{exp}) is the interpolated critical point for experimental data, and the green triangle (C_{RK}) is the calculated critical point [8]. *tie1* and *tie 2* in (**b**) are calculated tie-lines between two phases at constant pressures 6.891 and 6.036 MPa, respectively.

composition			$V_m(exp)$	*W*	*RK*	*S*	*PR*
CO_2	CH_4	N_2	$cm^3 \cdot mol^{-1}$				
0.8	0.1	0.1	56.64	64.61 (14.1%)	54.90 (-3.1%)	57.94 (2.3%)	53.59 (-5.4%)
0.8	0.2	0.2	58.92	65.81 (11.7%)	56.61 (-3.9%)	59.61 (1.2%)	56.93 (-6.1%)
0.4	0.3	0.3	61.08	67.08 (9.6%)	58.27 (-4.6%)	61.12 (0.1%)	56.93 (-6.8%)
0.2	0.4	0.4	62.90	68.28 (8.6%)	59.83 (-4.9%)	62.42 (-0.8%)	58.28 (-7.3%)

Table 4. Comparison of supercritical experimental molar volumes [23] at 100 MPa and 200 °C with two-constant cubic equations of state (abbreviations see Table 1). The percentage of deviation from experimentally obtained molar volumes is indicated in brackets.

Figure 3 and Table 4 illustrate that these modified two-constant cubic equations of state still need to be modified again to obtain a better model to reproduce fluid properties at sub- and supercritical conditions.

9. Modifications of modified equations of state

The number of publications that have modified the previously mentioned two-constant cubic equations of state are numerous, see also [11], and they developed highly complex, but purely empirical equations to define the parameters ζ_1 and ζ_2. A few examples are illustrated in the following paragraphs.

9.1. Chueh and Prausnitz [24]

The constant values in the definition of ζ_1 and ζ_2 (Eqs. 92 and 95) are modified for individual gases by Chueh and Prausnitz [24]. This equation is an arbitrary modification of the *RKeos*

[8]. Consequently, the calculation of the value of ζ_1 and ζ_2 is not any more defined by pseudo critical conditions, which give exact mathematical definition of these constants. Although the prediction of fluid properties of a variety of gas mixtures was improved by these modifications, the morphology of the Helmholtz energy equation in the *p-V-T-x* parameter space is not any more related to observed fluid properties. The theory of pseudo critical conditions is violated according to these modifications.

The mixing rules in Eqs. 106-108 were further refined by arbitrary definitions of critical temperature, pressure, volume and compressibility for fluid mixtures.

$$\zeta_2(i,j) = \frac{\Omega_i + \Omega_j}{2} \cdot \frac{R^2 T_{Cij}^2}{p_{Cij}} \quad (110)$$

$$a_{ij} = \frac{\Omega_i + \Omega_j}{2} \cdot \frac{R^2 T_{Cij}^{2.5}}{p_{Cij}} \quad (111)$$

where Ω_i and Ω_j are the newly defined constant values of component *i* and *j*, and T_{Cij} and p_{Cij} are defined according to complex mixing rules [see 24]. The values of T_{Cij} and p_{Cij} are not related to true critical temperatures and pressures of specific binary gas mixtures.

The prediction of the properties of homogeneous fluids at supercritical conditions (Table 5) is only slightly improved compared to *RKeos* [10], but it is not exceeding the accuracy of the *Seos* [11]. At sub-critical condition (Figure 4), the Chueh-Prausnitz equation is less accurate than the Redlich-Kwong equation (compare Figure 3) in the binary CO_2-CH_4 fluid mixture at 15 °C.

composition			$V_m(exp)$	*CP*	*H*	*B1*	*B2*
CO_2	CH_4	N_2	$cm^3 \cdot mol^{-1}$				
0.8	0.1	0.1	56.64	56.42 (-0.4%)	55.96 (-0.6%)	56.84 (0.4%)	56.53 (-0.2%)
0.8	0.2	0.2	58.92	57.85 (-1.8%)	57.68 (-2.1%)	59.43 (0.9%)	58.81 (-0.2%)
0.4	0.3	0.3	61.08	59.21 (-3.1%)	59.17 (-3.1%)	61.67 (1.0%)	60.79 (-0.5%)
0.2	0.4	0.4	62.90	60.44 (-3.9%)	60.38 (-4.0%)	63.45 (0.9%)	62.40 (-0.8%)

Table 5. The same experimental molar volumes as in Table 4 compared with two-constant equations of state according to Chueh and Prausnitz [24] (CP), Holloway [25, 26] (H), Bakker [27] [B1], and Bakker [28] (B2). The percentage of deviation from experimentally obtained molar volumes is indicated in brackets.

9.2. Holloway [25, 26] and Bakker [27]

The equation of Holloway [25, 26] is another modification of the *RKeos* [8]. The modification is mainly based on the improvement of predictions of homogenous fluid properties of H_2O and CO_2 mixtures, using calculated experimental data [29]. The value for ζ_1 and ζ_3 (both *b*) of H_2O is arbitrarily selected at 14.6 $cm^3 \cdot mol^{-1}$, whereas other pure gases are defined according to pseudo critical conditions. The definition of ζ_2 (i.e. *a*) for H_2O as a function of temperature was subjected to a variety of best-fit procedures [25, 26]. The fitting was improved from four experimental data points [25] to six [26] (Figure 5), but was restricted to temperatures above 350 °C. Bakker [27] improved the best-fit equation by including the entire data set [29], down to 50 °C (Eq. 112).

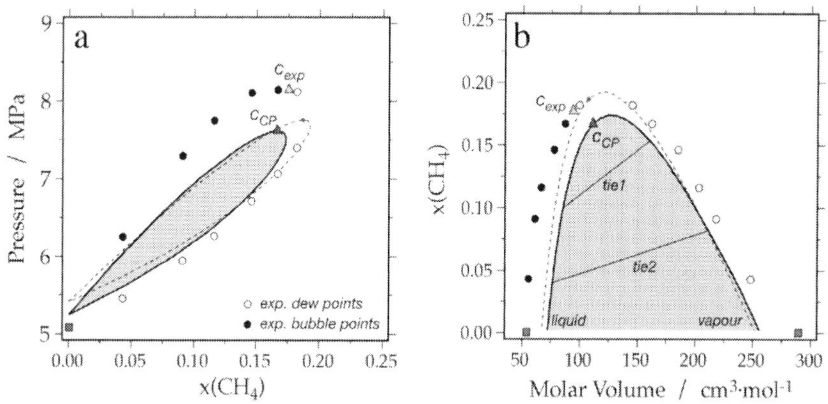

Figure 4. See Figure 3 for details. The *RKeos* is indicated by dashed lines in (**a**) and (**b**). The shaded areas are immiscibility conditions calculated with the Chueh-Prausnitz equation. *tie1* and *tie 2* in (**b**) are calculated tie-lines between two phases at constant pressures 6.944 and 5.984 MPa, respectively.

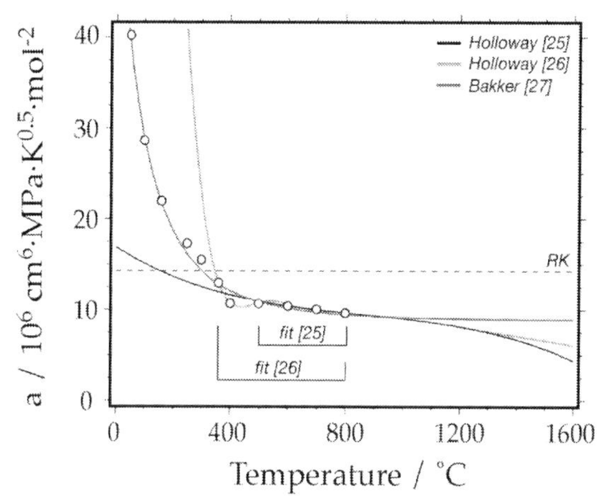

Figure 5. Temperature dependence of the *a* constant for pure H_2O in the modified cubic equation of state [25, 26]. The open circles are calculated experimental data [29]. *fit [25]* is the range of fitting in the definition of Holloway [25], and *fit [26]* of Holloway [26]. *RK* illustrates the constant value calculated from pseudo critical condition [8].

$$a_{H_2O} = \left(9.4654 - \frac{2.0246 \cdot 10^3}{T} + \frac{1.4928 \cdot 10^6}{T^2} + \frac{7.57 \cdot 10^8}{T^3}\right) \cdot 10^6 \quad (112)$$

where T is temperature in Kelvin, and the dimension of a is $cm^6 \cdot MPa \cdot K^{0.5} \cdot mol^{-2}$. The properties of homogeneous pure CO_2, CH_4 and N_2 fluids [27] were also used to obtain a temperature dependent a constant (Eqs. 113, 114, and 115, respectively).

$$a_{CO_2} = \left(-1.2887 + \frac{5.9363 \cdot 10^3}{T} - \frac{1.4124 \cdot 10^6}{T^2} + \frac{1.1767 \cdot 10^8}{T^3}\right) \cdot 10^6 \quad (113)$$

$$a_{CH_4} = \left(-1.1764 + \frac{3.5216 \cdot 10^3}{T} - \frac{1.155 \cdot 10^6}{T^2} + \frac{1.1767 \cdot 10^8}{T^3}\right) \cdot 10^6 \quad (114)$$

$$a_{N_2} = \left(0.060191 - \frac{0.20059 \cdot 10^3}{T} + \frac{0.15386 \cdot 10^6}{T^2}\right) \cdot 10^6 \quad (115)$$

The a_{ij} value of fluid mixtures with a H_2O and CO_2 component (as in Eqs. 106-108 and 110-111) is not defined by the value of pure H_2O and CO_2 (Eqs. 112 and 113), but from a temperature independent constant value (Eqs. 116 and 117, respectively). In addition, a correction factor is used only for binary H_2O-CO_2 mixtures, see [25, 29].

$$a_0(H_2O) = 3.5464 \cdot 10^6 \quad MPa \cdot cm^6 \cdot K^{0.5} \cdot mol^2 \quad (116)$$

$$a_0(CO_2) = 4.661 \cdot 10^6 \quad MPa \cdot cm^6 \cdot K^{0.5} \cdot mol^2 \quad (117)$$

Table 5 illustrates that the equation of Holloway [25] is not improving the accuracy of predicted properties of supercritical CO_2-CH_4-N_2 fluids, compared to Chueh-Prausnitz [24] or *Seos* [9], and it is only a small improvement compared to the *RKeos* [8]. The accuracy of this equation is highly improved by using the definitions of a constants according to Bakker [27] (see Eqs. 112-115), and result in a maximum deviation of only 1% from experimental data in Table 5.

Experimental data, including molar volumes of binary H_2O-CO_2 fluid mixtures at supercritical conditions [30, 31, 32] are used to estimate fugacities of H_2O and CO_2 according to Eq. 118 (compare Eq. 10).

$$RT \ln \varphi_i = \int_0^p \left[V_{m,i} - V_m^{ideal}\right] dp \quad (118)$$

where $V_{m,i} - V_m^{ideal}$ is the difference between the partial molar volume of component i and the molar volume of an ideal gas (see also Eq. 43). The difference between Eqs. 118 and 10 is the mathematical formulation and the use of different independent variables, which are temperature and pressure in Eq. 118. The integration to calculate the fugacity coefficient can be graphically obtained by measuring the surface of a diagram of the difference between the ideal molar volume and the partial molar volume (i.e. $V_{m,i}$ - V_m^{ideal}) as a function of pressure (Figure 6). The surface obtained from experimental data can be directly compared to calculated curves from equations of state, according to Eq. 10 (Table 6).

The dashed line in Figure 6 is calculated with another type of equation of state: a modification of the Lee-Kesler equation of state [33] that is not treated in this manuscript because it is not a two-constant cubic equation of state. Fugacity estimations of H_2O are similar according to both equations, and reveal only a minor improvement for the two-constant cubic equation of state [27]. The experimental data to determine fugacity of CO_2 in this fluid mixture is inconsistent at relative low pressures (< 100 MPa). The calculated fugacity [27] is approximately compatible with the experimental data from [31, 32].

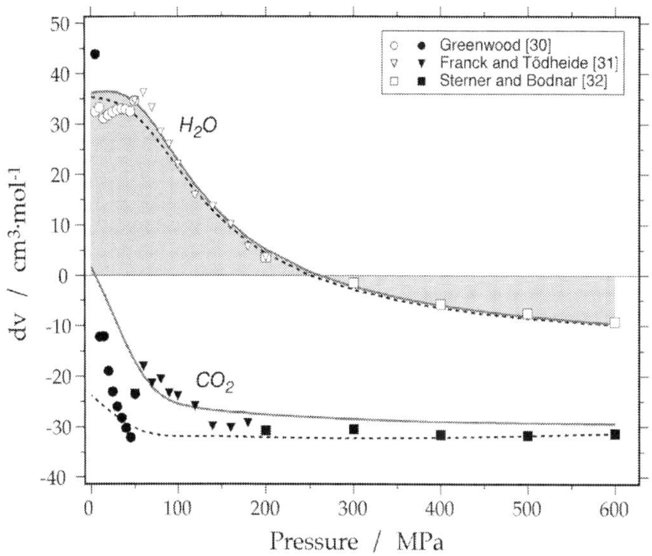

Figure 6. Fugacity estimation in a pressure - dv diagram at 873 K and a composition of $x(CO_2) = 0.3$ in the binary H_2O-CO_2 system, where dv is the molar volume difference of an ideal gas and the partial molar volume of either H_2O or CO_2 in binary mixtures. Experimental data are illustrated with circles, triangles and squares (solid for CO_2 and open for H_2O. The red lines are calculated with Bakker [27], and the shaded area is a measure for the fugacity coefficient of H_2O (Eq. 118).

Pressure (MPa)	*Exp. fugacity (MPa)*	*B1 fugacity (MPa)*
10	6.692	6.659 (-0.5%)
50	27.962	27.3061 (-2.3%)
100	45.341	44.6971 (-1.4%)
200	77.278	75.0515 (-2.9%)
300	114.221	111.072 (-2.8%)
400	160.105	157.145 (-1.8%)
500	219.252	216.817 (-1.1%)
600	295.350	294.216 (-04%)

Table 6. Fugacities of H_2O in H_2O-CO_2 fluid mixtures, $x(CO_2) = 0.3$, at 873.15 K and variable pressures. B1 fugacity is calculated with Bakker [27]. The deviation (in %) is illustrated in brackets.

9.3. Bowers and Helgeson [34] and Bakker [28]

Most natural occurring fluid phases in rock contain variable amounts of NaCl, which have an important influence on the fluid properties. Bowers and Helgeson [34] modified the *RKeos* [8] to be able to reproduce the properties of homogeneous supercritical fluids in the H_2O-CO_2-NaCl system, but only up to 35 mass% NaCl. The model is originally restricted between 350 and 600 °C and pressures above 50 MPa, according to the experimental data [35] that was used to design this equation. This model was modified by Bakker [28] including CH_4, N_2, and additionally any gas with a (ζ_2) and b (ζ_1) constants defined by the pseudo critical conditions (Eqs. 91-93 and 94-96). Experimental data in this multi-component fluid system with NaCl can be accurately reproduced up to 1000 MPa and 1300 K. Table 5 illustrates that this modification results in the best estimated molar volumes in the ternary CO_2-CH_4-N_2 fluid system at 100 MPa and 673 K. Similar to all modifications of the *RKeos* [8], this model cannot be used in and near the immiscibility conditions and critical points (i.e. sub-critical conditions).

10. Application to fluid inclusion research

Knowledge of the properties of fluid phases is of major importance in geological sciences. The interaction between rock and a fluid phase plays a role in many geological processes, such as development of magma [36], metamorphic reactions [37] and ore formation processes [38]. The fluid that is involved in these processes can be entrapped within single crystal of many minerals (e.g. quartz), which may be preserved over millions of years. The information obtained from fluid inclusions includes 1. fluid composition; 2. fluid density; 3. temperature and pressure condition of entrapment; and 4. a temporal evolution of the rock can be reconstructed from presence of various generation of fluid inclusions. An equation of state of fluid phases is the major tool to obtain this information. Microthermometry [39] is an analytical technique that directly uses equations of state to obtain fluid composition and density of fluid inclusions. For example, cooling and heating experiment may reveal fluid phase changes at specific temperatures, such as dissolution and homogenization, which can be transformed in composition and density by using the proper equations of state.

The calculation method of fluid properties is extensive and is susceptible to errors, which is obvious from the mathematics presented in the previous paragraphs. The computer package FLUIDS [6, 40, 41] was developed to facilitate calculations of fluid properties in fluid inclusions, and fluids in general. This package includes the group "Loners" that handles a large variety of equations of state according to individual publications. This group allows researchers to perform mathematical experiments with equations of state and to test the accuracy by comparison with experimental data.

The equations of state handled in this study can be downloaded from the web site **http://fluids.unileoben.ac.at** and include 1. "*LonerW*" [7]; 2. "*LonerRK*" [8]; 3. "*LonerS*" [9]; 4. "*LonerPR*" [10]; 5. "*LonerCP*" [24]; 6. "*LonerH*" [25, 26, 27]; and 7. "*LonerB*" [28, 34]. Each program has to possibility to calculate a variety of fluid properties, including pressure, temperature, molar volume, fugacity, activity, liquid-vapour equilibria, homogenization conditions, spinodal, critical point, entropy, internal energy, enthalpy, Helmholtz energy, Gibbs energy, chemical potentials of pure gases and fluid mixtures. In addition, isochores can be calculated and exported in a text file. The diagrams and tables presented in this study are all calculated with these programs.

Author details

Ronald J. Bakker
Department of Applied Geosciences and Geophysics, Resource Mineralogy, Montanuniversitaet, Leoben, Austria

11. References

[1] Dziewonski AM, Anderson DL (1981) Preliminary reference Earth model. Phys. Earth Planet. In. 25: 297-356.

[2] Press F, Siever R (1999) Understanding Earth. Freeman, New York, 679 p.

[3] Roedder E (1984) Fluid inclusions, Reviews in Mineralogy 12, Mineralogical Association of America, 646 p.

[4] Bakker RJ (2009) Reequilibration of fluid inclusions: Bulk diffusion. Lithos 112: 277-288.

[5] Prausnitz JM, Lichtenthaler RN, Gomes de Azevedo E (1986) Molecular thermodynamics of fluid-phase equilibria. Prentice-Hall, Englewood Cliffs, NJ, 600 p.

[6] Bakker RJ (2009) Package FLUIDS. Part 3: correlations between equations of state, thermodynamics and fluid inclusions. Geofluids 9: 63-74.

[7] Waals JD van der (1873) De continuiteit van den gas- en vloeistof-toestand. PhD Thesis, University Leiden, 134 p.

[8] Redlich OR, Kwong JNS (1949) On the thermodynamics of solutions, V: An equation of state, fugacities of gaseous solutions. Chem Rev. 44: 233-244.

[9] Soave G (1972) Equilibrium constants from a modified Redlich-Kwong equation of state. Chem. Eng. Sci. 27: 1197-1203.

[10] Peng DY, Robinson DB (1976) A new two constant equation of state. Ind. Eng. Chem. Fundam. 15: 59-64.

[11] Reid RC, Prausnitz JM, Poling BE (1989) The properties of gases and liquids. McGraw-Hill Book Company, NJ, 741 p.

[12] Levelt-Sengers J (2002) How fluids unmix, discoveries by the school of van der Waals and Kamerlingh Onnes. Koninklijke Nederlandse Akademie van Wetenschappen, Amsterdam, 302 p.

[13] Beyer WH (1991) CRC Standard mathematical tables and formulae. CRC Press, Boca Raton, Fl, 609 p.

[14] Horn RA, Johnson CR (1985) Matrix analysis. Cambridge University Press, Cambridge, 561 p.

[15] Prausnitz JM, Anderson TF, Grens EA, Eckert CA, Hsieh R, O'Connell JP (1980) Computer calculations for the multicomponent vapor-liquid and liquid-liquid equilibria. Prentice-Hall, Englewood Cliffs, NJ, 353 p.

[16] Donnelly HG, Katz DL (1954) Phase equilibria in the carbon dioxide - methane system. Ind. Eng. Chem. 46: 511-517.

[17] Baker LE, Luks KD (1980) Critical point and saturation pressure calculations for multipoint systems. Soc. Petrol. Eng. J. 20: 15-4.

[18] Konynenburg PH van, Scott RL (1980) Critical lines and phase equilibria in binary van der Waals mixtures. Philos. T. Roy. Soc. 298: 495-540.

[19] Arai Y, Kaminishi GI, Saito S (1971) The experimental determination of the P-V-T-X relations for carbon dioxide-nitrogen and carbon dioxide-methane systems. J. Chem. Eng. Japan 4: 113-122.

[20] Span R, Wagner W (1996) A new equation of state for carbon dioxide covering the fluid region from the triple point temperature to 1100 K at pressures up to 800 MPa. J. Phys. Chem. Ref. Data 25:1509-1596.

[21] Lorentz HA (1881) Über die Anwendung des Satzes vom Virial in den kinetischen Theorie der Gase. Ann. Phys. 12: 127-136.

[22] Waals JD van der (1890) Molekulartheorie eines Körpers, der aus zwei verschiedenen Stoffen besteht. Z. Ph. Chem. 5: 133-173

[23] Seitz JC, Blencoe JG, Joyce DB, Bodnar RJ (1994) Volumetric properties of CO_2-CH_4-N_2 fluids at 200 °C and 1000 bars: a comparison of equations of state and experimental data. Geochim. Cosmochim. Acta 58: 1065-1071.

[24] Chueh PL, Prausnitz JM (1967) Vapor-liquid equilibria at high pressures. Vapor-phase fugacity coefficients in non-polar and quantum-gas mixtures. Ind. Eng. Chem. Fundam. 6: 492-498.

[25] Holloway JR (1977) Fugacity and activity of molecular species in supercritical fluids. In: Fraser DG, editor. Thermodynamics in geology, pp 161-182.

[26] Holloway JR (1981) Composition and volumes of supercritical fluids in the earth's crust. In Hollister LS, Crawford MI, editors. Short course in fluid inclusions: Applications to petrology, pp. 13-38.

[27] Bakker RJ (1999a) Optimal interpretation of microthermometrical data from fluid inclusions: thermodynamic modelling and computer programming. Habilitation Thesis, University Heidelberg, 50 p.

[28] Bakker RJ (1999b) Adaptation of the Bowers and Helgeson (1983) equation of state to the H_2O-CO_2-CH_4-N_2-NaCl system. Chem. Geol. 154: 225-236.

[29] Santis R de, Breedveld GJF, Prausnitz JM (1974) Thermodynamic properties of aqueous gas mixtures at advanced pressures. Ind. Eng. Chem. Process, Dess, Develop. 13: 374-377.

[30] Greenwood HJ (1969) The compressibility of gaseous mixtures of carbon dioxide and water between 0 and 500 bars pressure and 450 and 800 °Centigrade. Am. J.Sci. 267A: 191-208.

[31] Franck EU, Tödheide K (1959) Thermische Eigenschaften überkritischer Mischungen von Kohlendioxyd und Wasser bis zu 750 °C und 2000 Atm. Z. Phys. Chem. Neue Fol. 22: 232-245.

[32] Sterner SM, Bodnar RJ (1991) Synthetic fluid inclusions X. Experimental determinations of the P-V-T-X properties in the CO_2-H_2O system to 6 kb and 700 °C. Am. J. Sci. 291: 1-54.

[33] Duan Z, Møller N, Weare JH (1996) A general equation of state for supercritical fluid mixtures and molecular simulation of mixtures PVTX properties. Geochim. Cosmochim. Acta 60: 1209-1216.

[34] Bowers TS, Helgeson HC (1983) Calculation of the thermodynamic and geochemical consequences of non-ideal mixing in the system H_2O-CO_2-NaCl on phase relations in geological systems: equation of state for H_2O-CO_2-NaCl fluids at high pressures and temperatures. Geochim. Cosmochim. Acta 47: 1247-1275.

[35] Gehrig M (1980) Phasengleichgewichte und pVT-daten ternärer Mischungen aus Wasser, Kohlendioxide und Natriumchlorid bis 3 kbar und 550 °C. University Karlsruhe, PhD-thesis, Hochschul Verlag, Freiburg, 109 p.

[36] Thompson JFH (1995) Magmas, fluids, and ore deposits. Short course 23, Mineralogical Association of Canada.

[37] Spear FS (1995) Metamorphic phase equilibria and pressure-temperature-time paths. Mineralogical Society of America, Monograph, 799 p.

[38] Wilkinson JJ (2001) Fluid inclusions in hydrothermal ore deposits. Lithos 55: 229-272.

[39] Shepherd TJ, Rankin AH, Alderton DHM (1985) A practical guide to fluid inclusion studies. Blackie, Glasgow, 239 p.

[40] Bakker RJ (2003) Package *FLUIDS* 1. Computer programs for analysis of fluid inclusion data and for modelling bulk fluid properties. Chem. Geol. 194: 3-23.

[41] Bakker RJ, Brown PE (2003) Computer modelling in fluid inclusion research. In: Samson I, Anderson A, Marshall D, editors. Short course 32, Mineralogical Association of Canada, pp. 175-212.

Chapter 8

Thermodynamics Simulations Applied to Gas-Solid Materials Fabrication Processes

Elisabeth Blanquet and Ioana Nuta

Additional information is available at the end of the chapter

http://dx.doi.org/10.5772/51377

1. Introduction

The development and the design of materials and/or the processes of their fabrication are generally very time consumer and with expensive operations. Various methods of development can be conceived. Often, "empirical" approaches are adopted: the choice of the experimental parameters is established either on technological or commercial criteria, the optimization being the results of a "trial and error" approach, or on the results of design of experiments (DOE) approach targeted at a property of a material or a parameter of a very particular process. Another approach is to use process modeling: to simulate the process by a more or less simplified model. The modeling of gas-solid materials fabrication processes brings together several physical and chemical fields with variable complexity, starting from thermodynamics and\or kinetics studies up to the mass and heat transport coupled with databases and with thermodynamic and/or kinetics transport properties.

The objective of this chapter is to illustrate the interest areas computer-aided materials design and of processes optimization based on the thermodynamic simulation and giving some interesting examples in different domains. Databases as well as their necessary tools for the implementation of the thermodynamic calculations will be described.

The thermodynamic simulations of multicomponent systems contribute at two important points: the selection of the material and the optimization of the conditions of fabrication. In order to obtain a finely targeted product which meet specific functionalities, it is necessary to answer the following questions:

- what type of composition, quantity, and microstructure of the material allow to obtain such properties?
- it is possible to elaborate the material? By what process, with which reagent/ species and which operating conditions?
- is stable this material during a treatment in temperature, and under a given atmosphere?

- does this material react with its environment (substrate, oven, atmosphere)?

These questions are connected because the properties of the material, essentially conditioned by the microstructure of the final material, are going to depend on its chemical composition, to the process, to the operating conditions and on dimensions of the fabrication equipment.

The answers to these questions can be provided and shown from calculation of phase diagrams, evaluation of chemical reactions, calculation of equilibrium pressures, and from reaction diagrams.

2. Description of the methodologies

To evaluate equilibrium state, two possible approaches exist. The first one is to choose *a priori* a limited number of species and the simple chemical reactions which are susceptible to represent the studied process, and to estimate one or several most favorable reactions. It can be reminded the very classic use of the diagrams of Ellingham for the synthesis of metals from their oxides.

The second approach, more complex, is based on the analysis and the consideration of all the species belonging to the chemical system in the studied process.

The optimization procedure must have the following stages:

a. The analysis of the system with the inventory of all the species reasonably susceptible to be present during the reactions taking place in the process.
b. The construction of a consistent set of thermodynamic data for these species.
c. The thermodynamic calculations at equilibrium of complex system
d. The best representation of the results for the users

Thus, the thermodynamic calculations often give satisfactory results for processes which use high temperatures and residence time or reaction but for processes at low temperature, the kinetic factors must be not neglected. That is why the recent developments of thermodynamic softwares tend to include descriptions of phenomena of diffusion and reaction kinetics.

The thermodynamic approach gives the superior limit of possibilities of process (considering the reaction rates as infinite). It can be the only way of modeling for a complex system where the mechanisms of reaction and the kinetic data are badly known.

To include some dynamic aspects (mass transport) in the modeling, an approach which takes into account the evaluation of flows will be presented. It concerns applications where the total pressure is low (<10 Pa).

2.1. The analysis of the system

The analysis of the system consists in listing the following points: the range of temperature T, range of pressure P or of volume V, the duration of the process, the list of the reagent species, the inventory of the components of the reactor and the nature of the atmosphere. For these last points, it means elaborating a list of all the compounds, the gaseous species, the elements and the solid solutions which result from the combination of the basic elements of the system.

This list is automatically generated thanks to interfaces with databases. However, it is advisable to make sure that the used database is very complete. As an example, the list corresponding to the chemical system Si-C-H-Ar (proceeded CVD (Chemical Vapor Deposition) contains about sixty species – excluding the hydrocarbons C_xH_y where x>3 [1].

2.2. Calculation of a thermochemical equilibrium

In a process reactor, at constant pressure, the balance is reached when the total free Gibbs energy function of the system is minimal (equation 1). To determine the nature and the proportion of the present phases at equilibrium, it is necessary to have the description of the energies of Gibbs of all these phases.

$$\frac{\Delta G}{RT} = \sum_{j=1}^{Ne} qj \frac{\Delta Gj}{RT} \tag{1}$$

where qj the number of moles of the species j, Gj the molar free energy of Gibbs of the species j, Ne total number of species.

The Gibbs energy can be described from the enthalpy (H) and the entropy (S):

$$G(T) = H(T) - T * S(T) \tag{2}$$

with

$$H(T) = \Delta H(298K) + \int_{298K}^{T} c_P(T)dT \tag{3}$$

$$S(T) = S(298K) + \int_{298K}^{T} \frac{c_P(T)}{T} dT \tag{4}$$

The necessary data are thus: Cp(T), ΔH(298K), S(298K) and the data of possible phases transitions Ttrans, ΔHtrans(T).

Various formalisms are adopted for the analytical expression of the function Cp (T). Among them, the formalisms of the SGTE (Scientific Group Thermodata Europe) [2] (equation 5) and of the NASA [3](equation 6) are :

$$c_P(T) = a + bT + cT^2 + \frac{d}{T^2} \tag{5}$$

$$c_P(T) = a + bT + cT^2 + dT^3 + eT^4 \tag{6}$$

where a, b, c, d, e are adjustable parameters.

So, it can be described analytically the Gibbs energy G for a stoichiometric compound (equation 7), for a gas (equation 8) and a solution phase (equation 9):

$$G(T) = A + BT + CT\ln T + DT^2 + ET^3 + \frac{F}{T} \tag{7}$$

$$G(T,P) = G(T) + RT\ln\frac{\bar{P}}{P_0} \tag{8}$$

$$\left.\begin{array}{l} G(T,x) = \sum_i G_i^{ref}(T) + G^{id}(T,x_i) + G^{excès}(T,x_i) \\ G^{id}(T,x_i) = RT\sum_i (x_i \ln x_i) \\ G^{excès}(T,x_i) \quad \textit{described from a} \text{ model} \end{array}\right\} \tag{9}$$

Besides, as neither the enthalpy nor the entropy can be described in an absolute way, a reference state must be used for these two functions of state. For the entropy, the adopted convention consists in taking a zero value at 0 K. In the case of the enthalpy, the most common convention is to choose the stable structure of the element at T = 298K, as standard reference state (e.g. Al cfc, Ti hcp, O_2 gas ...). For the reference state, ΔH(298K)=0 and S(0K)=0.

As the reliability of the results of the thermodynamics simulation depends widely on the quality and on the consistence of the necessary data, it is advisable to attach an importance to the consistence of the available information: thermodynamics measurements, theoretical calculations, characterizations (X-ray diffraction, Environmental Scanning Microscopy), balance of phases (diagrams). In the Table 1 are given some experimental and theoretical techniques usually used to obtain the thermodynamic data.

The thermodynamic information are accessible in compilations of binary phases diagrams (for example Hansen [5], Elliot [6], Massalski [7]), ternary (Ternary Alloys [8]), or specialized journals (CALPHAD, Journal of Phase Equilibria, Intermetallics...), or tables (JANAF Thermochemical Tables [9], Barin [10], Gurvich [11]).

Today, most of data are available in international electronic databases. In Europe, the economic interest group "Scientific Group Thermodata Europe [12]" proposes common data bases for compounds, pure substances and for solutions. Also let us quote the "Coach" data bank (more than 5000 listed species) proposed by Thermodata [1], well adapted to simulate gas/solid processes, the FACT bank (oxides/salts) proposed by the company GTT [13] and the Research Center in Calculation Thermodynamics [14], base TCRAS [15], bases NASA combustion [16], NIST [17].

	Experimental	Theory
ΔH(T)	Calorimetry (dissolution)	Ab initio calculations Estimations : Miedema [4], analogy
Cp(298K), S(298K)	Temperature measurement at low temperature	
Cp(T)		Estimations : Neumann-Kopp law(ΔCp=0 for a condensed compound)
Cp, S(T)	Differential Scanning Calorimetry (DSC)	volume-specific heat capacity calculations
G(T)	Electromotive force, Mass Spectrometry (activity data, partial pressures at equilibrium)	
T^{trans}, H^{trans} (T)	Differential Thermal Analysis (DTA) Thermogravimetric analysis (TGA)	

Table 1. The classically used techniques to obtain the thermodynamic data

2.3. Calculations of complex equilibrium

The software of complex equilibrium is based on the minimization at constant temperature T of the Gibbs energy and constant pressure P (equation 1) or Helmholtz energy (equation 10), at constant volume:

$$\frac{\Delta F}{RT} = \sum_{j=1}^{Ne} q_j \frac{\Delta G_j}{RT} - \frac{P.V}{RT} \tag{10}$$

q_j the number of moles of the species j .

The constraints of mass equilibrium of each present element in the chemical system expressed according to the number of atoms on the pure element i (C elements) are translated by the equation (11):

$$n_i = \sum_{e=1}^{Ne} q_e T_i^e \quad \text{with i} = 1 \ldots \text{C} \tag{11}$$

where T_i^e represents the stoichiometry of the species e for the element i.

These C equations can be translated under the matrix shape (equation 12):

$$\left[n\right] = \left[T\right] * \left[q\right] \text{ with Ne} > \text{C} \tag{12}$$

There are multiple algorithms allowing this minimization. Various classifications were given, the most exhaustive having been supplied by Smith and Missen [18]. In a simple way, two groups of algorithms can be distinguished: on one hand the methods of direct minimization, about zero order for the calculation of the function G, on the other hand, the methods of the first order based on the equality of the chemical potential which require the calculation of the function derivatives. These last ones also include the methods of second order, using among others the algorithm of Newton-Ralphson which is based on the second derivatives. It is necessary to note that the methods of the first order must be perfectly controlled because they can lead to a maximum instead of a minimum and consequently to a wrong result.

A method of the first group is described below: the matrix T is decomposed into a regular square matrix Tp of dimension C and a matrix Td of dimension (C, N_e-C) such as:

$$[n] = [T_p][q_p] + [T_d][q_d] \tag{13}$$

The C species which constitute the matrix column q_p are called the "main species" because they are chosen among the most important species and have by definition a linear independent stoichiometry. The N_e-C remaining species of the matrix T_d is called "derived species" although they are chosen as variables from the minimization. So the N_e-C values q_d are given by the procedure of minimization, C values q_p is calculated by resolving the linear system:

$$[q_p] = [T_p]^{-1}[n] - [T_p]^{-1}[T_d][q_d] \tag{14}$$

An iteration of this method is divided into two steps. Firstly, the phase of exploration, every variable is modified by a value + or - h.

If X_{n-1} is the vector representing the variables after n-1 iterations: the species i having a step h_i and G_i the value of the function

$$G_i^+ = \Delta G(x_1, x_2, ..., x_{i-1}, x_i + h_i, x_{i+1}, ..., x_v) \tag{15}$$

if $G_i^+ < G_i$ either,

$$G_i^{++} = \Delta G(x_1, x_2, ..., x_{i-1}, x_i + 2h_i, x_{i+1}, ..., x_v) \tag{16}$$

or,

$$G_i^- = \Delta G(x_1, x_2, ..., x_{i-1}, x_i - h_i, x_{i+1}, ..., x_v) \tag{17}$$

When the exploration phase is ended, the next step is to move to the second algorithm phase where from the values of G^+ and X_n^+ issued from the exploration phase, we calculate $X^{++} = X_{n-1} + (X_n^+ - X_{n-1})$ as well as the corresponding value G^{++} to obtain the optimal set [19]. To proceed the mimimization procedure, a certain number of more and more friendly softwares are available commercially. It can be listed as example:

- « Gemini1/Gemini2 » [1]
- « FactSage » [14]

- « MTDATA » [20]
- « Thermocalc » [21]

3. Applications

Thomas [22], Bernard [23] and Pons [24] present few examples on CVD processes to illustrate the use of an a priori thermodynamic analysis. In the following paragraphs, it was chosen to show other few examples which evidence the help of thermodynamic modeling in industrial bottlenecks:

- Thermal stability of Metal-Organic Precursors used in CVD and ALD processes.
- Stability of SiC in H_2 atmosphere
- HfO_2 plasma etching
- SiO_2 PVD evaporation-condensation deposition process.

In the last two examples which correspond to processes operating at low pressure (<1 Pa), in addition to pure thermodynamic approach, a dynamic approach was presented which includes calculations of the major species flows.

3.1. Microelectronics: Thermal stability of metal-organic precursors used in CVD and ALD processes

In the pursuit of smaller and faster devices manufacture, microelectronics industry scales down feature sizes and thus has to develop new materials and processes. Nowadays, organometallic precursors are widely used in ALD (Atomic Layer Deposition) and CVD (Chemical Vapor Deposition) deposition processes due to low deposition temperature (generally below 523 K). The objective of computational modeling for gaseous phase processes like ALD or CVD is to correlate the as-grown material quality (uniformity, growth rate, cristallinity, composition, etc) to general parameters such as growth conditions, reactor geometry, as well as local parameters that are actual flow, thermal fields and chemical kinetics at the solid/gas interface.

The gaseous precursors compounds used for the transport of the elements to be deposited by these processes have to meet several physicochemical properties requirements including relatively high volatility, convenient decomposition behavior and thermal stability. The tantalum organometallic precursor pentakis dimethylamino tantalum (PDMAT), remains an attractive solution for tantalum nitride films deposition. Unfortunately, information on physical and chemical behavior of this kind of precursor is scarce and namely species that are formed during vaporization and transported to the deposition chamber remain generally unknown. Thus, the knowledge of thermodynamics of these gaseous compounds could help in the understanding of the transport and growth mechanisms. Indeed, thanks to thermodynamics, it is possible to evaluate what evolves at equilibrium in the precursor source, in the input lines and in the deposition chamber where deposition reactions occur. To control, optimize and understand any ALD or CVD processes, thermodynamic simulations are very useful and therefore data should be primarily assessed.

3.1.1. Assessment of PDMAT thermodynamic data by mass spectrometry

In order to deposit thin layers of TaN using PDMAT, ALD experiments evidenced a cracking of this precursor in the ALD reactor [25].

Cracking reactions of PDMAT can be complex and occur at the same time. A quantitative interpretation of cracking reactions can be deduced from observed molecules by mass spectrometry [26, 27] with the condition that all products and reactants of the reaction are observed and measured by the mass spectrometer at the same time. Without additional hydrogen contribution, the two following cracking reactions of Ta $[N(CH_3)_2]_5(g)$ could occur:

i. either the Ta- $[N(CH_3)_2]$ bond breaks with the additional break of H-CH_2 to produce $HN(CH_3)_2$ (the so-called β substitution):

$$Ta\left[N(CH_3)_2\right]_5(g) \rightarrow Ta\left[N(CH_3)_2\right]_3\left[NCH_3CH_2\right](g) + HN(CH_3)_2(g), \quad (18)$$

ii. or the only bond break of Ta- $[N(CH_3)_2]$,

$$Ta\left[N(CH_3)_2\right]_5(g) \rightarrow Ta\left[N(CH_3)_2\right]_4(g) + N(CH_3)_2(g). \quad (19)$$

As no $N(CH_3)_2$ (g) radical was detected, the observed $HN(CH_3)_2$ (g) molecule could be formed by the following complete and rapid reaction,

$$N(CH_3)_2(g) + \frac{1}{2} H_2(g) \left(\text{ou } H(g)\right) \rightarrow HN(CH_3)_2(g). \quad (20)$$

Consequently, it could be assumed that the measured $HN(CH_3)_2$ (g) amount is the same as the initial produced amount of $N(CH_3)_2$ (g): this radical spontaneously reacts totally according to the reaction (20) after being produced by reaction (19). This mechanism could explain why $N(CH_3)_2$ (g) was not detected. So, in this study, the total cracking reaction was finally considered:

$$Ta\left(N(CH_3)_2\right)_5(g) + \frac{1}{2} H_2(g) \rightarrow Ta\left(N(CH_3)_2\right)_4(g) + HN(CH_3)_2(g) \quad (21)$$

Another cracking reaction could be noticed: $OTaN_4C_8H_{24}$ (g) molecule broke down into $OTaN_3C_6H_{18}$ (g) according to the following reaction,

$$OTaN_4C_8H_{24}(g) + \frac{1}{2} H_2(g) \rightarrow OTaN_3C_6H_{18}(g) + HN(CH_3)_2(g) \quad (22)$$

because the energy of the Ta-N bond is lower than the Ta-O bond.

The experimental study of these two reactions (21) and (22) requires to know or measure H_2(g) pressure. Hydrogen could come from either equilibrium with cracking cell deposits or either molecules losing one or more hydrogen atoms. In this last case, a new molecule should be present. In this study, as no hydrogen was detected or introduced intentionally in the cell, it was assumed that the amino radical is totally consumed and produces $HN(CH_3)_2$ (g) with just a sufficient hydrogen amount. So, it can be assumed that the partial pressure of

$p(HN(CH_3)_2)$ is quite equal to the partial pressure of $p(N(CH_3)_2)$. That allows us to calculate the equilibrium constant of reaction (19):

$$K_p(T) = \frac{p(N(CH_3)_2) \cdot p(Ta[N(CH_3)_2]_4}{p(Ta[N(CH_3)_2]_5} \tag{23}$$

Pressure measurements of these three molecules by mass spectrometry lead to the evaluation of standard enthalpy at 298 K from the third law of thermodynamics:

$$\Delta_r H^0_{298K} = -RT \ln K_p(T) - T \cdot \Delta_r fef^0_T \tag{24}$$

Measured partial pressures of Ta $[N(CH_3)_2]_5$ (g), Ta $[N(CH_3)_2]_4$ (g) and $HN(CH_3)_2$ (g) are elsewhere reported [26, 27].

From this, the average value of $\Delta_r H°_{298\ K}$ was evaluated to be equal to (85±5) kJ/mol.

3.1.2. *Thermodynamic simulation of PDMAT (thermal cracking)*

Thermodynamic simulations, based on the Gibbs free energy minimization of the Ta-C-N-H-(O)-(Ar) system were performed using GEMINI software [1] to provide the nature of the species that should be present at equilibrium under experimental conditions. The sets of thermodynamic data which have been used come from SGTE 2007 database [28] and from the mass spectrometry study for Ta $[N(CH_3)_2]_5$(g), Ta $[N(CH_3)_2]_4$(g),and NC_2H_6 (g) gaseous species [26]. Without any available literature data or any estimates, it cannot be considered any thermodynamic description of $OTaN_xC_yH_z$ (g) gaseous species and intermediate $TaN_xC_yH_z$ (g) species such as $TaN_3C_6H_{16}$ (g), even though these species are expected to appear as observed in mass spectrometric measurements and to play a role in PDMAT cracking and in Ta containing solid formation [26]. Two kinds of simulations have been performed within a temperature range from 400 to 750 K and at 10 Pa, which is our typical mass spectrometric total pressure in the cracking cell. First, homogeneous equilibrium was investigated - no solid phase is allowed to be formed - which corresponds to no deposition i.e. transport in gas lines held at temperature above the saturated one (Figure 1).

Second, a heterogeneous equilibrium - the solid phase is allowed to be formed - has also been simulated, which corresponds to the deposition process occurring in the ALD reactor and in the cracking cell.

In all these thermodynamic simulations, it appeared that Ta $[N(CH_3)_2]_5$(g), (PDMAT) is not stable. In Figure 1 the homogeneous equilibrium calculation show that Ta $[N(CH_3)_2]_4$(g) is stable but disappears after 450 K and Ta(g) is the only one main tantalum containing species after 415 K - but this species will soon be condensed due to large over saturation-. Added to Ta(g), a lot of cracking gaseous species such as N_2 (g), CH_4 (g), H_2 (g) originate from the complete amine decomposition and indeed among these species, NC_2H_7 (g) and NC_2H_6 (g) do not appear. The heterogeneous equilibrium calculations shows the formation of C solid that corresponds to the amine decomposition and this amount of free carbon increases with increasing temperature. Also, the formation of solid TaN was observed within the whole investigated temperature range and no gaseous tantalum containing species pertained contrary to mass spectrometric experiments.

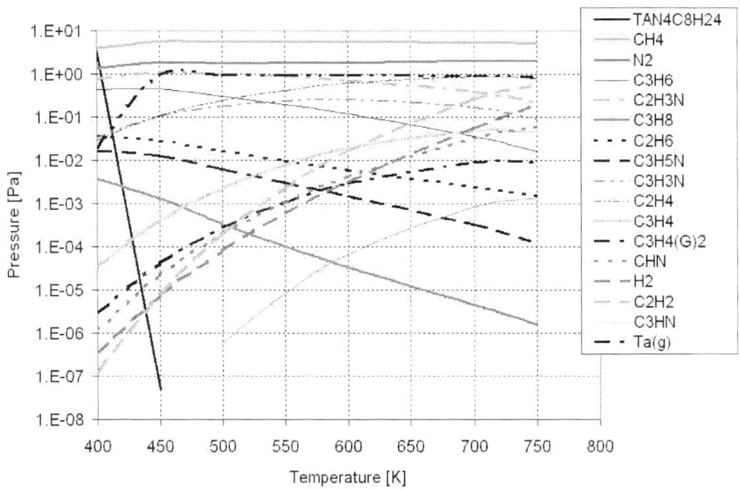

Figure 1. Homogeneous thermodynamic simulations performed starting from 1 mole of the compound Ta $[N(CH_3)_2]_5$(g) and 0.001 mol of Ar and for an applied total pressure equal to 10 Pa in order to be compared with the mass spectrometric experiments.

3.1.3. Conclusions

It is to be concluded that discrepancies exist between thermodynamic simulations and mass spectrometric experiments. Indeed, thermodynamics predicted total cracking of both Ta $[N(CH_3)_2]_5$(g) and amine molecules in the whole investigated temperature ranges, while in mass spectrometric experiments, Ta $[N(CH_3)_2]_5$(g) and NC_2H_7 (g) amine have been observed only in the low temperature below 623 K [26, 29]. The deviation vs. equilibrium could be analyzed experimentally by the use of various sizes of cracking cell as well as the deliberate and controlled introduction of H_2(g). However, despite these limitations, these results indicate the main features of the precursor thermal behavior which can be very useful in the first stages of the development of any new ALD or CVD (for precursor transport) processes.

3.2. High power electronics: Stability of SiC in H_2 atmosphere

Silicon carbide (SiC) possesses many favorable properties making it interesting for a multitude of applications, from high temperature to high frequency and high power device.

Among them, its excellent physico-chemical and electronic properties such as wide band gap and high breakdown field, together with the degree of maturity of technology, makes SiC a good candidate for mass production of Schottky diodes [30].

SiC device processing is conditioned to the fabrication of large area single crystal wafers with the lowest defect density associated to deposition of epitaxial thin films which present good structural quality and controlled doping level [31]. The most common processes used to develop SiC wafers and SiC thin films are the seeded sublimation growth technique so called the "Modified Lely method" and the Chemical Vapor Deposition technique from propane and silane, respectively.

Huge improvements for both processes have been observed in the last decades. They come mainly from extensive experimental effort, all over different groups in the world. However, macroscopic modeling has given valuable information to understand the impact of some growth parameters and propose new design of experiment to enlarge wafer size and deposition area.

On both processes [32-35], some modeling trends were largely reported combined with experimental results obtained in our research's groups.

Special emphasis is given to chemical related results. To carry out modeling, it was followed the different levels of complexity procedure described in the earlier paragraphs. Owing to the similarity of the two systems, studies on species and material databases have been naturally used for both processes.

CVD- grown SiC films can be obtained from a variety of precursors which are generally part of the Si-C-H system. However, to obtain high crystal quality of 4H and 6H SiC layers, which are the most interesting polytypes for the power devices applications, experimental investigations have demonstrated that silane ($SiH_4(g)$) – propane ($C_3H_8(g)$) gave the most stable growth, in the typical conditions (temperature higher than 1700K, pressure between 10 kPa to 100kPa, hydrogen as carrier gas) [36]. Operations are separated in two steps, first an in situ etching step to prevent epitaxy-induced defects, then the deposition step.

A great body of literature dealing with both theoretical and experimental results has been devoted to understanding chemistries relevant to the separate Si-H and C-H systems.

However, it appears that most is unknown about the chemical reactions in which organosilicon species, that include the three elements, can be involved. This is related to the difficulty to measure thermochemical properties of such reactive, short life time, species. Most of the thermodynamic data that have been used for these species come from ab initio electronic structure calculations combined with empiric bond additivity corrections [37, 38]. Mass spectrometry measurements have been carried out to estimate the thermodynamic data of the gaseous species $Si_2C(g)$, $SiC_2(g)$, $SiC(g)$ and the condensed $SiC(s)$ phase [39, 40].

With a purely thermodynamic approach, it was examined the preliminary operation of in situ etching. It was found that the $H_2(g)$ etching of SiC(s) at 1700 K under 100 kPa can lead to the formation of a condensed silicon phase, as shown on Figure 2. Thermodynamic study was made to understand the impact of the temperature, the pressure and the composition of the gas mixture [41].

Heterogeneous thermodynamic calculations show that the mixture ($H_2(g)$ + condensed SiC(s)) ends in the formation of gaseous species such as $CH_4(g)$ for the C-containing species and $SiH_2(g)$, $SiH_4(g)$, SiH(g), Si(g) and Si(s) in condensed phase for the Si-containing species. With the etching, the amount of the gaseous C-species formed (mainly $CH_4(g)$) is three times higher than the gaseous Si-species one. So, there is Si in excess which is condensed at the SiC(s) surface in a solid or liquid phase depending whether the etching temperature is higher or below the Si melting temperature. When the temperature is higher than 1800K, at atmospheric pressure, the quantity of formed gaseous Si-species becomes equal to the quantity of formed gaseous C-species. Consequently, the formation of liquid silicon is avoided.

3.2.1. *Conclusions*

Thermodynanic simulations have revealed the main phenomena and indicated some solutions. Reducing pressure would provide the same beneficial effect, though the etching rate decreases, as illustrated in Figure 3.

To compensate the formation of gaseous $CH_4(g)$, the addition of an hydrocarbon species such as propane in the initial gaseous mixture would prevent the formation of condensed silicon.

All these effects have been confirmed with experimental studies (Figure 2).

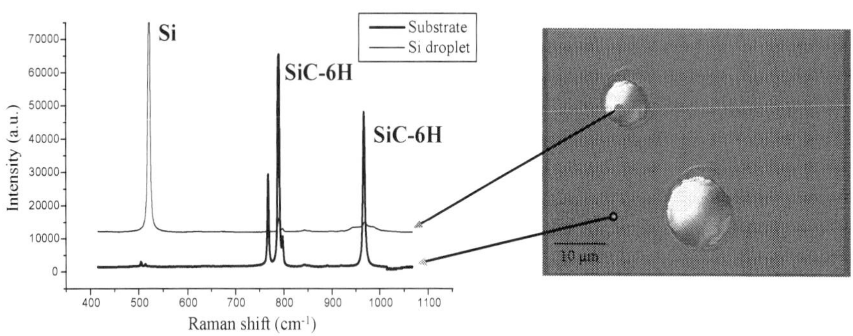

Figure 2. Optical Micrograph of SiC(s) surface etched with $H_2(g)$ at 1700 K under 100 kPa, showing silicon droplets (right). The silicon phase is identified by Raman spectroscopy (left) [41].

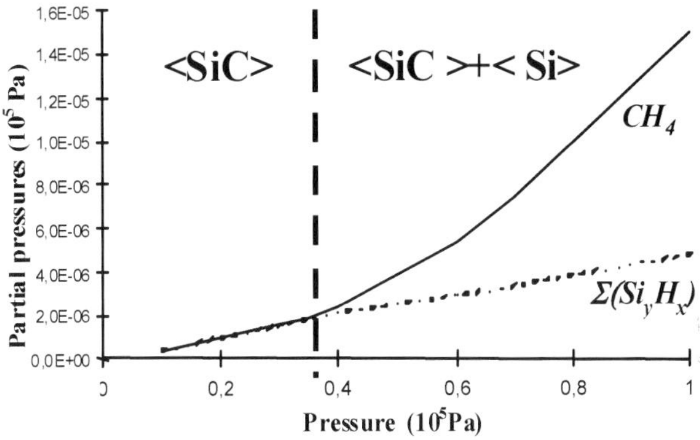

Figure 3. Heterogeneous equilibrium of the SiC(s) - H_2(g) system at 1685K. Gaseous species created by the etching are represented as a function of the pressure [41].

3.3. Thermodynamic analysis of plasma etching processes for microelectronics

With the constant downscaling of Complementary Metal-Oxide-Semiconductor (CMOS) devices and the consequent replacement of SiO_2 many high-*k* gate materials such as Al_2O_3, La_2O_3, Ta_2O_5, TiO_2, HfO_2, ZrO_2 and Y_2O_3 have been investigated. For each high-*k* material integration, the etch process has to be revisited.

In the case of the etching of HfO_2, one of the main issues is the low volatility of halogenated based etch by-products [42].

Compared to SiO_2 halide based etching process, thermodynamic data shows that, Hf based etch by-products ($HfCl_4$(g), $HfBr_4$(g), HfF_4(g)) are less volatile than Si etch by products ($SiCl_4$(g), $SiBr_4$(g), SiF_4(g)) [43]. Therefore, for HfO_2(s) etching, the choice of the halogenated based chemistry and substrate temperature are crucial parameters. In this work, thermodynamic studies have been carried out in the pressure (0.5 Pa) and temperature range (425 K to 625 K) conditions in order to select the most appropriate gas mixture and temperature leading to the formation of Hf and O based volatile products. Based on thermodynamic calculations in a closed system, the HfO_2(s) etching process has been simulated.

With this thermodynamic analyses, it is possible to determine an etch chemistry leading to volatile compounds and to estimate an etch rate under pure chemical etching conditions. It should be noted that the thermodynamic approach does not take into account the ion bombardment of the plasma.

3.3.1. Pure thermodynamic calculations of HfO_2 etching

For example, let's consider the etching of HfO_2(s) in CCl_4(g) plasma at 400 K and 0.5Pa. In such case, the thermodynamic system is composed by four elements Hf, O, C, and Cl. The thermodynamic calculation inputs are:each element of CCl_4(g) (C and Cl atoms) with

$HfO_2(s)$ as a solid phase. The main gaseous species are $CO_2(g)$ and $HfCl_4(g)$ and the main condensed species are $HfCl_4(s)$ on $HfO_2(s)$ in a solid phase. There are other gaseous species in very low amount so that they can be neglected as (CO, Cl_2, Cl). These results show that carbon and chlorine containing chemistries can lead to the etching of $HfO_2(s)$ by forming $CO_2(g)$ and $HfCl_4(g)$. Similar results have been obtained for the other halide chemistries $CCl_3F(g)$, $CCl_2F_2(g)$, $CCl_3F(g)$, $CCl_4(g)$.

3.3.2. *Thermodynamic analysis coupled to mass transport: evaluation of etching rate*

To point out the more promising chemistry among the usually adopted halogens precursors, the etch rate has been estimated from the flow calculations of each gaseous and condensed species under open conditions assuming molecular flow and the validity of the Hertz-Knudsen relation [44].

For these processes operating at low pressure (<10 Pa), it is possible to associate the incident and emitted flows from a given surface to the equilibrium partial pressures [45].

These calculations are based on the effusion calculations principles from the gas kinetic theory.

For a gaseous species e, the total flow Φ_e which is emitted from a vaporizing surface can be calculated according to the Hertz-Knudsen relation:

$$\Phi_e = \frac{p_e}{\sqrt{2\Pi M_e RT}} mol / s.m^2 \tag{25}$$

Where p_e et M_e are the partial pressure and molar mass of the species e, respectively.

For each etched or deposited element i, there is equality between the incident flow and the emitted or produced from reactions flows:

$$\Phi_i(incident) = \Phi_i(emitted) + \Phi_i(condensed) \tag{26}$$

on the deposited or etched surface

For example, in the case of $HfO_2(s)$ etching by $CHCl_3(g)$ with Ar(g), the system is Hf, O, C, H, Cl, Ar

For the previous thermodynamic calculations ($HfO_2(s)$ and $CHCl_3(g)$) the major species at equilibrium are:

- $CO_2(g)$, HCl(g), $HfCl_4(g)$, Ar(g) in the gaseous phase
- C(s), $HfO_2(s)$, $HfCl_4(s)$ in the solid phase

for a temperature of 300 K and a pressure of 5 Pa.

The flow equations are in this case (principle of mass conservation) :

- Flow of incident C = flow of evaporated C + flow of condensed C:

$$\frac{1}{\sqrt{2\pi.R.T}}\frac{P_{CO_2}}{\sqrt{M_{CO_2}}}+F\left(C(s)\right)=\phi\left(CHCl_3(g)\right) \tag{27}$$

- Flow of incident H = flow of evaporated H :

$$\frac{1}{\sqrt{2\pi.R.T}}\frac{P_{HCl}}{\sqrt{M_{HCl}}}=\phi\left(CHCl_3(g)\right) \tag{28}$$

- Flow of incident Cl = flow of evaporated Cl + flow of condensed Cl :

$$\frac{1}{\sqrt{2\pi.R.T}}\left(\frac{P_{HCl}}{\sqrt{M_{HCl}}}+4.\frac{P_{HfCl_4}}{\sqrt{M_{HfCl_4}}}\right)+F\left(HfCl_4(s)\right)=3.\phi\left(CHCl_3(g)\right) \tag{29}$$

with F(.(s)) the flow in solid phase species and $\phi\left(CHCl_3(g)\right)=\frac{p^0_{CHCl_3}}{\sqrt{2\pi.M_{CHCl_3}.R.T_0}}$ (T0 the temperature at the gases inlet and $p^0_{CHCl_3}$ the inlet partial pressure).

So five unknowns are obtained: $P_{CO_2}, F\left(C(s)\right), P_{HCl}, P_{HfCl_4}, F\left(HfCl_4(s)\right)$

To be able to solve this system, five equations are needed. Already, three equations with these ones of flow exist. For the both missing, it's enough to consider:

The equation of total pressure: $P_{CO_2}+P_{HCl}+P_{HfCl_4}+P_{Ar}=P_{totale}$

And the value of the equilibrium constant for the assessment of mass equation of the system, the fifth global equation is supplied by:

$$HfO_2(s)+2.CHCl_3(g)\rightarrow C(s)+\frac{1}{2}.HfCl_4(s)+CO_2(g)+2.HCl(g)+\frac{1}{2}.HfCl_4(g) \tag{30}$$

the equilibrium constant of this equation is :

$$K_p=\frac{P_{CO_2}.P^{\frac{1}{2}}_{HfCl_4}.P^2_{HCl}}{P^2_{CHCl_3}} \tag{31}$$

$$\text{with}:\begin{cases}P_{CO_2}=x_1\\P_{HCl}=x_2\\P_{HfCl_4}=x_3\\F\left(C(s)\right)=x_4\\F\left(HfCl_4(s)\right)=x_5\end{cases}. \tag{32}$$

The system of five equations and five unknowns is obtained:

$$\begin{cases} \dfrac{1}{\sqrt{2\pi.R.T}}\dfrac{x_1}{\sqrt{M_{CO_2}}} + x_4 = \phi(CHCl_3) \\ \dfrac{1}{\sqrt{2\pi.R.T}}\dfrac{x_2}{\sqrt{M_{HCl}}} = \phi(CHCl_3) \\ \dfrac{1}{\sqrt{2\pi.R.T}}\left(\dfrac{x_2}{\sqrt{M_{HCl}}} + 4.\dfrac{x_3}{\sqrt{M_{HfCl_4}}}\right) + x_5 = 3.\phi(CHCl_3) \\ x_1 + x_2 + x_3 + P_{Ar} = P_{Totale} \\ \dfrac{x_1.x_3^{\frac{1}{2}}.x_2^2}{P_{CHCl_3}^2} = K_p \end{cases} \quad (33)$$

The partial pressures of the main species are obtained. To determine the etch rate, it is needed to use the calculated values for the pressures of the gases containing the elements of material to etch.

In our example, the following gases $CO_2(g)$ and $HfCl_4(g)$ are considered.

The theoretical etch rate ER is given by the lowest value between:

ER = flow $HfCl_4(g)$.molar volume $HfCl_4(g)$ or flow $CO_2(g)$.molar volume $CO_2(g)$ (in m/s)

Where:

$$Flow_{CO_2} = \frac{\mathrm{N}}{\sqrt{2\pi.R.T}}.\frac{P_{CO_2}}{\sqrt{M_{CO_2}}} \quad (34)$$

and

$$Flow_{HfCl_4} = \frac{\mathrm{N}}{\sqrt{2\pi.R.T}}.\frac{P_{HfCl_4}}{\sqrt{M_{HfCl_4}}} \quad (\text{in molecules / s.m}^2) \quad (35)$$

with P_{CO_2} and P_{HfCl_4} determined by the resolution of the mathematical system and N the Avogadro number, P in Pa and M in kg.

Figure 4 shows the evolution of the calculated $HfO_2(s)$ etch rate as a function of temperature for different F/Cl ratios in $CCl_xF_y(g)$ based chemistries. The etch rate is lower when the F/Cl ratio increases in the gas mixture at temperature higher than 400 K. The decrease in the etch rate is explained by the non volatility of $HfF_4(g)$ in the investigated temperature range.

3.3.3. *Conclusions*

From these results, thermodynamic studies predict that a chlorocarbon gas mixture -such as CCl_4 -seems to be the most promising chemistry to etch $HfO_2(s)$ under pure chemical etching conditions.

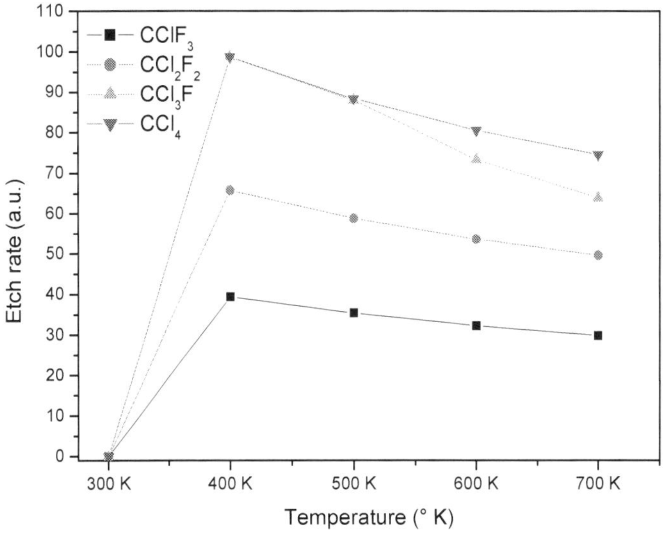

Figure 4. Evolution of the thermodynamically calculated $HfO_2(s)$ etch rate as a function of F-Cl ratio in $CCl_xF_y(g)$ based chemistries.

3.4. Optics: SiO_2 PVD deposition

From optics applications, the example of the evaporation/condensation process to obtain SiO_2 films is chosen. In that process, the surface of the evaporating source is heated by electronic bombardment, while the substrate is held at low temperature.

The control and the reproducibility of this type of process is based on the following points:

- Stability of the source with time (chemical composition, morphology of surface of the evaporating zone)
- Temperature and surface of the evaporated area (what is linked to the parameters of the electronic bombardment).

The object of this study is to simulate the evaporation of a source of glassy silica with the aim of depositing SiO_2. The heated zone is about 3-7 cm^2, the reactor has a volume about 1 m^3, the substrate is located at 1 m from the source.

The first paragraph is dedicated to the pure thermodynamic simulations to determine the major species originated from the evaporation. In the second one, the calculations of the flows of evaporation at equilibrium as well as exchanged flows between source and substrate surfaces are presented.

3.4.1. *Pure thermodynamic calculations of SiO_2 evaporation*

The thermodynamic simulations corresponding to the SiO_2(s) evaporation are realized by considering an excess of solid SiO_2(s) at a given temperature, in a constant volume. The range of tested temperature is 1600 - 2500 K. The results of the simulation indicate that the only solid present at equilibrium is SiO_2(s) and that there is no formation of solid silicon.

In the range of selected temperature, the evaporation of the silica is thus congruent (the ratio of the quantity of silicon and oxygen produced in the gaseous form is equal to 2). Figure 5 presents the nature and the partial pressures of the gaseous species formed at equilibrium.

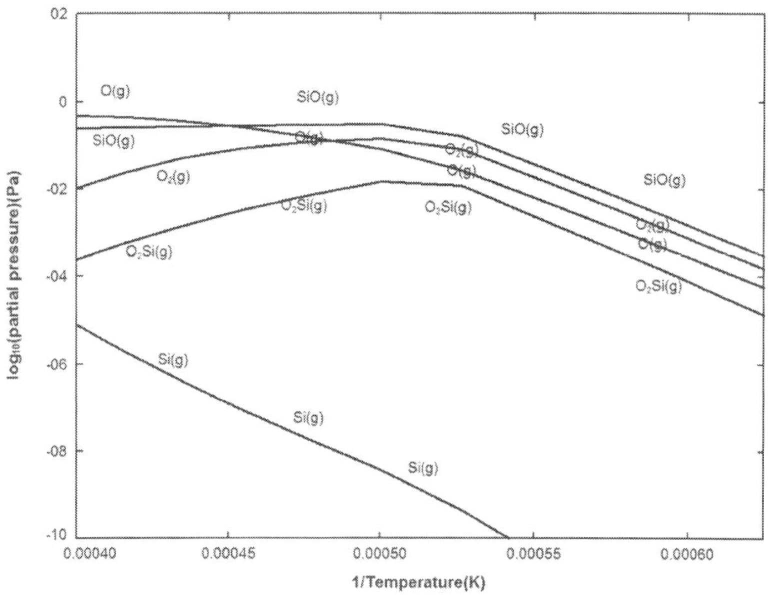

Figure 5. Calculated partial pressures for SiO_2 vaporization.

The major species in this range of temperature are SiO(g), O_2(g), O(g) and SiO_2(g), with trace of Si(g). It can be noted that the mainly evaporated species is not SiO_2(g) as it could believed to justify the stoichiometric composition of the deposits. From the results of the molar fractions calculated for various species, the gas phase reaction which takes place is globally the following:

$$SiO_2(s) = SiO(g) + 0.42\, O_2(g) + 0.15\, O(g) \tag{36}$$

These curves show that the evaporated material quantity and consequently the evaporation rate increases with temperature. That explains the best results obtained with sources carried beyond their melting point (besides the higher quality of the surface with regard to a solid source).

The calculated total pressure above silica is represented on the figure 5.

For the temperatures of evaporation above 1600 K, the total pressure over the silica is superior to 0.1 Pa. If the total pressure is fixed to a lower value, there is then complete consumption of the quantity of silica carried at the evaporation temperature from a thermodynamic point of view.

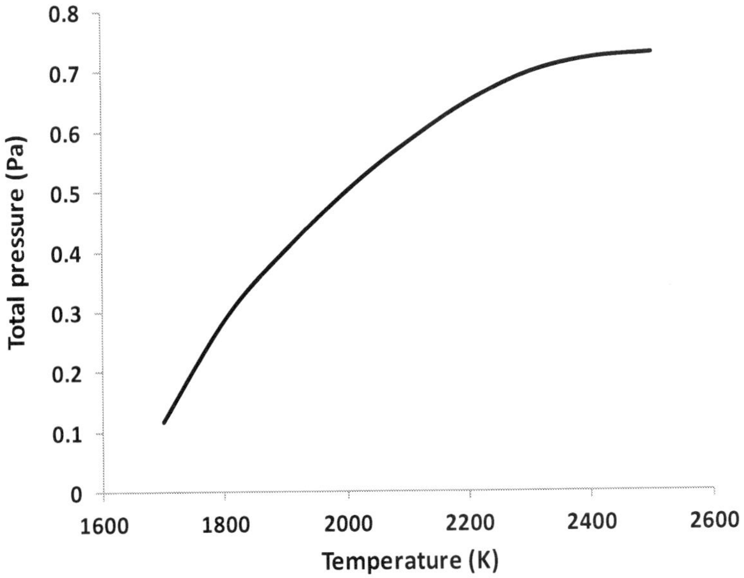

Figure 6. Calculated total pressure for $SiO_2(s)$ vaporization

3.4.2. *Thermodynamic analysis coupled to mass transport: evaluation of deposition rate*

The evaporation flows of the gaseous species originating from the $SiO_2(s)$ evaporation should respect the congruent vaporization relation (ratio Si/O in the gaseous phase = 2), demonstrated by the previous approach.

$$2 * \Phi_{O_2} + \Phi_{SiO} + \Phi_{O} + 2 * \Phi_{Si_2O_2} + 2 * \Phi_{SiO_2} = \Phi_{Si} + 2 * \Phi_{SiO} + \Phi_{O} + 4 * \Phi_{Si_2O_2} + 2 * \Phi_{SiO_2} \quad (37)$$

where Φ_e is the molecular flow of the species e, according to equation (25).

As illustrated by the previous etching case, it is possible to calculate all the species partial pressures (p_e) which verify the equation (37).

From the calculated flows from equation (37), the molar volume of $SiO_2(s)$, the reactor geometry, and the temperature conditions on the surface, it is possible to estimate the growth rate and the deposition profile (Figure 7). The growth rate on the substrate is given from the exchanged flows between the two surfaces source and substrate, from basic assumptions of molecular flows.

With coaxial source and substrate, the exchanged flow between the r_0 radius source and r_1, radius substrate, separated by a distance h is given by the relation (38).

$$\Phi_e = \frac{\alpha_e p_e}{\sqrt{2\Pi M_e RT}} * \frac{\Pi}{2} * [h^2 + r_0^2 + r_1^2 - \sqrt{(h^2 + r_0^2 + r_1^2)^2 - 4 * r_0^2 r_1^2}\,] mol / s.m^2 \tag{38}$$

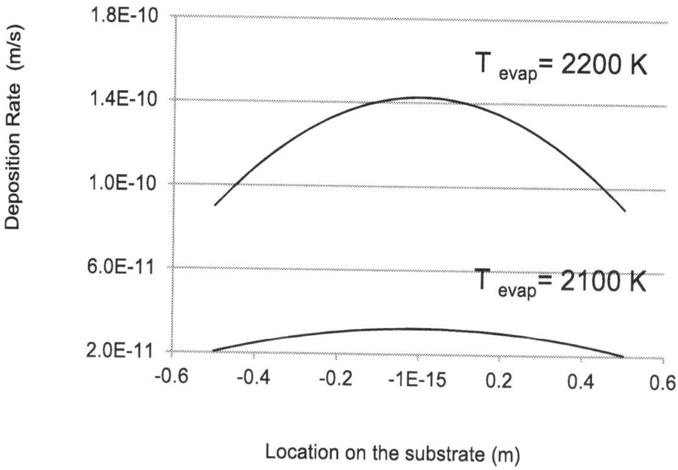

Figure 7. Deposition profile vs deposition temperature for two evaporation temperatures; Substrate radius r_0= 0.5 m; evaporating source surface= 3cm², h= 1m

The results of the simulations are the same order as the obtained experimental values.

3.4.3. *Conclusions*

The simulations of the evaporation of $SiO_2(s)$ show that it is congruent and that there is thus no evolution of the load in time. The mainly produced gaseous species are SiO(g), $O_2(g)$, and O(g). Their proportions remain constant but their quantities increase with the temperature of evaporation. Simulations of the evaporation/condensation process provide good estimations of the deposition rate.

4. Conclusions

This review illustrates the interest to operate a priori an thermodynamic approach to determine the feasibility and optimize a fabrication process, specially gas-solid fabrication process. Kinetic approaches will give rise to more realistic simulations but are often difficult to implement, for lack of reliable information. The classic pure thermodynamic can provide useful information. It can be the only approach in the case of complex chemical systems for which few kinetic data are available. To take into account the dynamic character of the processes, the approaches mixing thermodynamics simulations and calculations of exchanged flows are possible. In every case, the methodology has to contain continuous comparisons between experimental results and simulations.

Author details

Elisabeth Blanquet and Ioana Nuta
Laboratory "Science et Ingénierie des Matériaux et Procédés (SIMaP)"
Grenoble INP/CNRS/UJF, Saint Martin d'Hères, France

Acknowledgement

This paper was inspired by the collaborative works with the colleagues of the SIMaP (Laboratoire de Science et Ingénierie des Matériaux et Procédés): Jean-Noël Barbier, Claude Bernard, Christian Chatillon, Alexander Pisch, Arnaud Mantoux, Raphaël Boichot, Michel Pons.

5. References

[1] Cheynet, B., Chevalier, P.-Y. ,Fischer, E. (2002) Thermosuite. Calphad. 26: 167-174.
[2] Hack, K. (1996) The SGTE casebook: thermodynamics at work. Scientific Group Thermodata Europe and Institute of Materials.
[3] McBride, B. J., Zehe, M. J. ,Gordon, S. (2002) NASA Glenn Coefficients for Calculating Thermodynamic Properties of Individual Species NASA.
[4] Miedema, A., De Chatel, P. ,De Boer, F. R. (1980) Cohesion in alloys—fundamentals of a semi-empirical model. Physica B+ C. 100: 1-28.
[5] Hansen, M. ,Anderko, K. (1958) Constitution of Binary Alloys, Metallurgy and Metallurgical Engineering Series. New York: McGraw-Hill
[6] Eliott, R. (1965) Constitution of Binary Alloys, First Supplement. New York: McGraw-Hill.
[7] Massalski, T. B., Okamoto, H., Subramanian, P. R. ,Kacprazac, L. (1990) Binary alloy phase diagrams. Ohio: ASM International, Materials Park.
[8] Villars, P., Prince, A. ,Okamoto, H. (1995) Handbook of Ternary Alloy Phase Diagrams. Ohio: ASM International, Materials Park.
[9] Chase, M. (1998) NIST-JANAF thermochemical tables. Washington, D.C. and Woodbury, N.Y.: American Chemical Society.
[10] Barin, I. (1993) Thermochemical data of pure substances. New York: VCH.
[11] Gurvich, L. V., Veyts, I. ,Alcock, C. B. (1990) Thermodynamic Properties of Individual Substances. New York: Hemisphere Pub.

[12] SGTE, S. G. T. E. 38402 Saint Martin d'Heres, France.

[13] GTT-Technologies. 52134 Herzogenrath, Germany.

[14] Bale, C., Chartrand, P., Degterov, S., Eriksson, G., Hack, K., Ben Mahfoud, R., Melançon, J., Pelton, A. ,Petersen, S. (2002) FactSage thermochemical software and databases. Calphad. 26: 189-228.

[15] Belov, G. V., Iorish, V. S. ,Yungman, V. S. (1999) IVTANTHERMO for Windows--database on thermodynamic properties and related software. Calphad. 23: 173-180.

[16] Warnatz, J. (1984) Combustion chemistry. New York.

[17] (2005) NIST Chemistry WebBook, NIST Standard Reference Database. Gaithersburg MD.

[18] Smith, W. R. ,Missen, R. W. (1982) Chemical reaction equilibrium analysis: theory and algorithms. New York: Wiley

[19] Barbier, J. N. ,Bernard, C. (1986). In Calphad XV. Fulmer Grange, U.K.

[20] Davies, R., Dinsdale, A., Gisby, J., Robinson, J. ,Martin, S. (2002) MTDATA-thermodynamic and phase equilibrium software from the National Physical Laboratory. Calphad. 26: 229-271.

[21] Sundman, B., Jansson, B. ,Andersson, J. O. (1985) The thermo-calc databank system. Calphad. 9: 153-190.

[22] Thomas, N., Suryanarayana, P., Blanquet, E., Vahlas, C., Madar, R. ,Bernard, C. (1993) LPCVD WSi_2 Films Using Tungsten Chlorides and Silane. Journal of the Electrochemical Society. 140: 475-484.

[23] Bernard, C., Pons, M., Blanquet, E. ,Madar, R. (1999) Computer simulations from thermodynamic data: Materials productions and develeompment-Thermodynamic Calculations as the Basis for CVD Production of Silicide Coatings. MRS Bulletin-Materials Research Society. 24: 27-31.

[24] Pons, M., Bernard, C., Blanquet, E. ,Madar, R. (2000) Combined thermodynamic and mass transport modeling for material processing from the vapor phase. Thin Solid Films. 365: 264-274.

[25] Brize, V., Prieur, T., Violet, P., Artaud, L., Berthome, G., Blanquet, E., Boichot, R., Coindeau, S., Doisneau, B., Farcy, A., Mantoux, A., Nuta, I., Pons, M. ,Volpi, F. (2011) Developments of TaN ALD Process for 3D Conformal Coatings. Chem. Vapor Depos. 17: 284-295.

[26] Violet, P., Blanquet, E., Monnier, D., Nuta, I. ,Chatillon, C. (2009) Experimental thermodynamics for the evaluation of ALD growth processes. Surface and Coatings Technology. 204: 882-886.

[27] Violet, P., Nuta, I., Chatillon, C. ,Blanquet, E. (2007) Knudsen cell mass spectrometry applied to the investigation of organometallic precursors vapours. Surface and Coatings Technology. 201: 8813-8817.

[28] Dinsdale, A. (1991) SGTE data for pure elements. Calphad. 15: 317-425.

[29] Violet, P., Nuta, I., Artaud, L., Collas, H., Blanquet, E. ,Chatillon, C. (2009) A special reactor coupled with a high-temperature mass spectrometer for the investigation of the vaporization and cracking of organometallic compounds. Rapid Communications in Mass Spectrometry. 23: 793-800.

[30] Di Cioccio, L. ,Billon, T.(2002) Advances in SiC materials and technology for Schottky diode applications. In Silicon Carbide and Related Materials 2001, Pts 1 and 2, Proceedings, eds. Yoshida, S., Nishino, S., Harima, H. & Kimoto, T., 1119-1124.

[31] Matsunami, H. (2000) An overview of SiC growth. 125-130. Trans Tech Publ.

[32] Pons, M., Blanquet, E., Dedulle, J., Garcon, I., Madar, R. ,Bernard, C. (1996) Thermodynamic heat transfer and mass transport modeling of the sublimation growth of silicon carbide crystals. Journal of the Electrochemical Society. 143: 3727-3735.

[33] Pons, M., Anikin, M., Chourou, K., Dedulle, J., Madar, R., Blanquet, E., Pisch, A., Bernard, C., Grosse, P. ,Faure, C. (1999) State of the art in the modelling of SiC sublimation growth. Materials Science and Engineering: B. 61: 18-28.

[34] Pons, M., Baillet, F., Blanquet, E., Pernot, E., Madar, R., Chaussende, D., Mermoux, M., Di Coccio, L., Ferret, P., Feuillet, G., Faure, C. ,Billon, T. (2003) Vapor phase techniques for the fabrication of homoepitaxial layers of silicon carbide: process modeling and characterization. Applied Surface Science. 212–213: 177-183.

[35] Bernard, C., Blanquet, E. ,Pons, M. (2007) Chemical vapor deposition of thin films and coatings: Evaluation and process modeling. Surface and Coatings Technology. 202: 790-797.

[36] Hallin, C., Ivanov, I.G., Henry, A., Egilsson, T., Kordina, O., and Janzen, E. 1998, J. Crystal Growth, 183, 1-2, 163.

[37] Allendorf, M. D. ,Melius, C. F. (1992) Theoretical study of the thermochemistry of molecules in the silicon-carbon-hydrogen system. The Journal of Physical Chemistry. 96: 428-437.

[38] Allendorf, M. D. ,Melius, C. F. (1993) Theoretical study of thermochemistry of molecules in the silicon-carbon-chlorine-hydrogn system. The Journal of Physical Chemistry. 97: 720-728.

[39] Rocabois, P., Chatillon, C. ,Bernard, C. (1995) Thermodynamics of the Si-C system I. mass spectrometry studies of the condensed phases at high temperature. High Temperatures. High Pressures. 27: 3-23.

[40] Rocabois, P., Chatillon, C., Bernard, C. ,Genet, F. (1995) Thermodynamics of the Si-C system II. Mass spectrometric determination of the enthalpies of formation of molecules in the gaseous phase. High Temperatures. High Pressures. 27: 25-39.

[41] Neyret, E., Di Cioccio, L., Blanquet, E., Raffy, C., Pudda, C., Billon, T. ,J., C. (2000) SiC in situ pre-growth etching : a thermodynamic study. In Int. Conf. on Silicon Carbide and Related Materials, 1041-1044. Materials Science Forum.

[42] Helot, M., Chevolleau, T., Vallier, L., Joubert, O., Blanquet, E., Pisch, A., Mangiagalli, P. ,Lill, T. (2006) Plasma etching of HfO at elevated temperatures in chlorine-based chemistry. Journal of Vacuum Science & Technology A: Vacuum, Surfaces, and Films. 24: 30.

[43] Chen, J., Yoo, W. J., Tan, Z. Y. L., Wang, Y. ,Chan, D. S. H. (2004) Investigation of etching properties of HfO based high-K dielectrics using inductively coupled plasma. Journal of Vacuum Science & Technology A: Vacuum, Surfaces, and Films. 22: 1552-1558.

[44] Dumas, L., Chatillon, C. ,Quesnel, E. (2001) Thermodynamic calculations of congruent vaporization and interactions with residual water during magnesium fluoride vacuum deposition. Journal of Crystal Growth. 222: 215-234.
[45] Shen, J.-y. ,Chatillon, C. (1990) Thermodynamic calculations of congruent vaporization in III–V systems; Applications to the In-As, Ga-As and Ga-In-As systems. Journal of Crystal Growth. 106: 543-552.

Section 4

Material and Products

Chapter 9

Application of Thermodynamics and Kinetics in Materials Engineering

Lin Li

Additional information is available at the end of the chapter

http://dx.doi.org/10.5772/50335

1. Introduction

1.1. Background

During the last 30 years, there was a surge for coupling thermodynamics and phase diagrams in the field of material science. Many large programmers as well as the exploring of versatile solution models, the description of the magnetism energy of elements and solutions, etc. were developed. Thermo-Calc, as one of the most successful commercial software package for calculation of phase equilibrium from a broad database, offered appropriate guidance for the materials study and reduced the amount of experiments. For example, the slag database in this commercial software package could be used for predicting the composition of oxide and sulphide inclusion formed during the deoxidation and solidification process with a specific multiphase equilibrium calculation.

However, there was still far from much work in the field of prediction or estimation based on the principles of thermodynamics and kinetics for the practical processes in mass production or the designing of kinds of materials. This chapter aims at introducing some computational results on the designing of advanced materials.

2. Research and development on Transformation Induced Plasticity (TRIP) steel

2.1. TRIP steel containing Al

Low carbon low alloy TRIP steel receives significant interest in automobile industry since it offers excellent combination of high strength and formability for stretch forming or deep drawing applications. The high formability in this steel is attributed to the transformation induced plasticity of retained austenite. The amount and stability of retained austenite are the key factors, which influences the transformation behavior during straining.

As well known high Si content suppresses the formation of cementite during the bainitic transformation and it leads to an increase of the stability and the amount of retained austenite. However, high Si content might cause problems in steel production such as strong oxide layer, poor surface characteristics and low coat-ability. This leads to galvanizing problems.

Therefore, efforts are made to improve the composition concept, say, the substitution of Si by other elements which might not spoil surface quality. In the first phase, different amount of Al is added as an alternative to the steel to substitute for Si[1]. In order to understand the complex effects of Al, Si in TRIP steel, many samples of different composition are prepared. As an example, the overall compositions of five samples of them are listed in Table 1[2], where steel no.2, 4, 11 with high Al content are designed as the galvanizing base material with high requirement on elongation, steel no.5 and 10 are compared materials.

No.	C	Mn	Si	Al	P	Fe
2	0.18	1.56	0.02	1.73	0.017	Bal.
4	0.18	1.65	0.45	1.01	0.015	Bal.
5	0.21	1.41	1.07	0.32	0.017	Bal.
10	0.19	1.47	0.87	0.33	0.024	Bal.
11	0.19	1.47	0.22	0.94	0.024	Bal.

Table 1. Compositions of TRIP steels (in wt-%)[2]

Simulation of the inter-critical annealing at 780°C of steels is performed with Thermo-calc and its database. The Gibbs free energy of an arbitrary phase is listed as:

$$G_m = \sum_{I0} P_{I0}(Y) \cdot G_{I0}^{O} + RT \sum_{S} N^S \sum_{i} y_i^S \ln y_i^S + \sum_{Z>0} \sum_{IZ} P_{IZ}(Y) \cdot L_{IZ} \tag{1}$$

The first summation in equation (1) represented the standard energy, the second the mixing entropy and the third the excess Gibbs energy.

The results[2] (see in Table 2) show steel no.2 and 4 must have good plasticity and fairly good strength since both have high volume fraction of α phase (ferrite), though the γ phase (austenite) is less, it can still induce some bainite transformation in the low temperature and raise the strength of the steels. From Table 2[2], it seems steel no.11 may have good combination and highest value of strength and plasticity since it possesses much higher γ phase than steel no.2 and 4 and also the high α phase volume fraction which is favor to ductility.

Simulation of the continuous annealing process in the inter-critical temperature is performed with Dictra software package[2]. The concentration profiles in the equilibrium phases are estimated with the 'number-fixed frame of reference' with respect to the substitutional elements as follows:

$$\sum j_k^{\alpha} = \sum j_k^{\beta} = 0 \tag{2}$$

where j is flux and k was substitutional. For crystalline phases of iron based alloys, the vacancy exchange mechanism of diffusion was predominant, thus, in the lattice fixed frame of reference, the diffusion flux of component k could be written as:

$$\tilde{J}_k = -x_k \Omega_k \nabla \mu_k \quad (3)$$

where x_k was the mole fraction of k, Ω_k represented mobility and $\nabla\mu_k$ the gradient of chemical potential of element k.

Employing a coupled thermodynamic/kinetic method for the solution of the related equations, the concentration profile of the various elements is obtained. Some of the obtained diffusion data are used to guide the processes in mill.

Mechanical properties of the produced TRIP steels are shown in Fig.1[2], where all the steels are inter-critical annealed at 780°C for five minutes and then cooled to different temperatures in the bainite transformation zone.

steel no.	phase	C	Mn	Si	Al	P	α%	γ%
2	α	2.66E-4	1.18E-2	1.96E-4	2.00E-2	2.05E-4	73.4	26.6
	γ	6.07E-3	2.61E-2	2.10E-4	9.66E-3	7.19E-5		
4	α	1.71E-4	1.14E-2	4.58E-3	1.24E-2	1.98E-4	63.4	36.6
	γ	4.65E-3	2.55E-2	4.36E-3	6.00E-3	6.62E-5		
5	α	1.30E-4	8.57E-3	1.16E-2	4.22E-3	2.52E-4	51.1	48.9
	γ	4.18E-3	2.00E-2	9.71E-3	2.12E-3	8.32E-5		
10	α	1.20E-4	9.04E-3	9.40E-3	4.32E-3	3.52E-4	52.4	47.6
	γ	3.88E-3	2.10E-2	7.92E-3	2.17E-3	1.16E-4		
11	α	1.65E-4	9.77E-3	2.27E-3	1.21E-2	3.38E-4	54.5	45.5
	γ	3.99E-3	2.06E-2	2.12E-3	6.15E-3	1.22E-4		

Table 2. Calculated equilibrium compositions (in wt-%) at 780°C and corresponding vol.% of α and γ phases[2]

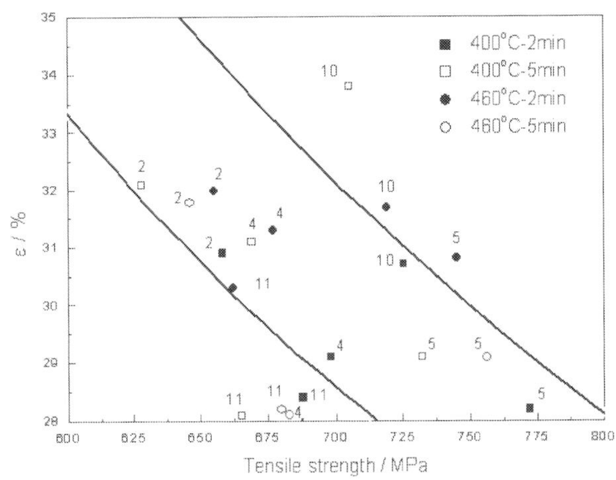

Figure 1. Mechanical properties of the various TRIP –aided steels after different heat treatment[2]

Mechanical properties of steel no.2 and 4 are in accordance with thermodynamic estimation except steel no.11 which does not show high strength and elongation rate as expected. Kinetic calculation results in Fig.2-5 explain that phenomena: after inter-critical annealing, the distribution of C and Mn in steels no.5, 10 and 11 does not exhibit substantial difference. However, high Si concentrates in the γ phase of steels no.5 and 10 while high Al in steel no.11. It follows from that result Al can substitute for Si to induce TRIP effect but its function is not as strong as Si.

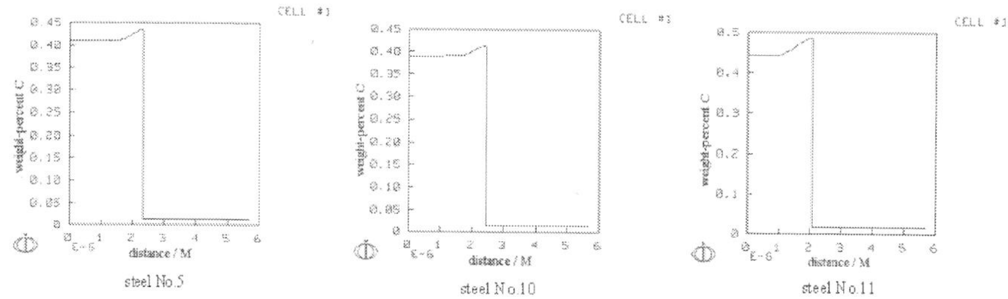

Figure 2. Concentration profile of C at the α/γ interface during inter-critical annealing at 780°C of the steel No.5, No.10 and No.11

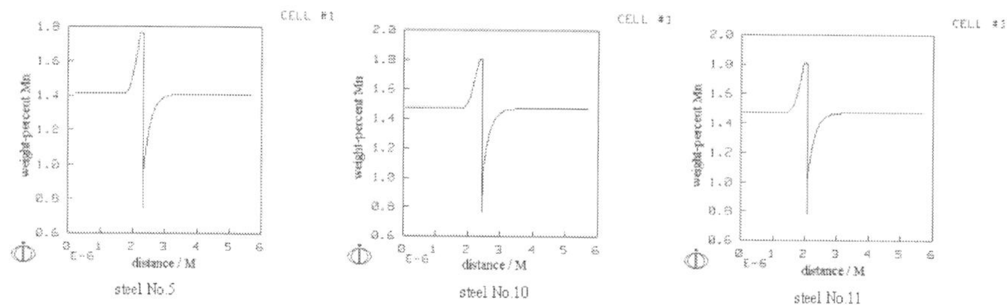

Figure 3. Concentration profile of Mn at the α/γ interface during inter-critical annealing at 780°C of the steel No.5, No.10 and No.11

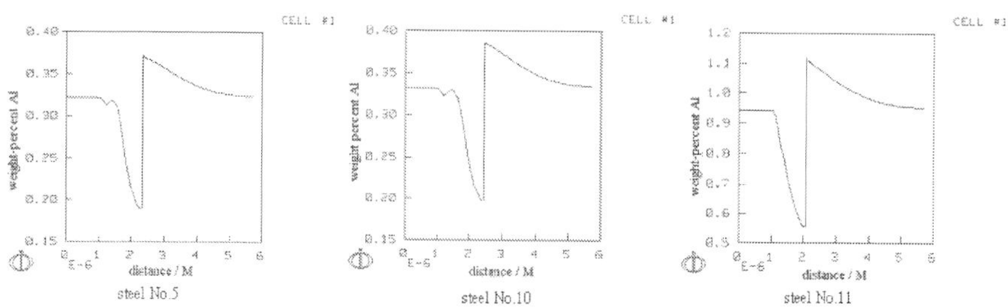

Figure 4. Concentration profile of Al at the α/γ interface during inter-critical annealing at 780°C of the steel No.5, No.10 and No.11

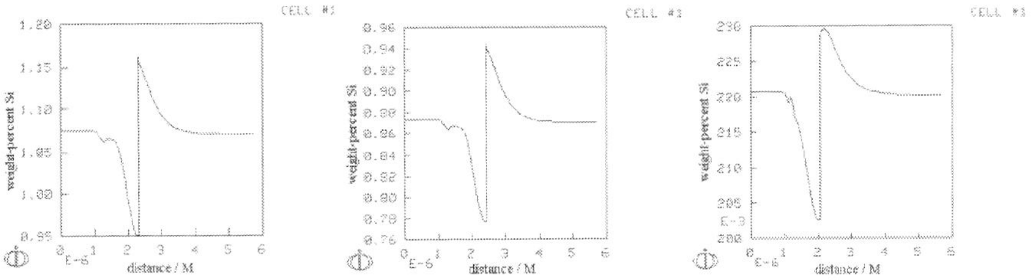

steel No.5 steel No.10 steel No.11

Figure 5. Concentration profile of Si at the α/γ interface during inter-critical annealing at 780°C of the steel No.5, No.10 and No.11

Different from steels no.5 and 10, steels no.2, 4 and 11 therefore possess both super coat-ability and satisfactory mechanical properties due to their high Al content. These three steels mentioned above were adopted and produced in ThyssenKrupp stahl and Voest-Alpine Arcelor and the sheet steels were manufactured as important structural parts, e.g., reinforcement A-pillars and reinforcement B-pillars in FIAT and PORSCHE, which are the pioneers in the organization of ULSAB (Ultra-Light Steel Auto Body).

2.2. TRIP steel containing P

As stated above, in order to replace Si partially and/or totally, efforts were made with Al instead[1,2]. The substitution was successful but only limited at low Al content. Side effect of high Al content in the steel is estimated that it may cause clogging during continuous casting due to precipitation of aluminum oxide[3]. Searching for the other effective element to substitute for Si was then triggered. In the periodic table of elements, Al, Si, and P are located nearby in the same period, these elements have similar electronic structures and may have similar properties. This comes true especially to Si and P since they both are semi-metals. It is then natural to consider P as another substitute for Si. P can, on one hand, depress the carbon activity in cementite, prevent cementite precipitation and increase the stability of austenite. On the other hand, this element has a strong tendency to segregate at grain boundary and to induce cold brittleness. Segregation formulation was developed by Guttmann and Mclean in a ternary system (Fe, M, I) [4] based on the sublattice model in melt salts and stoichiometric phases contributed by Hillert et al[5]. Li et al. extended the segregation equation into a five-element system[6], and generalized it into a multicomponent system as[7]:

$$y_I^g / y_V^g = (y_I^B / y_V^B)\exp\{[\Delta^0 G_I + 1/c^g \sum_{i\neq A} y_i^g \beta_{iI}^g - 1/c^B \sum_{i\neq A} y_i^B \beta_{iI}^B - 1/c^g \sum_{J\neq I} y_J^g L_{A:JI}^g + 1/c^g \sum_{J\neq I} y_J^g L_{A:JV}^g + 1/c^B \sum_{J\neq I} y_J^B L_{A:JI}^B - 1/c^B \sum_{J} y_J^B L_{A:JV}^B] / RT\} \quad (4)$$

$$y_M^g / y_A^g = (y_M^B / y_A^B)\exp\{[\Delta^0 G_M + 1/a^g \sum y_I^g \beta_{MI}^g - 1/a^B \sum y_I^B \beta_{MI}^B - 1/a^g \sum_{i \neq M} y_i^g L_{i:MV}^g + $$
$$+ 1/a^B \sum_{i \neq M} y_i^B L_{i:MV}^B + 1/a^B \sum_{i \neq A} y_i^g L_{Ai:V}^g - 1/a^B \sum_{i \neq A} y_i^B L_{Ai:V}^B] / RT\} \quad (5)$$

where variable y indicates site fraction in sublattice, g grain boundary, B matrix, V vacancy, A base metal, J and I represent impurity, i and M metal element, c and a fraction of sites available in grain boundary or matrix, $\Delta^0 G_I$ and $\Delta^0 G_M$ are the intrinsic segregation Gibbs free energy of impurity I and metal element M, $L_{A:JI}^{B(g)}$ stands for the interaction energy between J and I in one sublattice when another sublattice is fully occupied by element A, $\beta_{MI}^{B(g)}$ means the interaction energy between elements in different sublattices. Following references [7,8], assuming that interaction energy keeps the same on grain boundaries and in matrix and omitting the interaction between impurities and vacancy and between metal element and iron, a series of parameters[7,8] can be listed as follows: $a^g = c^g = 0.5$, $a^B = 0.75$, $c^B = 0.25$, $\Delta^0 G_P$ = 47KJ/mol, $\Delta^0 G_{Mn}$ = 8KJ/mol, β_{MnP} = 12.5 KJ/mol, $L_{Fe:PC}$ = -9KJ/mol. With all the data stated above and the composition of an assumed steel with enough high phosphorus (C = 0.15%, Mn = 1.6%, Si = 0.3%, P = 0.07%), the equilibrium segregation amount of phosphorus on grain boundaries at 400°C is calculated as 32%[9], which is a value high enough to cause temper brittleness.

The soaking temperature at the over-aging process of cold-rolled TRIP steel is set in the temperature range from 350 to 480°C. However, the soaking time is quite short, normally for 3~5 min, being far from equilibrium condition. Thus, the estimated equilibrium segregation amount of phosphorus is not suitable in describing the over-aging process of TRIP steels.

Based on Mclean contribution with his kinetic approach for a binary system[10], the amount of segregative element on grain boundaries can be expressed as:

$$(y_i^\phi(t) - y_i^\phi(0)) / (y_i^\phi - y_i^\phi(0)) \cong 2(X_i^\alpha / y_i^\phi d)(D_i t / \pi)^{1/2} \quad (6)$$

where $y_i^\phi(t)$ is the grain boundary coverage of element i at time t, $y_i^\phi(0)$ is its initial value and y_i^ϕ the equilibrium value. D_i is the bulk diffusion coefficient of i, d the grain boundary thickness and X_i^α the mole fraction of i in matrix. Taking d = 10^{-7}cm as usual, t = 5min = 300sec and T = 673K according to the actual process condition, X_P^α = 1.387x10^{-3} (i.e., wt = 0.07%), $y_i^\phi(0)$ = 0; according to Sundman et al, D_P = 3.53x10^{-18}[11].

The segregated amount of P at grain boundary after soaking at 400°C for 300 s is:

$$y_P^\phi(300\,\text{sec}) \cong 2(X_P^\alpha / d)(D_P t / \pi)^{1/2} = 0.051\% \text{[9]} \quad (7)$$

which is a small amount and would not be of any damage in grain boundary for the TRIP steel. Mechanical properties of the designed steel were measured at low temperatures (–20°C, –60°C) and no substantial difference was found between the results obtained at low temperatures and room temperature[9]. Moreover, cold fracture test was taken with the structural part assembles of automobiles and no sign of cold brittleness was detected. The new type TRIP steel containing P was manufactured according to design stated above in Ansteel company group in China.

2.3. Attractive effect of Al in TRIP steel

Aluminum, however, if the clogging of Al_2O_3 can be efficiently prohibited through careful control in the continuous casting process, is also a good promoter of TRIP effect. The reason has been studied thermodynamically and the authors found that with the existed database the phase diagram of a steel containing certain a amount of Al can not be described accurately[12] and must be revised. In their work[12], high purity Fe (99.99 wt%, Source), Al (99.97wt%, Source) and C (99.99wt%, Source) are used to prepare different composition Fe-Al-C alloys. Each alloy composition is checked by wet chemical analysis, as given in Table 3. Additional annealing of encapsulated ingots is carried out at 1100°C for 360 h to reach equilibrium states and to obtain a homogeneous element distribution. To check the homogeneity of the annealed alloys, an EDX line scanning of Al and C is applied on the cross-section of the alloys showing the elements are evenly distributed in the Fe matrix. The DIL experiments are carried out in a Formaster-D dilatometer to investigate the ferrite to austenite transformation during the heating and cooling cycles and the results are also listed in Table 3.

Alloy	Composition (wt%)			Experimental T (°C)		Calculated T (°C)	
	C	Al	Fe	A_1	A_3	A_1	A_3
C1	0.15	1.57	98.28	731	987	729	1006
C2	0.25	1.51	98.24	708	935	728	918
C3	0.41	1.47	98.12	739	892	727	850
C4	0.54	1.50	97.96	734	861	728	823
C5	0.75	1.53	97.72	732	810	729	796
C6	0.76	1.54	97.70	737	788	729	797
C7	0.91	1.42	97.67	726	827	726	769
C8	1.01	1.54	97.45	734	822	729	793
Al1	0.19	0.91	98.90	735	900	726	889
Al2	0.20	1.13	98.67	745	938	725	900
Al3	0.21	1.30	98.49	745	953	726	920

Table 3. Compositions of the Fe-rich alloys and the transformation temperatures of the alloys obtained by DIL analysis and thermodynamic calculation[12]

With the new set of experimental data, it is possible to optimize the Fe-Al-C system in Fe-rich corner. The standard element reference (SER) state of Fe (bcc), Al (fcc) and C (graphite) is used as the reference state of the Gibbs energy as usual. The molar Gibbs energy of the ternary liquid phase is described by the Redlich-Kister polynomial [13]:

$$G_m^{liquid} = \sum_{i=Fe,Al,C} x_i {}^0G_i^{liquid} + RT \sum_{i=Fe,Al,C} x_i \ln x_i + \\ + x_{Fe}x_{Al}L_{Fe,Al} + x_{Fe}x_C L_{Fe,C} + x_{Al}x_C L_{Al,C} + x_{Fe}x_{Al}x_C L_{Fe,Al,C} \quad (8)$$

Where x_i is the mole fraction of element i and ${}^0G_i^{liquid}$ is the Gibbs energy of liquid Fe, Al and C. The binary parameters $L_{i,j}$ are obtained from the thermodynamic evaluation of the limiting binaries, whereas the $L_{Fe,Al,C}$ is a ternary interaction parameter.

The Gibbs energy of solid phases, fcc (austenite), bcc (ferrite) and hcp in the Fe-Al-C system is described by a compound energy formalism consisting of substitutional and interstitial sublattices[14]. The Fe and Al substitute each other in the metal sublattice, while C and vacancies in the interstitial sublattice. The crystal structure is therefore modeled as $(Fe, Al)_a(C, Va)_c$. Symbols a and c denote the numbers of sites which is dependent on the crystal structure. Here the values of a and c are regarded as a = 1, c = 1, 3, and 0.5 for the fcc, bcc and hcp phases respectively. The molar Gibbs energy is expressed as:

$$\begin{aligned} G_m^{\phi} &= \sum_i y_i({}^{\circ}G_{i:C}^{\phi} y_C + {}^{\circ}G_{i:Va}^{\phi} y_{Va}) + aRT\sum_i y_i \ln y_i + cRT(y_C \ln y_C + y_{Va} \ln y_{Va}) \\ &+ \sum_i \sum_j y_i y_j (y_C L_{i,j:C} + y_{Va} L_{i,j:Va}) + y_C y_{Va} \sum_i y_i L_{i:C,Va} \\ &+ \sum_i \sum_j \sum_k y_i y_j y_k (y_C L_{i,j:C} + y_{Va} L_{i,j:Va}) + G_{mag} \end{aligned} \quad (9)$$

Where i, j = Fe, Al and K = C or Va. In Eq. 9, the variable y_i is the site fraction of component i in its sublattice. The components in different sublattices are separated by a colon and in one sublattice by a comma. ${}^{\circ}G_{i:Va}$ is the Gibbs energy of pure element i in a relevant non-magnetic state and ${}^{\circ}G_{i:C}$ is the Gibbs energy of a hypothetical non-magnetic state, where all interstitial sites are occupied by carbon. All G values are given relative to the reference state. G_{mag} is the magnetic part of the Gibbs energy.

In the Fe-Al-C system, only one ternary K phase was reported besides the liquid, fcc, bcc and hcp phases. In this optimization, the description of the K phase is after the thermodynamic assessment of Kumar et al [15].

Li et al[12] optimized the Fe-Al-C ternary system based on the previous thermodynamic description of the Fe-C, Fe-Al and Al-C systems[15, 16] as well as the obtained DIL data[12]. Furthermore, Li et al[12] extrapolated the thermodynamic properties of the limiting binary and ternary systems to higher order systems to construct the self-made database, especially for TRIP steels containing Mn, Si, Al, C, Nb and V as alloying elements. A calculated vertical section of the Fe-Mn-Si-Al-C system is presented in Fig.6 for the alloys meeting very well with the experimental results.

Phase diagram of a steel with the same composition as in Fig.6 but without Al is calculated with the self-made database and shown in Fig.7. It is obvious that the bcc+fcc field in Fig.6 is much wider than that in Fig.7. As calculated: when at 800℃, the carbon content in fcc phase in Fe-Mn-Si-Al-C steel is 0.476%, while at 802℃, the carbon content in fcc phase in Fe-Mn-Si-

C steel is only 0.206%. That means if soaking at the same temperature, the former dissolves much than two times carbon in the fcc phase resulting much higher stability of fcc phase and much better TRIP effect.

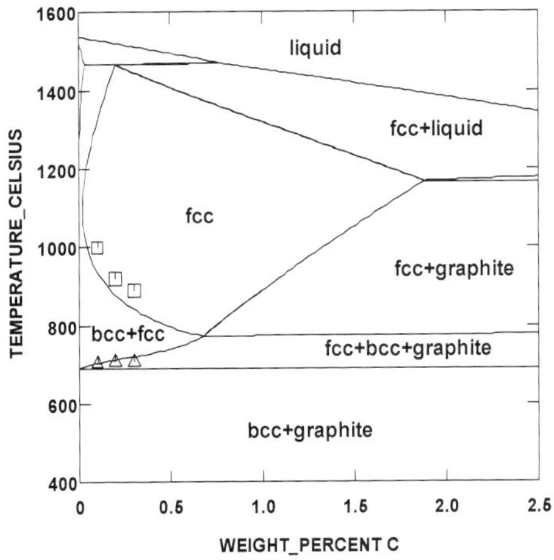

Figure 6. Extrapolated vertical section of the Fe-Mn-Si-Al-C steel with 1.60 wt% Mn, 0.37 wt% Si, and 1.31 wt% Al[12].

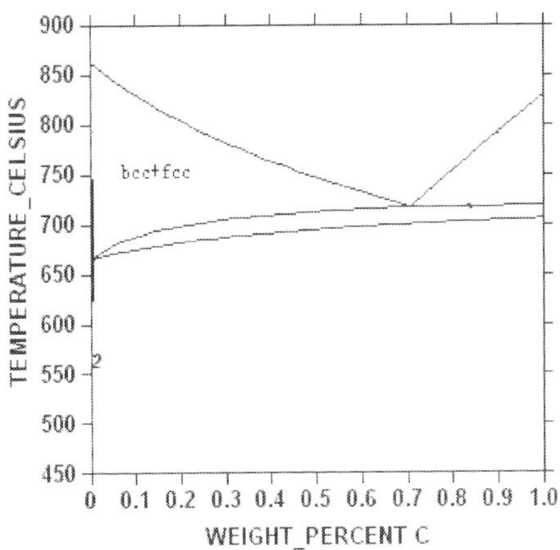

Figure 7. Phase diagram of the Fe-Mn-Si-Al-C steel with 1.60 wt% Mn, 0.37 wt% Si

This new finding brings substantial improvement for the mechanical properties of TRIP steel. Following this idea, i.e, keeping reasonably high amount Al in the steel and treating it in an ordinary Continuous Annealing Line (CAL), a high performance TRIP steel is manufactured in modern plant, whose strength and ductility product reaches 30,000MPa%. It is actually the property target of a third generation steel[17].

3. Research and development on large-section pre-hardened mold steel for plastic

In large-section pre-hardened plastic mold steel, the inclusion of high alumina content is detrimental to machinability, which is considered as the most important property of mold steel for plastic. It is known that non-metallic inclusion may appear in molten steel at various stages of the refining process or during solidification. One could not analyze in situ the precipitation of inclusions at every period of steel-making by traditional experimental method. However, the effective composition regulation and suitable heat treatment during production could influence inclusion precipitation and improve the quality of the mold steel for plastic. This paragraph is designed to produce a large-section pre-hardened plastic mold steel relying on the step by step thermodynamic calculation[18].

The SLAG database in Thermo-calc software package contributed by Gaye[19] is a useful tool which studied the slag mixture of m oxides ($(M_i)_{u_i} O_{v_i}$, i =1, 2, ..., m) including two kinds of sub-lattice. One is an anionic sub-lattice filled with oxygen ions and the other is cationic sub-lattice filled with the cations in the decreasing order of their charge (e.g Si^{4+}...Ca^{2+}...). The structure of the melt is described in term of symmetric (i-O-i) and asymmetric (i-O-j) cells, in numbers R_{ii},...$2R_{ij}$...; the energy parameters consist of parameters for asymmetric cells formation and parameters of interaction between cells. This simplified melt structure greatly and made it possible to describe the multi-component system in terms of only a few binary parameters. Because the formation of melt can be thought as the result of the following bond respond:

$$(i\text{-}O\text{-}i) + (j\text{-}O\text{-}j) = 2(i\text{-}O\text{-}j) \tag{10}$$

Then the expression for the mixing free energy of the melt can be derived by statistic thermodynamic method.

This paragraph applies the SLAG database of Thermo-calc, to predict a definite range of the composition, with which oxide and sulphide inclusion formed during the deoxidation treatment could be effectively controlled and inhibited.

Since Si and Mn in the steel affect the precipitation of alumina, the inter-relation curve of Si and Mn contents is calculated and plotted with a macro file which is input to the software. The calculated result can be used as a guide for the metallurgy engineer to select suitable composition of mold steel for plastic in a broad range. For Al contents of 100 ppm and 300 ppm at a given temperature of 1800K, the calculation results are shown in Fig.8 respectively. The curve in Fig.8 is the calculated critical line of alumina precipitation. When the composition of steel locates at the right side of the curve, alumina does not precipitate and the main inclusion in molten-steel is manganese-aluminate. However, if the composition locates at the left side of this curve, the alumina precipitates in molten steel. Thus, the mentioned results are used in the designing of mold steel.

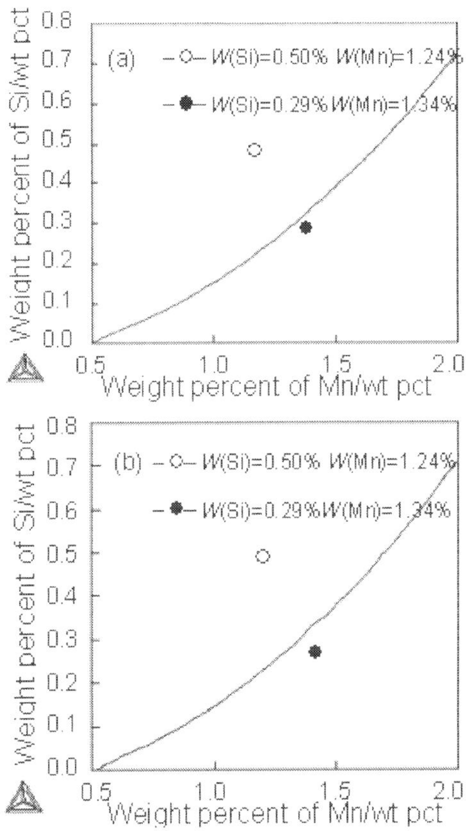

Figure 8. Effect of Si and Mn contents on alumina precipitation under 1800K for (a) Al content of 100ppm (b) Al content of 300ppm[18]

As stated above non-metallic inclusions may appear in molten steel at different stages. Therefore, in addition to the former considerations, the composition of the mold steel is designed in such a way that the oxide inclusion is fixed to be exogenous inclusion, which precipitates before solidification of liquid steel and can be eliminated after slag raking and gas rinsing, or be modified by injecting Ca-Si powder in the ladle. As calculation indicates in Fig.9a, the solidification temperature of the designed steel is 1760K while the lowest precipitation temperature of alumina is 1769K, 9K higher than the solidification point meeting well with the design requirement. The same result is obtained in Fig.9b, where the equilibrium phase diagram is calculated with another database of Thermo-calc, showing the solidification point of the steel is 1760K.

Moreover, the composition is carefully designed to let the driving force of the oxide precipitation be the biggest one following closely by the driving force of MnS. Then, when cooling, the alumina precipitates firstly and MnS precipitates afterwards to form the desired structure of good machinability where hard alumina is set in the core and soft MnS outside.

Furthermore, as known, good machining performance connects normally with high homogeneity of microstructures in the steel which is sometimes dependent on annealing process. With Thermo-calc, the main type of carbide which exists in the mold steel but can be dissolved during annealing is designed through the modification of steel composition considerately. The designed carbide is in the type of M_7C_3, and this prediction is proved by TEM detection. Soaking temperature for dissolving this carbide is estimated as 1220°C. Then suitable heat treatment process is taken to ensure the high homogeneity and super machinability of the steel after the annealing treatment in steel production.

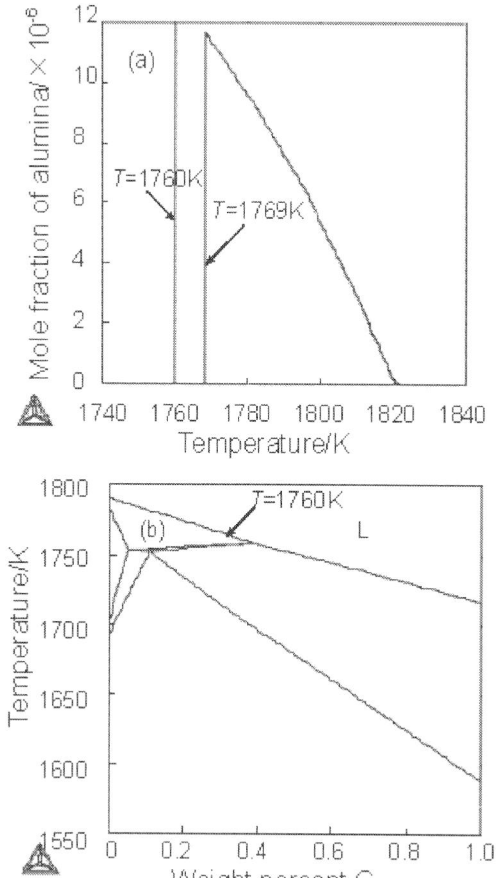

Figure 9. (a) Relationship between the precipitation amount of alumina and temperature in the designed mold steel, solidification temperature being shown (b) phase diagram of the same steel[18]

Measurement shows the inclusion number in the designed steel is reduced to be less than one tenth of the steel without composition regulation and its machinability reaches the world highest level of mold steel for plastic. The carefully designed mold steel is successfully put into mass production in Baosteel company.

4. Research and development on the high strength TWIP steel

Recently TWIP (TWin Induced Plasticity) steel, as one of the most promising advanced HSS (High Strength Steel), was developed showing the highest elongation rate (≧80%) and fairly good strength (≈600Mpa) due to the effect of TWIP. However, in practical use such as automobile anti-crash parts, the strength seems to be too low but ductility too high. With further consideration of alloy concept, much high strength can be obtained for the steel which comes from hcp martensite transformation. The equilibrium temperature between fcc phase and hcp phase in FeMnSi system and FeMnSiC system has been calculated thermodynamically[20]. It is found that the equilibrium temperature can be adjusted and controlled to be near to the room temperature through rational coordination of Si, Mn and C content in the steel. These results make it possible for realization of strain induced martensite transformation in the steel. However, the ε (hcp) martensite phase transformation is related to the stacking-fault energy (SFE) in austenite, which is affected by the special characteristics of alloy elements and their interactions and is discussed hereafter[20].

As known, the stacking-fault energy for pure metal can be considered as Gibbs free energy difference between γ atoms with two layer of fcc and ε atoms with two layer of hcp. While for alloy, its Gibbs free energy difference between fcc and hcp close-packed structure is not strictly equal to stacking-fault energy, because the element concentration at stacking faults is different from the average composition of alloy, especially the Suzuki Effect caused by C,N , which leads to the obvious difference of concentrations between stacking faults and base material. So according to the work of Olson and Cohen[21], the thermodynamic model of stacking-fault energy can be expressed as:

$$\gamma_{SF} = 2\rho_A \Delta G^{\gamma\to\varepsilon} + 2\sigma^{\gamma\to\varepsilon} \tag{11}$$

Where,

$$\rho_A = \frac{4}{\sqrt{3}} \frac{1}{a^2 N}$$

is the bulk density of atomic surface in face-centered cubic, a is the lattice constant of alloy, N is the avogadro's constant $N = 6.02\times10^{23}$. $\sigma^{\gamma\to\varepsilon}$ is the binding energy of phase interface, according to Allain[22], $\sigma^{\gamma\to\varepsilon} = (5\sim15)$mJ/m^2. For simplicity, an average, i.e., 10mJ/m^2 is taken by Li et al in their calculation[20]. $\Delta G^{\gamma\to\varepsilon}$ is the free energy difference between face-centered cubic (γ) and close-packed hexagonal (ε) . Referring to Dumay[23], this part can be expressed as:

$$\Delta G^{\gamma\to\varepsilon} = \Delta G^{\gamma\to\varepsilon}_{\text{FeMnSu}} + \Delta G^{\gamma\to\varepsilon}_{\text{FeMnSu/C}} + \Delta G^{\gamma\to\varepsilon}_{\text{mg}} \tag{12}$$

Where, the first item in the right part of the equation represents the effect of main elements such as Fe, Mn and Si on transformation free energy, denoted as:

$$\Delta G^{\gamma\to\varepsilon}_{\text{FeMnSu}} = \sum_i \Delta G_i^{\gamma\to\varepsilon} + \chi_{\text{Fe}}\chi_{\text{Mn}}[C + D(\chi_{\text{Fe}} - \chi_{\text{Mn}})] + \chi_{\text{Fe}}\chi_{\text{Si}}[E + F(\chi_{\text{Fe}} - \chi_{\text{Si}})] \tag{13}$$

Where $\Delta G_i^{\gamma\to\varepsilon}$ is the transformation free energy difference of $\gamma \to \varepsilon$ in pure components *i*. χ_i is the mole fraction of element *i*, T is temperature in Kelvin, the last two items in equation (13) represent the interaction between Fe-Mn and Fe-Si, where the interaction between Fe and other elements as well as the mutual action of alloy elements can be neglected. The parameters of C, D, E, F in equation (13) were obtained form the corresponding curve fitting and listed in Table 4 after Dumaya et al[23]. The $\Delta G^{\gamma\to\varepsilon}_{\text{FeMnSu/C}}$ in equation (12) represents effect of interaction between all the substitute elements and carbon on the transformation free energy, which is:

$$\Delta G^{\gamma\to\varepsilon}_{\text{FeMnSu/C}} = \frac{a}{X_{\text{C}}}(1 - e^{-bX_C}) + cX_{\text{Mn}} \tag{14}$$

While in equation (12), $\Delta G^{\gamma\to\varepsilon}_{\text{mg}}$ is the mole magnetic free energy difference between γ phase and ε phase, which is:

$$\Delta G^{\gamma\to\varepsilon}_{\text{mg}} = G^{\varepsilon}_{m} - G^{\gamma}_{m} \tag{15}$$

$G^{\varepsilon}_{m}, G^{\gamma}_{m}$ represents respectively the mole magnetic free energy[24-26] of γ phase and ε phase, which is:

$$G^{\varphi}_{m} = RT\ln(\frac{\beta^{\varphi}}{\mu_B} + 1)f^{\varphi}(\tau^{\varphi}) \tag{16}$$

Where:

$$f^{\varphi}(\tau^{\varphi}) = 1 - \frac{1}{D}\{\frac{79\tau^{-1}}{140p} + \frac{474}{497}[\frac{1}{p} - 1][\frac{\tau^3}{6} + \frac{\tau^9}{135} + \frac{\tau^{15}}{600}]\}(\tau = \frac{T}{T_C^{\varphi}} \le 1)$$

Or:

$$f^{\varphi}(\tau^{\varphi}) = -\frac{1}{D}[\frac{\tau^{-5}}{10} + \frac{\tau^{-15}}{315} + \frac{\tau^{-25}}{1500}](\tau = \frac{T}{T_C^{\varphi}} > 1), but D = \frac{518}{1125} + \frac{11692}{15975}(\frac{1}{p} - 1)$$

as for fcc and hcp crystal structure, p=0.28, for other structure, p=0.5.

$$T_N^{\gamma} = 250\ln(x_{Mn}) - 4750x_C x_{Mn} - 22x_{Cu} - 2.6x_{Cr} - 6.2x_{Si} + 720(K)$$
$$T_N^{\varepsilon} = 580x_{Mn}(K)$$

$\Delta G_{Fe}^{\gamma\to\varepsilon}$	-2243.38+4.309T $Jmol^{-1}$
$\Delta G_{Mn}^{\gamma\to\varepsilon}$	-1000.00+1.123T $Jmol^{-1}$
$\Delta G_{FeMn}^{\gamma\to\varepsilon}$	C=1246 $Jmol^{-1}$, D=-717 $Jmol^{-1}$
$\Delta G_{FeMnx/C}^{\gamma\to\varepsilon}$	a=1246 Jmol-1, b=24.29 Jmol-1,c=17175 Jmol-1
β^{γ}/μ_B	$0.7x_{Fe} + 0.62x_{Mn} - 0.64x_{Fe}x_{Mn} - 4x_C$ $Jmol^{-1}$
$\beta^{\varepsilon}/\mu_B$	$0.62x_{Mn} - 4x_C$ $Jmol^{-1}$
$\Delta G_{Al}^{\gamma\to\varepsilon}$	2800+5T $Jmol^{-1}$
$\Delta G_{Si}^{\gamma\to\varepsilon}$	-560-8T $Jmol^{-1}$
$\Delta G_{FeSi}^{\gamma\to\varepsilon}$	E=2850 $Jmol^{-1}$; F=3520 $Jmol^{-1}$
$\Delta G_{V}^{\gamma\to\varepsilon}$	-3500 $Jmol^{-1}$

Table 4. Parameters used in calculation of stacking-fault energy in Fe-Mn-X-C system(X=Si, Al, V)[23]

Y.K.Lee et al[27] considered that as for Fe-Mn alloy with 16-24% Mn, the phase transformation driving force of hcp martensite was -68~-120 J/mol. Allain et al[22] thought that for Fe-Mn-C system, if the stacking-fault energy (SFE) was above 12mJ/m^2, the austenite was easy to induce twins; if the SFE lower than 18 mJ/m^2, austenite easy to induce hcp martensite. The suggestion of Allain et al[22], say, the SFE value between 12~18 mJ/m^2 is adopted as the value which induces martensite transformation in the TWIP steel. The calculation result for a steel with 18% Mn but different Si and C is shown in Fig.10 where the composition range suitable for the assigned SFE is pointed.

Steels with different Mn,Si and C content but their SFE ranges between 12~18 mJ/m^2 is calculated and listed in Table 5 [20].

For designing of the test steel, a value of stacking fault energy in the range of 12~18mJ/m^2 is firstly adopted. Besides, the change of carbon content can be used to adjust the size of stacking fault energy while the content of Mn is to be set as low as possible to be favorable to the line process. Based on the above calculation the composition of the high strength high ductility steel is designed as 16~22%Mn, 0.3~0.6%C and 0.2~0.8%Si. The Gibbs free energy difference between fcc matrix phase (austenite) and hcp martensite in the steels is calculated with a self-made database[28] and, as an example, the difference of two phases in a steel of 18%Mn-0.528Si-0.35C is shown in Fig. 11. It can be seen that the T_0 temperature(at which the Gibbs energy of fcc phase equals to hcp phase) is 330K, if adding -68 J/mol as driving force[27] to the difference, the corresponding temperature is 293K, which means hcp martenisite can be obtained in room temperature.

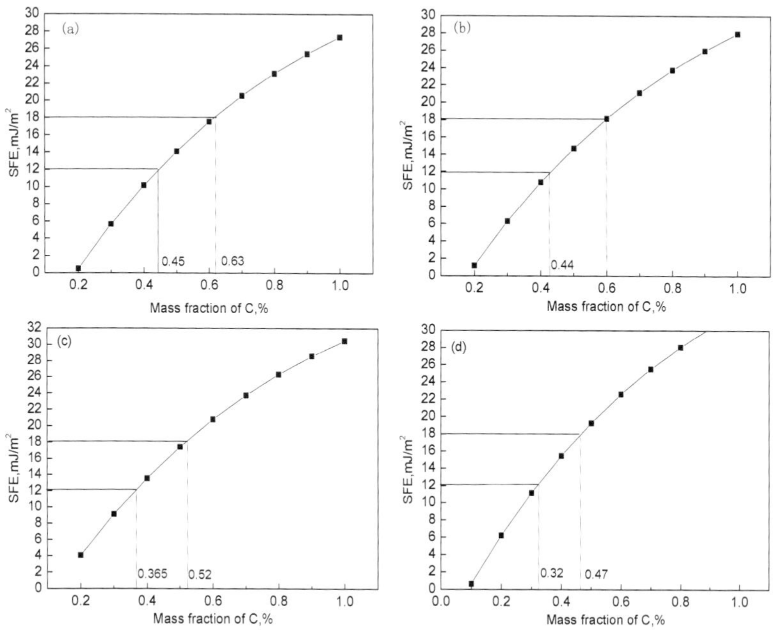

Figure 10. Estimated stacking fault energy(SFE) in Fe-18Mn-YSi-C alloys with Y = 0(a),Y= 0.3(b),Y= 2(c) and Y = 5(d) [20]

Fe-Mn	Si	C(SFE=12)	C(SFE=18)
Fe-22Mn	0	0.282	0.44
Fe-22Mn	0.3	0.28	0.43
Fe-22Mn	2	0.24	0.38
Fe-22Mn	5	0.23	0.37
Fe-18Mn	0	0.45	0.63
Fe-18Mn	0.3	0.44	0.6
Fe-18Mn	2	0.365	0.52
Fe-18Mn	5	0.32	0.47
Fe-15Mn	0	0.55	0.73
Fe-15Mn	0.3	0.53	0.7
Fe-15Mn	2	0.44	0.6
Fe-15Mn	5	0.37	0.52

Table 5. Chemical compositions with different SFE(mass fraction, %)

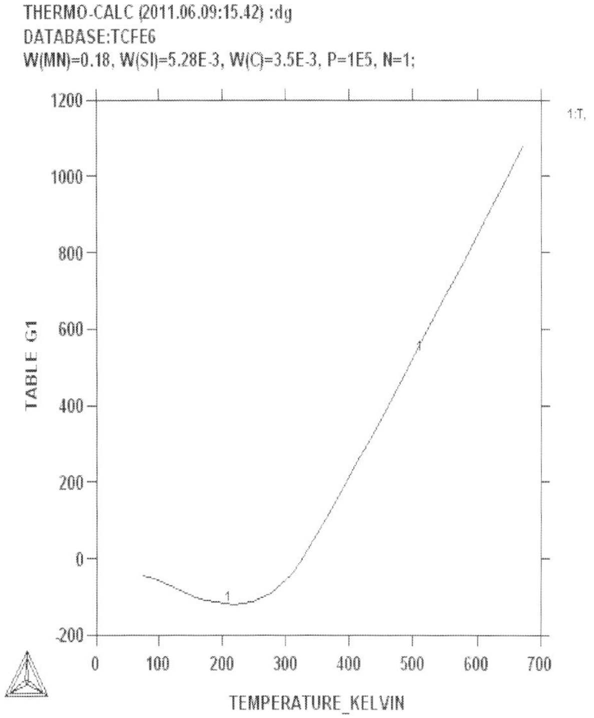

Figure 11. Gibbs free energy difference of two phases $\Delta G^{\gamma \to \varepsilon}$ in Fe-18Mn-0.528Si-0.35C(mass fraction, %)steel

The test steel is melt in the 35kg vacuum melting furnace and the measured stacking fault energy is 20mJ/m^2. The ingot is heated up to 1200°C and then forged into the stocks with a thickness of 25mm. The stocks are heated at 1100°C for 30min, finally hot rolled into the plates with a thickness of about 1.5mm and then quenched in oil. The final specimens are conducted with XRD, metallographic and SEM analyses. XRD spectrum shows weak ε-martensite peak in the steel. Optical metallography confirms there exists ε-martensite in the steel when the sample is chemical polished and the effect of mechanical grinding is removed. SEM topography also indicates thin strips with different orientations which is regarded primarily as ε-martensite. Therefore, as revealed by calculation, ε-martensite is formed in this high-Mn TWIP steels after oil quenching.

In order to raise the mechanical property of the steel, different heat treatment after hot rolling is applied. Table 6 shows different final rolling temperatures and cooling methods after hot rolling.

Steel No.	Final hot rolling temp./ °C	Cooling medium
Mn18_1	850	water
Mn18_2	750	water
Mn18_3	850	air

Table 6. Final rolling temperatures and cooling methods after hot rolling

Figure 12 shows the XRD spectrum of the test steels treated with different heat treatment methods after hot rolling processes. It can be seen that the microstructure is the combination of austenite and ε-martensite after hot rolling at 850°C and water quenching, but the microstructures are austenite alone after treatment by other processes. Figure 13 shows the strain-stress curves of the test steels treated with different processes. It can be seen that air cooling after hot rolling (as Mn18_3 steel) can lead to the best property such as strength above 1GPa and elongation rate above 60%. To examine the effect of structure on property, the fracture of the steels is analyzed with XRD as shown in Fig. 14. It can be found that ε-martensite peak is displayed in the fracture of three specimens. Among this peaks, ε-martensite peak in Mn18_3 specimen is the weakest. Thus, according to the XRD results before and after drawing and the result of tensile test, it can be concluded that ε-martensite existed before tensile test as well as a great amount of ε-martensite induced during tensile test may both raise the strength of the steel, while only a bit of induced ε-martensite during tension improves the property much effectively. All those steels exhibit strength higher than 800 MPa caused by martensite transformation.

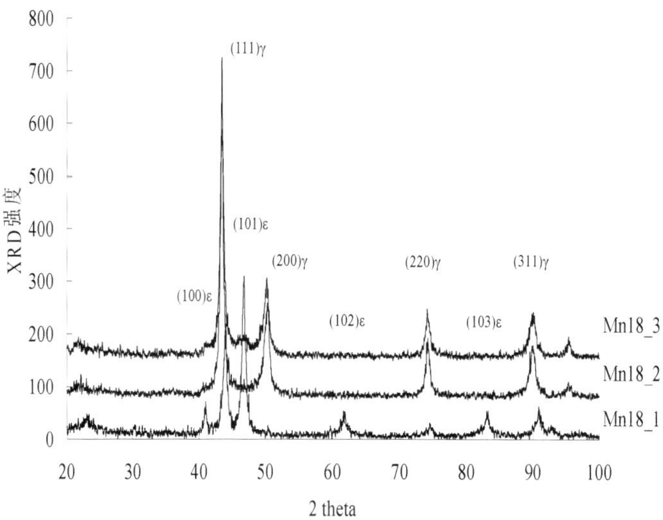

Figure 12. XRD spectra of hot rolling steel

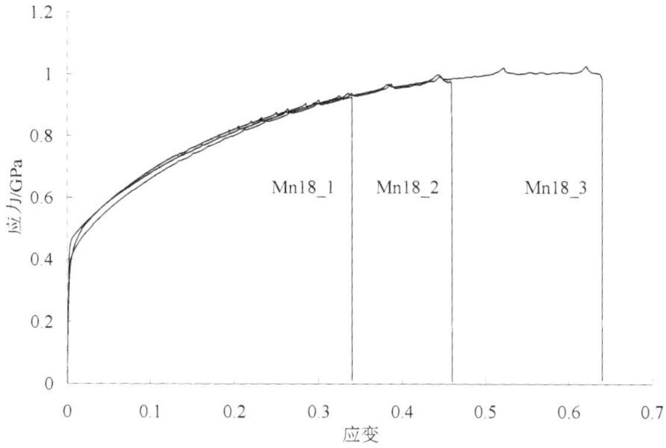

Figure 13. Stress-strain curves of hot rolling steels

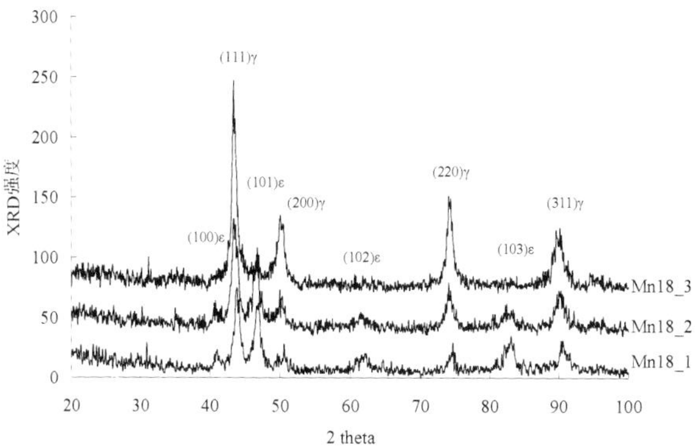

Figure 14. XRD spectra of hot rolling steel after deformation

Combining with the calculation results and experimental process stated above, new type of high strength and ductility TWIP steel is successfully produced in Ansteel group.

5. Research on high-tech ceramics

Ceria-doped zirconia ceramics found profound use in automobile emission system recently. When this ceramics is exposed to an inert atmosphere or low oxygen partial pressure at elevated temperature, the Ce^{+4} will be reduced to Ce^{+3}, a nonstoichiometric phase $Zr_{1-z}Ce_zO_{2-x}$ is then formed and the ZrO_2-CeO_2 system will convert to a quasi-ternary ZrO_2-CeO_2-$CeO_{1.5}$ system. It is therefore very useful to study the effect of oxygen partial pressure on the stability of the nonstoichiometric fluorite phase $Zr_{1-z}Ce_zO_{2-x}$ in the ZrO_2-CeO_2-$CeO_{1.5}$ system, which is directly connected to the ability to accumulate or release oxygen of a system, proved to be a guide in producing the high-tech ceramics in the modern industry especially in automobile manufacturing factory.

In this paragraph, the three limiting binary systems ZrO_2-CeO_2, ZrO_2-$CeO_{1.5}$ and $CeO_{1.5}$-CeO_2 are evaluated with compound energy model[29-31] and substitutional model[32]respectively and the calculation results met well with each other proving the substitutional model is still available to the complex oxide system[33,34].

In ZrO_2-$CeO_{1.5}$ system[33], compound energy model with structure $(Zr^{+4}, Ce^{+3})_1(O^{-2}, Va^0)_2$ is used for css phase. The Gibbs energy of one mole of formula unit can be expressed as follows:

$$\begin{aligned} G_m^{css} &= y_{Zr^{+4}} y_{O^{-2}} {}^0G_{Zr^{+4}:O^{-2}} + y_{Zr^{+4}} y_{Va^0} {}^0G_{Zr^{+4}:Va^0} \\ &+ y_{Ce^{+3}} y_{O^{-2}} {}^0G_{Ce^{+3}:O^{-2}} + y_{Ce^{+3}} y_{Va^0} {}^0G_{Ce^{+3}:Va^0} \\ &+ RT[(y_{Zr^{+4}} \ln y_{Zr^{+4}} + y_{Ce^{+3}} \ln y_{Ce^{+3}}) \\ &+ 2(y_{O^{-2}} \ln y_{O^{-2}} + y_{Va^0} \ln y_{Va^0})] + {}^EG_m \end{aligned} \tag{17}$$

The excess term ${}^EG_m$ is given by an equation of the form:

$$\begin{aligned} {}^EG_m &= y_{Zr^{+4}} y_{Ce^{+3}} [L^0_{Zr^{+4},Ce^{+3}:O^{-2}} \\ &+ (y_{Zr^{+4}} - y_{Ce^{+3}}) L^1_{Zr^{+4},Ce^{+3}:O^{-2}}] \end{aligned} \tag{18}$$

In the $CeO_{1.5}$-CeO_2 system[33], the nonstoichiometric phase <CeO_{2-y}> is described with structure $(Ce^{+4}, Ce^{+3})_1 (O^{-2}, Va^0)_2$. The Gibbs energy of one mole of formula unit can be expressed as follows:

$$\begin{aligned} G_m^{css} &= y_{Ce^{+4}} y_{O^{-2}} {}^0G_{Ce^{+4}:O^{-2}} + y_{Ce^{+4}} y_{Va^0} {}^0G_{Ce^{+4}:Va^0} \\ &+ y_{Ce^{+3}} y_{O^{-2}} {}^0G_{Ce^{+3}:O^{-2}} + y_{Ce^{+3}} y_{Va^0} {}^0G_{Ce^{+3}:Va^0} \\ &+ RT[(y_{Ce^{+4}} \ln y_{Ce^{+4}} + y_{Ce^{+3}} \ln y_{Ce^{+3}}) \\ &+ 2(y_{O^{-2}} \ln y_{O^{-2}} + y_{Va^0} \ln y_{Va^0})] + {}^EG_m \end{aligned} \tag{19}$$

The excess term ${}^EG_m$ is given as follows:

$$\begin{aligned} {}^EG_m &= y_{Ce^{+4}} y_{Ce^{+3}} [L^0_{Ce^{+4},Ce^{+3}:O^{-2}} \\ &+ (y_{Ce^{+4}} - y_{Ce^{+3}}) L^1_{Ce^{+4},Ce^{+3}:O^{-2}}] \end{aligned} \tag{20}$$

In the ZrO_2-CeO_2 system[33], the css phase is modeled with two sublattices, $(Zr^{+4}, Ce^{+4})_1(O^{-2})_2$. Because the valence of cations and anions can fit the condition of electroneutrality without the addition of any defects. All the limiting binaries are optimized and all the parameters are obtained[33].

When ZrO_2-CeO_2 system exposed in reduced atmosphere at high temperature, part of CeO_2 changes to $CeO_{1.5}$ in css phase. A new nonstoichiometric phase <$Zr_{1-z}Ce_zO_{2-x}$> is considered under the conditions. To describe the ionic characteristics of this phase, the compound energy model, i.e., $(Zr^{+4}, Ce^{+3}, Ce^{+4})_1(O^{-2}, Va^0)_2$ is also applied to this phase. Based on the parameters of phase css in the three sub-binary oxides system, The Gibbs free energy of one mole phase is[33]:

$$\begin{aligned} G_m^{css} &= y_{Zr^{+4}} y_{O^{-2}} {}^0G_{Zr^{+4}:O^{-2}} + y_{Zr^{+4}} y_{Va^0} {}^0G_{Zr^{+4}:Va^0} \\ &+ y_{Ce^{+3}} y_{O^{-2}} {}^0G_{Ce^{+3}:O^{-2}} + y_{Ce^{+3}} y_{Va^0} {}^0G_{Ce^{+3}:Va^0} \\ &+ y_{Ce^{+4}} y_{O^{-2}} {}^0G_{Ce^{+4}:O^{-2}} + y_{Ce^{+4}} y_{Va^0} {}^0G_{Ce^{+4}:Va^0} \\ &+ RT[(y_{Zr^{+4}} \ln y_{Zr^{+4}} + y_{Ce^{+3}} \ln y_{Ce^{+3}} \\ &+ y_{Ce^{+4}} \ln y_{Ce^{+4}}) + 2(y_{O^{-2}} \ln y_{O^{-2}} \\ &+ y_{Va^0} \ln y_{Va^0})] + {}^EG_m \end{aligned} \tag{21}$$

The excess term ${}^EG_m$ is represented by:

$$\begin{aligned} {}^EG_m &= y_{Zr^{+4}} y_{Ce^{+3}} [L^0_{Zr^{+4},Ce^{+3}:O^{-2}} + (y_{Zr^{+4}} - y_{Ce^{+3}}) \\ & L^1_{Zr^{+4},Ce^{+3}:O^{-2}}] + y_{Ce^{+4}} y_{Ce^{+3}} [L^0_{Ce^{+4},Ce^{+3}:O^{-2}} \\ &+ (y_{Ce^{+4}} - y_{Ce^{+3}}) L^1_{Ce^{+4},Ce^{+3}:O^{-2}}] + y_{Zr^{+4}} y_{Ce^{+4}} \\ & [L^0_{Ce^{+4},Zr^{+4}:O^{-2}} + (y_{Ce^{+4}} - y_{Zr^{+4}}) L^1_{Ce^{+4},Zr^{+4}:O^{-2}}] \end{aligned} \tag{22}$$

The calculated phase diagrams and corresponding experimental information of ZrO_2-CeO_2, ZrO_2-$CeO_{1.5}$ and $CeO_{1.5}$-CeO_2 systems are shown in Fig.15-17[33], where the little circles denote experimental data. In Fig.17, only the miscibility gap is considered in $CeO_{1.5}$-CeO_2 system. The calculated results met very well with experiments and keep consistent with each other.

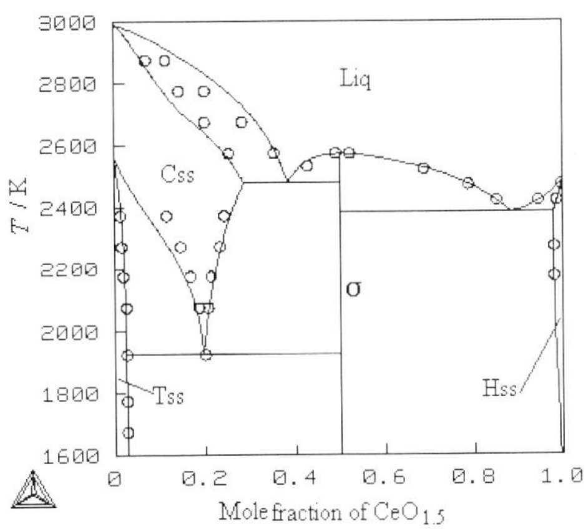

Figure 15. Calculated and experimental phase diagram of ZrO_2-$CeO_{1.5}$ system, circles showing the experimental result[33]

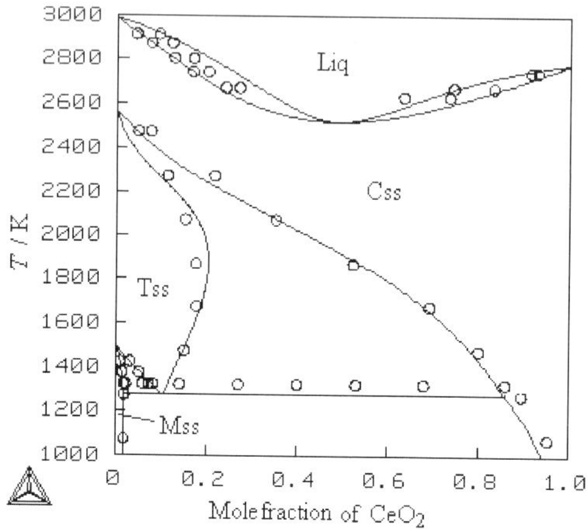

Figure 16. Calculated and experimental phase diagram of ZrO_2-CeO_2 system, cycles showing the experimental result[33]

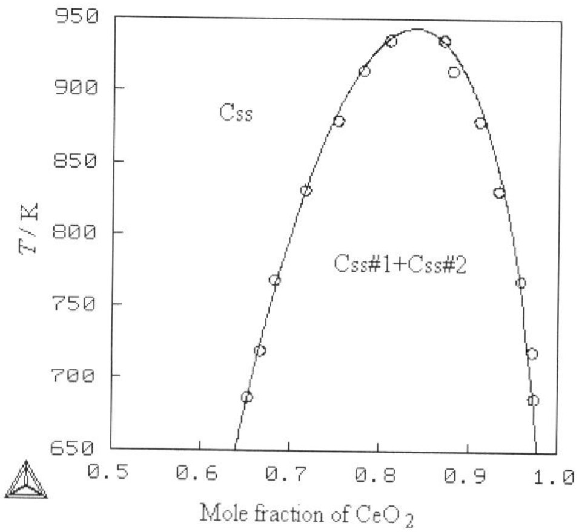

Figure 17. Calculated and experimental miscibility gap in $CeO_{1.5}$-CeO_2 system, circles showing the experimental result[33]

In order to understand thermodynamically the degasing of CeO_2, additional deduction is necessary as follows:

The chemical potential of oxygen in phase $<CeO_{2-y}>$ depends on the two mass variables, undefected species CeO_2 and defected $CeO_{1.5}$. In $CeO_{1.5}$-CeO_2 system, it would be natural to list the formula

$$4CeO_{1.5} + O_2 = 4CeO_2 \tag{23}$$

With the evaluated model parameters of phase <CeO_{2-y}> and Thermo-Calc, The authors[33] get the chemical potential of component CeO_2 and $CeO_{1.5}$ respectively. Therefore, the chemical potential of oxygen in this phase can be listed as:

$$\mu_{O_2} = {}^{0}\mu_{O_2} + RT\ln(PO_2) = 4\mu_{CeO_2} - 4\mu_{CeO_{1.5}} \tag{24}$$

Under reduced atmosphere, phase css in ZrO_2-CeO_2 system changes to <$Zr_{1-z}Ce_zO_{2-x}$>. Since no defects have been introduced in phase css in ZrO_2-CeO_2 system, The authors[33] also get oxygen potential for phase <Zr_{1-z} Ce_zO_{2-x}> from Eq. 24. To get the relation between nonstoichiometry y in <CeO_{2-y}> and oxygen potential in phase <$Zr_{1-z}Ce_zO_{2-x}$> in the ternary system at different temperatures, some manipulation has been made. It is supposed that phase <Zr_{1-z} Ce_zO_{2-x}> is composed of (1-z) mole ZrO_2 and z mole <CeO_{2-y}>. <CeO_{2-y}> is composed of m_2 mole CeO_2 and m_1 mole $CeO_{1.5}$. Then the O/Ce ratio is calculated from the ratio O/(Ce+Zr)

$$2 - y = [2 - x - 2(1 - z)] / z \tag{25}$$

To obtain the ln(y)-T-ln(PO_2) relation, one first need to determine the moles, m_i, of each components. These are calculated from the mass-balance for cerium and oxygen, which is

$$z = m_2 + m_1 \text{ and } z(2 - y) = 2m_2 + 1.5m_1 \tag{26}$$

Which lead to

$$m_1 = 2zy\,,\ m_2 = z(1 - 2y) \tag{27}$$

Where z =1 mole represents a system without component ZrO_2.

With the thermodynamic parameters of the three sub-binary systems[33] and Eq. 24, the relation ln(y)-ln(PO_2) in $CeO_{1.5}$-CeO_2 system, as well as in ZrO_2-$CeO_{1.5}$-CeO_2 system are calculated by Thermo-Calc[33].

Fig. 18-20 shows the calculated results. Fig. 18 represents the experiment data and the calculated relation ln(y)-ln(PO_2) at 1773K, 1573K, 1473K, where the little circles denote experiment data from Lindemer et al.[35]. It is obvious that the assessed parameters from CeO_2-$CeO_{1.5}$ system with compound energy model can represent the relations very well.

Fig. 19 reveals the relation between $CeO_{1.5}$ and ln(PO_2) in a system without component ZrO_2 (z=1) at different temperatures, showing temperature has great effect on oxygen partial pressure. In Fig. 20, different solid lines represent different z values (0.1, 0.2, 0.3, 1.0) corresponding to different mole fraction of ZrO_2 (0.9, 0.8, 0.7, 0 respectively) in the ternary system at 1473K. It is obvious that the same nonstoichiometry y corresponds to different oxygen partial pressure at different z. Also can be found is the significant effect of z value (the content of ZrO_2) on the relation between oxygen partial pressure and nonstoichiometry value.

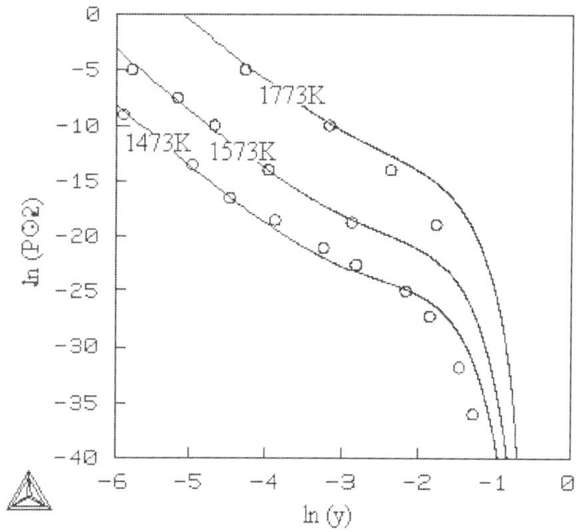

Figure 18. Calculated and experimental relationship ln(y)-ln(PO_2), circles showing the experimental result[33]

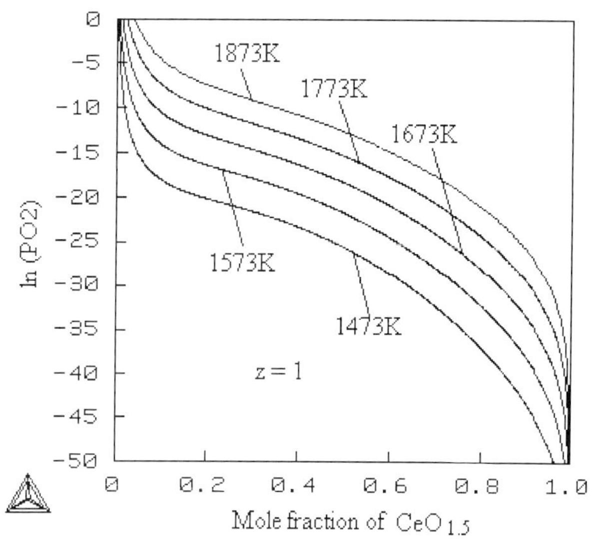

Figure 19. Relations between the different amounts of $CeO_{1.5}$ and oxygen partial pressure at different temperatures[33]

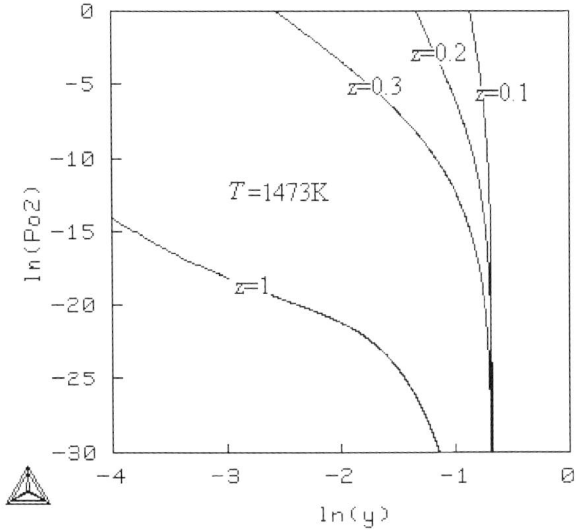

Figure 20. Predicted relations ln(y)-ln(PO_2) at 1473K with different z values [33]

A substitutional model was used to describe the quasibinary systems, which estimates the mixing entropy of complex ceramic systems in a very simple way. Kaufman et al[32] Du et al,[36] and Li et al.[37-39] already successfully evaluated oxide phase diagrams in this way. Li et al[40] pointed:

When calculating oxide systems, if there do not exist any short range ordering, shortage of elements or molecular-like associate in sublattice, the substitutional model is an effective alternative of sublattice model. Furthermore, if the valence of the cations are the same, the application of substitutional model can obtain the same formalism of Gibbs free energy and the same calculated result as that can be obtained by the subblatice model, though the former is more simple and easy to be extened into higher order symtems.

Then, the Gibbs free energy of one mole solution phase,Φ is described as:

$$G_m^{\Phi}(x,T) = \sum_{i=1}^{2} x_i.\Delta^{\circ}G_f^{\Phi-i}(T) + RT\sum x_i \ln x_i + x_1 x_2 \sum_{n=0}^{1} K^{n+3}(x_1 - x_2)^n \quad (28)$$

The first term on the right side accounts for the mechanical mixture of the pure components. $\Delta^{o}G_f^{\Phi-i}(T)$ is the standard Gibbs free energy , i.e., the formation energy of mass parameter i from the elements in their standard states. The second term refers to an ideal solution, and the third term is the excess Gibbs free energy function, which is a Redlich-Kister polynomial. The K^n factor in Eq. (29) is a temperature-related polynomial and is expressed as:

$$K^n = A^n + B^nT + C^nT\ln T + D^nT^2 + E^n / T + F^nT^3 \tag{29}$$

Gibbs free energy of one mole (cation) stoichiometric compound phase with pyrochlore structure is described as:

$$G(1/4)Zr_2Ce_2O_7 = (1/2)\Delta^O G_f^{css-ZrO_2} + (1/2)\Delta^\circ G_f^{css-CeO_{1.5}} + A + B^*T \tag{30}$$

Also with the relationship deduced in Eqs.(23-27), one can find in a system without ZrO_2 component i.e., z=1 (mole), the mole fraction of $CeO_{1.5}$ is $N_1=m_1/(m_1+m_2)$ and the mole fraction of CeO_2 is $N_2=m_2/(m_1+m_2)$. Moreover, $y=0.5–0.5N_2$.

Based on the thermodynamic information of the limiting quasi-binary systems, which is evaluated in reference [34,41] the molar Gibbs free energy of the nonstoichiometric phase $Zr_{1-z}Ce_zO_{2-x}$ can be written according to Maggianu[42] as:

$$\begin{aligned} &G'\langle Zr_{1-z}Ce_zO_{2-x}\rangle / (m_1 + m_2 + m_3) \\ &= N_1\Delta^\circ G_f^{css-ZrO_2} + N_2\Delta^\circ G_f^{css-CeO_2} \\ &+ N_3\Delta^\circ G_f^{css-ZrO_2} + N_1RT\ln(N_1) + N_2RT\ln(N_2) \\ &+ N_3RT\ln(N_3) + K_{1,2}^3N_1N_2 + K_{1,2}^4N_1N_2(N_1 - N_2) \\ &+ K_{1,3}^3N_1N_3 + K_{1,3}^4N_1N_3(N_1 - N_3) + K_{2,3}^3N_2N_3 \\ &+ K_{2,3}^4N_2N_3(N_2 - N_3) \end{aligned} \tag{31}$$

Where $\Delta^o G_f^{css-i}$ is the standard Gibbs free energy of formation of mass parameter i from the elements in their standard states. $\Delta^o G_f^{css-ZrO_2}$ is after the compilation of Pankratz, [43] whereas $\Delta^o G_f^{css-CeO_{1.5}}$ is retrieved from the quasi-binary phase diagrams [41]. $K_{i,j}^n$ is the interaction coefficient, where n is the order of the Redlich–Kister polynomial and the subscripts i,j represent the components, i.e., $CeO_{1.5}$=1, CeO_2=2, ZrO_2=3. N_i(i=1,2,3) is the molar fraction of $CeO_{1.5}$, CeO_2 and ZrO_2 respectively.

The partial molar free energy for mass parameter i equals to,

$$\partial[G' < Zr_{1-z}Ce_zO_{2-x} >] / \partial m_i = \Delta\overline{G}_i \tag{32}$$

With the equation similar to Eq. 24,

$$RT\ln(Po_2) = 4\Delta\overline{G}_{CeO_2} - 4\Delta\overline{G}_{CeO_{1.5}} \tag{33}$$

The relationship between ln(y) and ln(PO_2) at 1473K is calculated and shown in Fig. 21, which is exactly the same as shown in Fig. 20 calculated by CEM.

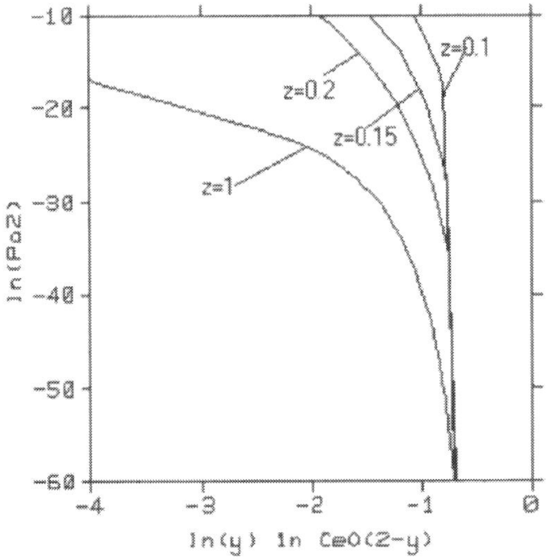

Figure 21. Predicted relationship between ln(y) and ln(PO2) at 1473 K at different CeO_{2-y} molar fractions in the ZrO_2–CeO_2–$CeO_{1.5}$ system[34]

The research staff members in Exhaust System of MERCEDES CAR GROUP checked independently the work stated above over years, they followed each of the published papers and calculated their system with the equations and model parameters in the publications. They were convinced that the calculation stated above predicted the mechanism in the automobile emission system and met very well with their detected data in all the gasoline engines of Mercedes Car.

6. Conclusion

An aspiration of engineers in materials field is to have the possibility to estimate in advance the composition, properties, processing parameters, etc., while they are planning to develop a new material. This chapter offers a basic sketch where one could find how thermodynamics, coupled with accurate experiment results, constructs the entrance for mass production. In order to limit the chapter length, only less examples were listed and most of them are stressed at steels, which include developing new type steel, improving the quality of existed steel and discussing the factors mainly influencing steel property. Calculation is applied in the light of the existed commercial software package though, in some cases, suitable derivation to reveal the relationship between thermodynamic properties of substances and self-made database exhibit also extreme importance in the designing of materials. Moreover, both compound energy model and substitutional model are used to describe the complex degassing behavior in the ZrO_2-CeO_2 system which reveals how and why the simple model could be successfully applied in complex systems.

Author details

Lin Li
School of Materials, Shanghai University, Shanghai, P.R.China

Acknowledgement

This work is financially supported by National Key Basic Research Development Project of China (973 Programme, No.2010CB630802), NSFC(No.50934011 and No.50971137) and School of Materials, Shanghai University.

7. References

[1] Meyer M.D., Vanderschueren D., De Cooman B.C., ISIJ Inter., 1999, 39(8), 813.
[2] Lin Li, B.C. De Cooman, P. Wollants, Yanlin He and Xiaodong Zhou, J Mat. Sci. & Technol., vol.20,No.2, 135-138, 2004.
[3] Li Lin, Liu Rendong, Shi Wen, Zhang Mei, He Yanlin, Liu Huachu, Fu Renyu, Development of a new type TRIP steel with good weldability and galvanizing property, Proc. 2009 Inter. Symp. automobile steel, Dalian, China, 2009, 332-336.
[4] Guttmann M., Mclean D., Grain Boundary Segregation in Multicomponent Systems, [A] eds. Johnson W.C., Blakely J.M., Interfacial Segregation[M]. Ohio: ASM, Metals Park, 1979. 261-348.
[5] M. Hillert, L.I. Staffansson, Acta Chemica, Scandinavica, 1970, 24, 3618
[6] Li L., Delaey L., Wollants P., O.Van Der Biest, Thermodynamic analysis of the segregation of multicomponent steels[J]. J.Chim. Phys., 1993, 90, 305-312.
[7] Li L., Delaey L., Wollants P., O.Van Der Biest, Thermodynamic Calculation of Segregation in Multicomponent Steels[J]. J.Mater.Sci.& Technol., 1996, 12(3), 238-240.
[8] Guttmann M., Dumoulin Ph., Wayman M., The thermodynamics of interactive co-segregation of phosphorus and alloying elements in iron and temper-brittle steels[J]. Metall. Trans., 1982, 13A, 1693-1711.
[9] Lin Li, Microstructure and property control of advanced high strength automotive steel, eds., Y.Q.Weng, H.Dong, Y.Gan, Metalluygical Industry Press and Springer, Beijing, 2010, 265
[10] Mclean D., Grain Boundaries in Metals, [M]Oxford: Clarendon Press, 1957.
[11] Sundman B., Jansson B., Andersson J., The thermo-calc databank system[J]. CALPHAD, 1985, 9(2), 153-190.
[12] L. Li, S.G. Huang, L. Wang, Y.L. He, J. Vleugels, O. Van der Biest, Thermodynamic re-assessment of the Fe-Al-C system based on the Fe-rich experimental data, Front. Mater. Sci., China, 3(1), 33-37, 2009.
[13] O. Redlich and A.T. Kister, Algebraic Representation of Thermodynamic Properties and the Classification of Solutions, Ind. Eng. Chem. 40 (1948), 345-348.

[14] Bo Sundman and John Ågren, A regular solution model for phases with several components and sublattices, suitable for computer applications, Journal of Physics and Chemistry of Solids, 42 (4) 1981, 297-301.

[15] K.C.H. Kumar and V. Raghavan, a thermodynamic analysis of the Al-C-Fe system, Journal of Phase Equilibria, 12 (3) (1991), 275-286.

[16] P. Gustafson, A thermodynamic evaluation of the Fe-C system, Scandinavian Journal of Metallurgy, 14 (1985), 259

[17] David K. Matlock, John G. Speer, Third Generation of AHSS, Microstructure Design Concepts,A. Haldar, S. Suwas, and D. Bhattacharjee (eds.), Microstructure and Texture in Steels, © Springer 2009,185-205

[18] Yanlin He, Lin Li, Xiaochun Wu, P. Wollants, J Mat. Sci. & Technol., vol.20, No.1, 71-74, 2004.

[19] H. Gaye, J. Welfringer,Proc.2nd Inter. Conf. Metallurgical Slags & Fluxes, Lake Tahoe, Nevada, Nov. 1984, 357

[20] Li Lin, Gao Yi, Shi Wen, Liu Rendong, He Yanlin, Fu Renyu, Zhang Mei, Martensite Transformation in High Mn Steels, HMnS2011, Soeal, Korea, A19

[21] G.B. Olson, M. Cohen: Metall. Trans. A, 7A (1976) p. 1897.

[22] S. Allain, J.-P. Chateau, O. Bouaziz, S. Migot, N. Guelton: Materials Science and Engineering A 387–389 (2004) p. 158.

[23] A. Dumaya, J.-P. Chateau, S. Allain, S. Migot, O. Bouaziz: Materials Science and Engineering A 483–484 (2008) p. 184.

[24] Inden G, Metallkde Z. 1977; 68: 529

[25] Hillert M, Jarl M. CALPHAD, 1978; 2: 227

[26] Dinsdale A T. CALPHAD, 1991; 15: 317

[27] Y.K. Lee, C.S.Choi: Metall. Mater. Trans., 31A(2000) p. 355.

[28] LIU Ren-dong, Shi Wen,He Yanlin,Li Lin, ZHANG Xiao-gang, WANG Fu,Designing and Investigating on the Hot Rolling TWIP Steel with TRIP Effect,Acta Metall. Sinica (in Chinese), 2012

[29] T.I. Barry, A.T. Dinsdale, J.A. Gisby, B. Hallstedt, M. Hillert, B. Jansson, S. Jonsson, B. Sundman and J.R. Taylor, J. Phase Equil., 1992, 13,459

[30] M. Hillert, J. Alloys Comp., 2001,320, 161

[31] B. Sundman, J. Alloys Comp., 1991, 12(1), 127

[32] Kaufman, L. and Nesor, H., Calculation of quasi-binary and quasi-ternary oxide systems I. CALPHAD, 1978, 2, 35–53.

[33] Shuigen Huang, Lin Li, O. V. D. Biest, J. Vleugels, Peiling Wang, J Mat. Sci. & Technol., vol.18, No.5, 422-426, 2002.

[34] Shuigen Huang, Lin Li, Jef Vleugels, Peiling Wang, Omer Van der Biest, Thermodynamic prediction of the nonstoichiometric phase $Zr_{1-z}Ce_zO_{2-x}$ in the ZrO_2–$CeO_{1.5}$–CeO_2 system, J. Eur. Ceram. Soc., 23 (2003) 99–106

[35] T.B. Lindemer,J. Brynestad,J. Am. Ceram. Soc., 1986, 69 (12), 867

[36] Du, Y., Yashima, M., Koura, T., Kakihana, M. and Yoshimura,M., Thermodynamic evaluation of the ZrO_2-CeO_2 system. Scripta Metall. Mater., 1994, 31, 327–332.

[37] Li, L., Xu, Z. Y. and Ao, Q., Optimization of the phase diagram of CeO_2-ZrO_2 system. J. Mater Sci. Tech., 1996, 12, 159–160.
[38] Li L., Van Der Biest O., Wang P. L., Vleugels J., Chen W. W. and Huang S. G., Estimation of the phase diagram in the ZrO_2-Y_2O_3-CeO_2 system. J. Eur. Ceram. Soc., 2001, 21(16), 2903–2910.
[39] Li L., Van Der Biest O., Xu L. P., Vleugels J. and Huang S. G., Prediction of the isothermal sections of the ZrO_2-Y_2O_3-CeO_2 system. J. Mater. Sci. Tech., 2001, 17(5), 529–534.
[40] Lin Li,O. Van Der Biest,Peiling Wang,Application of Substitutional Model in Oxide System, J.Mater. Sci. Technol., 19(1), 2003, 66
[41] Shuigen Huang, Lin Li, O.V.D. Biest, J. Vleugels, Peiling Wang, Thermodynamic Assessment of the ZrO_2–CeO_2 and ZrO_2–$CeO_{1.5}$ Binary System, J.Mater. Sci. Technol., 18(4), 2002, 325
[42] Muggianu, Y. M., Gambino, M. and Bros, J. P., Enthalpies de formation des alliages liquides bismuth-etain-gallium a 723 K. J. Chim. Phys., 1975, 22, 83–88.
[43] Pankratz, L. B., Thermodynamic Properties of the Elements and Oxides. US Dept. Interior, Bureau of Mines Bulletin 672, US Gov't Printing office, Washington, DC, 1982.

Chapter 10

Thermodynamics of Wool Dyeing

Rıza Atav

Additional information is available at the end of the chapter

http://dx.doi.org/10.5772/51618

1. Introduction

The pleasure derived from imparting color to clothing has existed since the time of the earliest civilizations; a world of fashion without color is impossible to imagine. Coloration processes produce the most visible results of all the finishing operations carried out during preparation of textile goods [1]. Dyes are synthetic aromatic compounds which can bring bright and firm color to other materials. It is reported that over 100,000 commercially available dyes exist and more than $7x10^5$ metric tons of dyes are produced worldwide annually [2]. When a textile fiber is immersed in a solution of dye under suitable conditions, the fiber becomes colored, the color of the solution decreases and dyeing has occurred [3].

The wool fiber is the hair of the sheep and forms the protecting covering of the animal [4]. Structurally, a wool fiber is an assembly of cuticle and cortical cells held together by the "cell membrane complex" (See Figure 1). The dyeing and diffusion properties of fibers are known as governed by this membranous structure, which is formed predominantly by internal wool lipids [5].

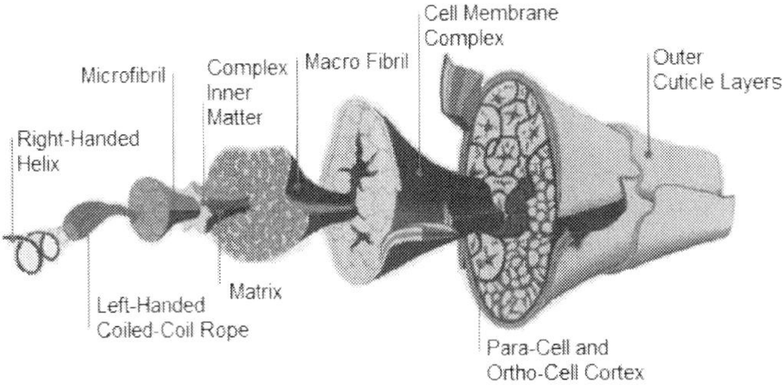

Figure 1. Cross-section diagram of a merino wool fiber [6]

Wool cuticle cells (overlapping cells that surround the cortex) are subdivided into two main layers, namely the exocuticle and endocuticle [7]. The outer surface of the scale of the cuticle is covered by a very thin membrane called the epicuticle. Below this hydrophobic epicuticle is the exocuticle, a cystine-rich part forming about two-thirds of the scale structure. The exocuticle just below the epicuticle is referred to as the "A" layer, having a distinctly higher cystine part than the rest of the exocuticle (known as the "B" layer). Below the exocuticle, forming the rest of the scale structure is the endocuticle and then a thin layer of intercellular cement [6].

For wool, its complexity is illustrated by the different important chemical groups it has and the intermolecular forces of attraction that are formed. First, there are the polar peptide groups (i.e., -CO-NH-). Second, the oxygen of the carbonyl groups (-CO-)is slightly negatively charged and, as a result, will form hydrogen bonds with the slightly positively charged hydrogen of the imino groups (-NH-) of other peptide groups. Third, cystine, the sulfur-containing amino acid capable of forming disulfide crosslinking tends towards greater chemical stability, resulting in less dye absorption [8].

Wool dyeing is a process involving high temperature for long periods in acidic to neutral pH medium to do good penetration, optimum fastness, and dyebath exhaustion. The shape of the conventional dye uptake curve is consistent with an initial dwell time (20 to 40°C) when dye is transported 1 through the medium, a primary exhaustion stage (40 to 60°C) when dye levels at the fiber surface and diffuses within, and a secondary stage (60 to 90°C and above) during which time the dye disperses and immobilizes within the fiber. There is much resistance to dye penetration from the epicuticle layer of the cuticle cell of the wool [9].

The surface of the cuticle cells is highly hydrophobic due to covalently bound fatty acids. The covalently bound fatty acids and the high amount of disulphide bridges make the outer wool surface highly hydrophobic. Especially in the printing and dyeing of wool, the hydrophobic character of the wool surface is disturbing. Diffusion of the hydrophilic dyes into the fibers is hindered [10].

Earlier workers identified the epicuticle with the barrier to dye penetration, thinking that this part forms a continuous membrane around the fiber. The barrier has also been ascribed to the cuticle and to the highly crosslinked A-layer of the exocuticle. All these suggestions on the nature of the barrier were based on a common belief that dyes must diffuse through the cuticle cells to reach the fiber cortex (i.e. the transcellular route shown in Figure 2) [11]. A continuous phase of membrane cells, the cell membrane complex underneath the cuticle, forms a network of penetrating canals to help mediate the impedance of the lipid barrier to dye diffusion, thereby facilitating dye penetration into the fiber interior [9].

In the manufacture of the almost infinite variety of wool textures, the dyeing of the wool itself is one of the series of many operations involved; and whilst of cardinal importance, the dyeing process must be subordinated to the particular final texture devised by the manufacturer [13]. For dyeing of wool fibers acid, metal complex 1:1 and 1:2, chrome

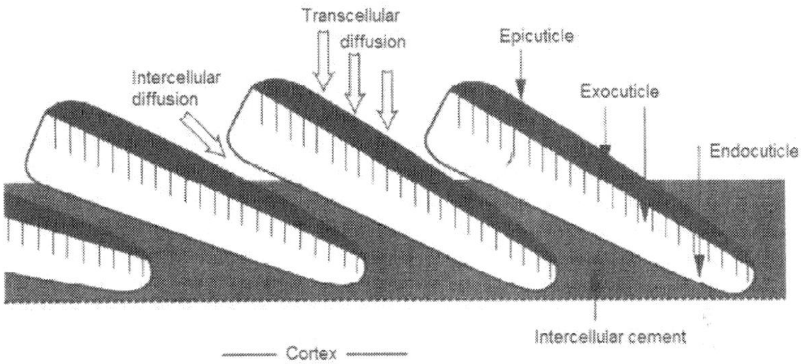

Figure 2. Diffusion pathways for dyes into wool [12]

and reactive dyes can be used [14]. In this chapter knowledge related to the thermodynamics of wool dyeing will be given.

2. Thermodynamics of wool dyeing

Dyeing systems can be classified as follows;

1. Non-ionic dyes applied to substrates which are non-ionic or ionic
a. Disperse dyes on essentially non-ionic substrates, such as the hydrophobic fibers produced from polyester, secondary cellulose acetate and cellulose triacetate, and
b. Disperse dyes on ionic substrates such as nylon.
2. Ionic dyes applied to substrates which themselves carry a charge
a. Ionic dyes applied to substrates which carry the same charge as the dye, as exemplified by anionic dyes on cellulose, and
b. Ionic dyes applied to substrates which carry the opposite charge as the dye, such as anionic dyes on wool and nylon and cationic dyes on acrylic fibers [3].

From above classification, it can be understood that the wool dyeing is a good example for dyeing systems described in 2b. In general, wool dyeing process occurs in four basic steps;

- Wetting and swelling of fibers in dyeing liquor
- Adsorption of dye molecules on the fibers
- Diffusion of dye molecules into the fibers
- Fixation of dye molecules in fibers via various physical and chemical forces [15]

Knowledge in detail related to the thermodynamics of wool dyeing such as aggregation and adsorption of dyes, rate, standard affinity, enthalpy and entropy of dyeing exc. is given below.

2.1. Aggregation of dyes

The nature of the dyeing process requires a complete understanding of the state of the dye in the dyebath, and many of the problems associated with the dyeing of wool can be explained in terms of the 'colloidal' or aggregated nature of the dye [16]. Aggregation can be defined as "Clustering of each particle of a substance to give it colloidal properties" [1]. A number of organic dyes form aggregates in solution via weak non-covalent interactions. The color, physical properties such as the solubility and photophysical behavior of dyes are affected due to aggregate formation [17]. In dyes, the aggregation principally forms dimers except at high concentrations in solvents with a high dielectric constant. For coloration of materials, it is the monomer adsorbed on the surface. There may be next build up of dye units through aggregation but the first layer must be composed of monomers [18].

Many studies have been carried out on the aggregation of dyes molecules or ions in aqueous and non aqueous solutions. The interactions between dye molecules and other molecular entities are of paramount importance in many industrial processes. The forces of attraction responsible for such interactions are also capable of causing physical interaction between dye molecules to produce molecular species ranging in size from dimmers to aggregates containing hundreds of molecules [19].

The dye aggregation is a function of temperature, electrolyte concentration, surfactant type, and dye concentration [20]. Aggregation is promoted by increasing the concentration of dye in solution [3]. The addition of electrolyte to a dyebath can increase the degree of aggregation, but fortunately the aggregation process is reversible and may be decreased by a rise in temperature. A certain degree of aggregation can be beneficial since it tends to increase the attraction of the dye for the fiber, but the large size of the dye aggregates can lead to a drastic reduction in the rate of fiber penetration, or in some cases to the precipitation (separation) of a dye from solution after prolonged storage [1]. Also chemical structure of dye molecule has an important effect on aggregation. For example; the tetrasulphonated dye C.I. Acid Red 41 has been shown to be monodisperse in aqueous solution by both diffusion-based and light scattering techniques. This result is to be expected, since the four sulphonate groups confer upon this molecule a high degree of hydrophilicity. When the number of sulphonate groups is reduced to one, as in C.I. Acid Red 88, then aggregation is expected to increase. The aggregation of C.I. Acid Red 88 has been determined by diffusion and by polarography at 25°C, and was found to range between 2 and 5 [16].

So far the changes in absorption spectra have been described qualitatively, yet the aggregation of dyes in aqueous and non-aqueous solutions may be studied quantitatively. The Maximum Slope has proved to be the simplest and the most convenient method to apply for such quantitative study. This method is based on the assumption that a simple equilibrium model between a monomer (m) and polymer $(m)_n$ is operative [19]:

$$n(m) \leftrightarrow (m)_n$$

First the following equation is used to estimate the values for degree of aggregation and aggregation constant:

$$\log C\,(\varepsilon_1 - \varepsilon) = n \log C\,(\varepsilon - \varepsilon'_n) + \log n\,K_n\,(\varepsilon_1 - \varepsilon'_n)^{1-n} \quad (1)$$

where $\varepsilon'_n = \varepsilon_n / n$

ε_1 = molar absorbativity of monomer

ε = experimentally measured molar absorptivities

ε_n = molar absorbativity of polymer

C = concentration of dye solution in mol/L

n = aggregation number

K_n = aggregation constant

By plotting logC ($\varepsilon_1 - \varepsilon$) against logC ($\varepsilon - \varepsilon'_n$), the points should lie on a straight line; the slope of which gives the degree of aggregation (n), and the aggregation constant (K_n) can be calculated from the intercept. Reliable results are very difficult to be obtained due to the assumption one has to make for ε_1 and ε_n. The above equation is then rearranged in the form [19]:

$$1/(n-1)\,\log C\,(\varepsilon_1 - \varepsilon) - n/(n-1)\,n\,\log C\,(\varepsilon - \varepsilon'_n) = \log \alpha C = X \quad (2)$$

where $\alpha = (nk)^{1/(n-1)}(\varepsilon_1 - \varepsilon'_n)^{-1}$

The Maximum Slope Method consists of correlating the curves of ε vs. logC and ε vs. X for different values of n and ε_n until a value of n is reached which gives the best fit between the experimental results and the theoretical curve [19].

2.2. Adsorption of dyes

Adsorption is a process that occurs when a gas or liquid solute accumulates on the surface of a solid or a liquid (adsorbent), forming a molecular or atomic film (the adsorbate). It is different from *absorption*, in which a substance diffuses into a liquid or solid to form a solution. The term *sorption* encompasses both processes, while *desorption* is the reverse process [21].

The strength of the adsorption depends on the kind of interactions taking place between the fiber surface and the dye molecules. Clearly, the highest strength results when a chemical interaction occurs. Unfortunately, in many fiber/dye systems such surface interaction cannot be realized. In the case of physical adsorption process, the strength is determined by the Lifshitz - van der Waals, hydrogen bonding, and electrostatic interactions. The nature of the process can be evaluated from the thermodynamic functions of the process [22]. As a result of its chemical structure, wool contains a considerable number of both acidic and basic groups and hence, it was natural to consider that the adsorption of ions by wool was directly concerned with the presence of these groups; this is in turn led to the idea that ions are adsorbed by wool on specific sites in the fiber [3].

Adsorption is usually described through isotherms, that is, functions which connect the amount of adsorbate on the adsorbent, with its pressure (if gas) or concentration (if liquid)

[21]. The Langmuir isotherm, the Freundlich isotherm, and the Nernst isotherm are all commonly used to classify dyeing isotherms even though they more correctly apply to the adsorption of gases on metal surfaces [16]. Their shapes are shown in Figure 3 [23].

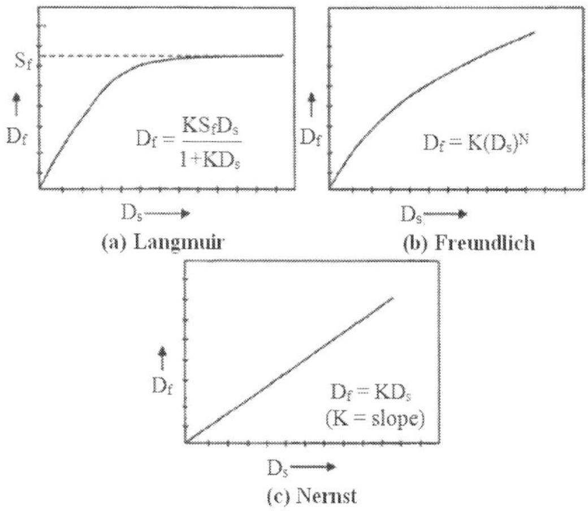

Figure 3. Specific isotherms of dye-fiber systems (D_s: Dye concentration in the dyebath, D_f: Dye concentration in the fiber) [23]

The simplest physically plausible isotherm is based on three assumptions:

1. Adsorption cannot proceed beyond monolayer coverage.
2. All sites are equivalent and the surface is uniform
3. There are no interactions between adsorbed molecules, so the ability of a molecule to adsorb at a given site is independent of the occupation of neighboring sites [24].

The Langmuir isotherm is well suited to describe dye adsorption by certain textile fibers. Considered in terms of dyeing, the basic postulate is that adsorption of dye takes place on specific sites in the fiber and that when a dye molecule occupies a site that site is saturated and incapable of further adsorption [3]. The Langmuir model is valid for monolayer sorption onto a surface with a finite number of identical sites. Langmuir isotherm theory is based on the assumption that adsorption on a homogeneous surface [25]. On the bases of these assumptions Equation 3 can be written [16]:

$$C_{f,a} = K'C_{s,m} / (1 + K'C_{s,m}) \tag{3}$$

$C_{f,a}$: concentration of adsorbed dye molecules on the fiber

$C_{s,m}$: concentration of mobile dye molecules in the bath

K': Adsorption constant

In the case of the Freundlich model (Equation 4) the dye is considered as being contained in an internal phase of volume V of the fiber.

$$\ln[C]_f = \ln K_F + x \times \ln[C]_s \tag{4}$$

where: $[C]_f$ represents the dye concentration in the fiber at equilibrium, in mol/kg dry fiber, and $[C]_s$ represents the dye concentration in solution at equilibrium, in mol/L; K_F is the equilibrium constant, and x is a sub-unitary power [26]. The Freundlich isotherm is commonly used to describe adsorption characteristics for heterogeneous surface [27]. The Freundlich model assumes a heterogeneous multilayer adsorption surface with sites that have different energies of adsorption which are not equally available [28].

The Nernst adsorption isotherm (Equation 5) is considered as a limit case of Freundlich and Langmuir adsorption isotherms [26].

$$[C]_f = K_P \times [C]_s \tag{5}$$

Nernst isotherm equation is a mathematical representation of the distribution law, which states that a dissolved substance, irrespective of its total amount, distributes itself between two layers or phases in a constant concentration ratio, at constant temperature; the ratio, equal to the constant in Equation 5, is referred to as the distribution or partition ratio. Theoretically the law is limited to describing the behavior of dilute solutions [3].

2.3. Rate of dyeing

Four steps can be envisaged in the process of dye uptake:

a. diffusion to the fiber surface,
b. transfer across that surface,
c. diffusion within the fiber to appropriate sites, and
d. binding at those sites.

In principle, any of these could be the slowest-and hence the rate determining-step. The theory of the kinetics of wool dyeing has generally been based on the premise that (c) is the rate determining step (diffusion control). Various algebraic expressions have been derived from Fick's laws of diffusion in an attempt to describe experimental dyeing rates. The most common treatment is the "parabolic", rate law, which was originally derived by Hill to explain the diffusion of phosphate and lactate in muscles. If a semi-infinite solid is brought into contact with a liquid containing a diffusible substance at concentration y_0, then the total amount diffused across a unit area (A) at time t is given by the following equation:

$$A = 2.y_0.(D_t/\pi)^{1/2} \tag{6}$$

where D is the diffusion coefficient of the substance in the solid [29]. The calculated diffusion coefficient appears to be exceedingly small when compared with values for other simpler ions and molecules. This may, of course, be due merely to the size of the dye anion, as diffusion within the solid is subject to greater restriction than aqueous diffusion, which suggests that the diffusing unit must force a path through more or less rigidly held molecules, so that a high activation energy is required. The apparently very low mobility of the dye ions may be due to interaction of ions with the fiber in such a way that the majority of ions within the fibers are held on specific sites; diffusion proceeds by way of a small fraction of mobile dye ions, thus giving an apparently very low value for diffusion coefficient [30].

The rate of diffusion of a dye in a fiber increases with increase in dyeing temperature. The effect of temperature on the rate of diffusion is given numerically by the activation energy of diffusion (E) according to the relationship shown in Equation 7:

$$D_T = D_0 e^{-E/RT} \tag{7}$$

Where D_T is the observed diffusion coefficient at absolute temperature T, D_0 is a constant and R is gas constant. Hence, if the logarithm of the observed diffusion coefficient is plotted against the reciprocal of T, a straight line of slope E/R should be obtained; from which E can be calculated [23].

Time of half-dyeing ($t_{1/2}$) is the time taken for a fiber to adsorb 50% of the dye it would adsorb at equilibrium. It provides an indirect measure of rate of diffusion and is useful for comparing the behavior of dyes applied under identical conditions. Because of the very gradual approach to equilibrium, the time taken to reach equilibrium cannot be measured accurately. However, equilibrium adsorption, C_α can be estimated readily by dyeing for a prolonged period and $t_{1/2}$ is then determined using a graph of exhaustion versus time of dyeing at $C_t = 0.5C_\alpha$ (Figure 4). In certain cases, it is preferable to plot exhaustion against logt since then the relevant portion tends to be a straight line and a more accurate estimate of $t_{1/2}$ can be obtained [23].

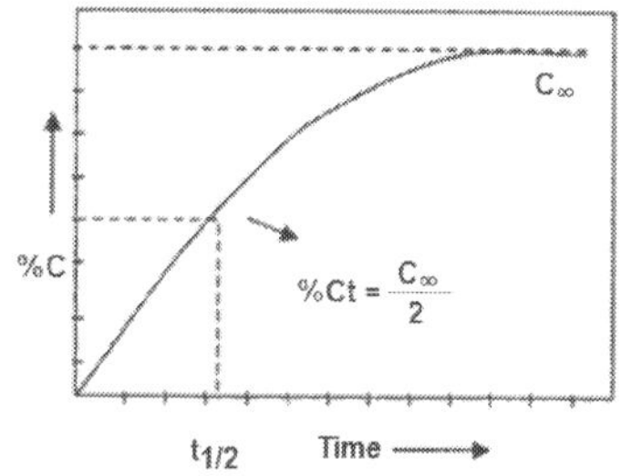

Figure 4. Exhaustion of dyes (%C) with time [23]

In practice, the experimental conditions must be chosen with care. If the conditions are such that the bulk of the dye is adsorbed on to the surface of the fibers quickly this will musk the effect of diffusion into the fiber [23, 31]. The conditions should be so chosen that the dye-fiber attraction is low and the final exhaustion is not very high (<80%). The time of half dyeing is also very dependent on the amount of dye used [23].

If the time of half dyeing is known for two different temperature activation energy of diffusion (E) can be calculated from the equation given below [31];

$$\log(t_{1/2})_{323} - \log(t_{1/2})_{363} = [E/2.3{*}R] * [(363-323)/(363{*}323)] \tag{8}$$

2.4. The standard affinity of dyeing

Thermal activation is only the first stage in the dyeing process. The driving force in the transfer of dyes from solution to the fiber is the difference in free energy between the two phases. In other words, the energy needed to compress the dye molecules from their freedom of movement in the solution to the restricted volume within the fiber. This is more conveniently expressed in the form of the chemical potential (μ) which is defined as the difference in free energy between two phases when a very small quantity of dye is added to the solution, all other variables kept constant. The standard potential is derived from the basic equation, applicable to ideal solution, namely Equation 9:

$$\mu = A + RT \ln C \quad (9)$$

where A is a constant, and C is the molar concentration [23].

A numerical parameter is needed to describe the behavior of individual dyes at equilibrium (thermodynamics) which is independent of variations in the dyeing conditions. Such a parameter would be standard affinity of the dye. Standard affinity of dyes has been defined as "the difference between the chemical potential of the dye in its standard state on the fiber and the corresponding chemical potential in its standard state in the dyebath" [32].

Affinity ($-\Delta\mu°$) is measurable according to the principles of thermodynamics and may be expressed in energy unit (kjoules/mole). It is the most basic thermodynamic parameter of the dye in dyeing solution towards fiber substrate. The greater the degree of exhaustion at equilibrium the greater is the affinity [33].

The standard affinity for the distribution of dye between the fiber and dyebath is proportional to the logarithm of the ratio of the absolute activities of the dye in the fiber and dyebath. Since the activity of the dye is assumed to be directly related to its concentration, one can write equation as follows [16];

$$-\Delta\mu^0 = RT[In([C]_F / [C]_L)] \quad (10)$$

$\Delta\mu°$: standard affinity (J mol^{-1})

R: gas constant (8.317 J K^{-1} mol^{-1})

T: absolute temperature (°K)

$[C]_F$: concentration of dye on the fiber at equilibrium (g/l)

$[C]_L$: concentration of dye remaining in the dyebath at equilibrium (g/l)

It was this point, where values had to be assigned to activities of the ions in the fiber, that a difference of opinion arose and this led to two different approaches being formulated, namely the Langmuir or Gilbert-Rideal method and the Donnan method. The differences between the two are described pictorially in Figure 5 for a simple, monobasic strong acid HX [3].

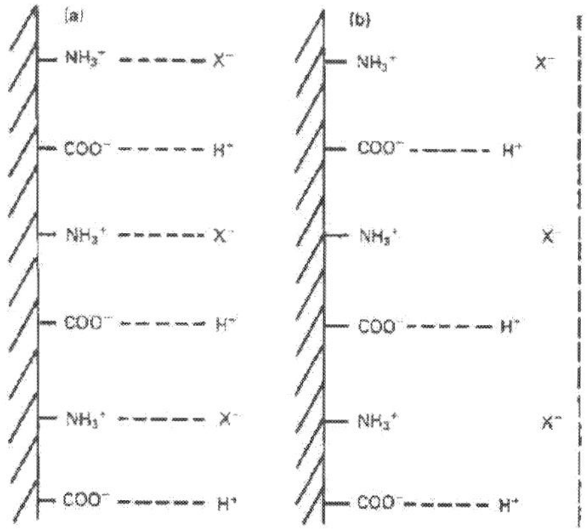

Figure 5. Pictorial representation of two approaches to wool dyeing theory: (a) Gilbert-Rideal, (b) Donnan [3]

a) Gilbert - Rideal theory: In that explanation, the activity coefficients of ions sorbed into the wool phase are reduced due to specific binding with sites on the wool, that is, the formation of ion pairs. Wool dyeing generally occurs in the presence of an acid as well as the dye (the acid is usually applied in a preliminary step). Gilbert and Rideal proposed that dye uptake is an anion exchange process, in which the dye molecules displace smaller anions [29].

b) Donnan theory: In this theory, the dye was considered to partition between the external solution and an internal solution phase in the wool. The latter phase is believed to contain a high concentration of fixed ionic groups, and hence solute molecules have reduced activity coefficients in that phase due to Coulombic interactions [29].

Derbyshire and Peters, proposed a unified interaction theory for all dye-fiber systems, postulating that the principal contribution to the standard affinity of the dyes, of all classical types, to all kinds of fibers arises from the nonpolar Van der Waals forces. They postulated that non-polar forces are mainly responsible for dye-fiber attachment and that in aqueous solution they occur between hydrophobic surfaces. The dyeing behavior of each type of fiber is predicted from the postulate that dye substantivity is due to non-polar forces. They suggested that the hydrophobic parts of the dye molecules have adequate affinity for the hydrophobic parts of the fiber. Accordingly the less hydrophobic dye molecules also have sufficient affinities for the more hydrophobic synthetic fibers, whereas dye molecules with larger hydrophobic parts will be required to dye a hydrophilic fiber such as wool [34].

Zollinger considered hydrophobic bonding in addition to the two well-known types of dye-fiber interaction in the adsorption of anionic dyes by wool such as coulombic attraction of dye anions to cationic groups on the fibers and Van der Waals (London) forces. The thermodynamic parameters such as standard enthalpies and standard entropies were measured for the adsorption of three suitable acid dyes on wool, and it was concluded that aliphatic side-chains contributed to the dye-fiber hydrophobic interaction. However, the detection of this contribution was difficult, because of the additional effect of dye-dye aggregation equilibria in solution [35].

Iyer et al. carried out a study using three different related acid dyes (Table 1); they calculated the heats and entropies of dyeing. They showed that with increasing hydrocarbon chain length, the affinity of the dye was increased, while the effect of a phenyl group was greater than that of a straight chain of four carbon atoms. It is surprising that the heat evolved on dyeing is decreased as the affinity increases. This is unusual, since higher affinity generally corresponds to larger heat of dyeing [36].

	R	$-\Delta\mu^{o}$ (kCal mol^{-1})	$-\Delta H^{0}$ (kCal mol^{-1})	ΔS^{0} (Cal k^{-1} mol^{-1})
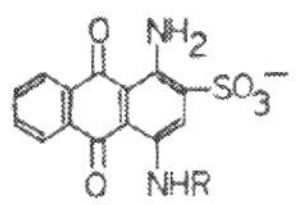	$-CH_3$	5.3	7.6	-7.2
	$-C_4H_{10}$	5.8	7.1	-4.1
	$-C_6H_6$	6.0	5.6	+1.2

Table 1. Thermodynamic parameters of acid dyes on wool (at 50°C and pH 4.6) [36]

The results of the study carried out by Ferrini et al. (Table 2) was also indicated that the phenyl group is more effective in reducing ΔH^{0} than is the butyl chain [37]. Asquith et al. claimed that the reason of the effect of a phenyl group to be greater than that of a straight chain was the effect of the phenyl group on dye aggregation to be greater than that of the butyl chain [38].

	Temp. (°C)	$-\Delta H^{0}$ (kCal mol^{-1})	ΔS^{0} (Cal k^{-1} mol^{-1})
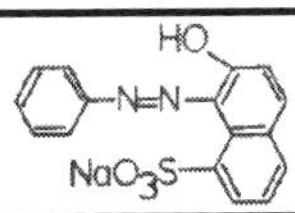	41-51	4.8	-2.2
	70-79	7.2	-9.3
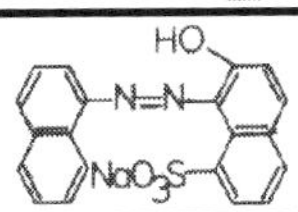	41-51	7.0	-4.9
	70-79	11.3	-17.7
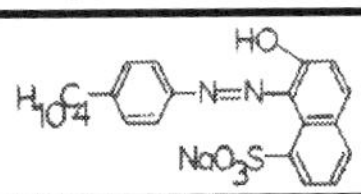	41-51	5.7	-0.8
	70-79	8.9	-10.5

Table 2. Heats of dyeing and entropies of acid dyes on wool [37]

2.5. The enthalpy of dyeing

The second thermodynamic quantity which is important in describing a dyeing process is the heat of dyeing or the change in enthalpy of the system when adsorption takes place [3]. As a result of dye molecules to be adsorbed on the fiber surface, free movement of these molecules gets restricted. So that entropy decreases when dye molecule is adsorbed. Adsorption also causes free energy to decrease.

$$\Delta G^0 = \Delta H^0 - T.\Delta S^0 \quad (11)$$

According to the Equation 11, enthalpy change should be negative, in other words adsorption is exothermic. For this reason adsorption amount will increase, when temperature is decreased [15]. Also the affinity of a dye is dependent on temperature. It decreases as the temperature rises, i.e. the equilibrium shifts in favor of the external phase. When $-\Delta\mu°/T$ is plotted against $1/T$ (T is the absolute temperature for a particular affinity value), the slope of the straight line obtained gives ΔH^0, the heat of dyeing [39]. However, the numerical value of $\Delta\mu°$, obtained from dye absorption, cannot be verified, and ΔS^0 is impossible to measure, while ΔH^0 is solely dependent on $\Delta\mu°$. Hence, using Equation 5, any error in $\Delta\mu°$ will accumulate in ΔH^0 and ΔS^0. One of the errors in determining $\Delta\mu°$ may be due to aggregation of the dye in solution under some conditions. Such an aggregation equilibrium could explain why a linear plot of $-\Delta\mu°/T$ against $1/T$ cannot always be obtained. In an attempt to clarify the latter point, Asquith et al. carried out a study on two dyes (C.I. Acid Orange 7 and 10), one known to be non-aggregating under the selected conditions of dyeing studied. Figure 6 shows the plots of $-\Delta\mu°/T$ against $1/T$ for these two dyes [38].

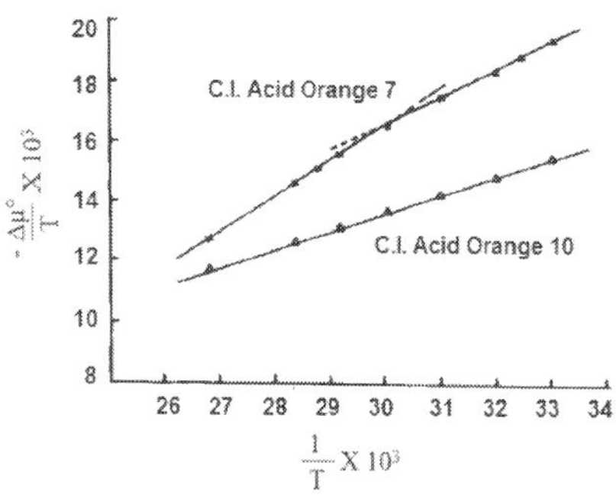

Figure 6. The relationship between $-\Delta\mu°/T$ and $1/T$ for various acid dyes [38]

According to these experimental results Asquith et al. denoted that for Orange 7, the plots each consist of two straight lines, intersecting at 60°C, which clearly indicates that this dye has two ΔH^0 values, corresponding to two temperature ranges-viz, below and above 60°C. The deviation from linearity in the above plots immediately leads to the speculation that the $\Delta\mu°$ and ΔH^0 values obtained from dyeing with Orange 7 has been influenced by dye aggregation equilibria in solution [38].

If the values of $\Delta\mu°$ are available for two absolute temperatures T_1 and T_2, it is more convenient to use the Equation 12 [39]:

$$\Delta H^0 = (T_2\Delta\mu_1° - T_1\Delta\mu_2°) / (T_2 - T_1) \quad (12)$$

where T_1 is the initial dyeing temperature in Kelvin; T_2, the final dyeing temperature in Kelvin, $\Delta\mu_1$, the affinity at T_1 °K, and $\Delta\mu_2$, the affinity at T_2 °K [40]. The value of ΔH^0

represents the difference in heat content between that needed to free one mole of dye from the solvent, and that required to free the same quantity of adsorbed molecules from the surface or from the interior of the solid by thermal agitation. It is thus the heat of sorption, i.e. the increase in heat content (enthalpy) of the system when the dye is sorbed. Large negative values of ΔH^0, therefore, correspond to high potential affinity [23].

2.6. The entropy of dyeing

The third thermodynamic parameter is the standard change in the entropy of the system (ΔS^0) [3]. The standard entropy of dyeing, ΔS^0, can be calculated from the Gibbs equation [26]:

$$\Delta\mu^\circ = \Delta H^0 - T.\Delta S^0 \tag{13}$$

Because there is no direct experimental means of measuring the entropy change, it is necessary to calculate its value from measurements of $\Delta\mu^\circ$ and ΔH^0 [3]. According to Boltzman's conception, the entropy is proportional to the probability of the system [26]. The change in entropy (ΔS^0) represents (on a logarithmic scale) the ratio of probability of sorption (β_1) to that of desorption (β_2) as shown in Equation 14:

$$\Delta S^0 = R\ln\beta_1 - R\ln\beta_2 = R\ln(\beta_1/\beta_2) \tag{14}$$

This entropy of sorption is the gain in entropy or disorder of the system when dye is adsorbed [3]. From the dyeing point of view the entropy change is of practical importance especially in two cases, in which either two dyes are compared on a common fiber or one dye is compared on two fibers. Therefore, dyeing entropy can almost be identified with dye/fiber orientation [26].

The entropy changes due to the dye sorption can be evaluated in a similar way to the enthalpies from the sorption isotherms determined at different temperatures via the free energy of adsorption ΔG^0.

$$\Delta S^0_{ad} = R(T_1 \ln C_1 - T_2 \ln C_2) / (T_2 - T_1) \tag{15}$$

The change in entropy upon the sorption process consists of two parts, one of which results from the dehydration of dye molecules before their adsorption (and dilution of the solution) and the other is connected with the actual interaction of the molecules with the solid surface. While the latter part is negative (loss of the molecular freedom), the former is positive. During a spontaneous adsorption process, the total entropy change is always positive [22].

3. Conclusion

Wool fibers have the most widely usage in the textile sector among the animal fibers and these fibers are dyed with acid, metal complex 1:1 and 1:2, chrome and reactive dyes. In order to understand the dyeing mechanism of these fibers better, it is needed to know thermodynamic and kinetic aspects of the dyeing such as aggregation and adsorption of dyes, rate, standard affinity of dye, enthalpy (standard heat) and entropy of dyeing exc.

Author details

Rıza Atav
Namık Kemal University, Textile Engineering Department, Tekirdağ, Turkey

4. References

[1] Ingamells W. Colour for Textiles a User's Handbook Society of Dyers and Colourists ISBN 0 901956 56 2 1993
[2] Gong R., Liu Y., Jiang Y., Li C. Isothermal, Kinetic and Thermodynamic Studies on Basic Dye Sorption onto Tartaric Acid Esterified Wheat Straw African Journal of Biotechnology 2009;8(24) 7138-7147.
[3] Johnson A. The Theory of Coloration of Textiles Second Edition Society of Dyers and Colorists 1989
[4] Von Bergen W., Krauss W. Textile Fiber Atlas: A Collextion of Photomicrographs of Common Textile Fibers American Wool Handbook Company New York 1942
[5] Marti M., Barsukov L.I., Fonollosa J., Parra J.L., Sukhanov S.V., Coderch L. Physicochemical Aspects of the Liposome-Wool Interaction in Wool Dyeing Langmuir 2004;20 3068-3073.
[6] Feughelman M. Introduction to the Physical Properties of Wool, Hair and Other α-Keratin Fibers in: Mechanical Properties and Structure Of Alpha-Keratin Fibers: Wool, Human Hair and Related Fibers UNSW Press 1997; 1-14.
[7] Silva C.J.S.M., Prabaharan M., Gubitz G., Cavaco-Paulo A. Treatment of wool fibers with subtilisin and subtilisin-PEG Enzyme and Microbial Technology 2005;36 917-922.
[8] Micheal M.N., El-Zaher N.A. Efficiency of Ultraviolet/Ozone Treatments in the Improvement of the Dyeability and Light Fastness of Wool Journal of Applied Polymer Science 2003;90 3668-3675.
[9] Cardamone J.M., Damert W.C. Low-temperature Dyeing of Wool Processed for Shrinkage Control Textile Research Journal 2006;76(1) 78–85.
[10] Onar N., Sarıışık M. Use of Enzymes and Chitosan Biopolymer in Wool Dyeing Fibers & Textiles in Eastern Europe 2005;13(1) 54-59.
[11] Marti M., Parra J.L., Coderch L. Natural Dyes: Chapter 5: Lipid Role in Wool Dyeing Ed. by Akcakoca Kumbasar E.P. InTech ISBN 978-953-307-783-3 2011.
[12] Simmonds D.H. The Amino Acid Composition of Keratins Australian Journal of Biological Sciences 1955;8(4) 537-540.
[13] Horsfall R.S., Lawrie L.G. The Dyeing of Textile Fibers Chapman&Hall LTD London 1946; 229.
[14] Yurdakul A., Atav R. Principles of Dyeing and Printing Ege University Engineering Faculty Department of Textile Engineering Bornova-Izmir 2006.
[15] Tarakçıoglu I., Textile Dyeing-I Notes, Ege University Engineering Faculty Press Izmir 1979-1980.
[16] Lewis D.M. Wool Dyeing Society of Dyers and Colourists 1992.

[17] Saha S.K. Dye-aggregation in Solution: Molecular Exciton Model 2010. http://www.scitopics.com/Dye_aggregation_in_solution_Molecular_exciton_model.htm l (accessed 22 April 2012)

[18] Walmsley F. Aggregation in Dyes: A Spectrophotemetric Study 1992;69(7) 583.

[19] Alarfaj N.A., El Khiate Z.M., Moussa E.A., Spectrophotometric Studies on Aggregation of Some Acid Dyes in Different Media Journal of King AbdulAziz University Science 2008; 99-110.

[20] Gabriela M.S., Elena G.M., Szabaday Z. Aggregation of Two Direct Dye Derived from 4,4′-Diaminostilbene-2,2′-Disulphonic Acid. Spectroscopic and Mathematical Studies, 14th International Electronic Conference on Synthetic Organic Chemistry, 1-30 November 2010.

[21] http://www.fpharm.uniba.sk/fileadmin/user_upload/english/Physical_Chemistry/5-Adsorption.pdf (accessed 25 April 2012)

[22] Espınosa-Jıménez M., Padılla-Weıgand R., Ontıveros-Ortega A., Perea-Carpio R., Ramos-Tejada M. M., Chıbowskı E. Investigation of the Polyamide 6,6 Dyeing Process with Acid Blue 45 dye. Part I. Thermodynamics of Acid Blue 45 Adsorption J. Adhesion Sci. Technol. 2002;16(3) 285-301.

[23] Kumar A., Choudhury R. Textile Preparation and Dyeing Science Publishers-USA ISBN 1-57808-402-4 2006

[24] Atkins P., Paula J. Elements of Physical Chemistry Oxford University Press Fourth edition ISBN: 0199271836 2005

[25] Tan L.S., Jain K., Rozaini C.A. Adsorption of Textile Dye From Aqueous Solution on Pretreated Mangrove Bark An Agricultural Waste: Equilibrium and Kinetic Studies 2010;5(3) 283-294.

[26] Simu G.M., Funar-Timofei S., Kurunczi L. Influence of Temperature and of the Electrolyte Concentration in the Dyeing Process of a Cellulosic Fiber with a Disazo Direct Dye Buletinul Stiinţific al Universităţii "Politehnica" din Timisoara ROMANIA 2008;53(67) 153-155.

[27] Zawani Z, Luqman Chuah A, Thomas S.Y. Choong Equilibrium, Kinetics and Thermodynamic Studies: Adsorption of Remazol Black 5 on the Palm Kernel Shell Activated Carbon (PKS-AC) European Journal of Scientific Research 2009;37(1) 63-71.

[28] Narayanachar K., Puttaswamya V., Gowdab K.N., Sudhakar R. Extraction of Natural Color Component from The Bark of Belleric Myrobalan (Terminalia Bellerica): Kinetic and Adsorption Studies European Journal of Chemistry 2010:1(3) 206-210.

[29] Bruce R.L., Broadwood N.V., King D.G. Kinetics of Wool Dyeing with Acid Dyes Textile Research Journal 2000; 70(6) 525-531.

[30] Alexander P., Hudson R.F. The Kinetics of Wool Dyeing: Part I: Simple Acid Dyes Textile Research Journal 1950:20 481-491.

[31] Dugg D.G., Sinclair S. Giles's Laboratory Course in Dyeing (Fourth Edition) ISBN-13: 978 0 901956 49 1 December 1989.

[32] Aspland J.R. Textile Dyeing and Coloration American Association of Textile Chemists and Colorists ISBN 0-9613350-1-7 1997.

[33] Adebayo G.B, Adekola F.A, Olatunji G.A, Bello I.A. Some Thermodynamic Parameters of Two Indigenous Mineral Dyes Applied on Wool Material Bulletin of Pure and Applied Sciences 2010:29C (2) 77-95.

[34] Derbyshire A.N., Peters R.H. An Explanation of Dyeing Mechanisms in Terms of Non-polar Bonding J. Soc. Dyers Colour. 1955:71 530-536.

[35] Zollinger H. The Dye and the Substrate: The Role of Hydrophobic Bonding in Dyeing Processes J. Soc. Dyers Colour. 1965;81 345-350.

[36] Iyer S.R.S., Ghanekar A.S., Singh G.S. The Chemistry of Synthetic Dyes Vol. VII Ed. by Venkataraman K. Academic Press New York 1974; 236

[37] Ferrini B., Kimura Y., Zollinger H. A Contribution to the Dyeing Mechanism of Acid Dyes on Wool Proc. Int. Wool Textile Res. Conf. Paris 1965; 291-300.

[38] Asquith R.S., Kwok W.F., Otterburn M.S. An Assessment of Some Thermodynamic Treatments of Wool Dyeing Systems Textile Research Journal 1980;50 333-336.

[39] Kumar A., Choudhury R. Textile Preparation and Dyeing Science Publishers ISBN 1-57808-402-4 2006.

[40] Samanta A.K., Agarwal P., Datta S., Physico-chemical Studies on Dyeing of Jute and Cotton Fabrics Using Jackfruit Wood Extract: Part II Dyeing Kinetics and Thermodynamic Studies Indian Journal of Fiber&Textile Research 2008;33 66-72.

Chapter 11

Mesomechanics and Thermodynamics of Nanostructural Transitions in Biological Membranes Under the Action of Steroid Hormones

L.E. Panin

Additional information is available at the end of the chapter

http://dx.doi.org/10.5772/51515

1. Introduction

Biological membranes are liquid heterocrystals with low shear stability. The main structure-forming bonds in biological membranes are covalent and hydrogen bonds, and hydrophobic and weak electrostatic interactions. These bonds are responsible for high membrane elasticity – a property of particularly importance to erythrocytes, which have to pass through blood capillaries of diameter equal to the erythrocyte one. Any structural changes that increase the erythrocyte membrane viscosity hamper the motion of erythrocytes through capillaries and may result in diffuse hypoxia. In this context, of great interest is the effect of stress hormones (cortisol, adrenaline, noradrenaline) on the behavior of erythrocyte membranes.

The nonspecific binding of stress hormones with erythrocyte membranes was studied earlier in [1]. It was shown that excess of these hormones in blood are capable for nonspecific binding with blood cells, primarily with erythrocytes, producing changes in rheological properties of the blood. It was found that CO, OH, and NH active groups incorporated in the structure of hormones can form hydrogen bonds with similar groups of proteins and phospholipids of erythrocyte membranes. Hydrophobic rings of hormones can participate in hydrophobic interactions with residues of phospholipid fatty acids, as a result of which complex domains arise in the membrane structure, the membrane microviscosity increases, and the motion of erythrocytes through capillaries becomes difficult. This effect is particularly dangerous to heart because it can leads to coronary syndrome X [2, 3]. Physicians still fail to understand the nature of this phenomenon, which shows up as exertional angina and ischemic ST segment depression on electrocardiograms with a normally functioning left ventricle.

Anabolic steroid hormones have been used for many decades, finding their most extensive use in sports medicine. Nowadays, it is impossible to train as an international class athlete without anabolic hormones. A coach's aspiration for high sporting results prompts that coach to use an ever increasing amount of anabolic steroids. Lacking a profound knowledge of sports medicine, such a coach cannot imagine all of the negative effects of anabolics on the body of an athlete. Moreover, sports medicine itself has no comprehensive information on the subject. As a consequence, the number of sudden and unexpected deaths of atheletes during the competitions has drastically increased in recent years [4].

Physicochemical analysis of the behavior of erythrocyte membranes as liquid heterocrystals makes it possible to disclose a link between structural changes in erythrocyte membranes and erythrocyte function. Of particular interest is activity of the Na+, K+-ATPhase that supports the transmembrane potential of cells and precludes their aggregation. It was previously supposed that the regulatory action of different ligands can be based on certain conformational changes of the Na^+, K^+-ATPhase [5]; however, the mechanism by which steroid hormones affect the activity of the Na^+, K^+-ATPase is poorly known.

In this work, the mechanism of testosterone, androsterone, dehydroepiandrosterone (DHEA), dehydroepiandrosterone sulfate (DHEAS) and cortisol interaction with structural components of erythrocyte membranes (mesomechanics and termodinamics of nanostructural transition) changes in their microviscosity and functional characteristics during the interaction have been studied. The results obtained could also shed light on the causes of cardiovascular catastrophes, which are often observed in sportsmen taking anabolic steroid hormones for a long time [6].

2. Materials and methods

The action of five hormones: testosterone, androsterone, dehydroepiandrosterone (DHEA), dehydroepiandrosterone sulfate (DHEAS) and cortisol (Amersham) is analyzed in the work (Fig. 1).

For this purpose, the following methods were used.

2.1. Atomic force microscopy (AFM) of erythrocytes

Erythrocytes were obtained from fresh blood after decapitation of Wistar rats under light nembutal narcosis. Blood was diluted twofold by isotonic phosphate buffer (pH 7.35) containing 0.043 M of KH_2PO_4 and 0.136 M of Na_2HPO_4. After precipitation of cells by centrifuging at 330 g for 10 min, supernatant liquor was decanted, and the washing procedure was repeated twice more.

All the procedures were performed at 4 °C [1]. The resulting erythrocyte suspension of 20 mcl volume was deposited onto a glass slide as a thin smear. The smear was predried for

10 min in air at 24 °C and humidity of 40%. After evaporation of excessive surface moisture, the smear was observed under a «Solver Bio» atomic force microscope (NT-MDT, Russia) at 24 °C using a semi-contact mode. An analogous procedure of obtaining red blood cells for the AFM examination was employed earlier by other authors [7]. In each experiment first a control specimen without hormones, and then the experimental one have been tested. Silicon cantilevers NSG11 (NT-MDT, Russia) with a resonant frequency between 120 and 180 kHz and spring constant ~ 6 N/m were used (all of these probe parameters were offered by manufacturer). Images of the surface relief of erythrocyte membrane after absorption of hormones were obtained with the scan size 1×1 μm² and 1.3×1.3 μm².

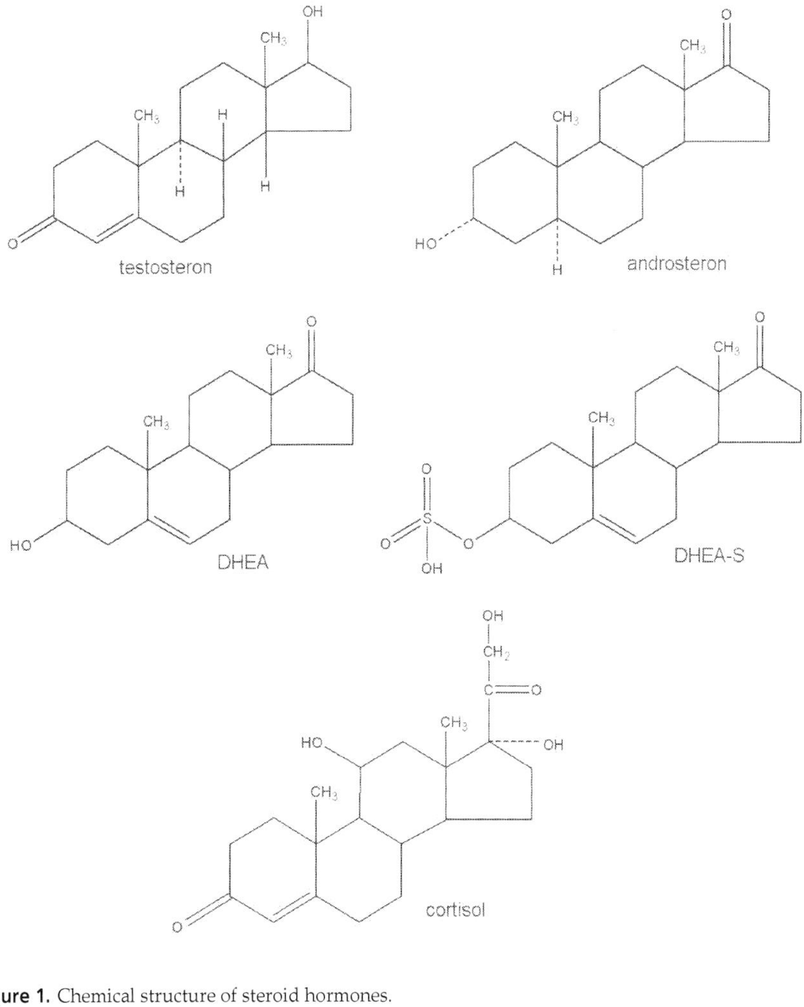

Figure 1. Chemical structure of steroid hormones.

2.2. IR spectroscopy of erythrocyte shadows

Erythrocyte shadows were obtained after their hemolysis in hypotonic phosphate buffer (pH 7.35) containing 2.75 mM of KH_2PO_4 and 8.5 mM of Na_2HPO_4. Shadows were precipitated by centrifuging at 5500 g, supernatant liquor was decanted. The washing procedure was repeated four more times [8]. All operations and further storage of shadows were performed at 4 °C.

A film for taking the IR spectra of erythrocyte shadows was prepared in a cuvette with fluorite backing via slow evaporation of water under weak vacuum at a pressure of ca. 0.1 atm (ca. $0.5 \cdot 10^4$ Pa) and temperature 4±1 °C [8]. Drying lasted 180 min. A suspension of erythrocyte shadows in a 0.001 M phosphate buffer with pH 7.35 and volume 60 mcl was introduced into a cuvette. This was supplemented with 30 mcl of the same buffer and 1.0 mcl of the hormone solution with concentration 10^{-6} M. Stirring and incubation lasted 10 min at 16-17 °C. The cuvette was placed horizontally on a special table of a vacuum unit.

When the film was prepared, the cuvette was transferred into an optical chamber and blown with dry air for 30 min, then the scanning unit was switched on. IR spectra were taken on a Specord-M80 spectrometer (Germany, Leipzig), sequentially experiment and control against the fluorite backing, or experiment and control to obtain a difference spectrum. Integration, determination of the spectrum band frequency, and mathematical processing were performed with special programs enclosed to the spectrometer. Erythrocyte suspensions were examined upon addition of cortisol using UV (Evolution 300, Thermo Scientific, USA). Merk or Sigma reagents were used in the work.

2.3. Fluorescence analysis of erythrocyte shadows

Fluorescence measurements were performed with a Shimadzu spectrofluorophotometer RF-5301(PC)SCE. 4 ml of hypotonic phosphate buffer containing 2.75 mM of KH_2PO_4 and 8.5 mM of Na_2HPO_4 (pH 7.35), and erythrocyte shadows were poured into a quartz cuvette of size $1 \times 1 \times 4$ cm^3. The concentration of shadow proteins was determined by the Warburg–Christian method from changes in the optical density of suspension [9]. On the average, it varied in the range of 0.100-0.250 mg/ml.

A cuvette with the shadow suspension was placed into a spectrofluorimeter thermostat for 1 hour. Getting a stationary temperature regime in the cuvette was controlled by an electronic thermometer. In all the experiments, temperature in the cuvette was 36 °C. After establishing a stationary temperature in the cuvette, intensity of the intrinsic fluorescence of tryptophan residues in protein membranes was measured. The tryptophan emission spectrum was taken in the range of 300 nm $\leq \lambda \leq$ 400 nm at the excitation wavelength 281 nm, with the maximum of emission intensity observed at 332 nm. The average value of maximum emission intensity was obtained graphically after its continuous measuring for 4 minutes. Intensity of tryptophan fluorescence fluctuated within 1%. The possible reasons include variation of temperature in the cuvette with suspension, instrumental error in determination of fluorescence intensity, and photochemical reactions occurring in the system. Spectral width of the slits was 1.5/10. The tryptophan absorption spectrum was recorded in the range of 220 nm ≤ λ ≤ 300 nm at the emission wavelength λ = 332 nm. Testosterone, androsterone, dehydroepiandrosterone (DHEA) and dehydroepiandrosterone sulfate (DHEAS) were dissolved in a mixture of dimethyl sulfoxide (DMS) and ethanol (1 : 1, V/V). Concentration of the hormone in the initial mother liquor was 10^{-3} M. If necessary, the solution was diluted with hypotonic phosphate buffer to obtain a desired concentration.

A solution of hormones with the concentration 10^{-6} M was prepared in hypotonic phosphate buffer. The time of hormone incubation with shadows was one hour. Absorption and emission spectra were taken, the average value of emission and absorption intensity was measured. For each hormone (testosterone, androsterone, dehydroepiandrosterone (DHEA) and dehydroepiandrosterone sulfate (DHEAS)), the binding constant K_b was calculated by the method [10] as well as the stoichiometric concentration of a bound hormone B_{max} and a change in free energy of the system ΔG. The interaction of hormone and erythrocyte membrane is described by the equation

$$B + nS = S_nB, \tag{1}$$

where B is a membrane protein, S is the hormone, and n is the number of moles of hormone per a mole of proteins. The binding constant K_b was calculated by the formula

$$K_c = \frac{\left[S_nB\right]}{\left[S\right]\cdot\left[B\right]}, \tag{2}$$

where $\left[S_nB\right]$ is the concentration of bound protein, $\left[B\right]$ is the concentration of free protein, and $\left[S\right]$ is the concentration of free hormone. It is supposed that hormone, upon binding to protein, completely quenches its fluorescence. Thus, the fluorescence intensity F will be proportional to the concentration of free protein. Let's write C for total concentration of protein in the cuvette, and x for concentration of the bound protein. Then,

$$\begin{aligned} F_{\max} &= \beta C \\ F &= \beta\left(C - x\right), \end{aligned} \tag{3}$$

where F is the intensity of tryptophan fluorescence at λ = 332 nm (the excitation wavelength Λ = 228 nm), $F_{\max}$ is the intensity of tryptophan fluorescence in the absence of hormone (when the entire protein is free), β is the proportionality factor, and A_S is the stoichiometric concentration of hormone. When concentration of hormone exceeds A_S, the fluorescence quenching does not increase. Dividing the first equation of set (2) by the second one gives

$$x = Q \cdot C, \text{ where } Q = \frac{F_{\max} - F}{F_{\max}} \tag{4}$$

$[S] = A - n \cdot x = A - n \cdot Q \cdot C$, where A is the total concentration of hormone; $n = \frac{A_S}{C}$; $[B] = C - x = C(1-Q)$. Substitution of (2) and (3) into expression for binding constant (1) gives

$$K_c = \frac{Q}{(1-Q)(A - nQC)} \tag{5}$$

In our case, molar mass of membrane proteins is unknown, so the concentration of proteins in cuvette C is determined in mg/ml, and concentration of hormones A in mol/l. The constant n is expressed in moles of molecules of hormone per milligram of protein (M/mg) and is a ratio of the maximum concentration of bound hormone to the concentration of membrane proteins. This can be written as

$$B_{MAX} = \frac{A_S}{C} \left[{}^{mole}\!/_{mg\ protein} \right] \tag{6}$$

Changes in Gibbs free energy ΔG of the system upon transition of hormone from aqueous medium to erythrocyte membrane are calculated by the formula

$$\Delta G = -RT \cdot \ln(K_c) \left[{}^{J}\!/_{mole} \right], \tag{7}$$

where $R = 8.314 \ \frac{J}{K \cdot mole}$, and T is the absolute thermodynamic temperature.

The measurement errors appeared due to inaccuracy in volumetric dosing of the shadow suspension specimens and their titration against hormones. Specimens were dosed using pipette dispensers DPAOF-1000 and DPAOF-50, their relative error at (20±2)°C being 1% and 2%, respectively. 4 ml of the buffer was taken once with a DPAOF-1000 pipette, and suspension of erythrocytes and hormones was dosed twice using a DPAOF-50 pipette. The fluorescence intensity of erythrocyte shadows F is directly proportional to the concentrations of proteins C and hormones A in the specimen. Relative error E_F in measuring the F value can be estimated by the formula

$$E_F = \sqrt{(1\%)^2 + (2\%)^2 + (2\%)^2} = 3\%. \tag{8}$$

Relative measurement errors for K_b and B_{MAX} can be obtained in the same way. They are equal to 10%.

In calculation, the values of fluorescence intensity F were corrected for dilution of suspension after the introduction of solution with hormone, for quenching of tryptophan emission by the solvent (a mixture of DMS and ethanol), for proper fluorescence of hormones, and evaporation of water from the cuvette.

2.4. Measurement of erythrocyte membrane microviscosity

Membrane microviscosity for translational diffusion of pyrene probe was calculated as a ratio of fluorescence intensity of the pyrene dimer to fluorescence intensity of the pyrene monomer. Microviscosity of erythrocyte membranes was measured also on a Shimadzu RF-5301(PC)SCE spectrofluorimeter. The experimental specimen was prepared as follows: 4 ml of hypotonic phosphate buffer containing 2.75 mM of KH_2PO_4 and 8.5 mM of Na_2HPO_4 (pH 7.35), a fluorescent pyrene probe, erythrocyte shadows and a specified amount of hormone were placed in a quartz cuvette of size $1 \times 1 \times 4$ cm^3. Before use, all the components were stored at 4 °C. The concentration of shadow protein in the cuvette was 0.100-0.250 mg/ml; that of pyrene, $7.76 \cdot 10^{-6}$ M. Pyrene was diluted in ethanol, its initial concentration being $1.5 \cdot 10^{-3}$ M. The cuvette was placed into the spectrofluorimeter thermostat for 10 min, then the fluorescence measurements were performed at 36 °C. Before placing the specimen into the spectrofluorimeter thermostat, it was shaken vigorously for 1 min. For fluorescence measurements of shadows upon their loading with a different amount of hormones, each time a new specimen was prepared by the same procedure. Such a procedure is necessary because pyrene favors fast degradation of erythrocyte membranes.

To measure microviscosity of a lipid bilayer near proteins (the region of protein-lipid interaction), we used the excitation wavelength $\lambda = 281$ nm and spectral slit width 1.5/5. Microviscosity of a lipid bilayer far from proteins (the region of lipid-lipid interaction) was measured with the excitation wavelength $\lambda = 337$ nm and spectral slit width 1.5/3. At that, the maxima of emission intensity were observed at $\lambda = 374$ nm and $\lambda = 393$ nm (the vibronic emission peaks of excited pyrene monomers), and $\lambda = 468$ nm (the emission maximum of excited pyrene dimer).

The relative microviscosity of membranes was determined as a ratio $\eta(A)/\eta(0)$, where $\eta(A)$ and $\eta(0)$ are microviscosities of membranes, respectively, with and without hormone added to the shadow suspension. For the region of lipid-lipid interaction, relative microviscosity was calculated by the formula

$$\frac{\eta(A)}{\eta(0)} = \frac{F_{468}(0)}{F_{468}(A)} \cdot \frac{F_{393}(A)}{F_{393}(0)} \tag{9}$$

where $F_{468}(A)$ is the fluorescence intensity of pyrene at wavelength $\lambda = 468$ nm in a specimen at the hormone concentration A in suspension; $F_{468}(0)$ is the fluorescence intensity of pyrene at wavelength $\lambda = 468$ nm in a specimen with no hormone in suspension. $F_{393}(A)$ and $F_{393}(0)$ are the fluorescence intensities of pyrene at wavelength $\lambda = 393$ nm at the hormone concentration A in suspension and without hormone in suspension, respectively. The excitation wavelength is 337 nm.

For the region of protein-lipid interaction, relative microviscosity was calculated by the formula

$$\frac{\eta(A)}{\eta(0)}=\frac{F_{468}(0)-I_{468}}{F_{468}(A)-I_{468}}\cdot\frac{F_{393}(A)-I_{393}}{F_{393}(0)-I_{393}} \tag{10}$$

where I_{393} and I_{468} are the fluorescence intensities of tryptophan at wavelength $\lambda = 393$ nm and $\lambda = 468$ nm, respectively. The excitation wavelength is $\lambda = 281$ nm. A relative measurement error for relative microviscosity was equal to 6%.

2.5. Change of the Na$^+$, K$^+$-ATPase activity

Erythrocytes were extracted from fresh blood of rats. Erythrocyte suspensions with hormones of differing concentration were analyzed to activity of the Na$^+$, K$^+$-ATPhase. The experimental procedures used in these investigations are described in [5].

3. Results

3.1. Atomic force microscopy

Under atomic force microscope, erythrocytes of healthy animals looked as large biconcave discs ca. 6 μm in diameter, which agrees with the results obtained by other authors [7]. At a higher magnification, their surface showed a slight nonuniformity caused most likely by the presence of membrane proteins. When the erythrocyte suspension was supplemented with DMS and ethanol (0.25% of the mixture volume), the surface nonuniformity increased, probably due to denaturating effect of solvent on the surface structural proteins (Fig. 2). Domains with the length 200-250 nm and height 2 nm are seen. The pattern changed upon addition of testosterone to erythrocyte suspension with the final concentration 10^{-7} M (Fig. 3). The interaction of testosterone with erythrocyte membranes leads to their restructuring. The surface is tuberous, there are domains of size 400 x 400 nm and height 20-25 nm, with smaller domains on the surface of large ones: size 50 x 50 μm^2 and height 10 nm. Between them, there are regions of loosened substance that form hollows. In this case, there are pronounced distortions in the primary structure of erythrocyte membranes.

Other structural changes of erythrocyte membranes were obtained in our study upon interaction with androsterone (Fig. 4) with the final concentration 10^{-6} M. The surface is flat, there are domains of size 100 x 100 nm and height 6-8 nm. In comparison with control specimens, domains decreased in area, but increased in height.

The surface of rat erythrocyte after adsorption of DHEA is depicted in Fig. 5. Concentration of the hormone is 10^{-7} M. The surface is tuberous, there are domains of size 220 x 220 nm and height 20 -25 nm. However, they are not separated into subdomains, as in the case of testosterone.

Of the four hormones, DHEAS has the weakest effect on the membrane morphology. The surface of rat erythrocyte after adsorption of DHEAS is shown in Fig. 6. Concentration of the hormone is 10^{-7} M. The surface is flat, there are domains of size 100 x 100 nm and height 3 - 4 nm. Changes are insignificant in comparison with control. It can be suggested that DHEA and DHEAS did not affect deep layers of the membranes, so the effect was much less pronounced. IR spectroscopy allowed us to reveal the nature of these structural transformations.

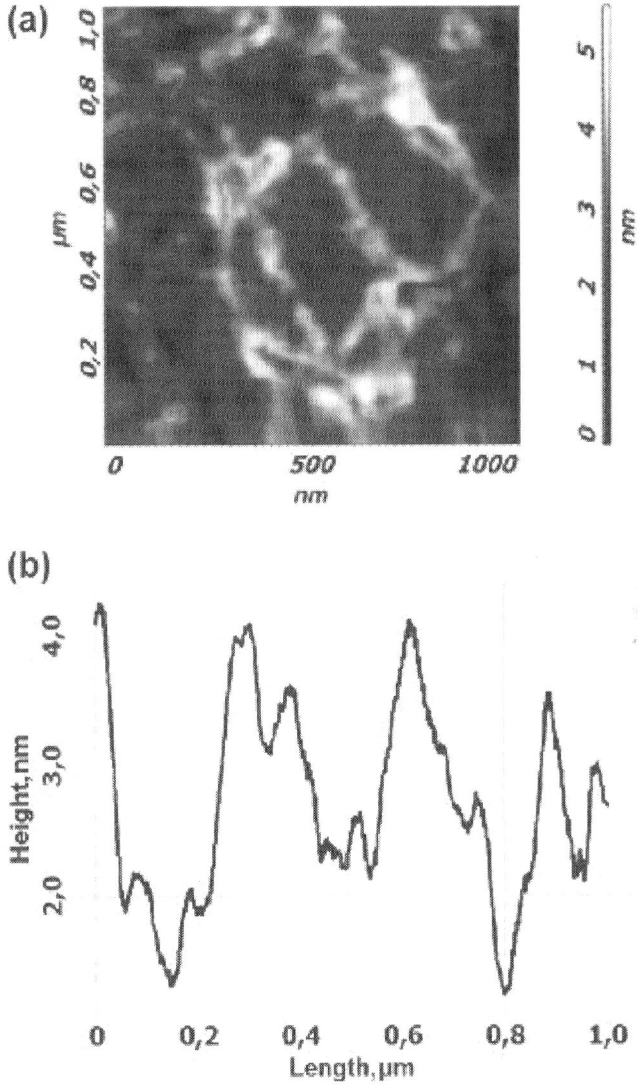

Figure 2. Control surface of rat erythrocyte. The erythrocyte suspension was supplemented with DMS and ethanol (0.25% of the mixture volume): (a) scan size 1 × 1 μm²; (b) center section of the surface.

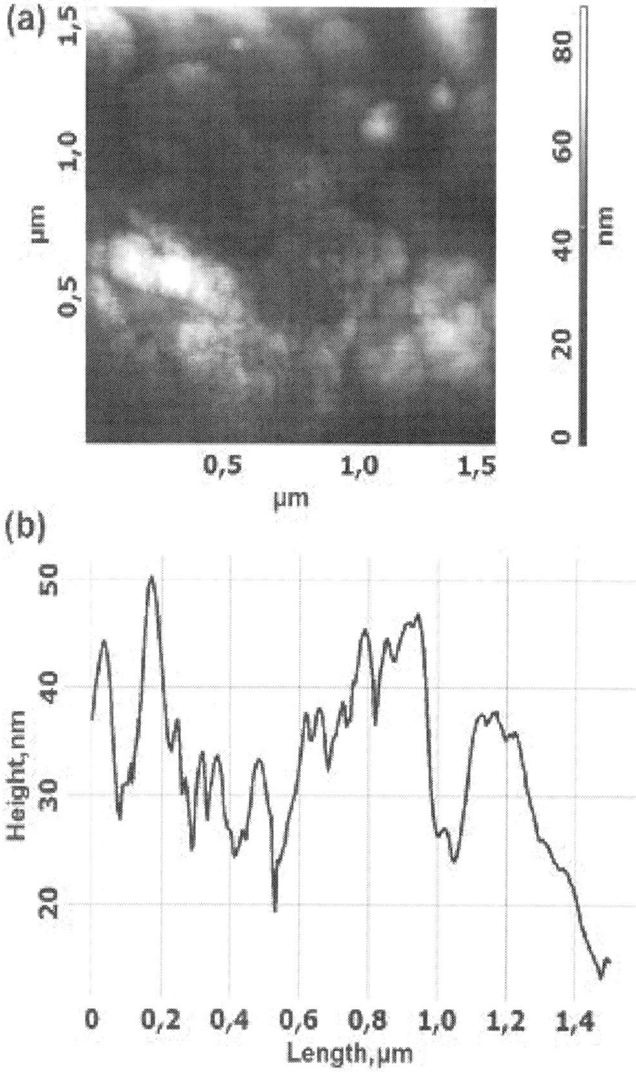

Figure 3. Surface of rat erythrocyte after adsorption of testosterone.
Concentration of the hormone is 10^{-7} M: (a) scan size 1.5×1.5 µm^2; (b) center section of the surface.

3.2. IR spectroscopy of erythrocyte shadows

Analysis of IR spectra of erythrocyte shadows obtained from rats with no hormone loading (Fig. 7) revealed in membrane-bound proteins not only a disordered structure, but also α-helix at 1650 and 1656 cm^{-1}, and β-structure at 1686 and 1520 cm^{-1}. Besides, we recorded NH stretching vibrations in proteins (3308 cm^{-1}), CH stretching vibrations in proteins and phospholipids (2948, 2930 and 2848 cm^{-1}), and a set of bands corresponding to phospholipids, in particular, C=O bond (1748 cm^{-1}), P=O bond (1236 cm^{-1}), CH_2 deformation vibrations (1460 and 1386 cm^{-1}), O_4C_4-C_5O_5 bond (1048 cm^{-1}) and C-C bond of deformation vibrations (978 cm^{-1}). It should be noted that C=O band (1736 cm^{-1}) is quite narrow, which gives grounds to suggest that phospholipids in membranes of normal erythrocytes are well ordered at a level of ester bonds of higher carboxylic acids and glycerol.

3.2.1. *Effect of testosterone*

Under the action of testosterone, intensity of absorption bands 1544, 1656 and 3292 cm^{-1} increased by 30% and more (Fig. 8, Table 1). Absorption band of NH bond showed a 3308 → 3272 cm^{-1} shift ($\Delta\nu$ = 36 cm^{-1}). The bands 2852 and 2932 cm^{-1} increased in intensity; the ratio of band intensities 2852/2932 cm^{-1} changed. The enhancement of integral intensity of the indicated absorption bands indicates an increased ordering of membrane proteins and, in particular, an increased fraction of α-helices.

The fraction of α-helices grows due to structural transition tangle → α-helix. A 3308 → 3272 cm^{-1} band shift of NH bond ($\Delta\nu$ = 36 cm^{-1}) is caused by the formation of hydrogen bond between keto group (C_3=O) in testosterone A-ring and NH bond of peptide group in membrane protein or indole ring in tryptophan. An increased intensity of the 2932 and 2852 cm^{-1} bands together with a growing intensity ratio 2852/2932 cm^{-1} confirm the rising orderliness of the entire membrane. Absorption band 1740 cm^{-1} (C=O bond of the ester group in phospholipids) increased in intensity and shifted to the short-wave region. The enhanced intensity of C=O bond reflects an increased ordering of phospholipids within domains and an increased interdomain ordering. The short-wave shift of this band is caused by the formation of hydrogen bond between OH group at C_{17} carbon atom in testosterone D-ring and C=O bond in phospholipids. Similar to segnetoelectrics a hysteresis phenomenon was observed in erythrocyte membranes [11, 12]. The spectrin-actin-ankyrin meshwork, which is connected both with membrane proteins and phospholipids, also contributes to the ordering of phospholipids. The 1088 → 1098 and 1236 → 1248 cm^{-1} shifts of absorption bands to the short-wave region result from dehydration of phospholipids due to increase in their orderliness, since the hydration process shifts these bands to the long-wave region [13] An increased intensity of bands 1098 and 1247 cm^{-1} (P-O-C and P=O bonds of phospholipids, respectively) in comparison with control specimens confirms an enhanced ordering of phospholipids under the action of the hormone.

Thus, the formation of complex domains in erythrocyte membranes upon their interaction with testosterone is caused by simultaneous interaction of CO and OH groups of the hormone with CO and NH groups both of proteins and phospholipids. In the process, water is displaced to adjacent regions, which is accompanied by membrane loosening.

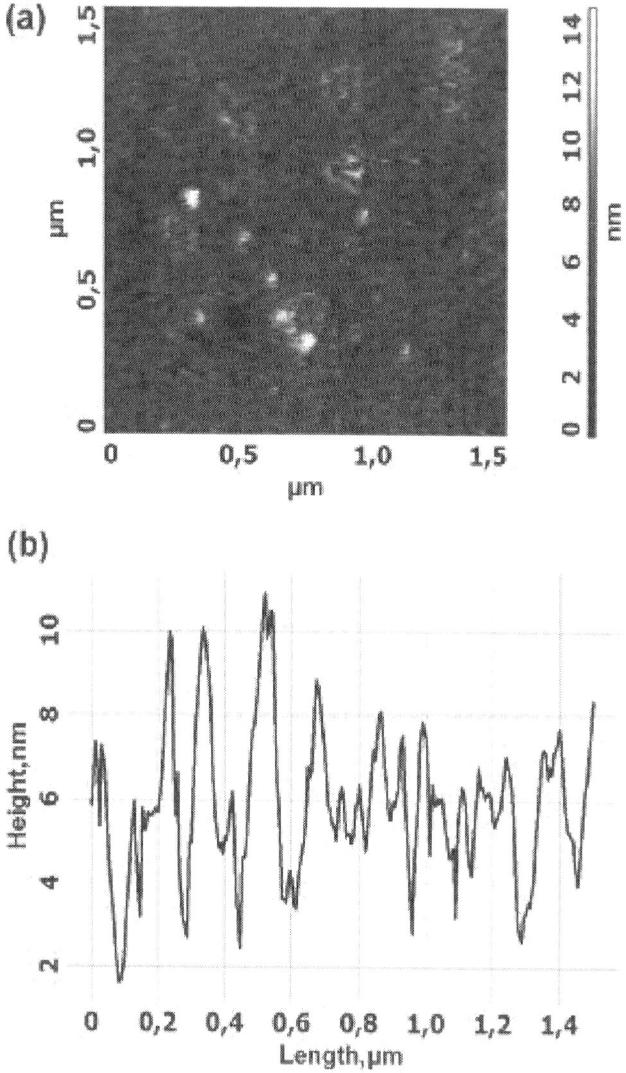

Figure 4. Surface of rat erythrocyte after adsorption of androsterone.
Concentration of the hormone is 10^{-6} M: (a) scan size 1.5 × 1.5 μm^2; (b) center section of the surface.

3.2.2. *Effect of androsterone*

Incubation of rat erythrocyte shadows with androsterone (C_C = 2.76 x 10^{-8} M) results in shifting the frequency of NH bonds (stretching vibrations of amide A) to the long-wave region by 38 cm^{-1} as well as shifting of NH bond of amide II ($\Delta\nu$ = 2 cm^{-1}) (Table 1). Integral intensity of absorption bands of CO (1654 cm^{-1}) and NH groups (3280 cm^{-1}) increased by 30% and more. There appeared a band at 1635 cm^{-1} corresponding to the β-structure. An increase in intensity of stretching vibrations of CH bonds at 2848 and 2930 cm^{-1} was observed. The frequency shift of NH bond is related with the formation of hydrogen bond with C_{17}=O group of the hormone D-ring. Androsterone has a more flexible structure as compared to cholesterol: its A, B and C-rings can take a more favorable conformation during the interaction with membrane proteins. Only D-ring has a flat structure, due to the presence of carbon C_{17} with sp_2 hybridization. Hydrophobic interaction with the membrane surface should also be taken into account. High conformational mobility of the molecule creates more advantageous steric conditions for hydrophobic interaction both with tryptophan, which fluorescence quenching was observed in our study, and hydrophobic regions on the membrane surface. This increases the constant of their binding to hormone and leads to more pronounced structural changes in the membranes. An increase in intensity of CO-peptide bond is related with the growing fraction of α-helices due to transition tangle →α-helix. An increase in intensity of absorption band 1620-1635 cm^{-1} is caused by structural transition tangle →β-structure. Of interest is a hypothesis stating that the indicated transitions may take place in contractile proteins, since their removal from the membrane results in a decrease or disappearance of transitions [12-13].

3.2.3. *Effect of dehydroepiandrosterone*

Incubation of DHEA with erythrocyte shadows showed that the frequency of stretching vibrations of NH peptide bond shifted by 20 cm^{-1} to the long-wave region (3308 →3288 cm^{-1}), whereas halfwidth of amide A decreased. An increase in the integral intensity of absorption bands at 1546, 1654.9 and 3288 cm^{-1} was observed (Table 1).

A 1236→1247.6 cm^{-1} band shift points to dehydration of phosphate groups in phospholipids. Shifting of the frequency of C=O bond in phospholipids (1748→1732 cm^{-1}) was observed; intensity of this band also increased. The 2930→2925.8 and 2848→2851 cm^{-1} shifts (CH stretching vibrations) took place, intensity of the bands increased. The intensity ratio 2852/2924 cm^{-1} changed.

3.2.4. *Effect of dehydroepiandrosterone sulfate*

Incubation of DHEAS with erythrocyte shadows resulted in the band shift 3308→3286 cm^{-1} (NH peptide bond) by 22 cm^{-1} (Table 1). Bands at 1548, 1656 and 3298 cm^{-1} increased in intensity with respect to control specimen; however, this was more pronounced upon addition of DHEA as compared to DHEAS. Absorption bands 1632 and 1684 cm^{-1} attributed to β-structure were observed. The band shift was recorded: 2930→2928 and 2848→2852 cm^{-1}, which was accompanied by a change in the 2852/2928 cm^{-1} ratio. The band at 1236 cm^{-1} (P=O bond) showed a strong splitting and had 3-4 bands in the region of 1236-1256 cm^{-1}. Bands at 1084 and 1100 cm^{-1} (P-O$^-$ bond) were observed. A 1748-1738 cm^{-1} shift was detected; however, it was less pronounced than in the case of DHEA addition. The DHEAS hormone has a stronger binding with hydrophilic heads of phospholipids as compared to DHEA, and a weaker binding with membrane proteins. This suggests that DHEAS molecules cannot penetrate deep into the membrane due to their higher hydrophilicity with respect to DHEA.

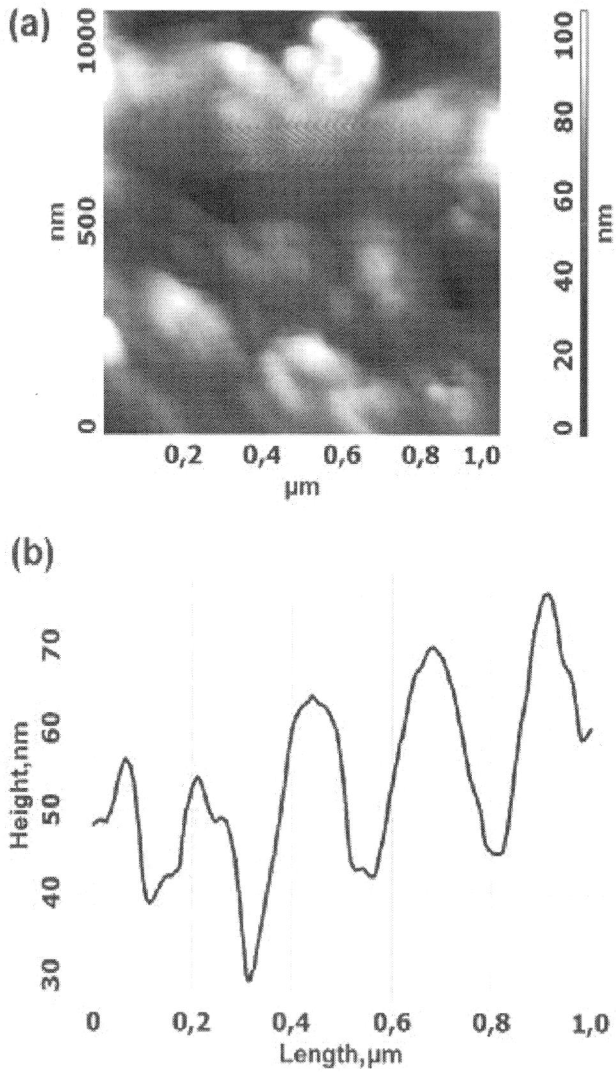

Figure 5. Surface of rat erythrocyte after adsorption of DHEA.
Concentration of the hormone is 10^{-7} M; (a) scan size 1 × 1 µm^2; (b) center section of the surface.

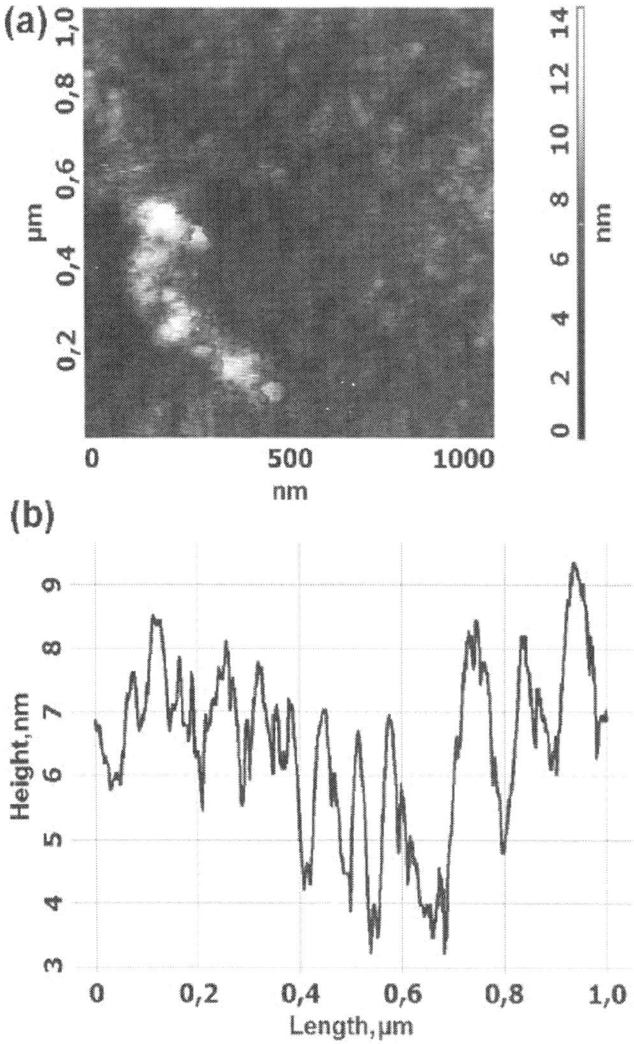

Figure 6. Surface of rat erythrocyte after adsorption of DHEAS.
Concentration of the hormone is 10^{-7} M; (a) scan size 1×1 μm^2; (b) center section of the surface.

Overall, it can be concluded that the interaction of DHEA and DHEAS with erythrocyte membranes is accompanied by the formation of hydrogen bonds between keto group (C_{17}=O) and NH group of proteins as well as between OH group at C_3 in the A-ring of the hormones and C=O group in biomembrane phospholipids. The formation of indicated hydrogen bonds leads to ordering of membrane proteins (transition tangle →α-helix) and phospholipids. Hydrophobic interactions of hormone with the surface of erythrocyte membranes also contribute to their structural rearrangement; however, they are much less pronounced for DHEAS as compared to DHEA. The reason is that substitution of OH group by SO_3 strongly diminishes the energy of hydrogen bond, since in OH group the unshared pair of electrons is located on the oxygen atom, whereas in SO_3 group it is delocalized over the entire π-conjugated bond.

4. Fluorescence analysis

In the study, absorption intensity (D) and emission intensity (F) of tryptophan were estimated at different wavelengths. Corrections were made for dilution of erythrocyte shadow suspension after the introduction of a hormone solution, for tryptophan emission quenching by a solvent (DMS : ethanol), intrinsic fluorescence of hormones, and evaporation of water from a cuvette. To obtain a correction for solvent, the erythrocyte shadow suspension was titrated with solvent.

It was shown that solvent decreases the intensity of tryptophan absorption at $\lambda = 227.8$ nm by 33% and results in its long-wave shift to $\lambda = 230.2$ nm. Absorption intensity at $\lambda = 281$ nm changed only by 1.3% without a long-wave shifting. A maximum of emission intensity differed from control specimen also at $\lambda = 332$ nm. It did not shift upon addition of solvent, but its intensity decreased by 1.3%.

Upon addition of testosterone with the final concentration $3{\cdot}10^{-6}$ M to erythrocyte shadows, the absorption intensity at 227 nm diminished by 19 a.u. or by 2.8%; this was accompanied by an upward shift of λ to 230.4 nm. As the hormone concentration increased to $6.05{\cdot}10^{-6}$ M, the absorption intensity at 227 nm decreased by 25 a.u., or 5.0%, which was accompanied by shifting the absorption maximum to 232 nm. In the region of 280 nm, addition of hormone caused only minor changes in fluorescence. Considerable changes in the spectrum were obtained upon addition of androsterone to the shadows. Even at a concentration of $6.92 \cdot 10^{-8}$ M, which is two orders of magnitude lower compared to the case of testosterone, the absorption intensity at 227 nm decreased by 90 a.u., or 122%. It means that this hormone penetrates deeper into erythrocyte membranes than testosterone and enhances the tangle → α-helix transition in proteins, thus increasing their ordering. When erythrocyte shadows were supplemented with DHEA or DHEAS, the hypochromic effect was weak or entirely absent. A decrease in absorption intensity and a long-wave shift observed in our study can be attributed to the effect of solvent.

Analysis of the tryptophan fluorescence quenching spectra testifies that all four hormones interact with membrane-bound proteins, although a degree of this interaction differs (Figs. 9-12). The most pronounced quenching was observed in the case of androsterone (Fig. 9). The maximum fluorescence quenching was observed at a concentration of $2.2{\cdot}10^{-8}$ M. Testosterone showed a lower fluorescence quenching (Fig. 10). The maximum quenching was observed at a concentration of $1.2{\cdot}10^{-7}$ M, which is 5.5 times higher as compared to androsterone. Fluorescence quenching was even less pronounced with DHEA (Fig. 11). In this case, the maximum quenching occurred at a concentration of $2.4 \cdot 10^{-6}$ M. And finally, the lowest fluorescence quenching was observed for DHEAS (Fig. 12). The maximum quenching took place at a concentration of $5.3 \cdot 10^{-6}$ M, which is 2.2 times higher as compared to DHEA.

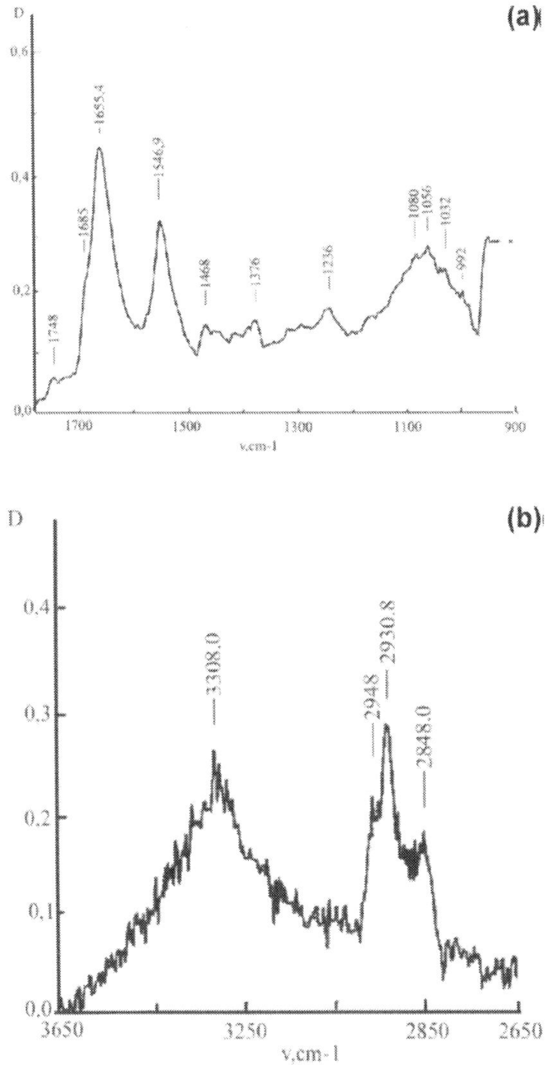

Figure 7. IR spectra of rat erythrocyte membranes (control) ($C_{phosph.buff.}$ = 0.01 M, pH 7.35, relative humidity 0%): (a) ν = 900-1800 cm^{-1}, (b) ν = 2600-3700 cm^{-1}.

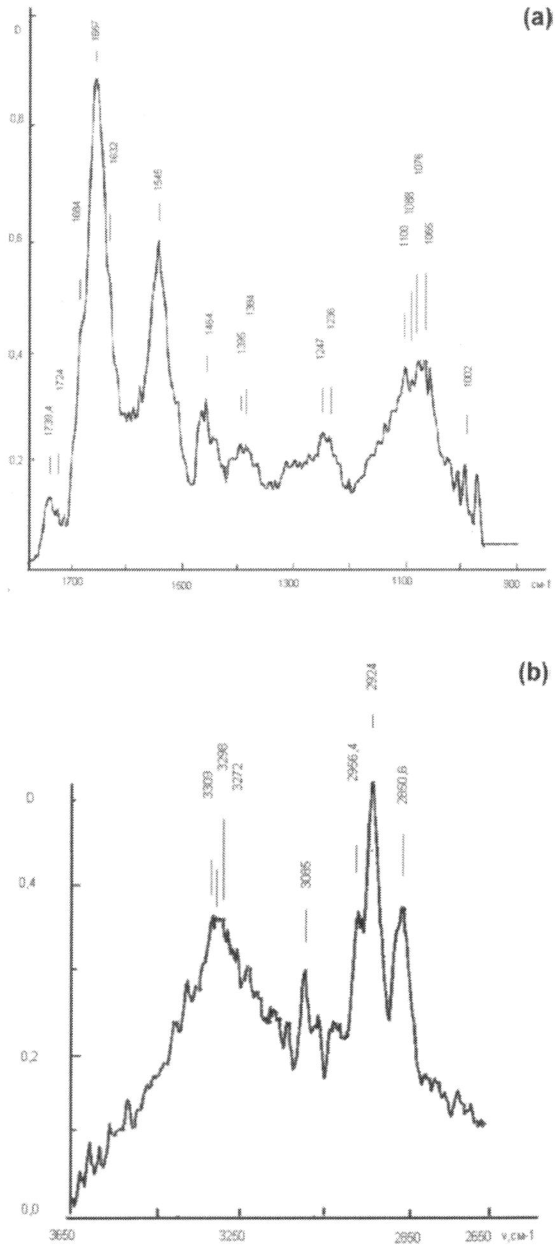

Figure 8. IR spectra of rat erythrocyte membranes incubated with testosterone ($C_C = 2.7\times 10^{-8}$ M, $C_{phosph.buff.} = 0.001$ M, pH 7.35, relative humidity 0%): (a) $\nu = 900\text{-}1800\ cm^{-1}$, (b) $\nu = 2600\text{-}3700\ cm^{-1}$.

According to the results obtained, testosterone and androsterone penetrate deeper into erythrocyte membrane and have a stronger effect on the structure of membrane-bound proteins toward their increased ordering. DHEA and DHEAS have some effect on erythrocyte membranes; these hormones adsorb on the membrane surface, but do not penetrate deep into hydrophobic layer of the membranes. These hormones have a weaker binding with proteins via hydrogen bonds.

This conclusion is confirmed by the calculated values of hormone binding constant (K_b), total amount of bound hormone (B_{max}), and changes in free energy (ΔG) upon hormone transition from free state to the membrane-bound one (Table 2). The highest values of K_b were obtained in our study for testosterone and androsterone, K_b for androsterone being higher by a factor of 4. Amount of the bound hormone (B_{max}) obeyed an inverse relationship: it was 2.4 times higher in the case of testosterone as compared to androsterone. Changes in free energy upon interaction of hormones with erythrocyte membranes were most pronounced

Compound	ν_{CO}	ν_{NH} stretch.	$\nu_{C=O}$	$\nu_{P=O}$	$\nu_{P\text{-}O\text{-}C}$	$\nu_{O5C4\text{-}C5O4}$	ν_{CH} stretch.	A_{CO}
Shadows (control)	1655.4 1686	3308	1748	1236	1080	1056	2948 2930 2848	1.2150 × 10
Shadows + androsterone (A = 2.76 x 10^{-8} M)	1656 1635 1620	3270 329 2		1260 1240	1098 1088		2958 2928 2848	
Shadows + testosterone (A = 2.7 x 10^{-8} M)	1657 1684 1632	3272 3298 3309	1739.4	1247 1236	1098 1088	1065 1076	2956.4 2924 2850	2.2433× 10
Shadows + DHEA (A = 2.64 x 10^{-8} M)	1654.9	3288.0	1732	1247.6	1088	1070.7	2956.3 2925.8 2851.8	2.1266 × 10
Shadows + DHEAS (A = 1.63 x 10^{-8} M)	1656.0 1680 1632.0	3286 3300 3312	1738	1248.0	1084	1070 1052.7	2952.0 2926.4 2852.0	1.2598 × 10

Table 1. IR spectroscopy. Frequency parameters of rat erythrocyte shadows after their interaction with hormones

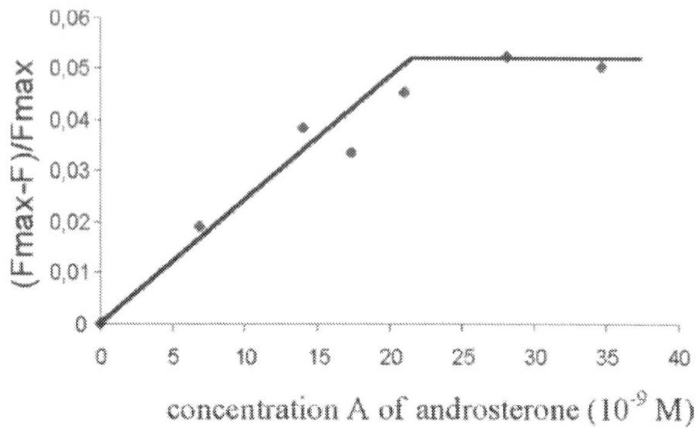

Figure 9. Q = (Fmax - F)/Fmax versus the concentration A of androsterone hormone introduced into a cuvette. Concentration of membrane protein C = 0.203 mg/mL.

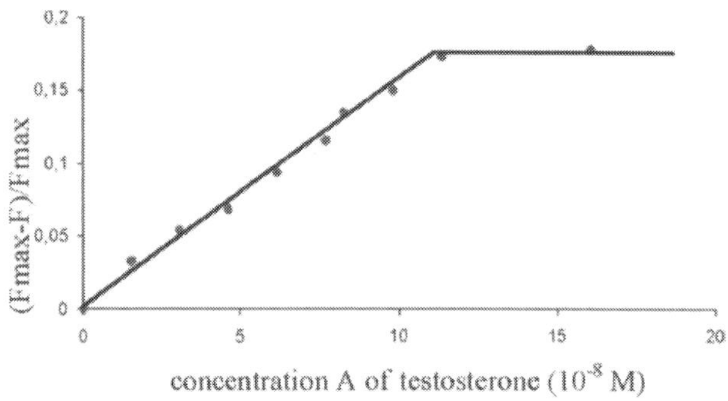

Figure 10. Q = (Fmax- F)/Fmax versus the concentration A of testosterone hormone introduced into a cuvette. Concentration of membrane protein C = 0.101 mg/mL.

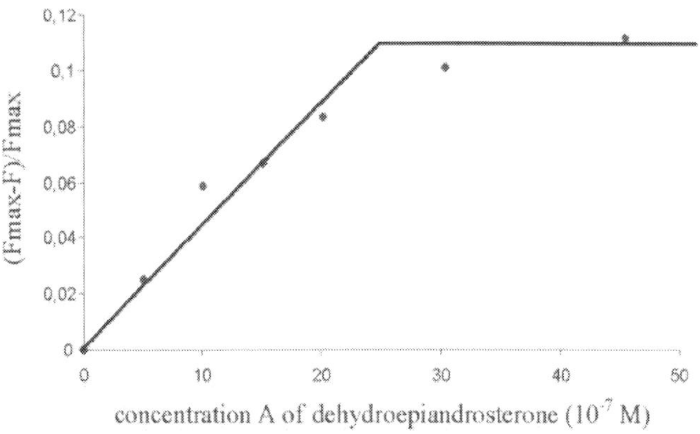

Figure 11. Q = (Fmax _ F)/Fmax versus the concentration A of dehydroepiandrosterone hormone introduced into a cuvette. Concentration of membrane protein C = 0.139 mg/mL.

All values for DHEA and DHEAS strongly differed from those listed above. Binding constants were nearly two orders of magnitude lower. Amount of the bound hormone (B_{max}) was much greater, indicating a low specificity of interaction with the membranes. Changes in free energy (ΔG) were low for both hormones (Table 2).

Thus, a comparison of two pairs of hormones demonstrated their considerable difference from each other. The higher is K_b, the greater is the binding specificity and the lower is the amount of bound hormone (B_{max}). Large negative values of ΔG for testosterone and androsterone testify that their interaction with erythrocyte membranes increases their ordering (negentropy). DHEA and DHEAS are characterized by a low specificity of binding to membranes.

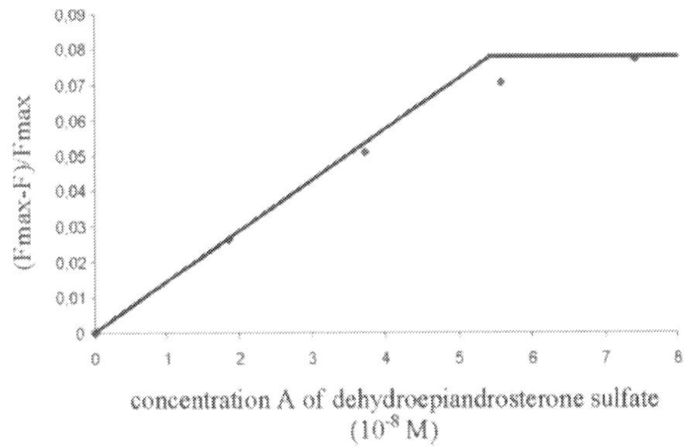

Figure 12. Q = (Fmax - F)/Fmax versus the concentration A of dehydroepiandrosterone sulfate hormone introduced into a cuvette. Concentration of membrane protein C = 0.139 mg/mL.

It shows up even when hormones are compared with each other. In DHEA, substitution of OH group by SO_3H in the 3rd position of A-ring decreases K_b by a factor of 3.8 and increases B_{max} by a factor of 2.2. A decrease in ΔG is also pronounced. The reason is that the presence of OH group and additionally of two keto groups and an S atom enhances the interaction of DHEAS with hydrophilic CO and NH groups of the surface proteins. DHEA and DHEAS cannot bind to the proteins residing in hydrophobic layer of the membrane. These two hormones do not change the conformational state of spectrin-actin-ankyrin meshwork and have only a slight effect on the morphology of membrane surface. DHEAS, being most hydrophilic among the four hormones, has the weakest effect. As hormone hydrophilicity increases, the amount of membrane-bound hormone rises and K_b decreases. During the interaction of testosterone and androsterone with erythrocyte membranes, both hydrogen bonds and hydrophobic interactions may strongly contribute to the growth of K_b. This is explained by a deeper penetration of hormones into hydrophobic layer of erythrocyte membrane, which increases the specificity of their interaction. The accompanying structural transitions in membrane proteins, tangle $\rightarrow \beta$-structure $\rightarrow \alpha$-helix, increase ordering of these proteins and substantially raise the ΔG value. Results obtained in the study agree well with changes in microviscosity of erythrocyte membranes.

Steroid hormone	Binding constant K_b (M^{-1})	Amount of bound hormone B_{max} (mol/mg protein)	Changes in free energy ΔG (kJ/mol)
testosterone	(2.24±0.22)x10^6	(1.09±0.11)x10^{-9}	-37.6
androsterone	(3.2±0.32)x10^6	(4.46±0.45)x10^{-10}	-38.5
DHEA	(5.99±0.60)x10^4	(1.80±0.18)x10^{-8}	-28.3
DHEAS	(1.56±0.16)x10^4	(4.03±0.40)x10^{-8}	-24.8

Table 2. Parameters of steroid binding to erythrocyte membrane based on tryptophan fluorescence quenching of membrane proteins

5. Changes in microviscosity

In erythrocyte membrane, a fluorescent pyrene probe is distributed in the lipid phase and can be a source of information on the state of its deeper layers. The rate of its migration and the ability to form excimers upon interaction with each other are estimated. This is the way to determine changes in microviscosity of the membranes.

In our study, an increase in microviscosity was most pronounced at the addition of androsterone to erythrocyte membranes. Microviscosity started to grow at a hormone concentration of 10^{-8} M, the growth proceeding up to $2.5 \cdot 10^{-8}$ M with subsequent saturation (Fig. 13). The S-shaped curve points to high cooperativity in changing the conformational state of the membrane. A microviscosity increment attained 50% with respect to the initial state. In the region of protein-lipid interactions it appeared earlier and reached a higher value as compared to the region of lipid-lipid interactions. The absorption intensity (D) and emission intensity (F) of tryptophan in membrane proteins started to decrease at the same concentrations and attained a maximum also at the same concentrations (Fig. 9). Thus, our results revealed a cooperative nature of changes in erythrocyte membranes under the action of androsterone.

Addition of testosterone produced similar changes in membrane microviscosity. In the region of protein-lipid interactions, microviscosity increased at lower concentrations of hormone and attained higher values as compared to the region of lipid-lipid interactions (Fig. 14). In both cases, the revealed structural changes were initiated in proteins and carried over to lipids by virtue of cooperativity.

The effect of DHEA and especially DHEAS on erythrocyte membranes is much less pronounced as compared to testosterone (Figs. 15, 16). DHEA and DHEAS increased microviscosity by 10% with respect to initial values. In these experiments, the concentration of DHEAS reached $8 \cdot 10^{-6}$ M. For DHEA, the growth started at a hormone concentration of $5 \cdot 10^{-7}$ M and attained its maximum at $1.5 \cdot 10^{-6}$ M (Fig. 15). Alteration of viscosity was described by S-curve and correlated with a decrease in fluorescence and absorption of tryptophan (Fig. 11). The latter processes started at the same hormone concentrations and reached their minima also at the same concentrations. Microviscosity in the region of protein-lipid interactions increased earlier, at lower concentrations of hormone, and was more pronounced than in the region of lipid-lipid interactions. Structural changes were initiated in proteins and involved lipids due to cooperativity.

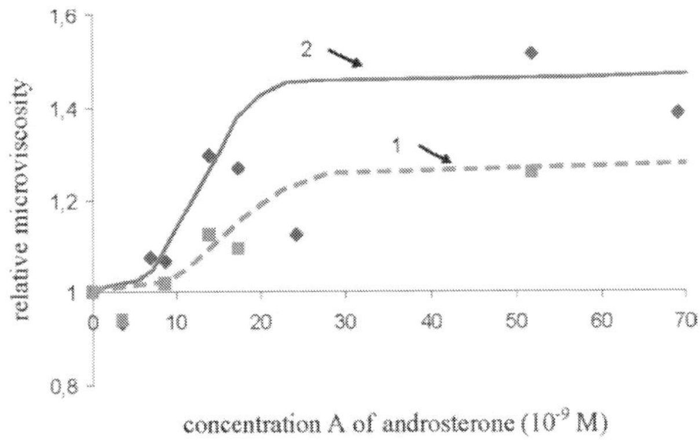

Figure 13. Changes in the relative microviscosity of membranes η(A)/η(0) of erythrocyte shadows at the concentration A of androsterone hormone. Concentration of shadows C = 0.133 mg protein/mL. Line 1 shows changes of the region of lipid-lipid interaction; line 2 – the region of protein-lipid interaction.

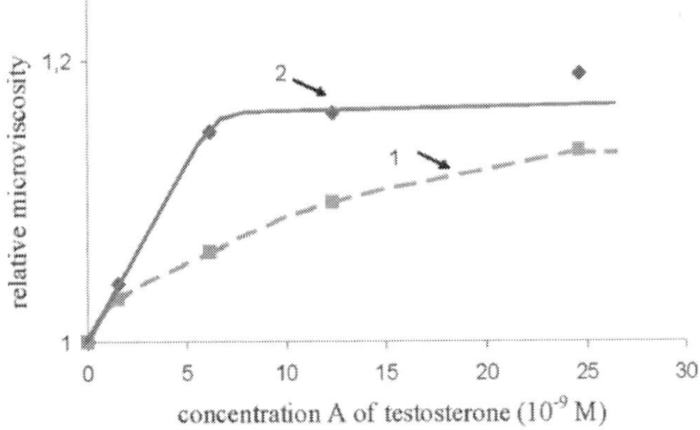

Figure 14. Changes in the relative microviscosity of membranes η(A)/η(0) of erythrocyte shadows at the concentration A of testosterone hormone. Concentration of shadows C = 0.117 mg protein/mL. 1 - the region of lipid-lipid interaction; 2 – the region of protein-lipid interaction.

The mechanism of changes in membrane microviscosity under the action of a more hydrophilic hormone DHEAS is quite different. Microviscosity goes to a constant value at a higher concentration of DHEAS in suspension as compared to that of DHEA ($5 \cdot 10^{-6}$ M versus $1.5 \cdot 10^{-6}$ M for DHEA). First changes of microviscosity appeared in the region of lipid-

lipid interactions (Fig. 16), which was followed by an increase of microviscosity in the region of protein-lipid interactions. DHEAS interacted with polar heads of phospholipids, then structural changes carried over to proteins due to cooperativity. Hydrophilic molecules of DHEAS cannot penetrate deep into hydrophobic layer of the membranes. There are only minor structural changes in the spectrin-actin-ankyrin meshwork and weak changes in membrane microviscosity.

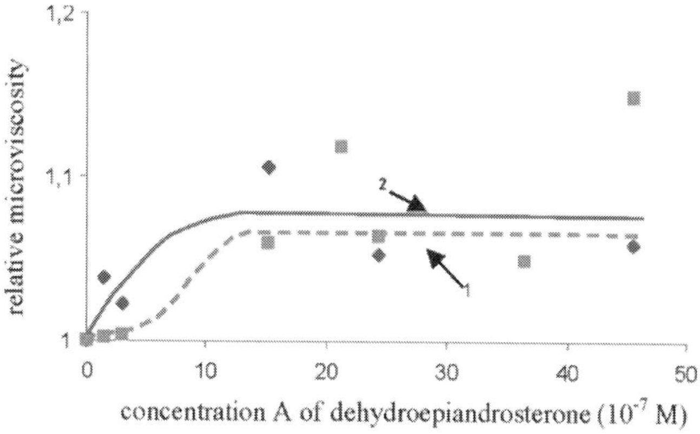

Figure 15. Changes in the relative microviscosity of membranes η(A)/η(0) of erythrocyte shadows at the concentration A of hormone DHEA. Concentration of shadows C = 0.113 mg protein/mL. 1 - the region of lipid-lipid interaction; 2 – the region of protein-lipid interaction.

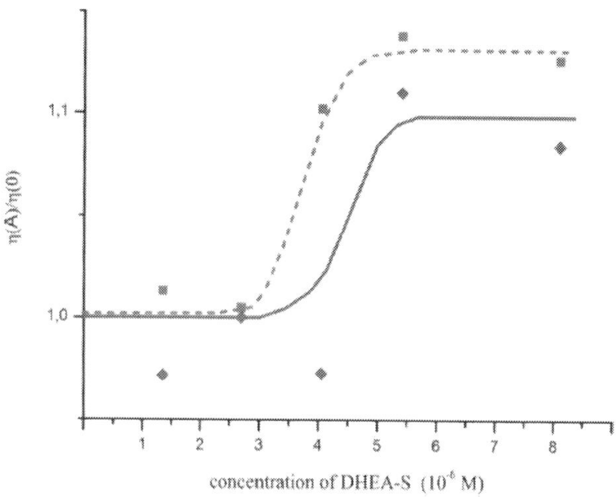

Figure 16. Changes in the relative microviscosity of membranes η(A)/η(0) of erythrocyte shadows at the concentration A of DHEAS hormone. Concentration of shadows C = 0.290 mg protein/mL. 1 - the region of lipid-lipid interaction; 2 – the region of protein-lipid interaction

6. Change of the membrane microviscosity, and Na^+, K^+-ATPase activity

Erythrocyte suspensions from fresh blood of rats with hormones of differing concentration were analyzed. For this purpose cortisol and adrenalin were used.

Under the action of hormones, the microviscosity of erythrocyte membranes increases following a saturation curve (Fig. 17). This increase depends on the hormone type and differs greatly for protein-lipid and lipid-lipid interactions. As can be seen from Fig. 17, protein-lipid interaction makes a decisive contribution to the increase in membrane microviscosity under the action of hormones. It is due to this contribution that the system of compaction domains is formed in the membrane, resulting in an increase in erythrocyte rigidity. The effect depends strongly on the hormone type. Adrenaline, which penetrates the entire erythrocyte bulk, rapidly increases the erythrocyte microviscosity, and the latter comes to saturation even at small hormone concentrations of $17 \cdot 10^{-9}$ M. Cortisol acts on an erythrocyte surface layer alone; hence, the microviscosity reaches saturation only at a cortisol concentration of $60 \cdot 10^{-9}$ M, and the increase in viscosity with cortisol is half as much as that with adrenaline.

The most important result concerns the influence of hormones on the activity of the Na^+, K^+-ATPase (Fig. 18). For both hormones analyzed, increasing the hormone concentration causes the quantity first to increase, reach its maximum, and then to decrease. The maximum of activity corresponds to the hormone concentration at which the microviscosity reaches saturation. A good correlation is found between the stages of variation in activity and in microviscosity with an increase in the concentration of different types of hormones.

Adrenaline, which is responsible for the rapid increase in erythrocyte microviscosity, is responsible as well for the rapid increase in the Na^+, K^+-ATPase activity and for its subsequent fast decline following the maximum (Fig. 18a). Increasing the cortisol concentration (Fig. 18b) causes a slow increase in microviscosity and Na^+, K^+-ATPase activity (γ), and then a slow decrease in γ whose value does remain high at a very high hormone concentration. At a $20 \cdot 10^{-9}$ hormone concentration, the activity γ is 0.05 and 0.03 μmol/h·mg protein for adrenaline and cortisol, respectively. At the stage of decline in γ at $60 \cdot 10^{-9}$ hormone concentration, γ ~ 0.02 and 0.035 μmol/h·mg protein for adrenaline and cortisol, respectively.

The maximum activity γ in the series of adrenaline and cortisol is observed at their respective concentrations of $3 \cdot 10^{-8}$ and $5 \cdot 10^{-8}$ M.

Important information on the nature of structural changes produced in erythrocytes by the analyzed hormones was obtained with infrared spectroscopy [14]. The increase in the absorption band intensity of CO- (1655.2 cm^{-1}) and NH-bonds (1548 and 3290 cm^{-1}) by about 20 % with cortisol points to enhanced ordering of membrane proteins due to the tangle $\rightarrow$ α-helix structural transition [15]. The shift 3308 $\rightarrow$ 3280 in stretching vibrations of the peptide NH-bond and the increase in its intensity owes to the formation of a hydrogen bond between cortisol and NH-bond of proteins. The increase in the absorption band intensity of the C=O-bond of phospholipids and its shift 1748 → 1740 points to enhancement of ordering of higher carboxylic acids and to a decrease in phospholipid entropy.

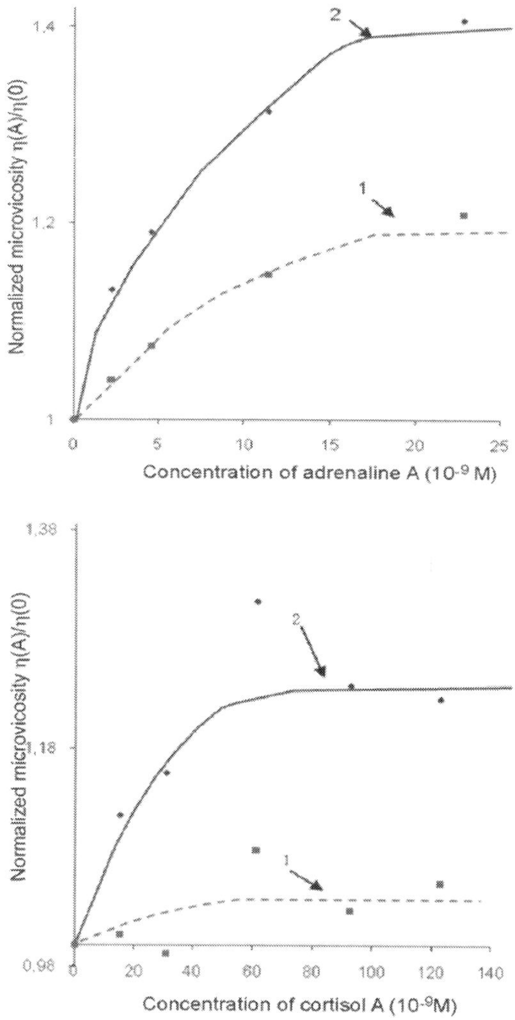

Figure 17. Changes in the relative microviscosity of membranes (L) of erythrocyte shadows at the concentration A for adrenaline (a) and cortisol (b) hormones added to the shadows suspension. Concentration of shadows C = 0.128 mg protein/ml. Concentration of pyrene in the suspension is $7.7 \cdot 10^{-6}$ M, temperature of the specimens 309.1±0.1 K (36°C), pH of the suspension 7.35. The measured value of L(A) exhibit an error of 6%. 1 - the region of lipid-lipid interaction; 2 – the region of protein-lipid interaction

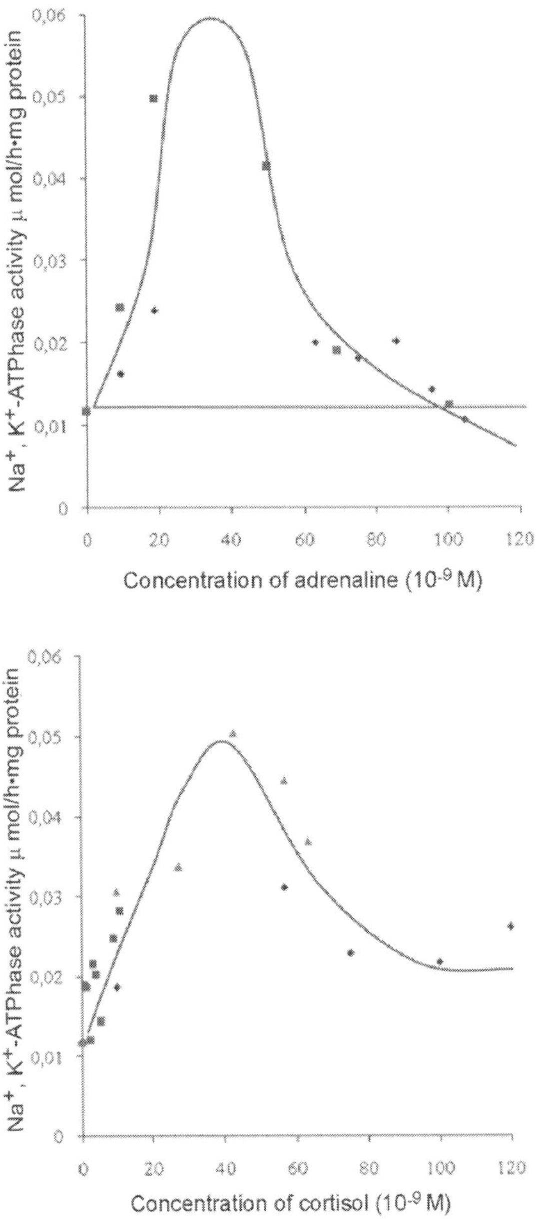

Figure 18. Changes in the activity of Na^{+}, K^{+}-ATPase of erythrocyte membranes as a function of a hormone concentration in suspension: a – adrenaline; b – cortisol.

The concurrent hormone interaction with protein and phospholipids enhances the protein-lipid interactions resulting in a complex domain structure in erythrocytes. The frequency shift of the P=O-bond toward the short-wave range and the increase in its intensity is associated with dehydration of membranes due to its hormone-induced compressive deformation. It is the loss of bound water that increases the frequency of the P = O-bond [16]. The displacement of water dipoles from protein-lipid domains to adjacent regions leads to the development of mesobands of localized deformation and discontinuities in them.

The obtained experimental results suggest that an important role in the action of hormones on erythrocytes belongs to two factors:

- collective effect of the increase in erythrocyte elastic modulus due to the formation of the domain structure in the region of protein-lipid and lipid-lipid interaction (macroscale);
- hindered structural transitions in mass transfer processes through a membrane (microscale).

The first factor causes an increase in energy quanta hν (phonone) required for structural transitions in mass transfer through erythrocyte membranes. This factor increases the activity of the Na^+, K^+-ATPase at the first stage of growth of hormone concentrations.

The second factor retards microscale structural transitions. Once the microviscosity ceases to increase (the formation of the domain structure is completed), the contribution of the first factor no longer grows, the contribution of the second factor continues to escalate, and the activity of the Na^+, K^+-ATPase decreases.

7. Effect of cortisol

Here we try to go beyond the influence of cortisol on the red cell membrane, and consider the action of the hormone on erythrocyte as a multi-layered liquid crystal system.

The mechanism of erythrocyte deformation and structural transformation of membranes and hemoglobin by the action of cortisol is still scantily investigated. The interaction of hemoglobin with contraction proteins and band 3 protein of erythrocytes is reported by Discher, D.E., Mohandas, N. & Evans, E.A [17]. These works imply that the disturbance and deformation of erythrocyte membrane caused by cortisol or other external factors can be transferred to hemoglobin by means of band 3 integral protein or contraction proteins. According to modern ideas, contraction proteins reside as at the inner as outside of membrane. Within this concept, a reverse response is also possible, i.e., the disturbance can be transferred from hemoglobin to the cell membrane.

The addition of cortisol to erythrocyte suspension with the hormone concentration of 10^{-8} to $6{\cdot}10^{-8}$ M produced a set of UV spectral curves. A maximum of absorption band at 418 nm was shown to decrease with increasing the hormone concentration. A decrease in the optical density was 22% as compared to erythrocyte suspension without hormone (control) (Fig. 19). The resulting set of curves was used to plot the dependences of optical density for band 418 nm on the concentration of hormones.

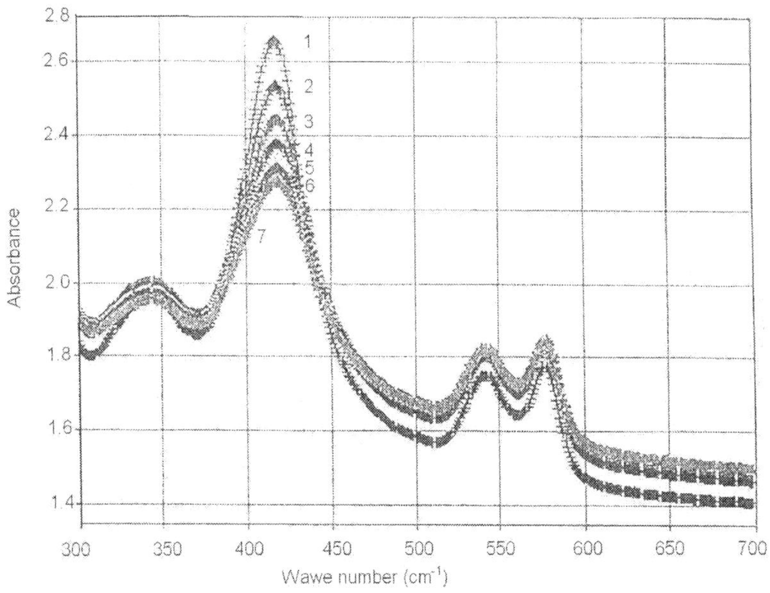

Figure 19. Changes in the optical density of absorption band 418 nm at the addition of cortisol to human erythrocyte suspensions (C = 10^{-8} – 6 · 10^{-8} M).

An increase in the optical density in the regions of 600-700 and 310 nm, which points to increasing diffusion of light has been also observed. In these regions, optical density changed by ~5.5%, which considerably exceeds the measurement error (0.5%) (Fig. 19). It should be noted that absorption band at 418 nm may shift spontaneously by ± 2-3 nm in different runs, generally to the short-wave region. Shifting of this band occurs either due to fluctuations in the structure of hemoglobin itself [18] or by the action of fluctuations in the structure of membrane and cell as a whole [19, 20].

The analysis of IR spectra of rat erythrocyte ghosts not loaded with hormone (control, Fig. 7) revealed not only a disordered structure, but also the presence of α-helix 1650-1656 cm^{-1} and β-structure (1686 and 1520 cm^{-1}) in the proteins of rat erythrocyte ghosts [1]. NH stretching vibrations of proteins (3308 cm^{-1}), CH stretching vibrations of hydrocarbon chains in proteins and phospholipids (2948, 2930 and 2848 cm^{-1}) as well as some bands typical of phospholipids, in particular, C=O bond (1748 cm^{-1}), P=O bond (1236 cm^{-1}), CH_2 deformation vibrations (1460 and 1386 cm^{-1}) of hydrocarbon chains, O_4-C_4-C_5-O_5 bond (1048 cm^{-1}) of monosaccharides in glycolipids and glycoproteins, and C-C deformation vibrations (978 cm^{-1}) have been recorded. Note that the C=O band (1736 cm^{-1}) is quite narrow; hence it follows that phospholipids are well ordered at the level of ester bonds in higher carboxylic acids and glycerol.

Analysis of the IR spectra of rat erythrocyte ghosts upon incubation with cortisol at its concentration of 4.4 · 10^{-8} M revealed a ca. 20% increase in intensity of the absorption bands of CO (1655.2 cm^{-1}) and NH bonds (1548 and 3290 cm^{-1}), the effect building up with an increase in the hormone concentration (Table 3, Fig. 20). A growing intensity of the band 1655.2 cm^{-1} testifies an increase in the fraction of α-helix [8]. The increasing fraction of α-helices in membrane proteins is related with the structural transition tangle → α-helix.

A shift of NH bond (stretching vibrations of peptide bond, 3308 →3280 cm^{-1}, $\Delta\nu = 28$ cm^{-1}) was accompanied by a growth of its intensity, which is related with the formation of hydrogen bond between cortisol and NH group. Hydrogen bond is likely to form between keto group of A-ring (C_3=O) and NH group of the membrane protein. Meanwhile, keto group (C_{20}=O) of D-ring and OH group at C_{11} in C-ring could also be involved in the formation of hydrogen bonds. The presence of several hydrophilic groups strongly changes the biological activity of cortisol and other steroid hormones, in distinction to cholesterol. Cholesterol binds to phospholipids mainly due to hydrophobic interaction (Van der Waals forces) with fatty acid residues [19]. Shifting of CH bond stretching vibrations 2848 → 2852 cm^{-1} ($\Delta\nu = 4$ cm^{-1}) and 2930 → 2925 cm^{-1} ($\Delta\nu = 5$ cm^{-1}) were observed. The latter increased in intensity under the action of hormone. Changes in intensity of this band confirm the presence of structural transition, but cannot differentiate the place where the transition occurs — in membrane proteins or in phospholipids, as CH bond is present both in proteins and phospholipids. However, as seen from our experimental data, this band reflects mainly the changes in phospholipid orderliness.

An increase in intensity of the absorption band of phospholipid C=O bond and its shift 1748 → 1740 cm^{-1} were observed. This increase of the band intensity indicates a growing orderliness of higher carboxylic acids and a decreasing entropy in phospholipids. Shift of the band is related with the formation of hydrogen bond between hormone, for example OH group at C_{21}, and CO bond of phospholipids. Such interaction of the hormone simultaneously with protein and phospholipids can occur at the interface between protein and phospholipids, i.e., in a near-boundary or annular layer of the band 3 integral protein, glycophorin and other proteins.

P=O bond shifted in frequency by 3 cm^{-1} to the short-wave region and increased in intensity. Shifting of P=O bond to the short-wave region is attributed to dehydration of membranes during their deformation under the action of hormone. A loss of bound water increases the frequency of P=O bond [1]. Deformation (contraction) occurs due to spectrin-actin and spectrin-ankyrin networks [19], since the extraction of spectrin from membrane relieves the deformation caused by hormones. It should be noted that 30% of membrane proteins is represented by spectrin. Overall, contraction proteins constitute 55-60% of all membrane proteins [21]. Steroids can attack either the spectrin-actin-ankyrin network located both on internal and external surfaces of the membrane or the integral proteins associated with contraction proteins [22].

Our FTIR spectroscopy study of the hormone effect on intact erythrocytes revealed considerable changes of the spectra in absorption regions both of proteins and phospholipids. In particular, cortisol gave rise to absorption band 1636 cm^{-1} corresponding to β-structure of membrane proteins, which indicates a transformation in the secondary structure of membrane proteins (tangle → β-structure transition) involving also the contraction proteins. Shifting of some other absorption bands attributed both to proteins and phospholipids was observed too (Fig. 21).

Noteworthy are the shift of absorption band 2870 cm^{-1} corresponding to stretching vibrations of CH bond in hemoglobin [23], and a more pronounced splitting in the region of stretching and deformation vibrations of phospholipid CH orderliness in and between the domains. A stronger splitting of CH bonds testifies the formation of new lipid-protein clusters as a result of intermolecular interaction, due to compaction of membrane elements caused by structural transformation of the contraction proteins network. In our earlier studies of high density lipoproteins (HDL), when calculating the enthalpy of structural transitions from experimental data, the occurrence of smectic A → smectic C transition in HDL phospholipids [8] has been suggested. Such a transitions may occur here, since it has a low enthalpy [8, 24].

Of interest is the appearance of the absorption band 2851.8 cm^{-1}, which is assigned to stretching vibrations of CH bond in phospholipids [8]. This band results from structural transition in membrane phospholipids.

No.	The object of measurement	ν_{CO}	$\nu_{NH\ val.}$	$\nu_{C=O}$	$\nu_{P=O}$	ν_{P-O-C}	$\nu_{05C4-C504}$	$\nu_{CH\ val.}$	A_{CO}
	Ghosts (control)	1655,4 1686	3308	1748	1236	1080	1056	2948 2930 2848	1,2150E + 01
2	Ghosts + cortisol (C= 4,4·10^{-8} M)	1655,2	3290,4 3308.0	1743 -	1236	1080	1051,2	2924,2 2848,9	1,5169E + 01
3	Ghosts + cortisol (C= 10^{-7} M)	1656,0 1630	3280 3300	1740	1239	1083,7	-	2962 2925 2852	1,5640E + 01
4	Erythrocytes (control)	1649,9	3282,1 3272,0		1245,7	1106,2		2956,8 2935,8 2871,7 3030,2	
5	Erythrocytes + cortisol (C= 10^{-8} M)	1642,7 1627,5	3285,3 3270,2 3247,3	1741,1 1707,5	1239,4 1201,3	1100,0 1089,5	1060,2	2956,6 2937,5 2872,3 3028,4 3052,4	

Note. A_{CO} is the integral intensity of absorption band ν_{CO} of the peptide bond in semilogarithmic form.

Table 3. Frequency characteristics of human erythrocytes and rat erythrocyte ghosts before and after their interaction with hormones

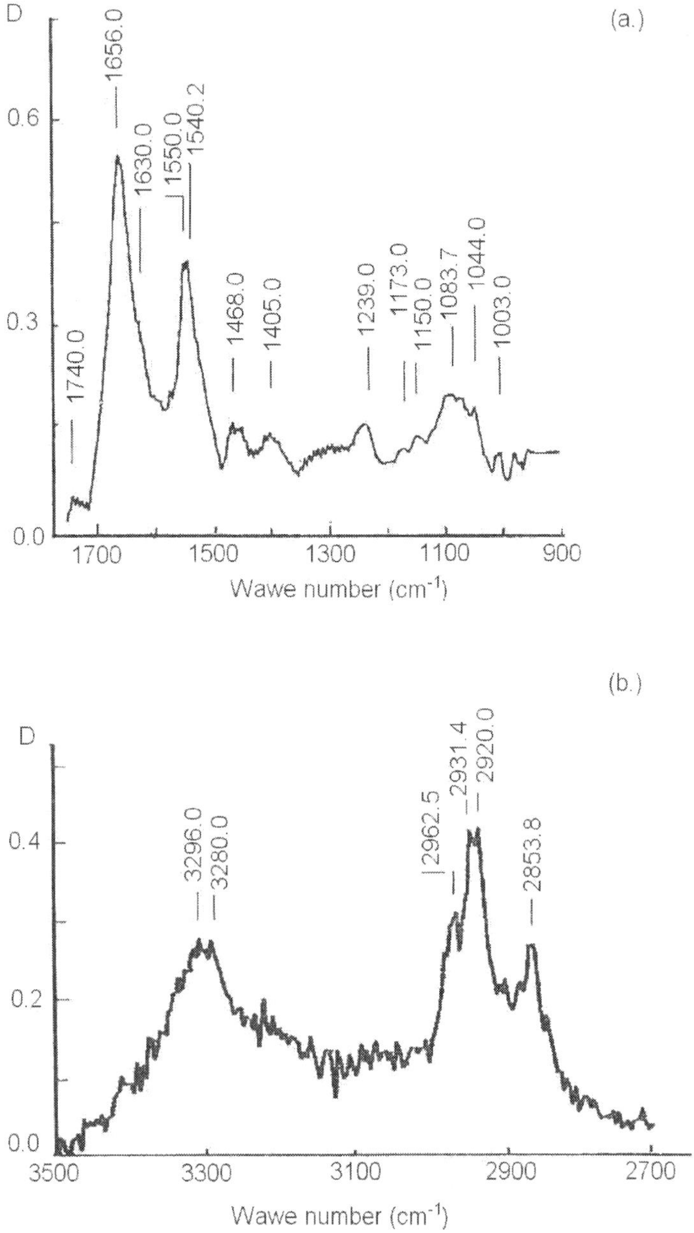

Figure 20. IR spectraof rat erythrocyte ghosts at the addition of cortisol (C = 4.4· 10^{-8} M): a. - ν = 1000 – 1800 cm^{-1}, b. - ν = 2600 – 3400 cm^{-1}.

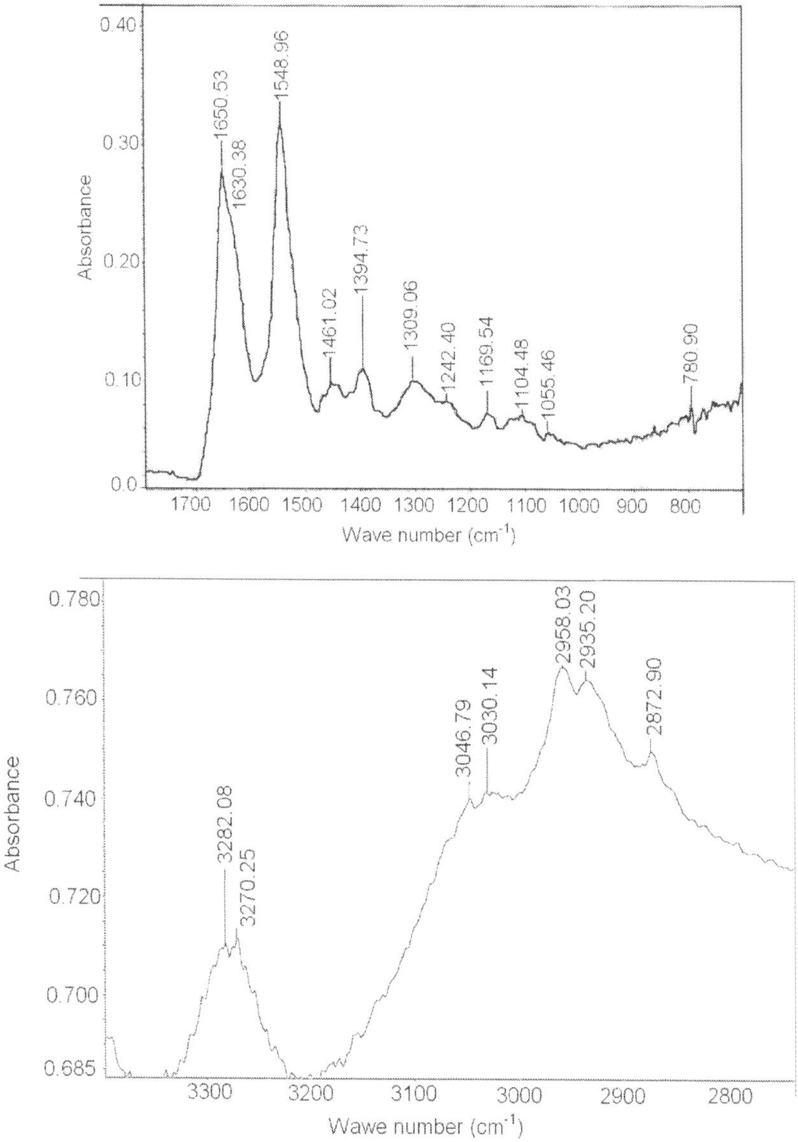

Figure 21. FTIR spectra of human erythrocytes at th addition of cortisol (C = 3 · 10^{-8} M): a. - ν = 1000 – 1800 cm^{-1}, b. - ν = 2600 – 3400 cm^{-1}.

Splitting in the region of 1088 (POC bond) and 3282 cm^{-1} (NH bond) was observed. Splitting of these bands indicates an increasing orderliness in phospholipids and membrane proteins, respectively. An increase in the fraction of β-structure points to the tangle → β-structure transitions; however, in this case more pronounced is the α-helix → β-structure transition, due to redistribution of intensity between absorption bands at 1650 and 1638 cm^{-1}. The first band corresponds to α-helices, the second one to β-structure [19].

Using the fourth derivative of the absorption band 1600-1700 cm^{-1} the maintenance of elements of the secondary structure in the erythrocyte membranes has been calculated. The results are given in Table 4. This table shows the considerable increase of β-structure under the action of adrenaline. In this case we observe the structural transition tangle → β-structure. However, the increase of α-helices and decrease of tangle under the action of cortisol has been seen. So it can be concluded the structural transition tangle → α-helix took place.

A comparison of IR spectra obtained from ghosts and intact erythrocytes revealed some general regularities: 1) splitting of absorption bands of NH peptide bonds, 2) an increased intensity of absorption bands corresponding to β-structure, 3) splitting of absorption bands corresponding to CH bonds of phospholipids, 4) a frequency shift of some bands (Table 4). However, there is also a distinction related with the appearance of absorption bands at 2870 and 1108 cm^{-1} corresponding to hemoglobin [23]. These bands are shifting when erythrocytes are subjected to the action of hormones.

Conformation	Erythrocyte (control)	Erythrocyte + cortisol $C_{cortis} = 3 \cdot 10^{-8}$ M
α-helical	25%	48%
β-structure	25%	30%
Random coil	25%	22%

Table 4. The quantitative definition of the elements of secondary structure in membrane proteins

Thus, these results suggest that the erythrocyte react to the effect of steroid hormones as a complex liquid-crystalline co-operative system in which nanostructured transitions are irreversible and are closely associated with the functional activity of cells.

8. Thermodynamics of nanostructural transitions in erythrocyte as a liquid crystal system, a relation with the cell function

The application of IR and UV spectroscopy showed that the interaction of steroid hormones with erythrocytes increases the ordering of both the membranes and hemoglobin, which means an increase in negentropy.

E. Schrodinger defined it as

$$-S = k \lg (1/D), \tag{11}$$

where $-S$ is the negative entropy, or negentropy; k is the Boltzmann's constant equal to $3.2983 \cdot 10^{-24}$ cal/deg; D is the quantitative measure of disorderliness of atoms in the system, $\lg (1/D)$ is the negative logarithm of D, and $1/D$ is the measure of orderliness.

However, of prime importance for us is that increasing negentropy is always supported by increasing amount of structural information. This can be expressed by the following equation:

$$-S = k\,\lg(1/D) + \sum p_i \cdot \log p_i \tag{12}$$

where p_i is the probability of individual events in the system. Thus, the informational component in this equation determines an increase of negentropy in the system and is related with acquisition of new properties.

Developing the concept about a correspondence between negentropy and structural information, we can present the following equality:

$$-S = k\,\lg(1/D) = \sum p_i \cdot \log p_i \tag{13}$$

Hence,

$$\lg(1/D) = (\sum p_i \cdot \log p_i)/k, \text{ and} \tag{14}$$

$$1/D = 10^{(\sum p_i \log p_i)/k}, \text{ then} \tag{15}$$

$$D = 10^{-(\sum p_i \log p_i)/k} \quad \text{or} \quad 1/10^{(\sum p_i \log p_i)/k} \tag{16}$$

If in the Helmholtz equation for free energy entropy is replaced by D, this gives the following expression:

$$F = U - \frac{T}{10^{\sum p_i \log p_i / k}} \tag{17}$$

Thus, F can be considered as a function of the amount of structural information in a system. This equation is essential for understanding the self-organization processes in living systems, so as the cell. An increase in the amount of structural information determines the transition from liquid crystal to crystal. This may incapacitate a cell from its functioning. It has been already shown that the interaction of steroid hormones increases microviscosity of erythrocyte membranes in the regions of lipid-lipid and protein-lipid interactions. At low concentrations of hormones in the incubation medium, the activity of erythrocyte Na^+,K^+-ATPase even increases, probably due to growing elasticity of the lipid microenvironment of the enzyme, which facilitates structural transitions in the enzyme itself. At high concentrations of hormones (the saturation phase), an increase in microviscosity in the region of lipid-protein interactions makes impossible structural transitions in the enzyme; so, its activity rapidly drops. This determines a dome shape of the enzyme activity curve. Since erythrocyte is a liquid crystal cooperative system, changes occur not only in the activity of Na^+,K^+-ATPase of erythrocyte membranes, but also in the state of cell hemoglobin, its ordering and ability to bind oxygen.

It seems interesting to compare changes in liquid crystals with those occurring in solid crystals in the fields of external action.

Destruction of solid and liquid crystals increases the molar volume [24, 25].

A dependence of the Gibbs thermodynamic potential F(υ) on the molar volume υ taking into account local zones of different scale stress concentrators is described by the equation:

$$F(\upsilon) = U - TS + p\upsilon - \Sigma\mu_i C_i, \tag{18}$$

where μ_i – chemical potential, C_i – concentration (Fig. 22, [26]) .

At critical values of molar volume υ_i = (1,2…6), the thermodynamic potential F(υ) has local minima. They reflect local nonequilibrium potentials in the zones of different scale hydrostatic tension. Critical values of υ_i correspond to different levels of homeostasis in a deformable solid:

υ_0 is an equilibrium crystal; the initial level of homeostasis;

υ_1 are the zones of stress microconcentrators where dislocation cores are generated; the next level of homeostasis;

υ_2, υ_3 are the zones of stress meso- and macroconcentrators where local structural-phase transitions with the formation of meso- and macrostripes of local plastic deformation take place; the next levels of homeostasis;

υ_4 corresponds to intersection of curve F(υ) with the abscissa. At a further increase of the local molar volume, changes of the Gibbs thermodynamic potential proceed under the conditions of F(υ) > 0, and the system becomes unstable. Various forms of material failure appear; solid crystal starts to behave as a liquid one.

$\upsilon > \upsilon_6$ – the existence of two phases is possible: at $\upsilon = \upsilon_5$ – the vacancy phase atom, at $\upsilon > \upsilon_6$ – different thermodynamic levels of the crystal lattice in a deformable solid, different levels of its homeostasis.

Thus, plastic deformation of solid and liquid heterocrystals in the fields of external action is a multilevel process of their destruction, with the corresponding levels of crystal lattice self-organization and levels of its homeostasis, i.e., the destruction via different phases of strengthening (self-organization). On solid crystals this decreases the orderliness and amount of structural information. In liquid crystals this increases the orderliness and amount of structural information i.e. liquid crystal → crystal transition.

Dependence of Gibbs thermodynamic potential on the molar volume υ and changes in the structural information (I), taking into account local zones of stress concentrators is determined by the expression:

$$F(\upsilon, I) = U - T / 10^{(\Sigma p_i \log p_i)/k} + p\upsilon - \Sigma\mu_i C_i \tag{19}$$

These quantitative interrelations underlie transition of the system to a new structural level of homeostasis.

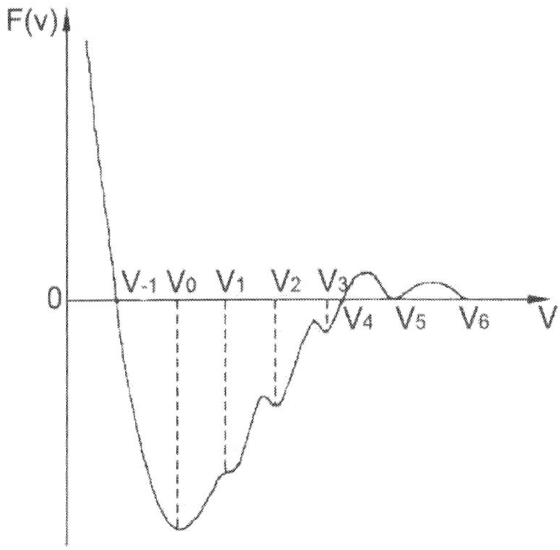

Figure 22. The dependence of the Gibbs thermodynamic potential F (v) from the molar volume v in the light of local zones of stress concentrators of different scales [26].

In biological membranes as liquid crystals, destruction is related with structural transitions and is generally accompanied by increasing structural orderliness (the order → order transition). Earlier [1], it was shown that the action of steroid hormones on erythrocyte membranes disturbs the mechanisms of self-organization that operate in the cells in normal functional condition. The active CO, NH and OH groups of stress hormones interact with CO and NH groups both of proteins and phospholipids in biological membranes. This leads to the formation of complex protein-lipid clusters, where "compressive" hydrophobic interactions are reinforced. Molecularly bound water is displaced to adjacent regions. Here, hydrostatic forces increase the "tensile" tangential stresses. Mobile nanostructural boundaries are formed, along which the biological membranes are destructed. This results in the formation of numerous pores and mesostrips of plastic deformation. In terms of physical mesomechanics, these transformations resemble those developing in solid crystals in the fields of external action (Fig. 23). However, in biological membranes such self-organization may be related even with increasing order and decreasing entropy, but this is incompatible with conditions that determine cell viability. Structural transitions cover the membrane-bound enzymes, transmembrane carriers and hormone receptors. It is reasonable to say that cell membranes go to a new level of homeostasis (self-organization) which is incompatible with life. The nature of life implies dynamics. The cell dies. Here, one can tell about thermodynamic features related with changes in the structure and function (properties) of solid crystals and biological membranes in the fields of external action.

Various structural transitions (phase transitions, nanostructural, etc.) strongly contribute to the functional activity of a cell. These are the transitions like smectic A → smectic C, smectic → cholesteric, and nematic → isotropic state; in proteins, the transitions tangle → β-structure and tangle → α-helix. They all affect the vital characteristics of a cell. I. Prigogine believed that there is "a wonderful analogy between instability of nonequilibrium origin and phase transitions" [27]. This problem is of great interest and deserves special examination.

Thus, on the curve of thermodynamic Gibbs potential versus molar volume F(υ), solid crystals fall in the region of strongly negative values, whereas liquid crystals are located near zero. Structural transformations taking place in the fields of external action draw together the positions of liquid and solid crystals on the functional curve. Morphologically, the destruction patterns of crystals are quite similar (Fig. 9a and b).

Thermodynamically, cells as hierarchic multilevel liquid crystal systems can function only near a zero value of thermodynamic Gibbs potential, i.e. in the region where reversible nanostructural transitions underlying life processes can occur.

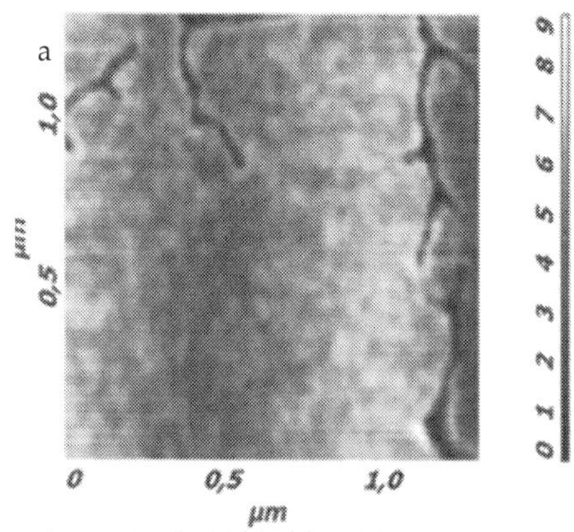

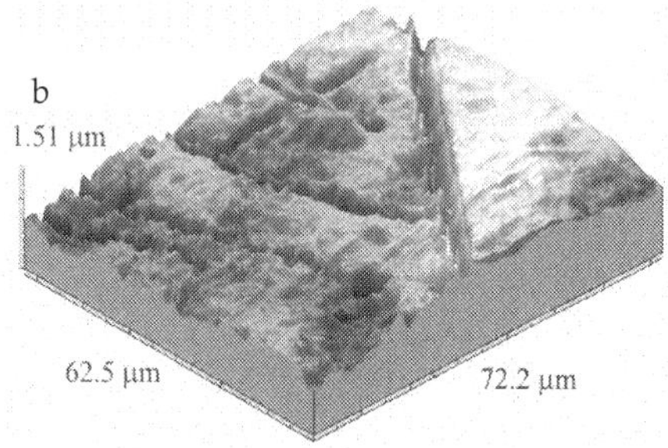

Figure 23. a – Atomic force microscopy. The surface of rat erythrocytes after adsorption of cortisol. Concentration of the hormone is 10^{-6}M. Deep meso-bands with bifurcation are seen; b – Formation of micropore chains along localized-deformation shear-bands. Plate of high-pure aluminum 180 nm thick glued on flat specimen of commercial Al. Alternative bending, T = 293 K; cycle number N = 17.55×10^{6} [26].

Biomembranes are of vital importance in the processes of self-organization of cell metabolism. These processes are transport of organic compounds through cell (plasma) membranes with delivery of nutrients to the cell and removal of their decay products (metabolism); diffusion of gases (O_2, CO_2) through a cell membrane; passive and active ion transport and production of electrochemical potential on the outer and inner surfaces of plasma membranes, and many others. All membranes are liquid crystals. Their behavior in an organism obeys physicochemical laws. The mechanism of their self-organization is the same as that of multilevel systems.

Multilevel systems mean the "hierarchy of scales of shear stability loss of the internal structure of a loaded material in local regions at the nano-, micro-, meso- and macrolevels" [26]. In liquid crystals, this is associated with lipid-lipid, protein-lipid, and protein-protein interactions, i.e. with cooperative behavior of a liquid crystal as a system. The ordering of these crystals as well as the nature of their components is determined by covalent and hydrogen bonds, weak electrostatic and hydrophobic interactions. Shear stability loss of natural liquid crystals depends on structural phase transitions such as the formation of smectic, cholesteric, nematic and isotropic structures and transitions between them. For membrane-bound proteins, states like α- helix, β-structure and chaotic coil are of significance. Structural transitions can be reversible and irreversible. In the latter case, defects are accumulated in liquid crystals, making some functions of the cell membranes unrealizable. The cell dies. These transitions, as a rule, arise on the surfaces of cell membranes and, because of cooperativeness, go deep into their lower levels. They can be also initiated at the inner membrane or particle interfaces and can be related to lipids and proteins. Thus, the case in point is different thermodynamic states of liquid crystals. The low transition enthalpy suggests that the transitions involve low-energy bonds, mainly hydrogen bonds, weak electrostatic and hydrophobic interactions. The external factors capable of changing (disrupting) the interactions are also physicochemical in nature. Among these factors are variations in temperature, pH, electrolyte content, etc. However, they are all of no fundamental character, and the behavior of liquid crystals under external actions fit in the same concept as the behavior of solid crystal does. It is very important for medicine.

As indicated in the report of World Health Organization, cardiovascular pathology, infections and oncological diseases are three main causes of human mortality all over the world [28]. Cardiovascular diseases stand first in this short list. In 2005 they killed 17.5 million people, which constitutes 30% of all deaths in the world. WHO predicts that in 2015 these diseases may take away the lives of 20 millions people. This will be caused mainly by infarctions and strokes, i.e. acute tissue hypoxia.

9. Conclusion

Especially dangerous is myocardial ischemia, which is related with the formation of atherosclerotic plaques within the coronary arteries and a considerable decrease in the blood flow rate. Such mechanism of tissue hypoxia development is typical of the older age groups. However, this pathology may develop also by a different mechanism. Nowadays acute myocardial ischemia and coronary deficiency are often observed in young people. There are many cases of sudden death that occur in young sportsmen during the competitions [29]. Coronary arteries of sportsmen are free of plaques, nevertheless, acute coronary deficiency develops somehow. Stress hormones and steroid anabolics may be a possible reason.

Today the world is changing very rapidly, and not everybody can adjust to these changes. Chronic stress becomes a wide-spread phenomenon [18]. Such reaction may be very pronounced in anxious persons. This is why cardiac syndrome X is often diagnosed now. Its clinical characteristics include angina chest pain with exertion and ischemic type ST segment depression on electrocardiogram, without angiographic signs of coronary artery stenosis and with normal left ventricle function. It means that acute coronary deficiency of obscure etiology occurs not only in sportsmen, but also among people at large [30].

Cortisol is the main stress hormone. It is a cholesterol derivative, and cholesterol is the essential component of all cell membranes. Of special interest is the erythrocyte membrane. Erythrocyte is a specialized cell that transfers oxygen from lungs to tissues by means of hemoglobin (Hb). In capillaries, oxyhemoglobin HbO_2 decomposes, and O_2 diffuses to the organ and tissue cells. The first obstacle to such diffusion is erythrocyte membrane. Changes in the properties of erythrocyte membrane determine the rate of oxygen diffusion across the membrane. Besides, capillary and erythrocyte may have comparable diameters. Sometimes the erythrocyte diameter happens to be even larger. To go through so small capillary, erythrocyte should have a high plasticity. Structural transformations in erythrocyte membrane under the action of stress hormones may be reflected not only in its plasticity, but also in the mechanism of gas exchange.

In this work, an attempt to elucidate the effect of steroid hormones on the structure of erythrocyte membranes and their physicochemical properties, i.e. the introduction of principles and regularities of physical mesomechanics in biology and medicine provides a deep insight into the mechanism interrelating structure and function of biological membranes, both in the norm and at systemic membrane pathology (upon variation of hormone concentration, temperature, pH, electrochemical potential, etc. has been made.

Thus, from a thermodynamic standpoint, life is the ability of cells to undergo reversible nanostructural transitions near a zero value of thermodynamic Gibbs potential. A loss of this ability leads to cell death and development of pathology.

Author details

L.E. Panin
Scientific Research Institute of Biochemistry SB RAMS, Russia

Acknowledgement

The author is grateful to Dr. Sci. (med.) V.G. Kunitsyn and Ph. D. (phys.-math.) P.V. Mokrushnikov for participation in the preparation of experimental material.

10. References

[1] Panin, L.E.; Mokrushnikov, P.V.; Kunitsyn, V.G. & Zaitsev, B.N. (2010). Interaction mechanism of cortisol and catecholamines with structural components of erythrocyte membranes. *J. Phys. Chem. B.* Vol. 114. No. 29, (Jul. 29, 2010) pp. 9462-9473. ISSN 1520-6106

[2] Hurst, T; Olson, T.H.; Olson, L.E. & Appleton, C.P. (2006). Cardiac syndrome X and endothelial dysfunction: new concepts in prognosis and treatment. *Am. J. Med.* Vol. 119. No. 7, (Jul. 2006) pp. 560-566. ISSN 0002-9343

[3] Rubart, M. & Zipes, D.P. (2005). Mechanisms of sudden cardiac death. *J. Clin. Invest.* Vol. 115. No. 9, (Sep. 2005) pp. 2305-2315. ISSN 0021-9738

[4] Courson, R. (2007). Preventing sudden death on the athletic field: the emergency action plan. *Curr Sports Med Rep.*, Vol. 6, No. 2, (Apr. 2007) pp. 93-100, ISSN 1537-890X

[5] Jorgensen, P.L.; Hakansson, K.O. & Karlish, S.J. (2003) Structure and mechanism of Na+,K+-ATPase. *Annu. Rev. Physiol.* Vol. 65, (May 2002) pp. 817–849. ISSN 0066-4278

[6] Golden, G.A.; Mason, P.E.; Rubin, R.T. & Mason, R.P. (1998) Biophysical membrane interactions of steroid hormones: a potential complementary mechanism of steroid action. *Clin. Neuropharmacol.* Vol. 21. No. 3, (May-Jun. 1998) pp. 181-189. ISSN 0362-5664

[7] Wu, Y.; Hu, Y.; Cai, J.; Ma, S.; Wang, X.; Chen, Y. & Pan, Y. (2009) Time-dependent surface adhesive force and morphology of RBC measured by AFM. *Micron.* Vol. 40. No. 3, (Apr. 2009) pp. 359-364. ISSN 0968-4328.

[8] Kunitsyn, V.G.; Panin, L.E. & Polyakov, L.M. (2001). Anomalous change of viscosity and conductivity in blood plasma lipoproteins in the physiological temperature range. *Int. J. Quantum Chem.*, Vol. 81, (Feb. 2001) pp. 348-369. ISSN 0020-7608

[9] Dawson, R.M.C.; Elliot, D.C.; Elliot, W.H. & Jones, K.M. (1986). *Data for biochemical research.* Clarendon Press, ISBN is absent, Oxford, United Kingdom.

[10] Attallah, N.A. & Lata, G.F. (1968). Steroid-protein interactions studies by fluorescence quenching. *Biochim. Biophys. Acta,* Vol. 168. Issue 2, pp. 321-333, ISSN 0005-2795

[11] Storozhok, S.A.; Sannikov, A.G. & Zakharov, Yu.M. (1997). *Molecular Structure of Erythrocyte Membranes and Their Mechanical Properties,* Tyumen University, Tyumen.

[12] Murray, R.K.; Granner D.K.; Mayes P.A. & Rodwell V.W. (2003). *Harper's Illustrated Biochemistry,* 26th ed., The McGraw-Hill Companies, Inc.

[13] Ooi, T.; Itsuka, E.; Onari, S. et al. (1988). *Biopolymers,* Imanisi Y. (Ed.), Mir, ISBN is absent: Moscow, Russia

[14] Panin L.E., Panin V.E. (2011). Thermodynamics and mesomechanics of nanostructural transitions in biological membranes under stress. *Int. J. Terraspace Science and Engineering.* V. 3. Iss. 1. pp. 3-12. ISSN:1943-3514

[15] Miyazawa, T. & Blout, E.R. (1961) The infrared spectra of polypeptide in various conformations: amid I and II bands. *J. Am. Chem. Soc.* Vol. 83, No. 3 (Febr. 1961) pp. 712-719. ISSN 0002-7863

[16] Semenov, M.A.; Gasan A.J.; Bol'bukh T.V.; et. al. (1996). Hydration and structural transitions of DNA from Micrococcus lysodeikticus in films. *Biophysics,* Vol. 41. No. 5, (Sept. – Oct. 1996) pp. 1007-1016. ISSN 0006-3509.

[17] Discher, D.E., Mohandas, N. & Evans, E.A. (1994). Molecular maps of red cell deformation: hidden elasticity and in situ connectivity. *Science.* Vol. 266. No. 5187, (Nov. 1994) pp. 1032-1035. ISSN 0036-8075.

[18] Shnol, S.E. (1979) *Physicochemical factors of biological evolution,* Nauka, Moscow (Rus.) ISBN 978-91-85917-06-8.

[19] Kunitsyn, V.G. (2002). *Structural phase transitions in erythrocyte membranes, lipoproteins and macromolecules.* Thesis, Dr. Sci. (biol.), Novosibirsk. (Rus.)

[20] Park, Y.; Best, C.A.; Badizadegan, K., Dasari, R.R.; Feld, M.S.; Kuriabova, T.; Henle, M.L.; Levine, A.J. & Popescu, G. (2010). Measurement of red blood cell mechanics during morphological changes. *Proc. Natl. Acad. Sci. U S A.* Vol. 107. No. 15, (Apr. 2010) pp. 6731- 6736. ISSN 0027-8424

[21] Bennett, V. (1984) In: *Cell Membranes: Methods and Reviews,* eds. Elson, E., Frazier, W. & Glaser, L. Vol. 2, Plenum, New York, pp. 149-195.

[22] Storozhok, S. A.; Sannikov, A. G. & Zakharov, Yu. M. (1997) *Molecular Structure of Erythrocyte Membranes and Their Mechanical Properties,* Tyumen Gos. University: Tyumen.

[23] Wolkers, W.F.; Crowe, L.M.; Tsvetkova, N.M.; Tablin, F. & Crowe, J.H. (2002). In situ assessment of erythrocyte membrane properties during cold storage. *Mol. Membr. Biol.* Vol. 19. No. 1, (Jan-Mar 2002) pp. 59-65. ISSN 0968-7688

[24] Panin, L.E. & Kunitsyn, V.G. (2009). Mechanism and thermodynamics of multilevel structural transitions in liquid crystals under external actions. *Physical Mesomechanics,* Vol. 12. No. 1-2, (Jan. – Apr. 2009) pp. 78-84. ISSN 1029-9599

[25] Panin, L.E. (2011). Thermodynamics and mesomechanics of nanostructural transitions in biological membranes under the action of male sex hormones. 13-th International conference on mesomechanics. Vicenza, Italy, 6-8 July 2011. Eds. G.Sih, P. Lazzarin, F. Berto. p. 48-51.

[26] Panin, V.E. & Egorushkin, V.E. (2008). Nonequilibrium thermodynamics of a deformed solid as a multiscale system. Corpuscular-wave dualism of plastic shear. *Physical Mesomechanics.* Vol. 11. No. 3-4, (May - August 2008) pp. 105 – 123. ISSN 1029-9599.

[27] Nicolis, G. & Prigogine, I. (1979). *Self-Organization in Non-Equilibrium Systems. From Dissipative structures to Order through Fluctuations,* Mir, Moscow, Russia.

[28] Vermel', A.E. (2006). Cardiac syndrome X. *Klin Med (Mosk).* Vol. 84. No. 6 (Nov.–Dec. 206) pp. 5-9. ISSN 0023-2149. Russian.

[29] Montagnana, M.; Lippi, G.; Franchini, M.; Banfi, G. & Guidi, G.C. (2008). Sudden cardiac death in young athletes. *Intern Med.* Vol. 47, No. 15, (Aug. 2008) pp. 1373-1378, ISSN 0918-2918

[30] Panin, L.E. & Usenko, G.A. (2004). *Anxiety, adaptation and prenosological clinical examination.* SB RAMS, Novosibirsk, ISBN 5-93239-050-6 (Rus.)

Chapter 12

Thermodynamics of Resulting Complexes Between Cyclodextrins and Bile Salts

Yu Liu and Kui Wang

Additional information is available at the end of the chapter

http://dx.doi.org/10.5772/51406

1. Introduction

Cyclodextrins (CDs), a class of macrocyclic oligosaccharides consisting of six, seven, or eight glucose units linked by α-1,4-glucose bonds, have been widely used as receptors in molecular recognition in the field of supramolecular chemistry because they are able to form inclusion complexes with hydrophobic guests in aqueous solution owing to their hydrophilic outer surface and their lipophilic cavity [1–3]. Therefore, much effort has been devoted to the design and synthesis of a wide variety of cyclodextrin (CD) derivatives to explore their binding behaviors for model substrates [4]. In order to further explore their inclusion complexation mechanism, most of these studies have been focused on the binding modes and complexation thermodynamics based on CDs and their derivatives in recent years [5]. Among the numerous guests researched, bile salts attracted much more attention because they are one kind of important surfactant-like biological amphipathic compounds possessing a steroid skeleton, which have distinctive detergent properties and play an important role in the metabolism and excretion of cholesterol in mammals [6]. For example: the thermodynamics and structure of inclusion compounds of glyco- and tauro-conjugated bile salts with CDs and their derivatives have been studied by Holm et al. during the last years [7–11]; the interactions of different kinds of bile salts with β-CD dimers linked through their secondary faces have been investigated by Reinhoudt and Vargas-Berenguel et al. [12–14]. It has been demonstrated that the formation of inclusion complexes between CDs and guest molecules is cooperatively governed by several weak forces, such as van der Waals interactions, hydrophobic interactions, hydrogen bonding, electrostatic interactions, and every weak force does its contribution to the complexation. In this chapter, the related investigations concerned on the binding modes, binding abilities, molecular selectivities and their thermodynamic origins of CDs and their derivatives with four typical bile salts (Cholate (CA), Deoxycholate (DCA), Glycocholate (GCA), and Taurocholate (TCA)) (Figure

1) have been summarized, which will be discussed from the aspect of the types of host molecules: (1) natural CD series; (2) modified CD series; (3) bridged CD series. This summary is helpful to improve understanding of the correlation between the structural features and molecular-recognition mechanism from thermodynamic viewpoints, and further guide its biological, medicinal and pharmaceutical applications in the future.

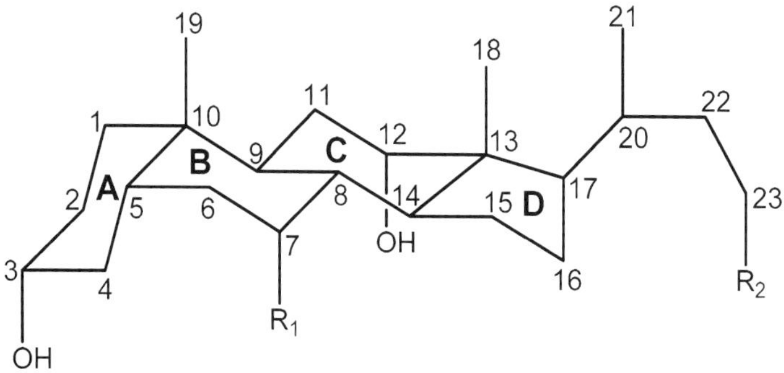

Guests	R1	R2
Cholate (CA)	OH	COONa
Deoxycholate (DCA)	H	COONa
Glycocholate (GCA)	OH	$CONHCH_2COONa$
Taurocholate (TCA)	OH	$CONHCH_2CH_2SO_3Na$

Figure 1. Molecular structures of four typical bile salts

2. Natural CD series

2.1. Binding modes for bile salts and natural CD series

Since two protons located closely in space (the corresponding internuclear distance is smaller than 3–4 Å) can produce NOE (Nuclear Overhauser Effect) cross-peaks between the relevant protons in NOESY (Nuclear Overhauser Effect Spectroscopy) or ROESY (Rotating Frame Overhauser Effect Spectroscopy) spectra, 2D NMR spectroscopy has become an important method for the investigation of the interaction between different kinds of CDs and guest molecules. It is well-known that only H3, H5, and H6 of CDs can give cross-peaks for analyzing host–guest interactions, as H2 and H4 are not facing to the inner cavity and H1 is affected by D_2O. For example, the ROESY study on the resulting complex of natural β-CD **1** (Figure 2) with CA has been reported by Tato et al. [15,16]. The results successfully indicated that in the 1:1 complex between **1** and CA the steroid body entered forward into the inner cavity of **1** by the side of the secondary hydroxyl groups, with the side chain folded toward the steroid body, i.e., rings D and C are totally and partially included, respectively. Therefore, the binding modes of bile salts with different kinds of CDs have been widely deduced by 2D NMR spectroscopy during the last years.

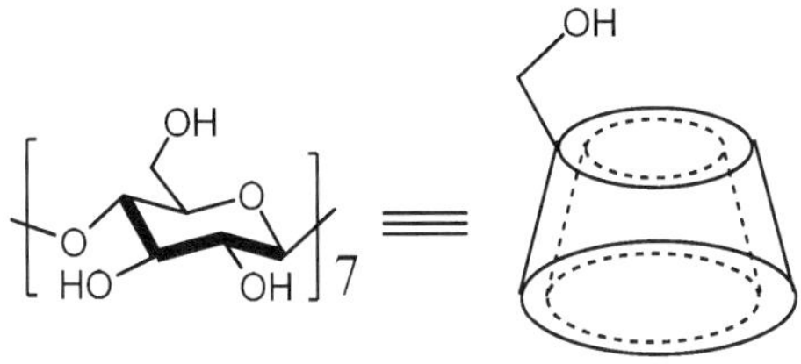

1

Figure 2. Structure of natural β-CD

2.2. Complexation thermodynamics for bile salts and natural CD series

The microcalorimetric titrations can be used to simultaneously determine the enthalpy and equilibrium constant from a single titration curve. Titrations were performed below the critical micelle concentration of bile salts. In each run, a solution of the host (guest) molecules in syringe was sequentially injected into the calorimeter sample cell containing a solution of guests (hosts). Each addition of hosts (guests) into the sample cell gave rise to a heat of reaction, caused by the formation of inclusion complexes between hosts and guests. The heats of reaction decrease after each injection because less and less molecules in cell are available to form inclusion complexes. A control experiment was performed to determine the heat of dilution by injecting a host (guest) solution into a pure solution containing no guest (host) molecules. The dilution enthalpy was subtracted from the apparent enthalpy obtained in each titration run, and the net reaction enthalpy was analyzed by using the "one set of binding sites" model. This model will work for any number of sites N if all sites have the same Ks and $\Delta H°$. In this case, the total heat Q was fitted via a nonlinear least-squares minimization method to the total host concentration in cell (M_t) using the following equation:

$$Q = (NX_t\Delta HV_o/2)\{1 + M_t/(NX_t) + 1/(NKsX_t) - \{[1 + M_t/(NX_t) + 1/(NKsX_t)]^2 - 4M_t/(NX_t)\}^{1/2}\} \quad (1)$$

where N is the number of binding sites of host, X_t is the total concentration of guests in cell and V_o is the cell volume. The value of Q above can be calculated (for any designated values of N, K, and ΔH) at the end of the ith injection and designated $Q(i)$. Then the correct expression for the heat released, $\Delta Q(i)$, from the ith injection is

$$\Delta Q(i) = Q(i) + \mathrm{d}V_i/(\text{<ital>}\ V_o)\{[Q(i) + Q(i-1)]/2\} - Q(i-1) \quad (2)$$

where $\mathrm{d}V_i$ is the volume of titrant added to the solution. Along with obtaining of Ks and $\Delta H°$ in this fitting program, the N value in eq 1 can also be obtained, which represents the numbers of guests bound to one host molecule.

The ORIGIN software (Microcal), used for the calculation of the binding constant (Ks) and standard molar reaction enthalpy ($\Delta H°$) from the titration curve, gave the relevant standard derivation on the basis of the scatter of data points in a single titration experiment. The binding stoichiometry was also given as a parameter when fitting the binding isotherm. Knowledge of the binding constant (Ks) and molar reaction enthalpy ($\Delta H°$) enabled the calculation of the standard free energy of binding ($\Delta G°$) and entropy change ($\Delta S°$) according to

$$\Delta G° = -RT \ln Ks = \Delta H° - T\Delta S° \tag{3}$$

where R is the gas constant and T is the absolute temperature.

The microcalorimetric experiments of natural β-CD **1** with bile salts (CA, DCA, GCA, and TCA) showed typical titration curves of 1:1 complex formation [17]. The stoichiometric ratios observed from curve-fitting results of the binding isotherm fell within the range of 0.9–1.1. This clearly indicated that the majority of the inclusion complexes had a 1:1 stoichiometry of bile salts and **1**.

Thermodynamically, the binding behaviors of bile salts by **1** were entirely driven by favorable enthalpy changes accompanied by small unfavorable entropy changes, which are attributed to the predominant contribution of the van der Waals interactions arising from the size/shape fit and geometrical complement between host and guest and to the accompanying decreases in translational and structural freedoms upon complexation.

As can be seen from Table 1, the enthalpy change for the complexation of **1** with DCA is more favorable than that with CA, which directly contributes to the increased complex stability. It is reasonable that DCA possesses a more hydrophobic structure due to the absence of C-7 hydroxyl group as compared with CA, as a result, it is easier to bind into the cavity of **1**, which leads to more favorable hydrophobic and van der Waals interactions and gives larger enthalpy and entropy changes. However, the enhanced favorable entropy gain by the desolvation effect may be canceled by the unfavorable entropy change caused by the structural freezing of the resulting complexes of **1** and DCA. Therefore, the stronger interaction between **1** and DCA only shows the larger negative enthalpy change, directly contributing the relatively larger complex stability constant. Meanwhile, **1** shows a lower binding ability upon complexation with GCA and TCA. Compared with **1** and CA, the complexation of **1** with GCA and TCA exhibit similar enthalpy changes but much more unfavorable entropy changes. The more polar side chains of GCA and TCA may be the reason for it.

3. Modified CD series

3.1. Binding modes for bile salts and modified CD series

3.1.1. Aminated β-CDs

The ROESY study on the resulting complex of **2** (Figure 3) with CA has been reported by Tato et al. [15]. The results exhibited different interactions of the side chain of CA with H5 and H6 of **2** from natural β-CD **1**. The facts indicated that the side chain was unfolded, with the negative carboxylate group moving toward the positive protonated amino group, and the side-chain elongation produced a deeper penetration of the steroid body in the inner cavity of **2**.

The ROESY experiments of modified β-CD **3** in the presence of CA or DCA have been performed in D_2O by Liu et al. [17]. The results indicate that the D-ring of CA is accommodated shallowly in the cavity and CA enters **3** from the second side of CD with the side chain and D-ring. At the same time, the side chain with the negative carboxylate group of CA moves toward the positive protonated amino group of **3**. For the resulting complex of DCA–**3**, the ROESY spectrum exhibits entirely different NOE cross-peaks and the D-ring of DCA is included within the cavity of CD from the primary side of CD. Meanwhile, the ethide protons of chiral tether interact with H6 of CD.

2D ROESY NMR experiment of **5** and CA has also been performed by Liu et al. in D_2O to investigate the binding mode between bile salt and CD [18]. The results show that steroid body enters the CD cavity from the second side with its tail and D-ring parts.

3.1.2. *Nucleobase-modified β-CDs*

In host **8**, the adenine group is deeply inserted into the β-CD cavity with an orientation parallel to the C7 axis of β-CD while the thymine and uracil groups are shallowly inserted in the β-CD cavity with an orientation perpendicular to the C7 axis of β-CD [19]. As a result, upon complexation with DCA guest, the deeply included adenine group in host **8** should be expelled from the cavity upon complexation with DCA guest, however, the shallowly included thymine and uracil groups in hosts **9** and **10** are hardly influenced by the inclusion of DCA guest.

3.1.3. *Tryptophan- and Tyrosine-modified β-CDs*

The binding modes of L/D-Trp-β-CD (**11** and **12**) with bile salts have been examined by Liu et al. by 2D ROESY NMR experiments [20]. For L-Trp-β-CD (**11**), the results show that in the absence of guest, L-Trp residue is only shallowly included or perching on the rim of the CD cavity. However, in the presence of DCA, the D-ring of DCA is close to the wide end of CD cavity, and the D-ring of DCA and the side chain is co-included in the same cavity from the primary side of **11**. For D-Trp-β-CD (**12**), the 2D NMR results indicate that the D-Trp residue attached to β-CD is more deeply self-included than the corresponding L-Trp residue in the absence of guest. However, in the presence of DCA, the carboxylate side chain and D-ring of DCA penetrate into the CD cavity from the secondary side shallowly.

The binding modes of L/D-Tyr-β-CD (**13** and **14**) with bile salts have further been examined by Liu et al. by 2D ROESY NMR experiments [21]. The results show that the L-tyrosine moiety was self-included in the β-CD cavity from the narrow opening. The DCA guest entered the β-CD cavity from the wide opening with the tail and the D ring and coexisted with the L-tyrosine substituent in the β-CD cavity to form a cooperative inclusion manner. For D-tyrosine-modified β-CD (**14**), the D-tyrosine substituent was deeply self-included in the β-CD cavity and might be located in the center of the β-CD cavity. Upon complexation with DCA, the D-tyrosine substituent of **14** would partially move out of the β-CD cavity. Compared with DCA + **13** complex, DCA penetrated into the β-CD cavity of **14** more deeply (Figure 4).

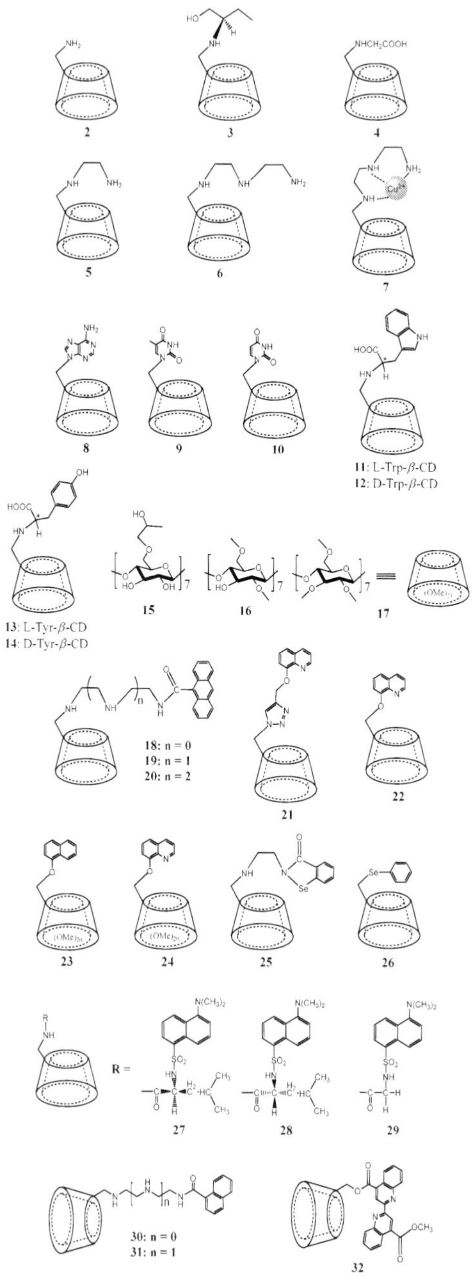

Figure 3. Structures of mono-modified β-CD derivatives

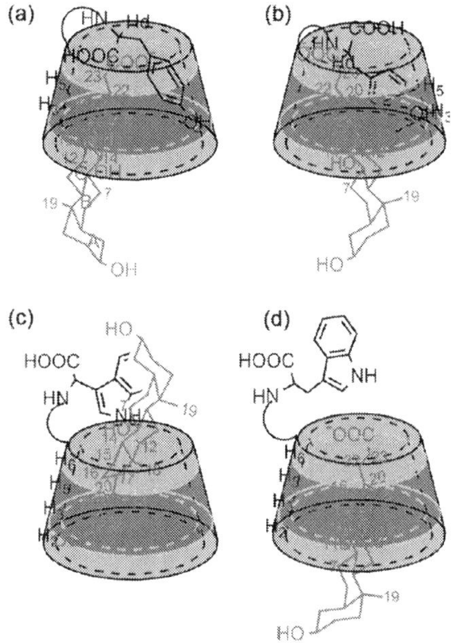

Figure 4. The possible binding modes of **11–14** (**11** (c); **12** (d); **13** (a); **14** (b)) with DCA

3.2. Complexation thermodynamics for bile salts and modified CD series

3.2.1. Aminated β-CDs

The microcalorimetric experiments of aminated β-CDs with bile salts clearly indicate that the majority of the inclusion complexes had a 1:1 stoichiometry [17]. Thermodynamically, the binding constants of **4** upon inclusion complexation with DCA, GCA, and TCA are less than that with natural β-CD **1**. It is reasonable that modified β-CD **4** decreased the microenvironment hydrophobicity of natural β-CD cavity due to the hydrophilic carboxylic group in the sidearm, and at the same time there is electrostatic repulsion between the anionic carboxylate at the sidearm of **4** and anionic carboxylate or sulfonate of bile salts. Unexpectedly, the resulting complex stability of aminated β-CD **4** with CA is higher than that of native β-CD **1**, which is mainly attributed to the more favorable enthalpy change. The possible reason may be the enhanced cooperative van der Waals, hydrogen-bonding, and electrostatic interactions exceeding the decreased hydrophobicity of the interior of β-CD **4**.

Positively charged monoamino-modified β-CD **2** and modified β-CD **3** possessing an additional binding site in the chiral arm evidently enhance the molecular binding ability and selectivity towards CA and DCA compared to those for native β-CD **1**, which is mainly attributed to the more favorable enthalpy change accompanied with unfavorable entropy change [17]. The more favorable enthalpy change most likely originates from effective electrostatic interactions and the additional binding site of hydroxyl group. In addition, the unfavorable entropy change is likely to originate from the conformation fixation of host and guest and the rigid complex formation upon complexation. β-CD derivatives **2** and **3** give a lower binding ability upon complexation with GCA and TCA as compared to the complexation with CA and DCA, which is similar to that for the complexation of β-CD **1** and derivative **4**. For the same reason, the more polar side chains at C23 for GCA and TCA remarkably affect their binding thermodynamics.

A study of ^{13}C chemical shifts as a function of concentration at different pH values has been performed by Tato et al., which shows a different behavior of complexation for CA and DCA with **5** resulting in 1:1 and 1:2 inclusion complexes [22]. However, the complexation phenomena do not depend on the pH of the solution. ^{13}C NMR chemical shifts of the host and guest molecules change on passing from the free to the complexed state. The side chains in **5** at position C-6 have a significant effect on the complexation process with the bile salts. The ROESY experiments confirm the overlap of the CA molecule with **5** resulting a 1:1 inclusion complex, while in the case of DCA molecule, the first molecule of **5** encapsulates the bile salt to a larger extent than the second molecule of **5**, resulting a 1:2 inclusion complex. Hence the most important factors for the formation of a stable inclusion complex are the relative size of **5** and the bile salt molecules, the nonpolar cavity of **5**, the hydrophobicity of the bile salts, and the presence of an electrostatic environment outside the toroidal cavity.

3.2.2. Nucleobase-modified β-CDs

The nucleobase-modified β-CDs **8–10** exhibit distinguishable binding abilities toward bile salts compared with parent β-CD **1** [19]. Host **10** shows increased binding of TCA/GCA. Host **9** exhibits increased binding of GCA while hosts **8–10** show less binding of the other bile salts. The inclusion complexation of hosts **8–10** is driven by favorable enthalpy changes, accompanied with unfavorable entropy changes. The driven forces are hydrogen-bonding and van der Waals interactions, simultaneously producing marked geometric configuration change. Host **8** displays weaker binding ability for every bile salt than hosts **9** and **10** owing to expelling adenine group from β-CD cavity to accommodate bile guests in hosts **8**, which is unfavorable to the host–guest complexation.

3.2.3. Tryptophan- and tyrosine-modified β-CDs

The microcalorimetric titrations of L/D-Trp-modified β-CD (**11** and **12**) with a series of bile acids, i.e., CA, DCA, GCA, and TCA, showed typical titration curves, which can be nicely analyzed by assuming the 1:1 complex stoichiometry [20]. Modified β-CDs **11** and **12** exhibited appreciably smaller binding abilities for GCA and TCA guests than those of native β-CD **1** since GCA and TCA, possessing a strongly hydrophilic and hydrated sulfonate tail, are not expected to deeply penetrate into the CD cavity by removing the originally included L/D-Trp group out of the hydrophobic cavity. In contrast, DCA and CA, possessing a less hydrophilic/hydrated carboxylate tail, showed comparable or even stronger binding and higher selectivities for host's chirality than TCA and GCA.

The ITC experiments of hosts **13** and **14** with bile salts (CA, DCA, GCA, and TCA) also showed the typical titration curves of the 1:1 complex formation [21]. The stoichiometric ratio "*N*" observed from the curve-fitting results was within the range 0.9 to 1.1, which clearly indicated that the majority of the inclusion complexes had a 1:1 binding mode. Thermodynamically, the binding of all CDs with the bile salts was entirely driven by the favorable enthalpy changes accompanied by the unfavorable entropy changes. **14** gave the higher bind ability toward CA and DCA than **1** and **12** due to the introduction of D-tyrosine substituent and the conformational difference between **12** and **14**. In addition, the bind constant of **14** for DCA was slightly bigger than that for CA. Possessing a more hydrophobic structure due to the absence of the C-7 hydroxyl group as compared with CA, DCA was easier to bind to the β-CD cavity than CA, which consequently led to the more favorable hydrophobic interactions between hosts and guests. Host **14** exhibited the obviously smaller binding abilities for GCA and TCA guests than **1** and **12**. Thermodynamically, the decreased binding affinities of host **14** toward GCA and TCA arose from the entropy change rather than the enthalpy change due to the weakened hydrophobic interactions and the relatively poor size-fit between host and guest. Compared with **1** and **11**, **13** showed clearly decreased binding abilities toward all four of the bile salts, especially for GCA and TCA. Thermodynamically, the inclusion complexation of **13** with four bile salts exhibited the favorable enthalpy changes and unfavorable entropy changes. The favorable enthalpy gain of **13** was slightly higher than those of **1** and **11**, but the entropy loss of **13** was much more than those of **1** and **11** toward corresponding guests.

3.2.4. *Methyl-β-CD and 2-hydroxypropyl-β-CD*

The interactions of CA, DCA, GCA, and TCA with **15** and **16** have been studied by Ollila et al. by means of isothermal titration calorimetry [23]. The results show that both CA and DCA bound to **15** and **16** with a 1:1 stoichiometry. The binding constant was significantly higher for DCA to **15** and **16** compared to CA. This difference in binding affinity is likely explained by the more hydrophobic nature of DCA due to the absence of the C-7 hydroxyl group, which is present in CA. The binding affinity was somewhat lower for CA binding to **15** compared to **16**, while DCA showed a markedly lower affinity for **15** compared to **16**. GCA and TCA have lower affinities to **15** and **16** compared to CA and DCA. TCA bound with lower affinity to **15** compared to GCA. Both GCA and TCA gave the same 1:1 stoichiometry for binding to **15** and **16** as did CA and DCA.

For **17**, all the hydroxyls are methylated, and the loss of hydrogen bonds for the resulting complexes is inevitable [24]. Therefore, host **17** only shows weak complex stability constants to bile salts, which are much lower than those of **1** and **16**. In addition, the release of higher energy water molecules in the cavity of β-CD upon complexation with guests makes the inclusion complexation more favorable, which cannot be obtained in the cases of **17** because almost no water molecule resides in the cavity of **17**. Besides that, **17** should need some conformational adjustment to accommodate bile guests, which is entropy-unfavorable for the inclusion complexation.

3.3. Binding modes for bile salts and chromophore-modified CD series

3.3.1. Anthryl-modified β-CDs

^{1}H ROESY experiment has been performed by Liu et al. to confirm the binding model of host **19** with CA [25]. The results indicate that CA molecule is included into the hydrophobic cavity from the secondary side of β-CD, with the side chain folded towards the steroid skeleton, and the anthracene group is excluded outside the cavity of β-CD. CA molecule and the tether of β-CD can be co-included into the cavity through the induced-fit interaction between host and guest.

3.3.2. Quinolinyl- and naphthyl-modified β-CDs

2D ROESY NMR experiments accompanied with molecular modeling studies have been performed by Liu et al. to investigate the binding modes of DCA with **21** and **22** [26]. The results show that the side chain and D-ring of bile salts were encapsulated in the β-CD cavity from the wide opening (Figure 5).

2D ROESY NMR experiment of complex of **23** with CA has also been performed to investigate the binding geometry between permethylated β-CDs and bile salts [24]. The results show that CA is deeply included into the cavity of host **23** with its ring A in the region of the narrow side and ring D in the region of the broad side. However, upon complexation with CA guest, the appended naphthalene group in **23** is not entirely expelled out of the cavity of permethylated β-CD but is removed from the central cavity to the region of the narrow torus rim. The cooperative inclusion manner of both guest molecule and substituent sidearm into the cavity is mainly benefited from the extended framework of permethylated β-CD.

3.4. Complexation thermodynamics for bile salts and chromophore-modified CD series

3.4.1. Anthryl-modified β-CDs

The stoichiometric ratios gotten from curve-fitting results of the binding isotherm fell within the range of 0.9–1.1, indicating that the resulting complexes of bile salts and CDs (**18–20**) are 1:1 [25]. As compared with parent β-CD **1**, modified β-CDs **18–20** with different chain length not only enhanced molecular binding ability but also significant molecular selectivity upon inclusion complexation with homologous steroids, except for resulting complex of **20** with TCA. The stability constants for the inclusion complexation of hosts and the each steroid molecule decreased in the following order: DCA > CA > GCA > TCA. The hydroxyl group at the C7 carbon atom of CA, GCA and TCA guests prevented deeper inclusion of the steroids in the β-CD cavity than that of DCA guest. On the other hand, the tether length of the host and induced-fit interactions also played crucial roles in the selective molecular binding process of modified β-CD **18–20** with guests. Host **19** possessing suitable tether length could encapsulate more tightly the steroid guests than the other, through the size/shape-matching and the induced-fit interactions between the host and guest.

Thermodynamically, the inclusion complexation of **18–20** with steroid guests is entirely driven by favorable enthalpy contribution with negative or minor positive entropy change [25]. The strong interaction between host and guest leads to the more favorable negative enthalpy change, which is counteracted by the relative more unfavorable negative entropy change. The introduction of anthracene group with different chain length, and additional binding site to CD rim can significantly enhance the binding ability of parent CD toward steroid guests.

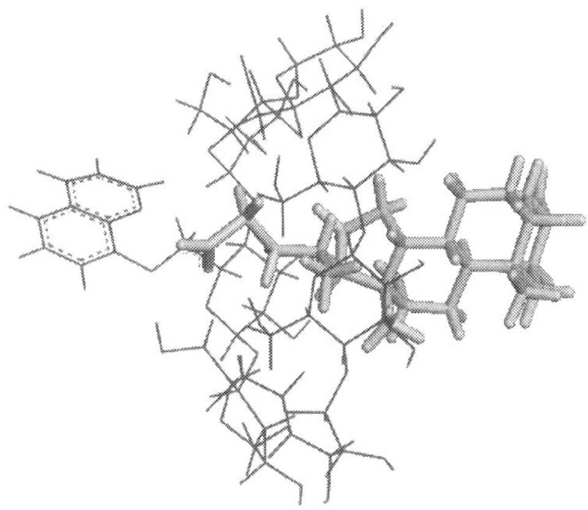

Figure 5. The possible binding mode of **22** with DCA

3.4.2. *Quinolinyl- and naphthyl-modified β-CDs*

The binding behaviors of two β-CD derivatives bearing 8-hydroxyquinolino and triazolylquinolino groups (**21** and **22**) with bile salts have been studied in aqueous buffer solution by means of microcalorimetrical titration [26]. The results showed that the host–guest binding behaviors were mainly driven by the favorable enthalpy changes, accompanied by the unfavorable entropy changes, and the hydrogen-bonding interactions and van der Waals interactions were the main driven forces governing the host–guest binding.

The binding stoichiometry of the permethylated β-CD derivatives **23** and **24** with bile salts has been determined by the Job's plot method, which showed that hosts and guests formed 1:1 complexes [24]. Thermodynamically, hosts **23** and **24** show much higher binding ability to bile salts than permethylated β-CD **17** when the naphthalene (or quinoline) sidearm is appended on it. The pronounced enhancement of complex stabilities for hosts **23** and **24** can be attributed to the cooperative complex interactions of both the cavity of permethylated β-CD and the chromophore sidearms. Furthermore, it should be mentioned that host **24** always forms more stable complexes with bile guests than host **23**, which indicates that the N atom on the quinoline ring plays a crucial role during the course of recognition of bile guests.

Hosts	Guests	pH	*Ks*	$\Delta H°$	$T\Delta S°$	Methods	Refs.
1	CA	7.2 (PBS)	4068	–22.98	–2.38	ITC	17
	DCA	7.2 (PBS)	4844	–25.79	–4.76	ITC	17
	GCA	7.2 (PBS)	2394	–22.99	–3.7	ITC	17
	TCA	7.2 (PBS)	2293	–23.77	–4.59	ITC	17
2	CA	7.2 (PBS)	11160	–25.53	–2.43	ITC	17
	DCA	7.2 (PBS)	7705	–32.16	–9.98	ITC	17
	GCA	7.2 (PBS)	2075	–25.90	–6.97	ITC	17
	TCA	7.2 (PBS)	2309	–26.89	–7.69	ITC	17
3	CA	7.2 (PBS)	16920	–28.11	–3.98	ITC	17
	DCA	7.2 (PBS)	9382	–35.78	–13.11	ITC	17
	GCA	7.2 (PBS)	3904	–24.74	–4.24	ITC	17
	TCA	7.2 (PBS)	2796	–20.37	–0.7	ITC	17
4	CA	7.2 (PBS)	4832	–24.90	–3.87	ITC	17
	DCA	7.2 (PBS)	4034	–38.91	–18.33	ITC	17
	GCA	7.2 (PBS)	2221	–19.75	–0.65	ITC	17
	TCA	7.2 (PBS)	1322	–32.75	–14.93	ITC	17
5	CA	7.2 (PBS)	11060	–36.44	–13.36	ITC	18
	DCA	7.2 (PBS)	11350	–41.15	–18.01	ITC	18
	GCA	7.2 (PBS)	3050	–25.48	–5.59	ITC	18
	TCA	7.2 (PBS)	3061	–18.43	1.47	ITC	18
6	CA	7.2 (PBS)	25315	–34.26	–9.13	ITC	18
	DCA	7.2 (PBS)	30300	–38.13	–12.55	ITC	18
	GCA	7.2 (PBS)	3098	–25.82	–5.89	ITC	18
	TCA	7.2 (PBS)	4659	–14.86	6.08	ITC	18
7	CA	7.2 (PBS)	25850	–23.53	1.65	ITC	18
	DCA	7.2 (PBS)	24785	–27.59	–2.51	ITC	18
	GCA	7.2 (PBS)	4722	–21.22	–0.25	ITC	18
	TCA	7.2 (PBS)	3022	–24.29	–4.43	ITC	18
8	CA	7.2 (PBS)	1726	–31.0	–13.3	ITC	19
	DCA	7.2 (PBS)	2839	–34.8	–14.9	ITC	19
	GCA	7.2 (PBS)	1032	–25.7	–8.5	ITC	19
	TCA	7.2 (PBS)	1003	–26.6	–9.5	ITC	19
9	CA	7.2 (PBS)	2567	–29.3	–9.9	ITC	19
	DCA	7.2 (PBS)	3137	–34.0	–14.0	ITC	19
	GCA	7.2 (PBS)	2898	–31.2	–11.4	ITC	19
	TCA	7.2 (PBS)	2284	–30.0	–10.8	ITC	19
10	CA	7.2 (PBS)	2605	–28.6	–9.1	ITC	19
	DCA	7.2 (PBS)	3813	–33.7	–13.3	ITC	19
	GCA	7.2 (PBS)	3140	–29.6	–9.7	ITC	19
	TCA	7.2 (PBS)	2402	–28.8	–9.5	ITC	19

Hosts	Guests	pH	*Ks*	Δ*H*°	*T*Δ*S*°	Methods	Refs.
11	CA	7.2 (PBS)	2020	–23.2	–4.3	ITC	20
	DCA	7.2 (PBS)	2310	–32.1	–12.9	ITC	20
	GCA	7.2 (PBS)	1110	–23.4	–6.0	ITC	20
	TCA	7.2 (PBS)	1060	–23.1	–5.8	ITC	20
12	CA	7.2 (PBS)	6680	–37.9	–14.5	ITC	20
	DCA	7.2 (PBS)	6770	–46.0	–24.1	ITC	20
	GCA	7.2 (PBS)	1760	–24.9	–6.4	ITC	20
	TCA	7.2 (PBS)	1470	–24.3	–6.2	ITC	20
13	CA	7.2 (PBS)	871	–26.7	–9.9	ITC	21
	DCA	7.2 (PBS)	1087	–33.1	–15.8	ITC	21
	GCA	7.2 (PBS)	428	–28.3	–13.3	ITC	21
	TCA	7.2 (PBS)	391	–25.7	–10.9	ITC	21
14	CA	7.2 (PBS)	8689	–41.7	–19.2	ITC	21
	DCA	7.2 (PBS)	9962	–50.5	–27.9	ITC	21
	GCA	7.2 (PBS)	1105	–30.5	–13.1	ITC	21
	TCA	7.2 (PBS)	809	–26.7	–10.1	ITC	21
15	CA	7.4 (Tris–NaCl)	2510	–7.9	38.6	ITC	23
	DCA	7.4 (Tris–NaCl)	4429	–10.65	34.0	ITC	23
	GCA	7.4 (Tris–NaCl)	1764	–8.2	34.5	ITC	23
	TCA	7.4 (Tris–NaCl)	1399	–8.75	31.0	ITC	23
16	CA	7.4 (Tris–NaCl)	2693	–5.7	46.6	ITC	23
	DCA	7.4 (Tris–NaCl)	6276	–6.8	49.9	ITC	23
	GCA	7.4 (Tris–NaCl)	1958	–7.9	36.6	ITC	23
	TCA	7.4 (Tris–NaCl)	2148	–7.2	39.6	ITC	23
17	CA	7.2 (PBS)	61			ITC	24
	DCA	7.2 (PBS)	774			ITC	24
	GCA	7.2 (PBS)	228			ITC	24
	TCA	7.2 (PBS)	162			ITC	24
18	CA	7.2 (PBS)	11760	–42.70	–19.47	ITC	25
	DCA	7.2 (PBS)	15030	–42.72	–18.87	ITC	25
	GCA	7.2 (PBS)	3870	–25.23	–4.75	ITC	25
	TCA	7.2 (PBS)	2647	–20.99	–1.47	ITC	25

Hosts	Guests	pH	*Ks*	Δ*H*°	*T*Δ*S*°	Methods	Refs.
19	CA	7.2 (PBS)	18965	–32.37	–7.95	ITC	25
	DCA	7.2 (PBS)	22485	–36.48	–11.46	ITC	25
	GCA	7.2 (PBS)	4888	–21.61	–0.56	ITC	25
	TCA	7.2 (PBS)	3755	–19.15	0.7	ITC	25
20	CA	7.2 (PBS)	11850	–33.23	–9.98	ITC	25
	DCA	7.2 (PBS)	13365	–39.57	–16.20	ITC	25
	GCA	7.2 (PBS)	4254	–20.07	0.65	ITC	25
	TCA	7.2 (PBS)	1833	–26.58	–7.96	ITC	25
21	CA	7.2 (PBS)	2216	–25.04	–5.94	ITC	26
	DCA	7.2 (PBS)	2007	–51.92	–33.07	ITC	26
	GCA	7.2 (PBS)	2434	–31.07	–11.74	ITC	26
	TCA	7.2 (PBS)	3478	–23.98	–3.76	ITC	26
22	CA	7.2 (PBS)	2443	–35.60	–16.25	ITC	26
	DCA	7.2 (PBS)	3177	–33.89	–13.90	ITC	26
	GCA	7.2 (PBS)	2811	–34.94	–15.24	ITC	26
	TCA	7.2 (PBS)	2809	–30.37	–10.68	ITC	26
23	CA	7.2 (Tris–HCl)	910			Fluorescence	24
	DCA	7.2 (Tris–HCl)	4320			Fluorescence	24
	GCA	7.2 (Tris–HCl)	4340			Fluorescence	24
	TCA	7.2 (Tris–HCl)	3820			Fluorescence	24
24	CA	7.2 (Tris–HCl)	3290			Fluorescence	24
	DCA	7.2 (Tris–HCl)	7460			Fluorescence	24
	GCA	7.2 (Tris–HCl)	10690			Fluorescence	24
	TCA	7.2 (Tris–HCl)	8710			Fluorescence	24
25	CA	7.4 (Tris–NaCl)	7400	–22.3	–0.2	ITC	27
	DCA	7.4 (Tris–NaCl)	6700	–32.1	–10.2	ITC	27
26	CA	7.4 (Tris–NaCl)	1280	–28.3	–10.5	ITC	27
	DCA	7.4 (Tris–NaCl)	2570	–33.3	–13.8	ITC	27

Hosts	Guests	pH	Ks	ΔH°	TΔS°	Methods	Refs.
27	CA	7.0 (PBS)	1650			Fluorescence	38
	DCA	7.0 (PBS)	2660			Fluorescence	38
28	CA	7.0 (PBS)	588			Fluorescence	38
	DCA	7.0 (PBS)	1520			Fluorescence	38
29	CA	7.0 (PBS)	60.4			Fluorescence	38
	DCA	7.0 (PBS)	1030			Fluorescence	38
30	CA	aqueous solution	–			Fluorescence	36
	DCA	aqueous solution	–			Fluorescence	36
	GCA	aqueous solution	–			Fluorescence	36
31	CA	aqueous solution	–			Fluorescence	36
	DCA	aqueous solution	–			Fluorescence	36
	GCA	aqueous solution	–			Fluorescence	36
32	CA	aqueous solution	–			Fluorescence	36
	DCA	aqueous solution	–			Fluorescence	36
	GCA	aqueous solution	–			Fluorescence	36

PBS: Phosphate Buffer Solution; ITC: Isothermal Titration Calorimetry;

Tris: Tris(hydroxymethyl)aminomethane;

–: The guest-induced variations in the excimer emission are too small for these values to be determined.

Table 1. Complex stability constants (Ks/M^{-1}), enthalpy ($\Delta H°/(kJ \cdot mol^{-1})$), and entropy changes ($T\Delta S°/(kJ \cdot mol^{-1})$) for intermolecular complexation of bile salts with natural β-CD and its mono-modified derivatives in aqueous solution

All the permethylated β-CD derivatives (**17**, **23** and **24**) present the weakest binding ability to CA guest because the cavity of permethylated β-CD possesses a broader hydrophobic region in comparison with **1**, and then permethylated β-CD is more suitable to include bile guests with longer tails (GCA and TCA) than **1** [24]. Moreover, there are similar structures between CA and DCA except for the difference of one hydroxyl in ring B. It is attractive that DCA can be included more tightly by **17**, **23** and **24** than CA. One reasonable explanation is that the absence of one hydroxyl in ring B makes the whole framework of DCA more hydrophobic than CA, and thereby DCA is more suitable to be immersed into the cavity of permethylated β-CDs.

4. Bridged CD series

4.1. Binding modes for bile salts and bridged CD series

4.1.1. Diseleno- and bipyridine-bridged β-CDs

ROESY experiments for the complexes of CDs (**25**, **26**, **33**, and **35**) and DCA have been performed to illustrate the binding modes between the CDs and bile salts [27]. The results show that the bridge linker does not interact with DCA and the bile salt molecule is not cooperatively bound by the two cavities of one dimer molecule. DCA is not included in the cavity of the dimer from the primary side (narrow open), but penetrates slightly into the cavity from the secondary side (wide open) using the side chain and D-ring moiety. For **33** (Figure 6), the A-ring moiety of DCA is simultaneously shallowly included in one of the cavities of another CD to form a liner structure. For monomer **25**, the D-ring moiety of DCA penetrates deep into the cavity of **25** from the secondary side. However, for monomer **26**, DCA is included in the cavity of **26** from the secondary side by its A-ring moiety, differing from other CDs (by D-ring moiety).

To further obtain the information about the binding modes of bile salts with diseleno- and bipyridine-bridged β-CDs, 2D ROESY spectra for typical host–guest pairs have also been determined by Liu et al. [28]. For dimer **35** and CA, the results indicate that the carboxylate side chain and D-ring of CA may penetrate into the CD cavity from the secondary side shallowly and two CA molecules are bound separately into two cavities of **35** from the secondary side, which is consistent with the 1:2 binding stoichiometry (Figure 7a). For dimer **39** and DCA, the results are quite different and show a 1:1 cooperative binding mode. The A-ring of DCA penetrates deeply into one CD cavity of **39**, attributing to the less steric hindrance and higher hydrophobicity of the substituent group on the C-7 position of DCA (Figure 7b). Under the same experiment using DCA as guest, host **38** adopts a different binding mode from **39**. The carboxylate side chain of two DCA molecules deeply penetrates into the CD cavity of **38** from the secondary side separately.

4.1.2. *Oligoethylenediamino-bridged β-CDs*

To obtain the information about the binding modes between bile salts and oligoethylenediamino-bridged β-CD dimers (**42–44**), 2D ROESY spectra for typical host–guest pairs have been determined by Liu et al. [29]. The results of ROESY experiments indicated that the D ring and side-chain of bile salt guest enter one β-CD cavity from the wide opening, and the linker group is partially self-included in the other β-CD cavity (Figure 8).

4.1.3. *Aromatic diamino- and sulfonyldianiline-bridged β-CDs*

From ROESY experiments, Zhao et al. found that the D-ring of CA is wholly included in the CD cavity of **45** from the wide opening, while the side-chain is located near the narrow opening of CD cavity and folded toward the steroid body and the phenyl moiety is not driven out of the CD cavity even after the guest inclusion [30]. Similar binding mode is also observed in other cases of **45**/bile salts complexes.

The binding modes between the aromatic diamino-bridged β-CDs **46–48** and bile salts have also been investigated by Zhao et al. via 2D ROESY experiments and the results show that the D-ring of CA is wholly included in the CD cavity with the wide opening, while the side chain is located near the narrow opening of the CD cavity and is folded toward the steroid body [31]. The phenyl moiety is not driven out of the CD cavity even after the guest inclusion.

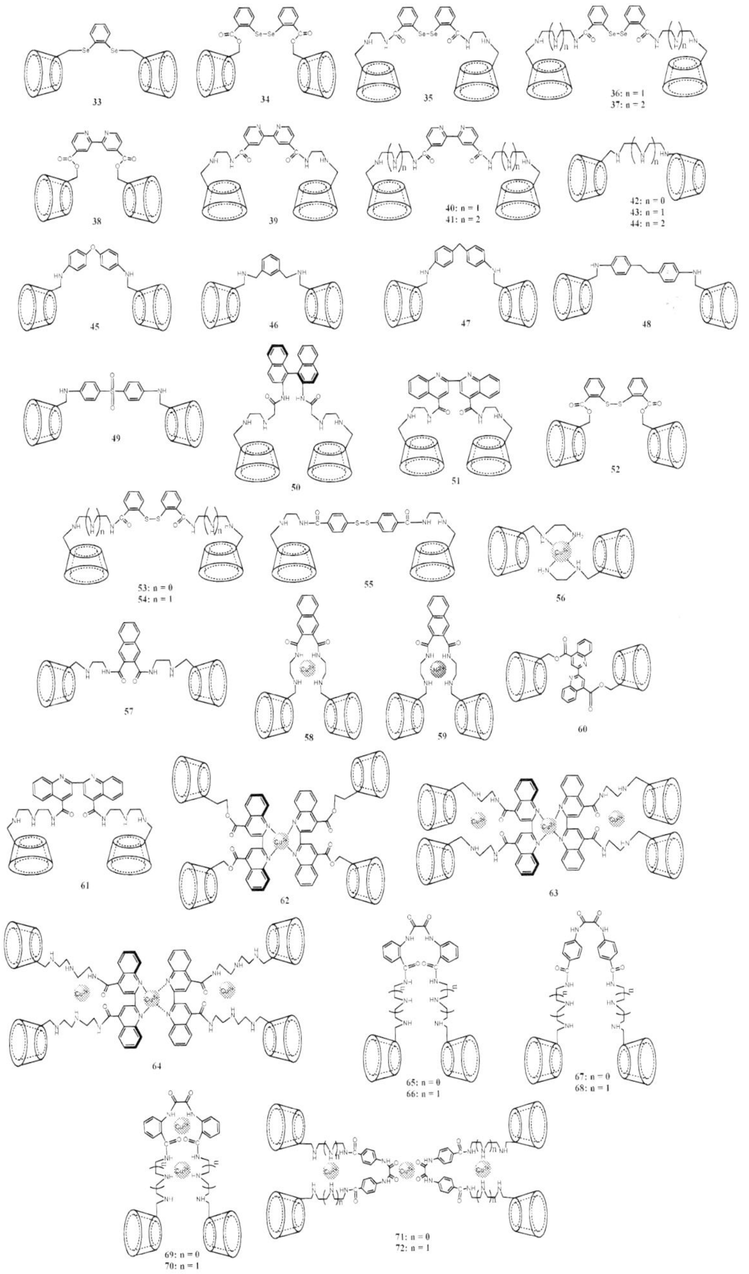

Figure 6. Structures of bridged β-CDs

To obtain the information about the binding modes between bile salts and sulfonyldianiline-bridged β-CD **49**, 2D ROESY spectra for typical host–guest pairs have further been determined by Zhao et al. [32]. The correlation signals, along with the 1:1 binding stoichiometry, jointly indicate a host-linker-guest binding mode between **49** and CA. That is, upon complexation with **49**, the carboxylate tail and the D ring of CA penetrate into one CD cavity of **49** from the wide opening deeply, while the phenyl moiety of the CD linker is partially self-included in the other β-CD cavity. Similar binding modes are also observed in other cases of **49**/bile salt complexes.

4.1.4. Binaphthyl-, biquinoline- and dithio-bridged β-CDs

The binding modes of binaphthyl-, biquinoline- and dithio-bridged β-CDs (**50–55**) and bile salts have been investigated by 2D ROESY experiments in aqueous solution [33]. The results show that CA enters the CD cavity of **53** from the second side of CD with the side chain and D-ring. The side chain with the negative carboxylate group of CA moves toward the positive protonated amino group of **53**. The other binaphthyl-, biquinoline- and dithio-bridged β-CDs/bile salts complexes show a similar binding mode as the complex **53**/CA, with only a slight degree of difference in the depth of guest insertion.

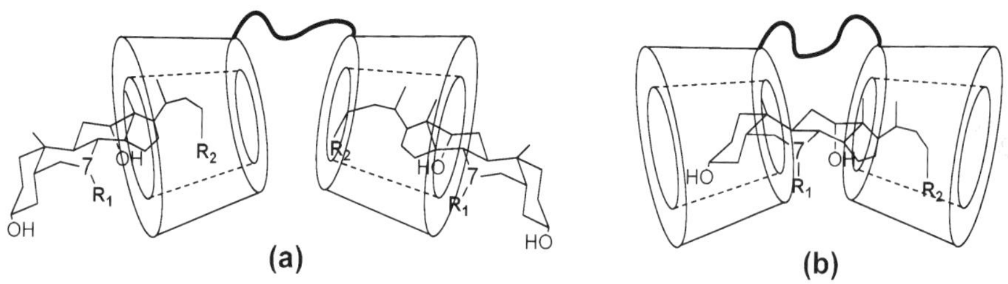

Figure 7. The possible binding modes of **35** with CA (a) and **39** with DCA (b)

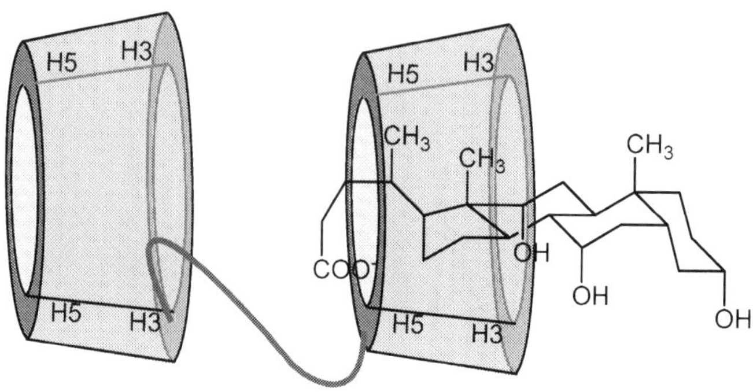

Figure 8. The possible binding mode of **42–44** with CA

4.2. Complexation thermodynamics for bile salts and bridged CD series

4.2.1. Diseleno- and bipyridine-bridged β-CDs

To elucidate the difference in binding behavior between the CD dimer and monomer, two CD dimers (**33** and **35**) and their monomer analogs (**25** and **26**) have been used for titration microcalorimetry with CA and DCA [27]. It is interesting that the results of the thermodynamic measurements show a 1:1 binding stoichiometry for hosts **25**, **26** and **33**, but 1:2 stoichiometry for host **35**. In addition, although the stability constants for the complexation between dimer **33** and the bile salts are much larger than those for monomer **26**, the long-linked dimer **35** unusually displays a lower cavity binding ability than its corresponding monomer **25** upon complexation with both guests CA and DCA. The enhancement of the binding ability of dimer **33** compared to monomer **26** could be ascribed not only to the cooperative binding but also partly to the peculiar self-inclusion conformation of **26** that leads to more unfavorable entropy changes, especially for the **26**–CA pair. For **35**, the two guest molecules are separately and independently included in the two cavities of **35** because the longer linker, especially the ethylenediamino moiety of dimer **35**, makes it possess a relatively large conformational freedom. As the considerable entropy loss cancels the advantage of enthalpy gain, dimer **35** displays relatively weak binding abilities. Both hosts **35** and **25** show similar binding ability for DCA and CA. The reason is that either binding with host **35** or host **25**, the two guest bile salts are included into the cavity of CDs by its D-ring and side-chain moiety, which reduces the influence of the substituent in C7. However, while binding with hosts **33** and **26**, the A-ring moiety participates in the binding process, so the more hydrophobic C7 substituent of DCA makes it bind more strongly with the host CDs, giving the higher binding constants than with CA, especially for host **26**.

Either for diseleno-bridged β-CDs (**34–37**) or for bipyridine-bridged β-CDs (**38–41**), the host–guest stoichiometry changes in the same order, that is, from 1:2 to 1:1 with the increase of spacer length [28]. For diseleno-bridged β-CDs, only **36** and **37** adopt the 1:1 binding mode. However, for bipyridine-bridged β-CDs, only host **38** adopts the 1:2 binding mode; the others all show the 1:1 cooperative binding mode. The thermodynamic results reveal that, with the longest spacer, **37** gives the largest stability constants in all diseleno-bridged β-CDs, while the largest stability constants of bipyridine-bridged β-CDs toward each guest molecule is obtained by the dimers **39** and **40** with the moderate spacer lengths, which suggests that only the CD dimers possessing the proper spacer length can give the perfect cooperative binding toward guests.

For the dimers adopting 1:1 cooperative binding mode, the enthalpy changes are not only the main contribution to the binding process but also the determining factor for the binding abilities [28]. Comparing the diseleno-bridged β-CDs with bipyridine-bridged β-CDs, all of the bipyridine-bridged β-CDs display much stronger binding abilities toward bile salts than corresponding diseleno-bridged β-CDs, which indicate that the presence of rigid spacer favors formation of a relatively fixed binding mode and results in the close contact between two CD cavities and guest molecule, leading to the stronger binding abilities. On the other hand, due to the presence of the bipyridine fragment, the hydrogen bond between the hydroxyl group of the bile salt and the nitrogen atom of bipyridine might also be taken as a plausible explanation for the strong binding abilities of bipyridine-bridged β-CDs as compared with diseleno-bridged β-CDs. Upon complexation with CA and DCA, all dimer hosts adopting a 1:1 binding mode show higher binding abilities than native β-CD **1** due to more favorable enthalpy changes, which perfectly confirms the advantage of cooperative binding of guests by two CD cavities.

4.2.2. Oligoethylenediamino-Bridged β-CDs

1:1 binding stoichiometry is observed for all the complexes between bile salts and oligoethylenediamino-bridged β-CDs (**42–44**) [29]. The inclusion complexation of bile salts with **42–44** is driven by favorable enthalpy changes, accompanied by slight to moderate entropy loss. Interestingly, the enthalpy changes for the inclusion complexation of **42–44** increased, while the entropic changes decreased, with the elongation of the linker group, giving a binding constant **42** > **43** > **44**. The stronger binding of bile salts by the short-linked β-CD dimer is not thermodynamically accomplished by an increase of the originally favorable enthalpy gain, but by a reduction of the unfavorable entropy loss. The short-linked β-CD dimer, with a better size and hydrophobicity match to bile salts, may experience more extensive desolvation upon complexation, and thus exhibits the less unfavorable entropy loss. With the elongation of linker group, the protonated amino group in the linker is located distant from the anionic carboxylate (or sulfonate) tail of bile salt, which consequently weakens the electrostatic interactions between the linker group and bile salt. Moreover, the increase of the number of -NH- fragments in the linker group decrease the hydrophobicity of β-CD dimer to some extent, which is also unfavorable to the hydrophobic interactions between host and guest.

The stability constants of the complexes formed by β-CD dimers **42–44** with bile salts are larger than those of the complexes formed by native β-CD **1** [29]. These enhanced binding abilities of β-CD dimers may be mainly attributed to the cooperative host-linker-guest binding mode between host and guest. The electrostatic interactions between the protonated amino groups in the linker and the anionic carboxylate (or sulfonate) tail of bile salt may strengthen the inclusion complexations of these β-CD dimers with bile salts. Moreover, the hydrogen bond interactions of the hydroxyl groups of β-CD and the -NH- fragments of the oligo(ethylenediamino) linker with the carboxylate (or sulfonate) tail of bile salt also contribute to the enhanced binding abilities of β-CD dimers **42–44**.

Compared with CA, GCA and TCA, DCA possesses a more hydrophobic structure due to the absence of C-7 hydroxyl group, which consequently leads to stronger hydrophobic interactions between host and guest. Therefore, DCA gives the highest binding abilities among the bile salts examined upon complexation with most CDs [29]. Possess more polar side-chains, GCA and TCA show weak binding abilities upon inclusion complexation

complexation with β-CD dimers due to the relatively poor hydrophobic interactions between host and guest.

4.2.3. *Aromatic diamino- and sulfonyldianiline-bridged β-CDs*

The stoichiometry for the inclusion complexation of **45** with bile salts were determined by the continuous variation method and the results showed a 1:1 inclusion complexation between **45** and bile salts [30]. The stability constants for the inclusion complexation of **45** with bile salts are much higher than those values for the native β-CD **1**. These enhanced binding abilities of **45** may be mainly attributed to the cooperative host-linker-guest binding mode between host and guest because the linker group provides some additional binding interactions towards the accommodate guest. Host **45** displays higher binding ability for CA than for DCA due to the hydrogen bond interactions between the 7-hydroxy group of CA and the 2- and 3-hydroxy group of CD. Host **45** shows the weaker binding abilities upon inclusion complexation with GCA and TCA than that of CA and DCA because GCA and TCA are unfavorable to insert into the cavity from the second side of CD cavity with their D ring attributing to the more hydrophilic tail attached to the end of the D ring.

The stoichiometries for inclusion complexation of aromatic diamino-bridged β-CDs **46–48** with bile salts were further determined by the continuous variation method and the results show that all the hosts and guests form 1:1 complexes [31]. β-CD dimers **46–48** also show enhanced binding ability toward bile salts as compared with β-CD **1**. The enhanced binding abilities of aromatic diamino-bridged β-CDs may be mainly attributed to the cooperative host-linker-guest binding mode between host and guest. In addition to the association of the CD cavity with a guest molecule, the linker group provides some additional binding interactions towards the accommodated guest.

Unlike the β-CD **1**, the bridged β-CDs **46–48** show larger binding constants for CA than for DCA [31]. Among them, the host **47** gave the highest stability constant for inclusion complexation with CA. One possible reason for the stronger affinity for CA may involve hydrogen-bond interactions between the 7-hydroxy group of CA and the 2- and 3-hydroxy groups of CD, which subsequently strengthen the host–guest association. Moreover, all the hosts show lower binding ability for complexation with GCA and TCA as compared with complexation with CA and DCA. The highest binding constants towards GCA and TCA are with host **47**. The universally decreased binding ability toward GCA and TCA must be related to structure differences between CA and DCA. Attributing to the more hydrophilic tail, which is attached to the end of the D ring, GCA and TCA are unfavorable for insertion into the cavity from the second side of the β-CD cavity with their D ring.

The binding constants for the complexation of each bile salt by hosts **46–48** increases in the following order: **47** > **48** > **46** [31]. That is, host **47** with a tether of moderate length and rigidity among the β-CD dimers studied is the most suitable for inclusion complexation with bile salts. This may be attributable to the strict size fit between these bile salts and the moderate length-tethered β-CD dimer **47**, which consequently exhibits strong van der Waals and hydrophobic interactions between host and guest.

The stoichiometry for the inclusion complexation of sulfonyldianiline-bridged β-CD **49** with bile salts has also been determined by the "continuous variation" method and the results indicate that all the bile salts can form 1:1 complexes with **49** [32]. Thermodynamically, the binding constants of **49** with bile salts are larger than those of native β-CD **1**. The enhanced binding abilities of **49** may be also mainly attributed to the cooperative host-linker-guest binding mode between host and guest. In addition to the association of the CD cavity with a guest molecule, the linker group provides some additional binding interactions towards the accommodate guest. Distinctly, the binding constant is significantly higher for DCA compared to CA by native β-CD **1**. However, different from native β-CD **1**, sulfonyldianiline-bridged β-CD **49** reverses this binding selectivity, showing larger binding constants for CA than DCA. One possible reason for the stronger affinity for CA may involve H-bond interactions between CA and CD, which subsequently strengthen the host-guest association. Moreover, all the hosts show a weaker binding ability upon complexation with GCA and TCA than with CA and DCA. The universal decreased binding ability toward GCA and TCA must relate to the structure differences from CA and DCA. Attributing to the more hydrophilic tail, which is attached to the end of the D ring, GCA and TCA are unfavorable to insert into the cavity from the second side of β-CD cavity with their D ring. It is worthy to note that the binding ability of **49** is significantly larger for TCA than for GCA, which leads to a relatively strong molecular selectivity.

4.2.4. Binaphthyl-, biquinoline- and dithio-bridged β-CDs

The stoichiometric ratios from the binding patterns for the titrations of steroids with binaphthyl-, biquinoline- and dithio-bridged β-CDs **50–55** fell within the range of 1.8-2.1, which clearly indicates that the majority of the inclusion complexes have a 1:2 stoichiometry of steroids and bridged β-CDs [33]. Thermodynamically, bridged β-CD **52**, possessing a relatively short and rigid tether without amino groups, still gives an enhanced binding ability upon complexation with steroids, except TCA, when compared its one single unit of cavity with that of native β-CD **1**. The enthalpy changes for the inclusion complexation of bridged β-CD **52** with DCA and CA are more negative than that of native β-CD **1**, resulting in the relatively stronger binding. On the other hand, the enthalpy change for the complexation of **52** with DCA is higher than that with CA, which directly contributes to the increased complex stability. It is reasonable that, possessing the more hydrophobic structure due to the absence of C-7 hydroxyl group as compared with CA, DCA is easier to bind into the β-CD cavity than CA, which should lead to the more favorable van der Waals interactions.

All the complexation of aminated bridged β-CDs (**50**, **51**, and **53–55**) toward DCA and CA give more negative enthalpy changes as compared with that of neutral bridged β-CD **52**, validating the contribution of the attractive electrostatic interactions between positively charged protonated amino group of β-CD tethers and negatively charged carboxylate group of DCA and CA [33]. Accompanied with the more exothermic reaction enthalpies, the inclusion complexation of DCA and CA by aminated bridged β-CDs (**50**, **51**, and **53–55**) exhibits more unfavorable entropy changes compared to that for neutral bridged bis(β-CD)

52, which possibly originates from the conformation fixation of host and guest and the rigid complex formation upon complexation.

Mostly, bridged β-CDs **50**, **51**, and **53–55** give the lower binding ability upon complexation with GCA and TCA as compared with the complexation with CA and DCA, which is similar as the complexation of β-CD **1** and bridged β-CD **52** [33]. The universal decreased binding ability toward GCA and TCA must relate to the structure differences from CA and DCA. The more polar side chains at C23 for GCA and TCA remarkably affect their binding thermodynamics.

4.3. Binding modes for bile salts and metallobridged CD series

4.3.1. Metallobridged β-CDs with naphthalenecarboxyl linkers

2D ROESY NMR and circular dichroism spectroscopy experiments for the complexes of bile salts with bridged and metallobridged CDs with naphthalenecarboxyl linkers have been performed by Liu et al. to investigate the binding modes between host and guests [34]. The result of **57**/DCA complex showed that the guest DCA was included in the β-CD cavity with the D-ring and the carboxylic tail located near the narrow opening but the B-ring located near the wide opening and the naphthyl group was excluded from the β-CD cavity upon inclusion complexation. Moreover, the result of 2D ROESY NMR showed that the ethylenediamino moiety of the linker group was also partially self-included in the β-CD cavity from the narrow opening. Similar results were also found in other ROESY experiments of hosts **57** and **59** with bile salts.

4.3.2. Metallobridged β-CDs with biquinoline linkers

2D NMR experiments in D_2O and molecular modeling studies for the complexes of bridged and metallobridged β-CDs with biquinoline linkers and bile salts have been performed by Liu et al. to deduce the binding modes between the bile salts and β-CD dimers [35,36]. The results show that a cooperative "host-tether-guest" binding mode is operative in the association of β-CD dimers with a guest molecule; upon complexation with β-CD dimers, the guest steroid is embedded into one hydrophobic β-CD cavity from the primary side, while the tether group is partly self-included in the other cavity. In the metallobridged β-CDs, the tether group is entirely excluded from the β-CD cavities as a result of metal coordination. This arrangement allows two side groups of the guest molecule to be embedded into the hydrophobic β-CD cavities from the primary side of the β-CD to form a sandwich host–guest inclusion complex.

4.3.3. Metallobridged β-CDs with oxamidobisbenzoyl linkers

^1H ROESY experiments have been performed in D_2O to investigate the binding modes between bridged and metallobridged β-CDs with oxamidobisbenzoyl linkers and bile salts [37]. The results show a "host-linker-guest" binding mode between **66** and CA. That is, upon inclusion complexation with β-CD dimer, the carboxylate tail and the D-ring of CA enter into one CD cavity of **66** from the wide opening, while the linker group of **66** is partially self-included in the other CD cavity (Figure 9a). A similar binding mode is also observed for the inclusion complexation of **66** with DCA.

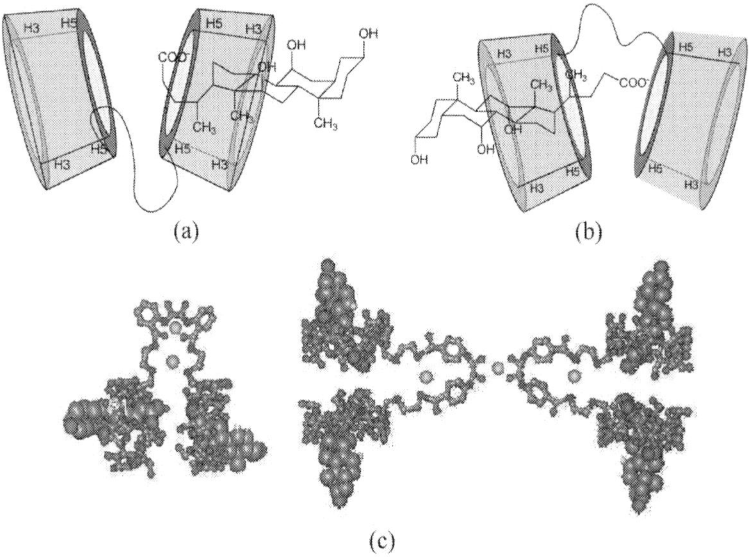

Figure 9. The possible binding modes of **66** (a) and **67** (b) with CA, and the possible binding modes of **69–72** with bile salts

With a shallowly self-included conformation, β-CD dimers **65**, **67**, and **68** show a binding mode different from that of **66**. For example, for **67**/CA complex, the carboxylate tail and D-ring of CA enter the CD cavity from the wide opening, and the carboxylate tail is located close to the linker group. On the other hand, the linker group is mostly moved out from the CD cavity after complexation with CA (Figure 9b). **65**/CA, **65**/DCA, **67**/DCA, **68**/CA, and **68**/DCA complexes show a similar binding mode to the **67**/CA complex.

In the cases of the metallobridged β-CDs **69–72**, the strong electrostatic attraction from the coordinated Cu^{II} ions in the linker group may also favor the penetration of the carboxylate tail of bile salt into the CD cavity through the wide opening. Moreover, the 1:2 or 2:4 binding stoichiometry indicates that each CD cavity of a metallobridged β-CD is occupied by a bile salt (Figure 9c).

4.4. Complexation thermodynamics for bile salts and metallobridged CD series

4.4.1. Metallobridged β-CDs with naphthalenecarboxyl linkers

The interactions between hosts (**27–29**, **57–59**) and bile salts have been studied by Liu and Ueno et al. by the method of fluorescence [34,38]. The results show that all the hosts (**27–29**,

57–59) can form 1:1 complexes with bile salts CA and DCA. Thermodynamically, bridged β-CDs possess much stronger binding abilities compared with mono-modified β-CDs. These enhanced binding abilities should be attributed to cooperative binding of the β-CD cavity and the linker group towards the guest molecule, leading to greatly strengthened van der Waals and hydrophobic interactions between host and guest when compared with mono-modified β-CDs. Furthermore, after metal coordination, the metallobridged bis(β-CD)s **58** and **59** significantly enhance the original binding ability of native β-CD **1**, mono-naphthyl-modified β-CDs **27–29** and even parent bridged β-CD **57**. This enhancement may be subjected to a multiple recognition mechanism of metallobridged β-CDs towards model substrates. On one hand, the coordination of a metal ion to the linker group shortens the effective distance of two β-CD cavities to some extent and thus improves the size-fit degree between host and guest. On the other hand, the electrostatic attraction between the anionic carboxyl group of guest bile salt and the coordinated metal ion of metallobridged β-CD may also favour the host–guest binding to some extent.

4.4.2. Metallobridged β-CDs with biquinoline linkers

The interactions between host **51** and bile salts have been investigated by Liu et al. by the method of fluorescence [35]. The results show that all the bile salts can form 1:1 complexes with **51**. Thermodynamically, the binding constants obtained for CA and DCA are much larger than those reported for mono-modified β-CDs by Ueno et al. under practically the same experimental conditions [38]. This enhancement is probably due to the cooperative binding of the steroids by **51**. The complex stability decreases in the order: DCA > CA > GCA > TCA. The highest affinity for DCA is likely to arise from its more hydrophobic steroid skeleton. Host **51** shows comparable affinities toward CA and GCA, whereas TCA, possessing a highly polar anionic tail gives the lowest binding constant.

The stoichiometry for the inclusion complexation of hosts **60–64** with bile salts has also been determined by Job's method [36]. The results show that the stoichiometry of the inclusion complex formed by the **63**/CA system is likely to be 2:2, with intramolecular complexation. Stoichiometries of 1:1 (for bridged β-CD) or 2:2 (for metallobridged β-CD) were obtained in other similar cases of host–guest inclusion complexation. Thermodynamically, the stability constants of the complexes of bridged β-CDs **51**, **60** and **61** with bile salts are larger than those of the complexes formed by mono-modified β-CDs **27–29** by a factor of about 1.1 to around 200 benefitting from cooperative binding. In addition to inclusion complexation of the guest molecule within one hydrophobic CD cavity, the tether group located near the accommodated guest provides some additional interactions with the guest. In control experiments, the changes in the fluorescence spectra of **30–32** upon addition of guest steroids were too small to allow calculation of the stability constants, which may be attributed to strong self-inclusion of the substituted group preventing penetration of the guest into the CD cavity. Except **60**, the mono- and bridged-β-CDs display higher binding affinities for DCA than for CA. This stronger affinity for DCA is likely to arise from the more hydrophobic steroid skeleton of this compound compared with that of CA. The abilities of both the short-tethered compound **60** and the long-tethered host **61** to bind CA and DCA are unexpectedly limited compared to the binding abilities of mono-modified CDs **27–29** due to the self-inclusion of the tether group for the short-tethered β-CD dimer **60** and the steric hinderance from the relatively large 5-amino-3-azapentyl-2-quinoline-4-carboxyamide fragment on the exterior of the CD cavity for the long-tethered β-CD dimer **61**, respectively.

The metal-ligated oligomeric β-CDs **62–64** have significantly enhanced (around 50–4100 higher) binding affinities for the tested guest molecules compared with those of the mono-modified β-CDs [36]. These results can be explained by considering a mechanism involving an uncommon multiple recognition behavior of metallobridged β-CDs. A metallobridged β-CD affords four hydrophobic binding sites (four CD cavities) and one (or three) metal coordination center(s), which jointly contribute to the cooperative binding of the oligomeric host with the guest molecule upon inclusion complexation. In addition, ligation of a Cu^{II} ion shortens the effective length of the tether to some extent and thus improves the size fit of the host with the guest. The cumulative result of these factors is that the metal-ligated β-CD oligomers have binding abilities around 6–200 times higher than those of their parent bridged β-CDs.

4.4.3. Metallobridged β-CDs with oxamidobisbenzoyl linkers

The stoichiometry for the inclusion complexation of hosts **65–72** with bile salts has been determined by Job's method [37]. The results indicate that each of the Job's plots for the inclusion complexation of **65–68** with bile salts shows a maximum at a β-CD dimer molar fraction of 0.5, confirming the 1:1 binding stoichiometry between host and guest. For the inclusion complexation of metallobridged β-CDs **69–72** with bile salts, however, each of the Job's plots shows a maximum at a bridged β-CD unit molar fraction of 0.33, which indicates 1:2 stoichiometry between each bridged β-CD unit and guest. The metallobridged β-CDs **69** and **70** may only bind two bile salts to form a stoichiometric 1:2 inclusion complex. However, the metallobridged β-CDs **71** and **72** may adopt intramolecular 2:4 stoichiometry upon inclusion complexation with bile salts. Thermodynamically, the binding constants for the inclusion complexation of CA and DCA with bridged β-CDs **65–68** are higher than the K_S values reported for the inclusion complexation of these bile salts with native or mono-modified β-CDs [33,38]. These enhanced binding abilities highlight the inherent advantage of the cooperative "host-linker-guest" binding mode of bridged β-CDs **65–68**. In addition to the association of the CD cavity with a guest molecule, the linker group provides some additional binding interactions towards the accommodated guest.

The bile salts CA and DCA are better bound by bridged β-CD **65**, which possesses the shortest linker group, than by the long-linked bridged β-CDs [37]. This may be attributable to the strict size-fit between these bile salts and the short-linked bridged β-CD **65**, which consequently gives strong van der Waals and hydrophobic interactions between host and guest.

Significantly, metallobridged β-CDs **69–72** show greatly enhanced binding abilities with regard to the bridged β-CDs **65–68** [37]. These significant enhancements in the binding abilities may be attributable to a more complicated multiple recognition mechanism involving the cooperative binding of several CD cavities, conformation adjustment by the metal coordination, and additional binding interactions between the metal-coordinated linker group and the accommodated guest molecules.

Except for **66**, all of the hosts examined display higher binding abilities for CA than for DCA [37]. One possible reason for the stronger affinities for CA may involve hydrogen bond interactions between the 7-hydroxy group of CA and the 2- and 3-hydroxy groups of CD.

4.4.4. Metallobridged β-CDs with aminated linkers

The microcalorimetric experiments of β-CD **1** and modified β-CDs **2**, **5**, **6**, **7** with bile acids showed typical titration curves of 1:1 complex formation [18]. However, metallobridged β-CD **56** displays a 1:2 host–guest binding stoichiometry. Thermodynamically, as compared with native β-CD **1**, most oligo(ethylenediamino)-β-CDs **2**, **5** **6**, and their Cu^{II} complexes **7** and **56** show enhanced molecular binding abilities and guest selectivities towards bile acids. The inclusion complexation of bile acids with native β-CD **1** and their derivatives (**2**, **5**, **6**, **7**, and **56**) is absolutely driven by favorable enthalpy changes accompanying with moderate unfavorable or slightly favorable entropy changes. The favorable enthalpy change is attributed to the dominant contribution of the hydrophobic interactions. Meanwhile, the unfavorable entropy given by most of the complexes is due to the decrease of rotational and structural freedom upon complex construction.

As compared with native β-CD **1**, **5** shows increased binding abilities toward negatively charged bile acids guest molecules, which should be mainly due to the additional electrostatic interactions between the amino tether moiety of hosts and anionic carboxylate or sulfonate tail of guests [18]. Moreover, β-CD dimer **56** shows a larger binding constant upon inclusion complexation with CA and DCA than **5**. This may be attributed to that the coordination of copper ion onto the amino tether of CD affords a more positive charged environment as compared with its precursor **5**. Compared with **5**, host **6** also shows stronger binding abilities toward guest molecules. However, the introduction of copper actually decreases the original binding ability of **6** towards DCA and gives comparable stability constant upon complexation with CA. All the hosts, including native β-CD **1** and modified β-CDs **2**, **5**, **6**, **7**, and **56**, show the weaker binding abilities upon inclusion complexation with GCA and TCA than those of CA and DCA. It is also found that complexes stabilities enhance with the extended length of spacer for the same guest except for **2**/CA to **5**/CA. It is reasonable to believe that the increased stability is due to the enlarged hydrogen binding interactions.

Hosts	Guests	pH	*Ks*	Δ*H*°	*T*Δ*S*°	Methods	Refs.
33	CA	7.4 (Tris–NaCl)	6860	–30.5	–8.6	ITC	27
	DCA	7.4 (Tris–NaCl)	9700	–37.0	–14.3	ITC	27
35	CA	7.4 (Tris–NaCl)	2700	–27.1	–7.5	ITC	27
	DCA	7.4 (Tris–NaCl)	3300	–35.7	–15.7	ITC	27
36	CA	7.4 (Tris–NaCl)	4100	–24.9	–4.3	ITC	28
	DCA	7.4 (Tris–NaCl)	5400	–35.0	–13.7	ITC	28
37	CA	7.4 (Tris–NaCl)	5030	–29.1	–8.0	ITC	28
	DCA	7.4 (Tris–NaCl)	6100	–40.2	–18.6	ITC	28
39	CA	7.4 (Tris–NaCl)	12700	–32.4	–9.0	ITC	28
	DCA	7.4 (Tris–NaCl)	12400	–45.4	–22.0	ITC	28
40	CA	7.4 (Tris–NaCl)	12400	–25.5	–2.2	ITC	28
	DCA	7.4 (Tris–NaCl)	13100	–31.9	–8.3	ITC	28
41	CA	7.4 (Tris–NaCl)	6800	–25.4	–3.5	ITC	28
	DCA	7.4 (Tris–NaCl)	7500	–35.2	–13.1	ITC	28
42	CA	7.2 (PBS)	21065	–32.8	–8.1	ITC	29
	DCA	7.2 (PBS)	22780	–42.7	–17.9	ITC	29
	GCA	7.2 (PBS)	9707	–23.0	–0.3	ITC	29
	TCA	7.2 (PBS)	6848	–22.4	–0.5	ITC	29
43	CA	7.2 (PBS)	5868	–39.3	–17.7	ITC	29
	DCA	7.2 (PBS)	7017	–47.4	–25.5	ITC	29
	GCA	7.2 (PBS)	4031	–25.8	–5.2	ITC	29
	TCA	7.2 (PBS)	2947	–26.9	–7.1	ITC	29
44	CA	7.2 (PBS)	5606	–41.0	–19.6	ITC	29
	DCA	7.2 (PBS)	5511	–52.1	–30.7	ITC	29
	GCA	7.2 (PBS)	2847	–26.5	–6.9	ITC	29
	TCA	7.2 (PBS)	1877	–29.0	–10.3	ITC	29
45	CA	7.2 (PBS)	27050			Fluorescence	30
	DCA	7.2 (PBS)	22930			Fluorescence	30

Hosts	Guests	pH	*Ks*	Δ*H*°	*T*Δ*S*°	Methods	Refs.
	GCA	7.2 (PBS)	7200			Fluorescence	30
	TCA	7.2 (PBS)	17610			Fluorescence	30
46	CA	7.2 (PBS)	15310			Fluorescence	31
	DCA	7.2 (PBS)	8790			Fluorescence	31
	GCA	7.2 (PBS)	3040			Fluorescence	31
	TCA	7.2 (PBS)	4100			Fluorescence	31
47	CA	7.2 (PBS)	39900			Fluorescence	31
	DCA	7.2 (PBS)	31880			Fluorescence	31
	GCA	7.2 (PBS)	10400			Fluorescence	31
	TCA	7.2 (PBS)	5360			Fluorescence	31
48	CA	7.2 (PBS)	25930			Fluorescence	31
	DCA	7.2 (PBS)	14330			Fluorescence	31
	GCA	7.2 (PBS)	7950			Fluorescence	31
	TCA	7.2 (PBS)	4590			Fluorescence	31
49	CA	7.2 (PBS)	26200			Fluorescence	32
	DCA	7.2 (PBS)	10140			Fluorescence	32
	GCA	7.2 (PBS)	3150			Fluorescence	32
	TCA	7.2 (PBS)	7730			Fluorescence	32
50	CA	7.2 (PBS)	7351	–33.0	–10.9	ITC	33
	DCA	7.2 (PBS)	5504	–42.7	–21.4	ITC	33
	GCA	7.2 (PBS)	5936	–15.1	6.4	ITC	33
	TCA	7.2 (PBS)	3058	–24.5	–4.6	ITC	33
51	CA	7.2 (PBS)	5559	–49.3	–27.9	ITC	33
	CA	7.2 (PBS)	11300			Fluorescence	35
	DCA	7.2 (PBS)	8372	–48.1	–25.7	ITC	33
	DCA	7.2 (PBS)	21730			Fluorescence	35
	GCA	7.2 (PBS)	2979	–18.1	4.2	ITC	33
	GCA	7.2 (PBS)	11040			Fluorescence	35
	TCA	7.2 (PBS)	4441	–19.7	1.1	ITC	33
	TCA	7.2 (PBS)	6040			Fluorescence	35
52	CA	7.2 (PBS)	5039	–28.2	–7.1	ITC	33
	DCA	7.2 (PBS)	7900	–31.6	–9.4	ITC	33
	GCA	7.2 (PBS)	4262	–21.5	–0.8	ITC	33
	TCA	7.2 (PBS)	1975	–22.0	–3.2	ITC	33
53	CA	7.2 (PBS)	10700	–30.6	–7.6	ITC	33
	DCA	7.2 (PBS)	8912	–38.1	–15.6	ITC	33
	GCA	7.2 (PBS)	5689	–22.7	–1.3	ITC	33
	TCA	7.2 (PBS)	2762	–37.3	–17.6	ITC	33
54	CA	7.2 (PBS)	9899	–37.5	–14.7	ITC	33
	DCA	7.2 (PBS)	11150	–39.9	–16.8	ITC	33
	GCA	7.2 (PBS)	4061	–23.5	–2.9	ITC	33
	TCA	7.2 (PBS)	2502	–20.2	0.8	ITC	33

Hosts	Guests	pH	*Ks*	Δ*H*°	*T*Δ*S*°	Methods	Refs.
55	CA	7.2 (PBS)	6196	–39.3	–17.6	ITC	33
	DCA	7.2 (PBS)	10325	–39.4	–16.5	ITC	33
	GCA	7.2 (PBS)	2891	–23.3	–3.5	ITC	33
	TCA	7.2 (PBS)	2189	–20.0	–0.9	ITC	33
56	CA	7.2 (PBS)	13330	–29.77	–6.23	ITC	18
	DCA	7.2 (PBS)	12065	–34.02	–10.72	ITC	18
	GCA	7.2 (PBS)	2925	–23.36	–3.58	ITC	18
	TCA	7.2 (PBS)	2478	–21.46	–2.09	ITC	18
57	CA	7.4 (Tris–HCl)	10540			Fluorescence	34
	DCA	7.4 (Tris–HCl)	12400			Fluorescence	34
58	CA	7.4 (Tris–HCl)	15500			Fluorescence	34
	DCA	7.4 (Tris–HCl)	15700			Fluorescence	34
59	CA	7.4 (Tris–HCl)	31400			Fluorescence	34
	DCA	7.4 (Tris–HCl)	95900			Fluorescence	34
60	CA	aqueous solution	5380			Fluorescence	36
	DCA	aqueous solution	2790			Fluorescence	36
	GCA	aqueous solution	–			Fluorescence	36
61	CA	aqueous solution	3380			Fluorescence	36
	DCA	aqueous solution	3710			Fluorescence	36
	GCA	aqueous solution	–			Fluorescence	36
62	CA	aqueous solution	30500			Fluorescence	36
	DCA	aqueous solution	529000			Fluorescence	36
	GCA	aqueous solution	1745000			Fluorescence	36
63	CA	aqueous solution	196000			Fluorescence	36
	DCA	aqueous solution	283700			Fluorescence	36
	GCA	aqueous solution	13000			Fluorescence	36
64	CA	aqueous solution	246000			Fluorescence	36
	DCA	aqueous solution	51000			Fluorescence	36
	GCA	aqueous solution	891000			Fluorescence	36
65	CA	7.2 (Tris–HCl)	18500			Fluorescence	37
	DCA	7.2 (Tris–HCl)	12200			Fluorescence	37
66	CA	7.2 (Tris–HCl)	8130			Fluorescence	37
	DCA	7.2	–			Fluorescence	37

Hosts	Guests	pH	Ks	ΔH°	TΔS°	Methods	Refs.
		(Tris–HCl)					
67	CA	7.2 (Tris–HCl)	11900			Fluorescence	37
	DCA	7.2 (Tris–HCl)	11500			Fluorescence	37
68	CA	7.2 (Tris–HCl)	8820			Fluorescence	37
	DCA	7.2 (Tris–HCl)	1870			Fluorescence	37
69	CA[a]	7.2 (Tris–HCl)	5.73×10^7			Fluorescence	37
	DCA[a]	7.2 (Tris–HCl)	2.03×10^7			Fluorescence	37
70	CA[a]	7.2 (Tris–HCl)	9.93×10^7			Fluorescence	37
	DCA[a]	7.2 (Tris–HCl)	3.47×10^7			Fluorescence	37
71	CA[a]	7.2 (Tris–HCl)	3.96×10^7			Fluorescence	37
	DCA[a]	7.2 (Tris–HCl)	3.78×10^7			Fluorescence	37
72	CA[a]	7.2 (Tris–HCl)	2.95×10^7			Fluorescence	37
	DCA[a]	7.2 (Tris–HCl)	6.2×10^6			Fluorescence	37

PBS: Phosphate Buffer Solution; ITC: Isothermal Titration Calorimetry;

Tris: Tris(hydroxymethyl)aminomethane;

–: The guest-induced variations in the fluorescence intensities are too small for these values to be determined.

[a]: Unit of *Ks* is in M^{-2}.

Table 2. Complex stability constants (Ks/M^{-1}), enthalpy ($\Delta H°/(kJ \cdot mol^{-1})$), and entropy changes ($T\Delta S°/(kJ \cdot mol^{-1})$) for intermolecular complexation of bile salts with bridged β-CDs in aqueous solution

5. Conclusion

In conclusion, the binding modes, binding abilities, and molecular selectivities of four typical bile salts (CA, DCA, GCA, and TCA) upon complexation with CDs and their derivatives are summarized in this chapter from thermodynamic viewpoints. Generally, native and mono-modified CDs display relatively limited binding ability towards guest molecules, probably because of weak interactions between hosts and guests, which would result in a relative small negative enthalpy change, and then, a relative weak binding. However, bridged and metallobridged CDs have greatly enhanced the binding abilities in relation to the parent CDs, owing to a multiple recognition mechanism, which would lead to a relative large negative enthalpy change, and then a strong binding. This summary of the binding modes and thermodynamic data for the complexation of bile salts with CDs and their derivatives is quite important to improve the understanding of molecular recognition mechanism in supramolecular systems and further guide the design and synthesis of new supramolecular systems based on different kinds of CDs in the future.

Author details

Yu Liu* and Kui Wang
Department of Chemistry, State Key Laboratory of Elemento-Organic Chemistry, Nankai University, Tianjin, P. R. China

Acknowledgement

This work was supported by the 973 Program (2011CB932502) and NSFC (20932004), which are gratefully acknowledged.

6. References

[1] Szejtli J (1998) Introduction and General Overview of Cyclodextrin Chemistry. Chem. Rev. 98: 1743–1753.
[2] Connors K.A (1997) The Stability of Cyclodextrin Complexes in Solution. Chem. Rev. 97: 1325–1357.
[3] Wallimann P, Marti T, Fürer A, Diederich F (1997) Steroids in Molecular Recognition. Chem. Rev. 97: 1567–1608.
[4] Khan A.R, Forgo P, Stine K.J, D'Souza V.T (1998) Methods for Selective Modifications of Cyclodextrins. Chem. Rev. 98: 1977–1996.
[5] Rekharsky M.V, Inoue Y (1998) Complexation Thermodynamics of Cyclodextrins. Chem. Rev. 98: 1875–1917.
[6] Danielsson H, Sjövall J (1985) Sterols and Bile Acids. Amsterdam: Elsevier.
[7] Holm R, Hartvig R.A, Nicolajsen H.V, Westh P, Østergaard J (2008) Characterization of the Complexation of Tauro- and Glyco-Conjugated Bile Salts with γ-Cyclodextrin and 2-Hydroxypropyl-γ-Cyclodextrin Using Affinity Capillary Electrophoresis. J. Incl. Phenom. Macrocycl. Chem. 61: 161–169.
[8] Holm R, Shi W, Hartvig R.A, Askjær S, Madsen J.C, Westh P (2009) Thermodynamics and Structure of Inclusion Compounds of Tauro- and Glyco-Conjugated Bile Salts and β-Cyclodextrin. Phys. Chem. Chem. Phys. 11: 5070–5078.
[9] Schönbeck C, Westh P, Madsen J.C, Larsen K.L, Städe L.W, Holm R (2010) Hydroxypropyl-Substituted β-Cyclodextrins: Influence of Degree of Substitution on the Thermodynamics of Complexation with Tauroconjugated and Glycoconjugated Bile Salts. Langmuir 26: 17949–17957.

* Corresponding Author

[10] Holm R, Madsen J.C, Shi W, Larsen K.L, Städe L.W, Westh P (2011) Thermodynamics of Complexation of Tauro- and Glyco-Conjugated Bile Salts with Two Modified β-Cyclodextrins. J. Incl. Phenom. Macrocycl. Chem. 69: 201–211.

[11] Schönbeck C, Westh P, Madsen J.C, Larsen K.L, Städe L.W, Holm R (2011) Methylated β-Cyclodextrins: Influence of Degree and Pattern of Substitution on the Thermodynamics of Complexation with Tauro- and Glyco-Conjugated Bile Salts. Langmuir 27: 5832–5841.

[12] Jong M.R.de, Engbersen J.F.J, Huskens J, Reinhoudt D.N (2000) Cyclodextrin Dimers as Receptor Molecules for Steroid Sensors. Chem. Eur. J. 6: 4034–4040.

[13] Casas-Solvas J.M, Ortiz-Salmerón E, Fernández I, García-Fuentes L, Santoyo-González F, Vargas-Berenguel A (2009) Ferrocene–β-Cyclodextrin Conjugates: Synthesis, Supramolecular Behavior, and Use as Electrochemical Sensors. Chem. Eur. J. 15: 8146–8162.

[14] Casas-Solvas J.M, Quesada-Soriano I, Carreño-Gázquez D, Giménez-Martínez J.J, García-Fuentes L, Vargas-Berenguel A (2011) β-Cyclodextrin Dimers Linked through Their Secondary Faces with Rigid Spacer Arms as Hosts for Bile Salts. Langmuir 27: 9729–9737.

[15] Cabrer P.R, Alvarez-Parrilla E, Meijide F, Seijas J.A, Rodríguez Núñez E, Vázquez Tato J (1999) Complexation of Sodium Cholate and Sodium Deoxycholate by β-Cyclodextrin and Derivatives. Langmuir 15: 5489–5495.

[16] Cabrer P.R, Álvarez-Parrilla E, AlSoufi W, Meijide F, Rodríguez Núñez E, Vázquez Tato J (2003) Complexation of Bile Salts by Natural Cyclodextrins. Supramol. Chem. 15: 33–43.

[17] Liu Y, Yang Y.-W, Cao R, Song S.-H, Zhang H.-Y, Wang L.-H (2003) Thermodynamic Origin of Molecular Selective Binding of Bile Salts by Aminated β-Cyclodextrins. J. Phys. Chem. B 107: 14130–14139.

[18] Liu Y, Cao R, Yang Y.-W (2006) Selective Binding Thermodynamics of Bile Acids by Oligo(ethylenediamino)-β-Cyclodextrins and Their Copper (II) Complexes. J. Incl. Phenom. Macrocycl. Chem. 56: 183–191.

[19] Liu Y, Zhang Q, Guo D.-S, Zhuang R.-J, Wang L.-H (2008) Thermodynamics of Complexes between Nucleobase-Modified β-Cyclodextrins and Bile Salts. Thermochim. Acta 470: 108–112.

[20] Wang H, Cao R, Ke C.-F, Liu Y, Wada T, Inoue Y (2005) Diastereomeric Molecular Recognition and Binding Behavior of Bile Acids by L/D-Tryptophan-Modified β-Cyclodextrins. J. Org. Chem. 70: 8703–8711.

[21] Chen Y, Li F, Liu B.-W, Jiang, B.-P, Zhang H.-Y, Wang L.-H, Liu Y (2010) Thermodynamic Origin of Selective Binding of β-Cyclodextrin Derivatives with Chiral Chromophoric Substituents toward Steroids. J. Phys. Chem. B 114: 16147–16155.

[22] Singh A.P, Cabrer P.R, Alvarez-Parrilla E, Meijide F, Tato J.V (1999) Complexation of 6-Deoxy-6-(aminoethyl)Amino-β-Cyclodextrin with Sodium Cholate and Sodium Deoxycholate. J. Incl. Phenom. Macrocycl. Chem. 35: 335–348.

[23] Ollila F, Pentikäinen O.T, Forss S, Johnson M.S, Slotte J.P (2001) Characterization of Bile Salt/Cyclodextrin Interactions Using Isothermal Titration Calorimetry. Langmuir 17: 7107–7111.

[24] Liu Y, Shi J, Guo D.-S (2007) Novel Permethylated β-Cyclodextrin Derivatives Appended with Chromophores as Efficient Fluorescent Sensors for the Molecular Recognition of Bile Salts. J. Org. Chem. 72: 8227–8234.

[25] Liu Y, Zhao Y.-L, Yang E.-C, Zhang H.-Y (2004) Molecular Recognition Thermodynamics of Steroids by Novel Oligo(aminoethylamino)-β-Cyclodextrins Bearing Anthryl: Enhanced Molecular Binding Ability by Co-inclusion Complexation. J. Incl. Phenom. Macrocycl. Chem. 50: 3–11.

[26] Li N, Chen Y, Zhang Y.-M, Li Z.-Q, Liu Y (2010) Binding Thermodynamics of Quinolinyl-Modified β-Cyclodextrins with Bile Salts. SCIENTIA SINICA Chimica 40: 1355–1362.

[27] Liu Y, Li L, Zhang H.-Y, Yang Y.-W, Ding F (2004) Correlation between Thermodynamic Behavior and Structure in the Complexation of Modified β-Cyclodextrins and Bile Salts. Supramol. Chem. 16: 371–379.

[28] Liu Y, Li L, Chen Y, Yu L, Fan Z, Ding F (2005) Molecular Recognition Thermodynamics of Bile Salts by β-Cyclodextrin Dimers: Factors Governing the Cooperative Binding of Cyclodextrin Dimers. J. Phys. Chem. B 109: 4129–4134.

[29] Liu Y, Kang S, Chen Y, Cao R, Shi J (2007) Thermodynamics of Molecular Recognition of Bile Salts by 3,6′-(Oligoethylenediamino-Bridged) β-Cyclodextrin Dimers. Comb. Chem. High T. Scr. 10: 350–357.

[30] Zhao Y, Gu J, Chi S.-M, Yang Y.-C, Zhu H.-Y, Wang Y.-F, Liu J.-H, Huang R (2008) Aromatic Diamino-bridged Bis(β-Cyclodextrin) as Fluorescent Sensor for the Molecular Recognition of Bile Salts. Bull. Korean Chem. Soc. 29: 2119–2124.

[31] Zhao Y, Gu J, Chi S.-M, Yang Y.-C, Zhu H.-Y, Huang R, Wang Y.-F (2009) Bis(β-Cyclodextrin)s Linked with an Aromatic Diamine as Fluorescent Sensor for the Molecular Recognition of Bile Salts. J. Solution Chem. 38: 417–428.

[32] Zhao Y, Yang Y.-C, Chi S.-M, Shi H, Zhao Y, Zhu H.-Y, Li Q.-L, Wang Y.-F (2010) Fluorescence Sensing and Selective Binding of a Novel 4,4′-Sulfonyldianiline-Bridged Bis(β-Cyclodextrin) for Bile Salts. Helv. Chim. Acta 93: 999–1011.

[33] Liu Y, Yang Y.-W, Yang E.-C, Guan X.-D (2004) Molecular Recognition Thermodynamics and Structural Elucidation of Interactions between Steroids and Bridged Bis(β-Cyclodextrin)s. J. Org. Chem. 69: 6590–6602.

[34] Liu Y, Wu H.-X, Chen Y, Chen G.-S (2009) Molecular Binding Behaviours of Bile Salts by Bridged and Metallobridged Bis(β-Cyclodextrin)s with Naphthalenecarboxyl Linkers. Supramol. Chem. 21: 409–415.

[35] Liu Y, Song Y, Wang H, Zhang H.-Y, Wada T, Inoue Y (2003) Selective Binding of Steroids by 2,2′-Biquinoline-4,4′-dicarboxamide-Bridged Bis(β-cyclodextrin): Fluorescence Enhancement by Guest Inclusion. J. Org. Chem. 68: 3687–3690.

[36] Liu Y, Song Y, Chen Y, Li X.-Q, Ding F, Zhong R.-Q (2004) Biquinolino-Modified β-Cyclodextrin Dimers and Their Metal Complexes as Efficient Fluorescent Sensors for the Molecular Recognition of Steroids. Chem. Eur. J. 10: 3685–3696.

[37] Liu Y, Yu H.-M, Chen Y, Zhao Y.-L (2006) Synthesis of Bridged and Metallobridged Bis(β-cyclodextrin)s Containing Fluorescent Oxamidobisbenzoyl Linkers and Their Selective Binding towards Bile Salts. Chem. Eur. J. 12: 3858–3868.

[38] Ikeda H, Nakamura M, Ise N, Oguma N, Nakamura A, Ikeda T, Toda F, Ueno A (1996) Fluorescent Cyclodextrins for Molecule Sensing: Fluorescent Properties, NMR Characterization, and Inclusion Phenomena of *N*-Dansylleucine-Modified Cyclodextrins. J. Am. Chem. Soc. 118: 10980–10988.

Chapter 13

Thermodynamics of Hydration in Minerals: How to Predict These Entities

Philippe Vieillard

Additional information is available at the end of the chapter

http://dx.doi.org/10.5772/51567

1. Introduction

Many papers have suggested that several specific "types" of H_2O exist. These have been labelled variously as " zeolitic", " loosely held", "structural", "crystal", "tightly bound" and "external" [1-3]. These labels suggest both the differing apparent energies of H_2O as well as their differing apparent structural roles. Three useful distinctions can be made about H_2O in compounds:

1. H_2O that varies in content as a continuous function of temperature and pressure
2. H_2O that changes discontinuously at a unique temperature for a given pressure
3. H_2O that is sorbed to external surfaces

The first type (continuously varying H_2O) is characteristic of the zeolites and clay minerals and considers that most of the zeolites and clay minerals lose or gain H_2O in response to small changes in temperature and pressure over an extended temperature, relative humidity or the nature of cations in exchanged sites.

The second type of H_2O is similar in nature to that found in hydrates of salts and considers that dehydration occurs over narrow temperature intervals in some compounds. Some zeolites, like analcime and laumontite, do not exchange water at room temperature. This type can be called "hydrate" H_2O and has a specific position in the crystal structure. Hydroxides are excluded from this chapter because the H_2O molecule is not identified but is only virtually present as a hydroxyl OH.

The third type of H_2O is externally sorbed to the crystal and may be referred to as external. This type is present in quantities much smaller than the H_2O present within the structure of any zeolite grain size. In the clay minerals, whose specific surfaces are greater than those of zeolites, the water located in inter-particle spaces, in inter-aggregate spaces and at the surface of uncharged clays minerals indicates that a special status of H_2O may be different in the regimes of adsorption or of capillary condensation, and this is not considered in this chapter.

2. Definition of thermodynamic entities of hydration

In evaluating the thermodynamic quantities, it is important to define the standard state chosen for a given constituent. In the present study, the standard state for water is chosen as pure water at a temperature of 25°C with a vapour pressure of P_o, and for the minerals, the dry compound is at a temperature of 25°C with a vapour pressure of P=0.

The reaction of hydration may be expressed by the following reaction:

$$X_{(c)} + n * H_2O_{(l)} \rightarrow X - nH_2O_{(c)} \tag{1}$$

If we consider any standard thermodynamic entity of hydration $\Xi^{\circ}_{hyd,298}$ (enthalpy $\Delta H^{\circ}_{hyd,298}$ free energy $\Delta G^{\circ}_{hyd,298}$, entropy $\Delta S^{\circ}_{hyd,298}$, heat capacity $Cp_{Hyd.}$ and molar volume $V_{Hyd.}$) for a given mineral having n bounded molecules, the entity of hydration is related to the total formation entity $\Xi^{\circ}_{f,298}$ (enthalpy $\Delta H^{\circ}_{f,298.15}$ free energy $\Delta G^{\circ}_{f,298.15}$, entropy S°_{298}, heat capacity $Cp_{298.15}$ and molar volume V_{298}) of the mineral by the following:

$$\Xi^{o}_{hyd,298} = \Xi^{o}_{f,298}\left(X - nH_2O\right) - \Xi^{0}_{f,298}\left(X\right) - n * \Xi^{o}_{f,298}(H_2O_{liq}) \tag{2}$$

where $\Xi^{o}_{f,298}\left(X - nH_2O\right)$ and $\Xi^{0}_{f,298}\left(X\right)$ are the thermodynamic entity of formation of the mineral in the hydrated and dehydrated states, respectively, and $\Xi^{0}_{f,298}\left(H_2O_{liq}\right)$ stands for the thermodynamic entity of formation of bulk water. The unit of the entity $\Xi^{o}_{hyd,298}$ is based per n moles of bound water or per mole of an anhydrous compound X.

The standard thermodynamic parameters of the hydration water can be calculated if the thermodynamic entities of formation (enthalpy $\Delta H^{o}_{f,298.15}$ free energy $\Delta G^{o}_{f,298.15}$, entropy S°_{298}, heat capacity $Cp_{298.15}$. and molar volume $V_{298.}$) of the anhydrous and hydrous end members are available in thermochemical tables or calculated from calorimetric measurements. However, the thermodynamic properties of the hydration water can be estimated by considering the hypothetical intra-crystalline reaction:

$$X_{(c)} + n * H_2O_{(hydr)} \rightarrow X - nH_2O_{(c)} \tag{3}$$

where $H_2O_{(hyd)}$ represents the hydration water. The thermodynamics of the intra-crystalline reaction can be written as follows:

$$\Xi^{o}_{r,298} = \Xi^{o}_{f,298}\left(X - nH_2O\right) - \Xi^{0}_{f,298}\left(X\right) - n * \Xi^{o}_{f,298}(H_2O_{hyd.}) \tag{4}$$

By assuming that the standard thermodynamic entities of the intra-crystalline reaction are equal to zero, *i.e.*, $\Xi^{o}_{r,298} = 0$, the standard values of the molar thermodynamic entity of the hydration water, $\Xi^{0}_{f,298}\left(H_2O_{Hyd}\right)$, is finally obtained from eqs (2) and (4):

The values of $\alpha_{A^{z+}}$ and β for the different salts of cation $A^{z_1^+}$ are given in table 3 with the respective standard deviations and % errors of the predicted enthalpy of formation of the hydrated salts.

An example is given by considering the data of all of the sodium salts displayed in Figure 1A and shows the great variation of the enthalpy of hydration having the same number as the crystalline water. The different straight lines are obtained from Eq. (10) with values of $\alpha_{A^{z+}}$ and β for Na^+ and are plotted for the different values of the number of crystalline water molecules and the nature of the crystalline salts ($\Delta_H(B^{z_2^-})A^{z_1^+}$). For a given anhydrous salt (same value of $\Delta_H(B^{z_2^-})A^{z_1^+}$), the enthalpy of hydration per mole of water appears to not be constant but decreases when the number of molecules of the crystalline water increases. Figure 1B displays the enthalpy of crystalline water versus the nature of the anhydrous salt showing the great disparity and shows that the enthalpy of the crystalline water in all of the sodic salts is more negative than those of the hexagonal ice. When the number of crystalline water molecules of any salt increases, the enthalpy of formation of the crystalline water decreases from large values (NaOH. H_2O; $\Delta H^\circ_{f,298}$ H_2O= -309.25 kJ mol^{-1}) to values close to those of ice (Na_2Se. 16 H_2O; $\Delta H^\circ_{f,298}$ = -297.22 kJ mol^{-1}) and are different from the average value for 61 Na salts ($\Delta H^\circ_{f,298}$ = -297.74 kJ mol^{-1}).

Cations	$\alpha_{A^{z+}}$	β	Nb. of data	% error[1]	Std dev. (kJ/mol H_2O)
Li^+	-0.277	-0.341	19	0.34	± 3.58
Na^+	-0.294	1.759	61	0.29	± 3.44
K^+	-0.115	-9.168	15	0.66	± 10.70
Rb^+	-0.360	-6.769	6	0.29	± 3.38
Cs^+	-0.577	18.954	4	0.48	± 6.35
Be^{+2}	-0.485	-3.068	7	0.25	± 3.48
Mg^{+2}	-0.312	-10.708	21	0.43	± 5.39
Ca^{+2}	-0.270	-3.892	25	0.44	± 4.64
Sr^{+2}	-0.349	-1.894	20	0.32	± 5.93
Ba^{+2}	-0.327	-4.637	19	0.80	± 5.31
Zn^{2+}	-0.211	-16.461	12	0.44	± 6.57
Cd^{2+}	-0.322	2.266	8	0.33	± 4.29
Cu^{+2}	-0.433	-1.791	11	0.65	± 4.22
Ni^{2+}	-0.176	-18.056	10	0.46	± 2.81
Co^{2+}	-0.260	-2.296	11	0.36	± 3.60
Fe^{2+}	-0.136	-17.806	6	0.18	± 1.58
Mn^{2+}	-0.238	-5.144	15	0.25	± 6.14
Cr^{+2}	-0.478	31.459	3	0.28	± 1.77
UO_2^{+2}	-0.247	-7.945	18	0.35	± 4.80

1: the % error weighs the difference (in %) between the measured and predicted enthalpy of the formation of the hydrated salts.

Table 3. Values of $\alpha_{A^{z+}}$, β, number of data, % error and statistical error per moles of crystalline water.

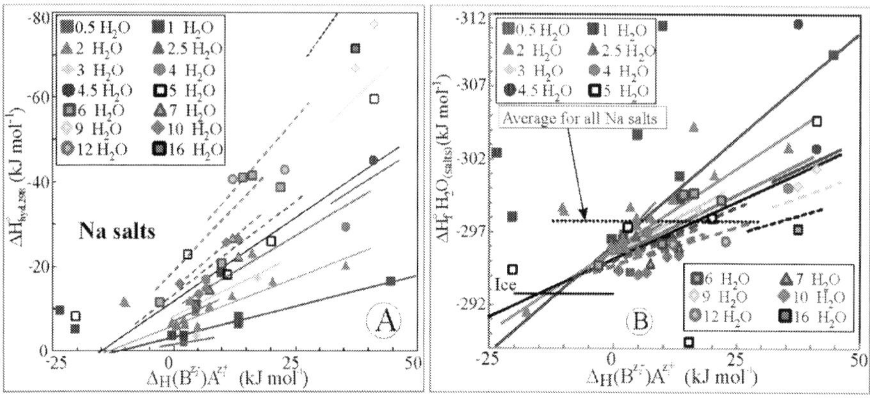

Figure 1. A – Hydration enthalpy of sodium salts versus the parameter $\Delta_H(B^{z_2^-})A^{z_1^+}$; B – Enthalpy of the formation of the crystalline water versus the parameter $\Delta_H(B^{z_2^-})A^{z_1^+}$.

Zeolites: The anhydrous compound X (reaction 1) is a zeolite with the following formula: $\left(\sum_{i=1}^{i=n_c} A_i\right)\left(Al_{n_{Al}} Fe_{n_{Fe}} Si_{n_{Si}}\right)O_N$, in which Al, Fe and Si represent the framework with different channels occupied by various exchangeable cations present, which are represented by A^{z+}. A compilation of the average hydration enthalpies per mole of water of 145 diversely originating zeolites measured using different technical methods [54] provided 76 data points from the transposed-temperature drop calorimetry (TTDC), 57 data points from the immersion calorimetry (IC), 6 data points from the phase equilibriums (PE), 5 data points from the gas-adsorption calorimetry (GAC) and 3 data points from the hydrofluoric acid calorimetry (HF).

From the selected average enthalpies of hydration of the different zeolites with their chemical formulas (145 data points), a regression equation for the enthalpy of hydration per mole of water is proposed [54]:

$$\Delta H_{Hyd-W} = -\left(Al/Si\right) * e^{\left[a + b * WP\right]} \tag{11}$$

where coefficients a and b are

$$a = 4.664 - 2.787 * \left(Al/\left(Al + Si\right)\right) + 0.056 * FD_{anh.} - 0.003 * \left(\Delta_H O^{=} site\ A\right) \tag{12}$$

$$b = -1.52 + 0.0047 * \left(\Delta_H O^{=} site\ A\right) * \left(Al/Si\right) \tag{13}$$

Coefficients of Eqs. (11) (12) and (13), such as the ratio Al/Si, Al/(Al+Si), $FD_{anh.}$, $\Delta_H O^{=}(site\ A)_{(aq)}$ and WP, represent the extra-framework charge, the ratio of tetrahedral substitution, the framework density of the anhydrous zeolite, the average cation electronegativity in the exchange site and the intracrystalline water porosity, respectively. All of these parameters can be calculated from the chemical formula and the unit cell volume for any zeolite.

The framework density (FD) [55, 56] represents the number of tetrahedral atoms per 1000 Å^3 and are obtained as:

$$FD_{anh.} = \left[\left(n_{Al} + n_{Fe} + n_{Si} \right) / Unit\ cell\ Vol \right] * 1000 \tag{14}$$

The electronegativity of site A^{z+}, defined by $\Delta_H O^=$ (site A), represents the weighed average of $\Delta_H O^= M^{z+}(aq)$ of n_c different cations in the exchanged site A:

$$\Delta_H O^=(site.A) = \frac{\sum_{i=1}^{i=n_c} n_i * x_i . \left(\Delta_H O^= M_{aq}^{z_i +} \right)}{x_A} \tag{15}$$

The number of oxygens balancing site Mz+ (in extra-framework sites) is then:

$$x_A = \sum_{i=1}^{i=n_c} n_i * x_i \tag{16}$$

The parameter ΔHO= Mz+ (aq) characterises the electronegativity for a given cation M^{z+} and is defined as the difference between the enthalpy of formation of the corresponding oxides $\Delta H_f^{\circ}\left(M_i O_{x_i}\right)_{(c)}$ and the enthalpy of formation of the corresponding aqueous cation $\Delta H_f^{\circ}\left(M_i^{z_i +}\right)_{(aq)}$:

$$\Delta_H O^= M_{aq}^{z_i +} = \frac{1}{x_i} \left[\Delta H_f^{\circ}\left(M_i O_{x_i}\right)_{(c)} - \Delta H_f^{\circ}\left(M_i^{z_i +}\right)_{(aq)} \right] \tag{17}$$

where z is the charge of the cation M^{z+} and x is the number of oxygen atoms combined with one atom of M in the oxide (x = z/2), such that the difference in Eq. (17) refers to one oxygen atom. A set of internally consistent values was presented by Vieillard & Mathieu [54] and is given in Table 4.

The total intracrystalline pore volume, WP, was introduced by Barrer [57] for determining the volume of liquid water that can be recovered thorough the outgassing of the fully hydrated zeolite. Assuming that the unit cell volume of zeolitic water is the same as that of liquid water ($V_{u.c.\,H_2O}$ = 29.89 Å^3), during the hydration-dehydration processes, the water porosity (WP) can be calculated as the volume of liquid water in 1Å^3 of crystal and is expressed as follows:

$$WP = \frac{29.89 * n_{H_2O}}{\left(V_{u.c.\,hydr.}\right)} \tag{18}$$

where $V_{u.c.\ hyd.}$ represents the unit cell volume of a hydrated zeolite containing n_{H2O} water molecules. Assuming that the number of water molecules n_{H2O} varies from 0 to $n_{H2O\ max}$, the unit cell volume of a zeolite can be directly related to the number of water molecules. Although such variations in the unit -cell volume do occur, let us assume a linear variation in the unit-cell volume as a function of the number of water molecules:

$$V_{u.c.hyd} = V_{u.c.anh.} + k * n_{H_2O} \tag{19}$$

where $V_{u.c.\ anh.}$ represents the unit-cell volume of the anhydrous zeolite. Parameter k weights the variation in the unit-cell volume between an anhydrous zeolite and a fully hydrated zeolite per one water molecule. This parameter can be calculated from the available unit-cell volumes of anhydrous and fully hydrated zeolites. Thus, knowledge of the number of water molecules and the unit-cell volume of anhydrous and hydrated zeolites is required for the calculation of the water porosity:

$$WP = \frac{29.89 * n_{H_2O}}{\left(V_{u.c.anh.} + k * n_{H_2O}\right)} \tag{20}$$

Ions	$\Delta_H O^= M^{z+}(aq)$ (kJ mol^{-1})	Ions	$\Delta_H O^= M^{z+}(aq)$ (kJ mol^{-1})
$Li^+(aq)$	-40.90	$Ca^{+2}(aq)$	-92.1
$Na^+(aq)$	65.80	$Mg^{+2}(aq)$	-134.6
$K^+(aq)$	141.00	$Fe^{+2}(aq)$	-181.58
$Rb^+(aq)$	163.24	$Mn^{+2}(aq)$	-164.4
$Cs^+(aq)$	170.0	$Zn^{+2}(aq)$	-197.07
$Ba^{+2}(aq)$	-15.6	$La^{+3}(aq)$	-124.8
$Sr^{+2}(aq)$	-40.4	$H^+(aq)$	-285.83

Table 4. Parameter $\Delta_H O^= M^{z+}$ (aq) of selected cations [54].

By eliminating 9 erroneous data points [54], the regression coefficient for 136 data is R2= 0.880, with a standard error of ± 3.46 kJ mol-1 for all data, regardless of the nature of the experimental data. Figure 2 displays the predicted enthalpy of hydration of zeolitic water calculated from Eqs. (11) (12) and (13) versus the experimental enthalpy of hydration coming from the different technical measurements. Regression coefficients and standard errors are different within the two main groups: R2= 0.88 and std. err. = ± 3.41 kJ mol-1 for TTDC data (67 data points), R2= 0.658 and std err. = ± 2.66 kJ mol-1 for the "IC" data (57 data points).

When Al/Si and Al/(Al+Si) = 0, the hypothetical integral enthalpy of water in zeolites obtained by the extrapolation of Eq. (11) is 0, which corresponds to the enthalpy of hydration of water in cordierite (2.2 ± 1.6 kJ mol^{-1} from Carey [58]). Unlike zeolites, the water in cordierite is not coordinated by cations, so the molecular environment of H_2O in cordierite is similar to that of H_2O in water.

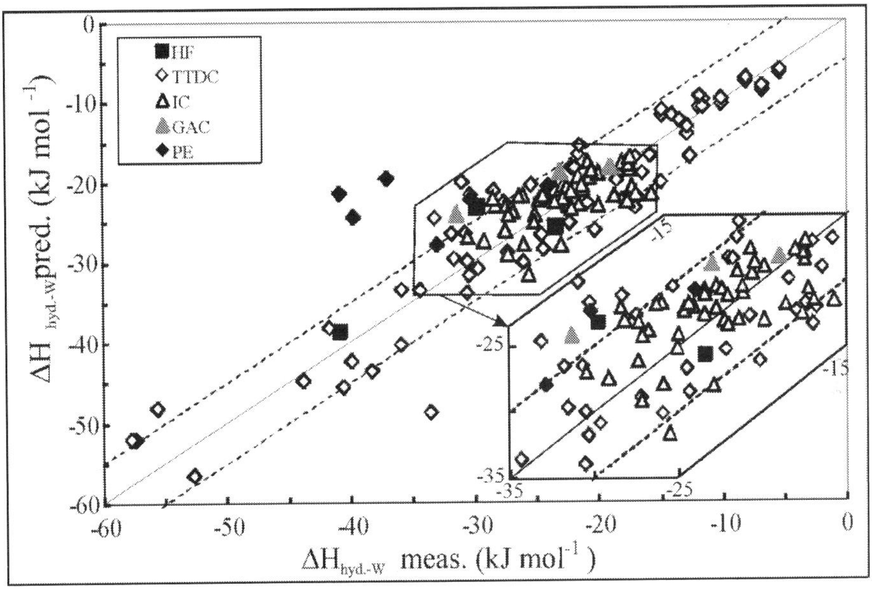

Figure 2. Predicted enthalpy of the hydration of zeolitic water calculated from Eq. (11) versus the experimental enthalpy of hydration using different technical measurements.

A detailed computation of the enthalpy of hydration is given as an example for three natural clinoptilolites (-Ca, -Na and –K) with the following chemical formula given by [22]:

- Clinoptilolite-Na: $(Ca_{0.12}Mg_{0.29}K_{0.75}Na_{5.21})(Al_{6.78}Fe_{0.06}Si_{29.2})O_{72.05}-21.3H_2O$
- Clinoptilolite-K: $(Ca_{0.13}Mg_{0.25}K_{5.84}Na_{0.27})(Al_{6.85}Fe_{0.1}Si_{29})O_{71..9}-18.5H_2O$
- Clinoptilolite-Ca: $(Ca_{2.34}Mg_{0.57}K_{0.9}Na_{0.18})(Al_{6.7}Fe_{0.017}Si_{29})O_{71..8}-21.9H_2O$

The values of the ratio Al/Si yield a constant value of 0.235. The electronegativity of the exchangeable site A^{z+} for clinoptilolite-Na is, for example, the average electronegativity of four cations (Na^+, K^+, Ca^{2+} and Mg^{2+} (table 4)) as follows:

$$\Delta_H O^{=}(site.A) = \frac{2.607^*\left(\Delta_H O^{=}Na^{+}_{aq}\right) + 0.375^*\left(\Delta_H O^{=}K^{+}_{aq}\right) + 0.12^*\left(\Delta_H O^{=}Ca^{2+}_{aq}\right) + 0.29^*\left(\Delta_H O^{=}Mg^{2+}_{aq}\right)}{3.39} \quad (21)$$

i.e., $\Delta_H O^{=}(site.A)$ =51.42 kJ mol^{-1}.

The unit-cell volume of the anhydrous clinoptilolite–K (V u.c.) is 2019.51 Å^3 [59]. This means that there are 35.95 tetrahedral atoms (6.95+0.1+29) in 2019.51 Å^3; which indicates that $FD_{anh.}$= 17.8 tetrahedral atoms in 1000Å^3. The unit-cell volume of the hydrated clinoptilolite-K (at saturation or fully hydrated) with 18.5 moles of zeolitic waters is 2089.50 Å^3 [54], which is slightly greater than the anhydrous form and allows the setting of the relationship of water porosity versus the number of hydration water moles:

$$WP = \frac{29.89 * n_{H_2O}}{\left(2019.51 + 3.78 * n_{H_2O}\right)} \qquad (22)$$

From these previous examples, the parameters requested for the prediction of the enthalpy of hydration are displayed in table 5 for the three clinoptilolites (-Na, -K and –Ca).

Zeolite minerals	V u.c. anh. (Å^3)	Vu.c. hydr. (Å^3)	k	$\Delta_H O^= $ Site A kJmol^{-1}	FD. Anh.	WP	Al/Si
Clinoptilolite -Na	2000.00 [1]	2140.00 [1]	6.35	51.39	18.02	0.31	0.23
Clinoptilolite -K	2019.51 [2]	2089.50 [2]	3.78	109.16	17.80	0.26	0.24
Clinoptilolite -Ca	2019.51 [2]	2089.50 [2]	3.20	-64.60	17.78	0.31	0.23

1 - [60]; 2 - [59]

Table 5. Unit-cell volume of hydrated and anhydrous clinoptilolite, factor k (eq. 20), mean $\Delta_H O^= A^{z+}$ (site A), framework density and ratio Al/Si for 3 clinoptilolites (-Na, K and –Ca).

From the values given in table 5, the enthalpy of hydration of the zeolitic water, ΔH_{Hyd-W}, can be computed with Eqs. (11), (12) and (13) and are close to experimental values from [22] (Table 6).

The model of the computation of the enthalpy of hydration of the hydration water represents a very useful tool and contributes to the knowledge of enthalpies of formation of hydrated zeolites from anhydrous ones [61].

A useful method for avoiding complications at the outset of a thermodynamic analysis is to emphasise the H_2O rather than the zeolite structure. Using such an approach, the equilibriums can be considered as the equilibrium between the H_2O in the fluid-vapour phase and the H_2O in the zeolite. According to the definition of equilibrium, the chemical potential of H_2O in the vapour phase must equal the chemical potential of the H_2O component in the zeolite. Any series of measurements of the amount of H_2O in zeolite at a known fugacity (or partial pressure of H_2O under ideal gas conditions) provides the basis for the thermodynamic description of the system. The most elegant approach to develop a thermodynamic formulation is to know the partial molar enthalpy of hydration, $\Delta\overline{H}$, and the partial molar entropy $\Delta\overline{S}$. A constant partial molar enthalpy of hydration indicates an ideal mixing of the H_2O.

Zeolite minerals	ΔH_{Hyd-W} predicted (kJ mol^{-1} H_2O)	ΔH_{Hyd-W} measured (kJmol^{-1} H_2O)	References
Clinoptilolite.-Na	-22.20 -22.20	-30.17	[22]
Clinoptilolite -K	-20.24	-23.97	[22]
Clinoptilolite-Ca	-29.03	-32.87	[22]

Table 6. Predicted and experimental enthalpy of hydration, ΔH_{Hyd-W} of clinoptilolites –Na, -K and –Ca.

The bulk enthalpy of hydration or the integral enthalpy of hydration can be used as the product of the enthalpy of hydration of the water times the number of zeolitic water:

$$\Delta\tilde{H}_{Hyd-Z} = \left(\Delta H_{Hyd-W}\right) * n_{H_2O} = \left[-\left(Al/_{Si}\right) * e^{\left(a + b * WP\right)}\right] * n_{H_2O} \tag{23}$$

As the integral hydration enthalpy $\Delta\tilde{H}_{Hyd-Z}$ is the integral function of the partial molar enthalpy hydration from $n_{H_2O} = 0$ (corresponding to an anhydrous zeolite) to a maximum value of $n_{H_2O\max}$ (corresponding to the fully hydrated zeolite), the partial molar enthalpy of hydration relative to the liquid water, $\Delta\overline{H}$, which is the derivative function of $\Delta\tilde{H}_{Hyd-Z}$, becomes:

$$\Delta\tilde{H}_{Hyd-Z} = \frac{1}{n_{H_2O}} \int_{n=0}^{n=n_{H_2O}} \Delta\overline{H} * dn \tag{24}$$

$$\Delta\overline{H} = \left[-\left(Al/_{Si}\right) * e^{\left(a + b * \left(\frac{29.89 * n_{H_2O}}{\left(V_{u.c.anh.} + k * n_{H_2O}\right)}\right)\right)}\right] * \left(1 + \frac{29.89 * b * V_{u.c.anh.} * n_{H_2O}}{\left(V_{u.c.anh.} + k * n_{H_2O}\right)^2}\right) \tag{25}$$

The fractional water content can be obtained if the maximum number of zeolitic waters is known at the saturation state and is equal to:

$$\theta = \frac{n_{H_2O}}{n_{H_2O,\max.}} \tag{26}$$

From values of table 5, the partial molar enthalpy of clinoptilolites K, Na and Ca can be calculated versus the fractional water content and plotted in Figure 3 with the experimental partial molar enthalpies measured from immersion calorimetry [62] and from thermogravimetry [22]. The modelled values of the partial molar enthalpy for the three clinoptilolites appear to be closer to the data from the immersion calorimetry than from thermogravimetry. The second important point is the fact that the calculated and experimental partial molar enthalpy of hydration has a similar behaviour.

The partial molar enthalpy of hydration per mole of H_2O increases smoothly from low water content to high θ and indicates that the H_2O in clinoptilolites occupies a continuum of energetic states. This is not the case for chabazite-Ca [23], which exhibits three energetically distinct types of H_2O.

This shows that, within different zeolites with the same exchangeable cation, the partial molar enthalpy as a function of the hydration degree may display different functions. This is the reason why the fundamental relationship verifying the enthalpy of hydration by means of the chemical composition and the accurate knowledge of the unit-cell parameters of anhydrous and hydrated zeolites needs to be improved with new data of hydration enthalpy of the partially hydrated zeolites on the one hand and a better fit of the effective water porosity as a function of the hydration degree on the other hand. Then, the thermodynamic description of the hydration-dehydration process can be modelled as a function of pressure and temperature with the contribution of predicted enthalpies of hydration and entropies of hydration.

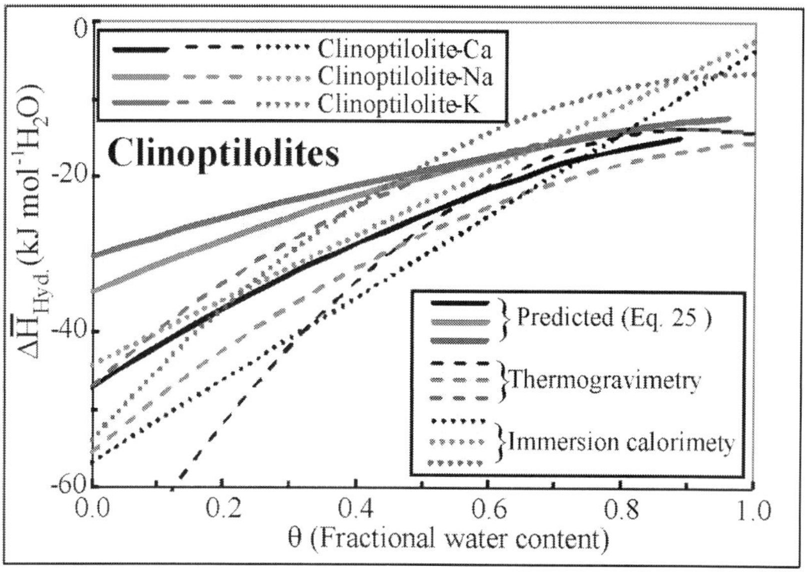

Figure 3. Partial molar enthalpy of hydration of K-, Na- and Ca- clinoptilolites derived from a model in [54], immersion calorimetry [62] and thermogravimetry results [22]

Clay minerals:

Unlike zeolites, smectites are clearly not inert phases, as the particles size increases with the relative humidity [15]. Adsorbed water is distributed throughout the interlayer space, the outer surfaces of particles and the open pores space in the sample. [15-17] provided measured values of the surface covering waters for both the hydration and dehydration reactions of a set of eight homo-ionic SWy-1 montmorillonites saturated by alkali and alkaline-earth cations from their BET specific surface area (Table 1) and their basal spacing variations with relative humidity. To quantify the effective amount of water involved in the hydration reaction, the amount of poral H_2O must be quantified and subtracted from the total amount of H_2O taken up by the clay sample. The number of external surface water molecules (expressed in mmol/g dry clay) can be expressed as a function of relative humidity [63]. There are numerous papers about the adsorption-desorption isotherms performed on various clays minerals, but few are devoted to the acquisition of enthalpy of hydration-dehydration on SWy-1 [15-17, 64] [18, 19, 30] [20] [34] [24, 26, 65] [25] and on vermiculites [21].

All recent predictive models of the hydration of smectites are based on an approach that uses the solid solution initiated by Ransom & Helgeson [66]. The hydration of a smectite is considered through the following reaction between hydrated and dehydrated end-members:

$$Smectite + n_m * H_2O_{(l)} \rightarrow Smectite - n_m * H_2O_{(i)} \quad (27)$$

where n_m represents the maximal number of moles of water that can be included in the smectite on the basis of a half-cell (i.e., $O_{10}(OH)_2$). One considers the interlayer water $H_2O(i)$ (where subscript i stands for interlayer) as a variable weighing the hydration ratio. The amount of interlayer water in a smectite is proportional to the mole fraction of hydrated end-member: $x*H_2O= n_m*x_{hs}=(1-n_m)*x_{as}$; to the mole fraction of the hydrous end-member, x_{hs} or to the anhydrous end-members x_{as}. Ransom & Helgeson [66] considered the system hydrated smectite – anhydrous smectite + bulk water as a strictly regular binary solid solution by considering the free energy only. Thus, the excess free energy of mixing can be expressed as follows:

$$G_{xs} = x_{as} * (1 - x_{as}) * W_G \quad (28)$$

where W_G is an excess mixing constant. The integral Gibbs free energy of hydration is:

$$\Delta\tilde{G}_{Hyd} = x_{hs} * \Delta G^{\circ}_{Hyd,298} + x_{hs} * (1 - x_{hs}) * W_G \quad (29)$$

in which $\Delta G^{\circ}_{Hyd,298}$ is the standard free energy of hydration. The previous parameter and W_G are determined only from adsorption isotherms by assuming a maximal number of moles of interlayer water fixed to 4.5 moles. As there are no experimental measurements of heat of adsorption, the standard enthalpy of hydration, $\Delta H^{\circ}_{Hyd,298}$, is determined from $\Delta G^{\circ}_{Hyd,298}$ by assuming a constant value of S°_{298} $H_2O_{(i)}$ equal to 55.02 J K^{-1} mol^{-1} [67]. From calculated $\Delta H^{\circ}_{Hyd,298}$, the standard enthalpy of formation of the interlayer water, $\Delta H^{\circ}_{f,298}(H_2O)_i$, is given in table 7.

Vidal et al. [68-69] have assumed that a smectite could be considered strictly regular solid solutions between four end-members with 0.7 H_2O, 2 H_2O, 4 H_2O, and 7 H_2O. Those compositions correspond to four different hydrated states (with 0, 1, 2, and 3 water sheets, respectively). For any strictly regular solid solution, the integral enthalpy of hydration is expressed as the following:

$$\Delta\tilde{H}_{Hyd} = x_{hs} * \Delta H^{\circ}_{Hyd,298} + x_{hs} * (1 - x_{hs}) * W_H \quad (30)$$

with

$$\Delta H^0_{hyd,298} = n_m * \left[\Delta H^0_{f,298}\left(H_2O_{(i)}\right) - \Delta H^0_{f,298}\left(H_2O_{liq}\right)\right] \quad (31)$$

The hydration enthalpy is retrieved from data of [24] for montmorillonite –Na and –K and derived from differential heats of adsorption obtained from the measurements of the heat of immersion [18, 20] for Na-, Ca- and Mg- montmorillonite. Thus, an integral enthalpy of hydration $\Delta H^{\circ}_{f,298}(H_2O)_i$ is provided for each of the four solid solutions with n_m equal to 0.7, 2, 4 and 7 and is given in table 7. Vieillard et al. [63] consider the hydration of a smectite to be an asymmetrical regular solid solution between anhydrous and hydrated smectite.

Units (kJ mol^{-1})	H_2O	Cs	Rb	K	Na	Ba	Sr	Ca	Mg
$\Delta H^{\circ}_{f,298}(H_2O)_i$ [1]	4.5	-289.93	-290.23	-290.45	-291.25	-294.14	-294.57	-294.86	-295.71
$\Delta H^{\circ}_{f,298}(H_2O)_i$ [2]	0-0.7			-295.50	-299.71			-301.43	-302.86
W_H [2]				0.	0.			0.	0.
$\Delta H^{\circ}_{f,298}(H_2O)_i$ [2]	0-2			-293.58	-295.90			-298.60	-300.05
W_H [2]				-10.	-10.			-10.	-10.
$\Delta H^{\circ}_{f,298}(H_2O)_i$ [2]	0-4			-291.96	-293.83			-296.25	-297.62
W_H [2]				-10.	-10.			-10.	-10.
$\Delta H^{\circ}_{f,298}(H_2O)_i$ [2]	0-7			-291.05	-292.33			-293.86	-294.69
W_H [2]				-10.	-10.			-10.	-10.
$\Delta H^{\circ}_{f,298}(H_2O)_i$ [3]	5.5	-287.47	-288.01	-288.16	-290.01	-292.10	-292.65	-294.72	-296.56
W_{H_1} [3]		16.0	19.0	15.0	8.0	-60.0	-55.0	-60.0	-49.0
W_{H_2} [3]		-90.0	-95.0	-68.0	-33.0	-75.0	-90.0	-100.0	-129.0

1 –[66]; 2 – [68]; 3 - [63].

Table 7. Standard thermodynamic enthalpy $\Delta H^{\circ}_{f,298}(H_2O)_i$ (in kJ mol^{-1}) of interlayer water and Margules interaction parameters for the system H_2O – smectite.

The integral hydration enthalpy is obtained by the following relationships:

$$\Delta \tilde{H}_{Hyd} = x_{hs} * \Delta H^{\circ}_{Hyd,298} + x_{hs} * (1 - x_{hs}) * \left[W_{H_1} * x_{hs} + W_{H_2} * (1 - x_{hs}) \right] \quad (32)$$

in which $\Delta H^{\circ}_{Hyd,298}$ and Margules parameters W_{H_1} and W_{H_2} are determined by a minimisation procedure applied to the difference between computed and calorimetric integral enthalpies on eight homoionic montmorillonite SWy-1 [15-17] and given in table 7. The maximum number of moles of interlayer water was set to 5.5 moles. Figure 4A displays the comparison of integral enthalpy of hydration of a smectite- Na computed from the three models developed previously with different experimental values.

The curve 1 in Figure 4A (yellow dotted line) comes from model of Ransom & Helgeson [66], is linear and is obtained from the following equation of integral enthalpy of hydration:

$$\Delta \tilde{H}_{Hyd,298} = -24.39 * x_{hs} \quad (33)$$

where the coefficient -24.39 is computed from $\Delta H^{\circ}_{f,298}(H_2O)_i$ given in Table 7. The curves 2b, 2c and 2d (green dotted lines) displayed in Figure 4A come together from the model of Vidal et al. [68] and are graphically represented by four truncated lines with their respective equations:

$$0 \text{ layer } \left(0 < x_{hs} < 0.7\right):\ \Delta\tilde{H}_{Hyd,298} = -9.716 * x_{hs} \tag{34}$$

$$1 \text{ layer } \left(0 < x_{hs} < 2.0\right):\ \Delta\tilde{H}_{Hyd,298} = -20.14 * x_{hs} + x_{hs} * \left(1 - x_{hs}\right) * 10.0 \tag{35}$$

$$2 \text{ layers } \left(0 < x_{hs} < 4.0\right):\ \Delta\tilde{H}_{Hyd,298} = -32.0 * x_{hs} + x_{hs} * \left(1 - x_{hs}\right) * 10.0 \tag{36}$$

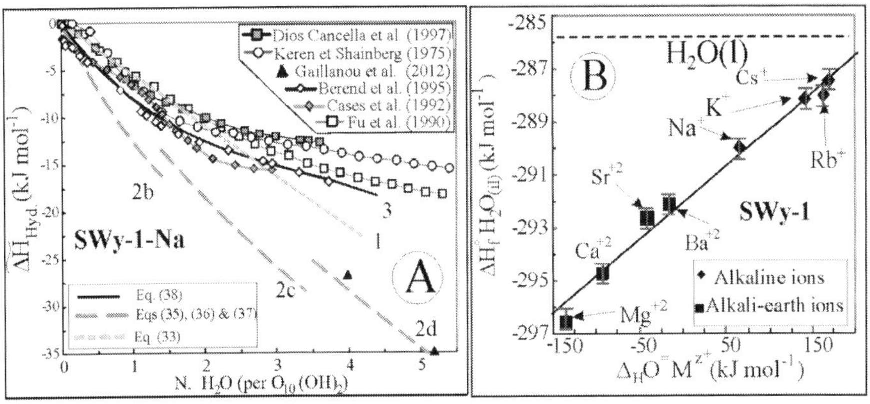

Figure 4. A – Integral enthalpy of hydration versus the number of interlayer cation; B – Standard enthalpy of formation of interlayer water versus the nature of the interlayer cation.

$$3 \text{ layers } \left(0 < x_{hs} < 7.0\right):\ \Delta\tilde{H}_{Hyd,298} = -45.5 * x_{hs} + x_{hs} * \left(1 - x_{hs}\right) * 10.0 \tag{37}$$

The curve 3 (black line) in Figure 4A comes from the model of Vieillard et al. [63] and is represented by the following equation:

$$\Delta\tilde{H}_{Hyd} = -23.0 * x_{hs} + x_{hs} * \left(1 - x_{hs}\right) * \left[8.0 * x_{hs} - 33.0 * \left(1 - x_{hs}\right)\right] \tag{38}$$

It appears that integral enthalpy of hydration modelled by Vieillard et al. [63] (black line) based on experimental measurements of heat of adsorption from [15-17] encompasses nearly all experimental data. As the standard entropy of hydration is assumed constant in [66] and [68] models, the integral hydration enthalpies provided by these two previous models are nearly linear. Those obtained by [68] exhibit three sections of curves corresponding to the three states of hydration (1, 2 and 3 water layers).

In Figure 4B, a relationship between the standard state enthalpy of interlayer water with the nature of the interlayer cation located in the SWy-1 montmorillonite with a constant layer charge of 0.38, characterised by its electronegativity ($\Delta_H O^{=}M^{z+}$ (aq), Table 4) is proposed [63] and given as follows:

$$\Delta H^{\circ}_{f,298}(H_2O)_i - \Delta H^{\circ}_{f,298}(H_2O)_{(l)} = 0.02784 * \Delta_H O^{=} M^{z+} - 6.275 \quad (39)$$

Lowering the electronegativity of cation in the interlayer sites stabilises the enthalpy of formation of the interlayer water. It should be kept in mind that Eq. (39) has been settled for the same support, *i.e.,* a constant layer charge. For natural montmorillonite-Na, most enthalpies of hydration of the interlayer water are consistent within a narrow range of interlayer charge, IC (0.31<IC<0.38) [63]. Due to the lack of calorimetric measurements of enthalpy of hydration for high charge (IC>0.5) and low charge (IC<0.3) montmorillonites, the variation of the standard hydration enthalpy as a function of the layer charge remains highly questionable. An important remaining matter of discussion is the extent to which the model is able to predict hydration properties of dioctahedral and trioctahedral smectites with different layer charges and compositions.

5.2. Entropy

Salts: Average values of the entropy of hydration water were obtained [39] and [35] within chlorides, sulphates and salts (table 2). These average values of salts appear very similar to the value of entropy for crystalline water, S°=40.17 J/K^{-1} Mol^{-1} [70]. To improve the accuracy of prediction, statistical relationships between entropy and the molar volume of hydration should be investigated.

Zeolites: Initially, an average value of entropy of zeolitic water was proposed by Helgeson et al. [70] and was equal to $S°H_2O$ (zeol) = 58.99 J/K^{-1} Mol^{-1}. From very few available calorimetric measurements of anhydrous and hydrated zeolites [4, 6, 71], it has been shown that the entropy of hydration water, at saturation, remains constant at approximately 52.0 J/K^{-1} Mol^{-1} [72]. Some authors provide slight deviations of entropy of hydration depending of the nature of cations in clinoptilolite [22].

Clay minerals: Entropies of interlayer water in clay minerals have never been measured directly, but have been evaluated from indirect measurements such as a contribution of experimental enthalpies of immersion and adsorption-desorption isotherms [63] [34] or from the Clausius-Clapeyron rules with adsorption isotherms performed at different temperatures [31], [30]. A constant value 55.0 J/K^{-1} Mol^{-1} for the entropy of interlayer water has been chosen [67] and [68] regardless of the nature of the interlayer water and the numbers of interlayer waters layers. By considering the hydration of a smectite always as a symmetrical regular solid solution, the integral entropy is expressed as in [63]:

$$\begin{aligned}\Delta\tilde{S}_{Hyd} &= x_{hs} * \Delta S^{\circ}_{Hyd} - R\left[x_{hs} * \ln(x_{hs}) + (1 - x_{hs}) * \ln(1 - x_{hs})\right] \\ &+ x_{hs} * (1 - x_{hs}) * \left[W_{S_1} * x_{hs} + W_{S_2} * (1 - x_{hs})\right]\end{aligned} \quad (40)$$

$$\text{with} \quad \Delta S^0_{hyd,298} = n_m * \left[S^0_{298}\left(H_2O_{(i)}\right) - S^0_{298}\left(H_2O_{liq}\right) \right] \tag{41}$$

where ΔS°_{Hyd}, W_{S_1}, and W_{S_2} represent the standard entropy of hydration and Margules parameters of the excess entropy of mixing, respectively. The integral entropy is expressed in J K-1 mol-1 of smectite based on O10(OH)2. The determination of these parameters is retrieved from values of ΔH°_{Hyd}, W_{H_1}, and W_{H_2} obtained previously in the minimisation procedure of integral hydration enthalpy and from the experimental adsorption-desorption isotherm interlayer water versus relative humidity. This minimisation procedure uses a least square method and provides ΔS°_{Hyd}, W_{S_1} and W_{S_2} (Table 8).

	$S^{\circ}_{298}H_2O_{(i)}$ [1] J/K/mole	W_{S_1} [1] J/K/mole	W_{S_2} [1] J/K/mole	$S^{\circ}_{Hyd}M^{z+}_{(aq)}$ [2] J/K/mole
Cs^+	64.6	163.5	-348.0	-78.
Rb^+	63.8	142.0	-355.0	-84.
K^+	63.7	104.0	-267.5	-93.
Na+	59.8	49.1	-94.0	-130.
Ba^{2+}	55.7	-116.0	-211.5	-224.
Sr^{2+}	54.3	-117.8	-244.5	-261.
Ca^{2+}	51.5	-93.5	-270.5	-271.
Mg^{2+}	45.4	-42.8	-380.0	-350.

1 –[63]; 2 - [73].

Table 8. Standard thermodynamic enthalpy ΔS°_{Hyd} of interlayer water, Margules interactions parameters and hydration entropy of the ion under its hydrated aqueous state for the eight cations.

Figure 5A displays the comparison of the integral entropy of hydration of a smectite- Na computed from the three models [67] [68] and [63] developed previously with different experimental values.

As indicated in the enthalpy section, integral hydration entropies modelled by [66] (yellow dotted line, N°1) and [68] (green dotted line, N°2a, 2b and 2c corresponding to 1st, 2nd and 3rd layer) are linear and merge together. In the model proposed by Vieillard et al. [63], three calculated integral entropies of SWy1-Na –water system were plotted in Figure 5A and correspond to the adsorption (red line), the desorption (blue line) and the theoretical equilibrium water – Swy1-Na (black line, N° 3), whose equation is given:

$$\begin{aligned} \Delta \tilde{S}_{Hyd} &= -55.8 * x_{hs} - R\left[x_{hs} * \ln\left(x_{hs}\right) + \left(1 - x_{hs}\right) * \ln\left(1 - x_{hs}\right) \right] \\ &+ x_{hs} * \left(1 - x_{hs}\right) * \left[49.05 * x_{hs} - 94.0 * \left(1 - x_{hs}\right) \right] \end{aligned} \tag{42}$$

The comparison of hydration-dehydration curves calculated by [63] with those provided by Fu et al. [34] show an opposite interpretation. The data from [34] show that integral entropies during adsorption are less negative than those during desorption. Data from our model show the opposite. This difference is explained by the fact that from experimental works of [34], adsorption follows desorption, while with our data, desorption follows adsorption. Thus, a maximum entropy difference between hydration and dehydration functions can be depicted and is equal to 18 J/K/mole. These observations show the importance of movement of exchangeable cations from ditrigonal cavities and the rotation of tetrahedrals in the tetrahedral sheets when dry collapsed layers are progressively exposed to water vapour.

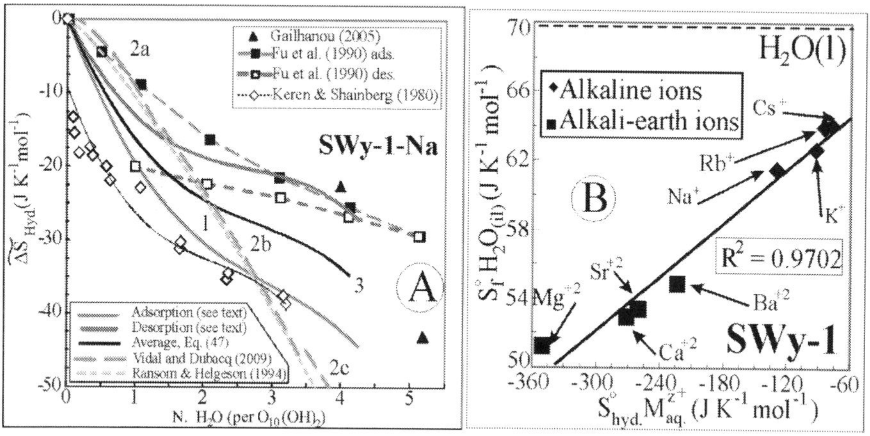

Figure 5. A – Integral entropy of hydration versus the number of interlayer cation; B – Standard entropy of formation of interlayer water versus the nature of the interlayer cation.

A correlation between the entropy of interlayer cation $S^{\circ}_{298}(H_2O)_i$ with the theoretical hydration entropy, S°hyd M^{z+} [73] (Table 8), of the interlayer cation [63] is displayed in Figure 5B with the following equation:

$$S^{\circ}_{298}(H_2O)_i - S^{\circ}_{298}(H_2O)_{(l)} = 0.06542 * S^{\circ}_{hyd}M^{z+} - 0.413 \quad (43)$$

Negative values of the hydration entropy of the hydrated ion lower the hydration entropy of the interlayer water.

5.3. Heat capacity

Salts: An acceptable average heat capacity ranging from 39.0±5.3 J/K^{-1}mol^{-1} for sulphates and sulphites to 42.0±2.7 J/K^{-1}mol^{-1}for chlorates and chlorides has been proposed [35] and seems to be close to the average value of heat capacity for crystalline water Cp=40.04 J/K^{-1}mol^{-1} [70]. Only the average heat capacity function of crystalline water was proposed [70]:

$$Cp(H_2O)_{salts} = 29.75 - 3.448 * 10^{-2} * T \quad (44)$$

Zeolites: Three equations of heat capacities ([41], [74] [45]) were proposed for a set of silicate minerals, including analcime and natrolite, and are displayed in table 9.

	Heat capacity function (298.15- 500 K)	
Eq. A	$Cp(H_2O)_{sil} = 56.61 - 2.64 * 10^2 * T^{-0.5}$	[41]
Eq. B	$Cp(H_2O)_{sil} = 87.62 + 7.58 * 10^2 * T^{-0.5} + 0.53 * 107 * T^{-3}$	[74]
Eq. C	$Cp(H_2O)_{sil} = 39.65 + 32.5 * 10^{-3} * T - 4.694 * 10^5 * T^{-2}$	[45]
Eq. D	$Cp(H_2O)_{zeol.} = 23.999 + 66.295 * 10^{-3} * T - 0.3948 * 10^5 * T^{-2}$	[72]

Table 9. Heat capacity functions for zeolites minerals

Heat capacities of the hydration in zeolites have been performed by Neuhoff & Wang [75] on three zeolites (analcime, natrolite and wairakite) and exhibit marked variations in the heat capacity of hydration with temperature. Four zeolites (mordenite, wairakite and two different values from different sources for analcime and natrolite) for which the heat capacity of zeolitic water can be obtained by a difference between the heat capacities measured on both the anhydrous and the hydrated phases [72] are represented in Figure 6A. Also displayed are the heat capacity values of water in all its states (ice, liquid and vapour) for comparison [35]. The heat capacity values of zeolitic water obtained by minimisation techniques (Eq. D of table 9, turquish line) matches with the heat capacity function equations of [74] and [45].

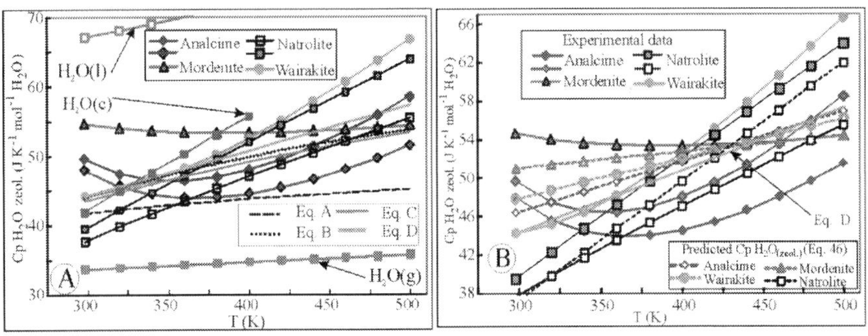

Figure 6. Heat capacity of zeolitic water versus temperature: A – Heat capacities of zeolitic water in zeolites, in different prediction models and of water in its various states (l for liquid, c for ice and g for vapour); B – Experimental heat capacities of zeolitic waters in natural zeolites and predicted heat capacities from Eq. (46).

Eq. D of table 9 provides a statistical error of 4.51%, 2.35% and 4.18% for hydrated zeolites at 298.15 K, 400 K and 500 K, respectively. To improve the accuracy of the prediction of the heat capacity of hydrated zeolites, one considers the normalised hydration volume, which is the difference of unit-cell volume ($V_{u.c.}$) between the hydrated and dehydrated forms for any zeolite per zeolitic water molecule [72]:

$$\Delta V_{Hy-Deh} = \left(V_{u.c.hydr.} - V_{u.c.dehyd.}\right) / n_{H_2O} \tag{45}$$

$$Cp(H_2O)_{zeol.} = \left[55.85 - 13.073 * \sqrt{\left(\Delta V_{Hy-Deh}\right)}\right] + \left[0.0062 + 0.0305 * \sqrt{\left(\Delta V_{Hy-Deh}\right)}\right] * T + \left[0.008 + 0.0048 * \sqrt{\left(\Delta V_{Hy-Deh}\right)}\right] * T^{-2} \quad (46)$$

The heat capacities of hydration water given in Figure 6B have been calculated for the four hydrated zeolites (mordenite, analcime, wairakite and natrolite) from the unit-cell volumes of their dehydrated and hydrated forms [72]. It can be observed that the heat capacities of hydration water calculated in this way are much better than those obtained by Eq. D of table 9 (indicated in Fig. 6B as a turquish full line). With this improvement, the errors made on hydrated zeolites decrease to 3.5%, 2.35% and 3.89% at 298.15 K, 400 K and 500 K, respectively.

Clay minerals: The heat capacity function comes initially from [67] and is assumed to be independent of the nature of the interlayer cations:

$$Cp(H_2O)_{clays} = 37.84 + 51.631 * 10^{-3} * T - 4.0959 * 10^{5} * T^{-2} \quad (47)$$

This heat capacity equation has also been used by [68] and by [63] in the behaviour of thermodynamic entities versus temperature.

5.4. Molar volume

Salts: Only [35] provided an average value for different salts: 14.5±1.2 cm³/mole for sulphates and sulphites and 14.1±2.3 cm³/mole for chlorides and chlorates. These average values appear to be similar to the molar volume of crystalline water ($V(H_2O)_{(salts)}$= 13.7 cm³/mole [76]).

Zeolites: No study on the variation of the molar volumes of the zeolitic water has been performed. Numerous values of molar volumes of hydrated and anhydrous zeolites have been provided and gathered in such calculations for enthalpy and entropy of hydration [77], [61], [72]. In the study of the relationship between entropy and molar volume within zeolites [72], an average value of the molar volume of hydration water has been calculated and is $V(H_2O)_{(zeol)}$=2.06 cm³/mole, which is not significantly different from those proposed by Helgeson ($V(H_2O)_{(zeol)}$= 8 cm³/mole) estimated from standard molal volumes of zeolites. The weak value of the molar volume of the hydration water is explained by the fact that the anhydrous zeolites exhibit a rigid (Si, Al) framework with different channels or cavities occupied by exchangeable cations. During the hydration process, the water molecules are bound to cations inside the cavities, do not affect the framework and have a weak impact on lattice parameters.

Clay minerals: The molar volume of the interlayer water was initially determined by Ransom [67] and also used by [68] and by [63] in thermochemical calculation. Its value is $V(H_2O)_I$ = 17.22 cm³/mole. This value, greater than those of salts, is explained by the fact that, during hydration, the molar volume of clay minerals increases in one dimension characterised by basal spacing (Miller index 001).

6. Discussion of the Gibbs free energy of the formation behaviour of the hydration water in environmental conditions

The relationships among the free energy change, enthalpy of reaction and entropy for an isothermal process is:

$$\Delta G = \Delta H - T * \Delta S \tag{48}$$

where ΔG stands for free energy change, ΔH for enthalpy of reaction, ΔS for the change in entropy and T for the absolute temperature. This relationship holds true for both the integral entities that are represented by $\Delta\tilde{G}$, $\Delta\tilde{H}$ and $\Delta\tilde{S}$ and for the partial quantities that are represented by $\Delta\bar{G}$, $\Delta\bar{H}$ and $\Delta\bar{S}$.

Salts: The free energy of formation of the hydration water of hydrated salts can be calculated from the enthalpy and entropy of hydration given previously. A relationship has been demonstrated by the statistical regressions [35] between the enthalpy and Gibbs free energy of crystalline water in the following relation:

$$\Delta G^{\circ}_{f,298}\left(H_2O_{salts}\right) = 1.008 * \Delta H^{\circ}_{f,298}\left(H_2O_{salts}\right) + 59.406 \tag{49}$$

The predicted values contribute to the variations of solubility products and the cation selectivity constants among the series of salts. An example is given in [78] for two series of hydrated salts of $MgCl_2$ and $SrCl_2$.

For zeolites and clays, the adsorbent (a single smectite or zeolite phase) is supposed to be inert so that the derived enthalpy and entropy functions formally represent the energy variations between the adsorbed and the bulk water.

The integral free energy of hydration $\Delta\tilde{G}_{hyd}$ may be determined from the gravimetric isotherms of adsorption of the water vapour by the following expression [79]:

$$\Delta\tilde{G}_{hyd} = R * T * \int_{n=0}^{n=n_w} Ln\left(\frac{P}{P_0}\right) * dn \tag{50}$$

where R = 8.314 $J.mol^{-1}K^{-1}$ is the ideal gas constant, n is the amount of adsorbed water, P is the partial water vapour pressure, and P_0 is the saturated water vapour pressure at temperature T (P/P_0 = Relative Humidity/100). This value can be determined by plotting R*T*Ln(P/Po) versus the water content n (adsorbed or desorbed) and determining the area under the curve from 0 to n_w. When n_w approaches 0, R*T*Ln(P/Po) approaches infinity, making integration somewhat difficult.

Zeolites: Eq. (50) has been used on three clinoptilolites (-Na, -K and -Ca) [22] and show a good correlation between the Gibbs free energies calculated from the adsorption-desorption isotherms and from the contribution of the enthalpy of hydration and the entropy of hydration extracted from the equilibrium at high temperature. The model described by Carey & Bish [62] used a solution solid model between anhydrous and hydrated zeolite, and the variation of the integral Gibbs free energies of hydration is expressed as a function of a parameter θ characterising the ratio of H_2O/(maximum H_2O). The main difficulty in these calculations is the determination of the maximal amount of hydration water at the saturation state ($p/p^{\circ}=1$).

Clay minerals: Among the numerous measurements of adsorption-desorption isotherms on various clays, very few works provided a consistent Gibbs free energy of adsorption-desorption due to the difficulty of measurements in the low relative humidity domains. By choosing the system of water – SWy-1-Na, Figure 7 displays the adsorption-desorption isotherms (Figure 7A), the variation of the integral free energy of hydration versus the number of moles of the interlayer water (Figure 7B) and versus the relative humidity (Figure 7C) and the variation of the free energy of formation of the interlayer water versus the relative humidity (Figure 7D).

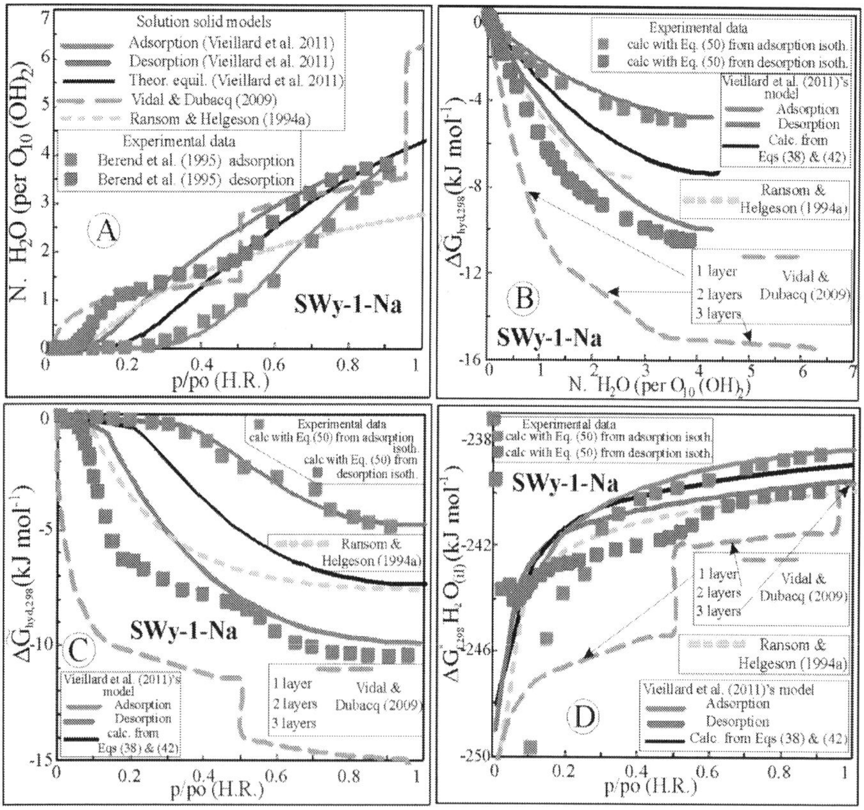

Figure 7. A - Adsorption-desorption isotherms; B - Integral free energy of hydration versus the number of moles of interlayer water; C - Integral free energy of hydration versus the relative humidity; D - Free energy of formation of the interlayer water versus the relative humidity.

In each figure, the data of the free energy of hydration (Eq. 50) from the experimental isotherms [15, 16] (blue square for adsorption and red square for desorption) and the free energy of hydration are computed and are reported for comparison of three different works: the Vieillard et al. model [63] (blue, red and black lines for adsorption, desorption and theoretical equilibrium, respectively), the Ransom & Helgeson model [66] (dotted yellow line), and the Vidal & Dubacq model [68] (dotted green line). For the adsorption and desorption reactions, the modelled isotherm calculated by the Vieillard et al. model [63] (blue and red lines, respectively) agrees well with the experimental data acquired by [15, 16] in high relative humidity.

At low relative humidity, the uncertainty appears very important between the experimental desorption data and the desorption curve from the model in [63] (3 kJ/mol for integral free energy, Fig C) due to the analytical limits and mathematical constraints.

The theoretical equilibrium (black line) suggests that the hydration of a Na-SWy-1 begins at only R.H.= 0.15 (Figure 7A), which correlates well with the experimental adsorption but not with the desorption. The integral Gibbs free energy of hydration (figure 7B and 7C) and the Gibbs free energy of the interlayer water (Figure 7D) of the theoretical equilibrium between the water and Na-Smectite (black line) are plotted and correspond to the average values of calculated adsorption-desorption isotherms. The theoretical curve of equilibrium between the water and Na-smectite provided by Ransom & Helgeson [66] displays a small number of moles of the interlayer water close to saturation, while those of Vidal & Dubacq [68] reproduce this stepwise behaviour fairly well because it assumes three solid solutions with 3 end-members (0.7 H_2O, 2 H_2O, 4.5 H_2O). However, it does not match the behaviour of water adsorption versus relative humidity, which is not a step function. The advantage of this approach is that it allows reproducing the observed stepwise evolution of the smectite volume and water content with varying T and relative humidity. At low relative humidity, as the hydration of a Na-SWy-1 begins at only R.H.= 0.15, the computations from [68] and [66] (green and yellow dotted lines) do not match the observed isotherms because there is a strong hydration in the RH= 0.0- 0.18 range. During the interval of R.H.=0.0-0.2, the adsorption of water molecules occurs predominantly on the external surface of tactoids (basal external faces and edges of tactoids). This explains why we observe a very negative integral free energy in the low relative humidity domain for these two models (green dotted line and in a less extent, yellow dotted lines). The necessity to take into account only the hydration water, sensu-stricto, is justified. This procedure has been incompletely performed by Ransom & Helgeson [66] with the external surface of a kaolinite to estimate the amount of surface covering water, while Vidal & Dubacq [68] made no correction on the gravimetric waters. At high relative humidity, Vidal & Dubacq [68] proposed a third step of hydration, with the number of water molecules increasing up to 7 moles per O10(OH)2. All of the experimental isotherms performed by [15, 16] exhibit an increase in the water contents beyond RH = 0.9, identified as osmotic water. The osmotic swelling of the sodium montmorillonite corresponds to an iso-enthalpic effect, with the heat of adsorption being constant. The theoretical equilibrium between water and Na-SWy-1 (black line) matches the experimental dehydration near saturation.

The cationic exchange and hydration reactions are intimately related, and the cationic exchange reactions not only modify the interlayer composition but also imply changes in the interlayer water content. These changes can be quite significant and can be evaluated using the Vieillard et al. model [63], which provides a theoretical number of moles of interlayer water and integral thermodynamic properties (H, S and G) for the SWy-1 montmorillonites. Thus, the cationic exchange reactions may depend on an external parameter, such as the relative humidity.

7. Conclusions and perspectives

This chapter demonstrated some useful relationships in the prediction of thermodynamic entities. Some important areas are still lacking and deserve further study.

Among the anhydrous and hydrated salts, the relationships between the entropy and molar volume may exist and will improve the prediction of the entropy of hydration. The function of the average heat capacities for the hydration water is lacking and will help in characterising the behaviour of the crystalline water versus temperature.

In the zeolite field, the relationships among the molar volume, water content and relative humidity may exist within zeolites families with a reversibility in the hydration-dehydration processes. These relationships will contribute to a better understanding of the behaviour of partial molar enthalpy versus the degree of saturation or versus the relative humidity.

In the clay Minerals, three following recommendations are needed: 1) to dissociate the hydration water from the gravimetric water; 2) to measure the heat of hydration among the dioctahedral and trioctahedral clays; and 3) to find a relationship between the hysteresis and the entropy of the hydration-dehydration by the molecular modellings. With these three aims, the knowledge of mechanisms of the hydration within clay minerals will advance our understanding of cationic exchange in soils under various physical and chemical constraints.

Author details

Philippe Vieillard
CNRS IC2MP, UMR 7285, Hydrasa, France

Acknowledgement

Financial support from the French National Radioactive Waste Management Agency (ANDRA), from the French Geological Survey (BRGM) and from the national council for scientific Research (CNRS) is gratefully acknowledged. Special thanks to Ph. Blanc, H. Gailhanou and Cl. Fialips for their effective and decisive contribution to this document.

8. References

[1] Reeuwijk van LP (1974) The thermal dehydration of natural zeolites. In: H. V, B.V. Z, editors. Wageningen 1-88.

[2] Tardy Y, Mercury L, Roquin C, Vieillard P (1999) Le concept d'eau *ice-like*: hydratation-deshydratation des sels, hydroxydes, zéolites, argiles et matières organiques vivantes ou inertes. Compt. Rendus Académie Science Paris 329: 377-388.

[3] Mercury L, Vieillard P, Tardy Y (2001) "Thermodynamics of Icepolymorphs and "Ice-like" water in hydrates and hydroxides. Appl. Geochem. 16: 161-181.

[4] Johnson GK, Flotow HE, O'Hare PAG, Wise WS (1982) Thermodynamic studies of zeolites : analcime and dehydrated analcime. Amer. Mineral. 67: 736-748.

[5] Johnson GK, Flotow HE, O'Hare PAG (1983) Thermodynamic studies of zeolites:natrolite mesolite and scolecite. Amer. Mineral. 68: 1137-1145.

[6] Johnson GK, Tasker IR, Flotow HE, O'Hare PAG (1992) Thermodynamic studies of mordenite, dehydrated mordenite and gibbsite. Am. Mineral. 77: 85-93.

[7] Gailhanou H, Blanc P, Rogez J, Mikaelian G, Kawaji H, Olives J, et al. (2012) Thermodynamic properties of illite IMt-2, smectite MX-80 and beidellite SBld-1 by calorimetric methods: Enthalpies of formation, heat capacities , entropies and Gibbs free energies of formation. . Am. Mineral. (submitted).

[8] Kiseleva I, Navrotsky A, Belitskii IA, Fursenko BA (1996) Thermochemistry and phase equilibria in calcium zeolites. Am. Mineral. 81: 658-667.

[9] Sun P, Deore S, Navrotsky A (2007) Enthalpy of formation and dehydration of lithium and sodium zeolite beta. Microporous Mesoporous Mat. 98: 29-40.

[10] Sun P, Navrotsky A (2008) Enthalpy of formation and dehydration of alkaline earth cation exchanged zeolite beta. Microporous Mesoporous Mat. 109: 147-155.

[11] Yang SY, Navrotsky A (2000) Energetics of formation and hydration of ion-exchanged zeolite Y. Microporous Mesoporous Materials 37: 175-186.

[12] Shim SH, Navrotsky A, Gaffney TR, Macdougall JE (1999) Chabazite: Energetics of hydration, enthalpy of formation, and effect of cations on stability. Amer. Mineral. 84: 1870-1882.

[13] Gailhanou H, van Miltenburg JC, Rogez J, Olives J, Amouric M, Gaucher EC, et al. (2007) Thermodynamic properties of anhydrous smectite MX-80, illite IMt-2 and mixed-layer illite-smectite ISCz-1 as determined by calorimetric methods. Part I: Heat capacities, heat contents and entropies. Geochim. Cosmochim. Acta 71: 5463-5473.

[14] Barrer RM, Cram PJ (1971) Heats of immersion of outgassed and ion-exchanged zeolites. In: L.B. FEMS, editor. Molecular sieves zeolites. Washington D.C.: American Chemical Society. pp. 105-131.

[15] Cases JM, Bérend I, Besson G, François M, Uriot JP, Thomas F, et al. (1992) Mechanism of adsorptionand desorption of water vapor by homoionic montmorillonite. I.The sodium-exchanged form. Langmuir 8: 2730-2739.

[16] Bérend I, Cases JM, François M, Uriot JP, Michot L, Masion A, et al. (1995) Mechanism of adsorption and desorptionof water vaporby homoionic montmorillonites:2. The Li+, Na+, K+, Rb+ and Cs+ exchanged forms. Clays Clay Minerals 43: 324-336.

[17] Cases JM, Bérend I, François M, Uriot JP, Michot L, Thomas F (1997) Mechanism of adsorption and desorption of water vapor by homoionic montmorillonite: 3. the Mg^{2+}, Ca^{2+},Sr^{2+} and Ba^{2+} exchanged forms. Clays Clay Minerals 45: 8-22.

[18] Keren R, Shainberg I (1975) Water vapor isotherms and heat of immersion of Na/Ca-Montmorillonite systems. I Homoionic clay. Clays Clay Minerals 23: 193-200.

[19] Keren R, Shainberg I (1979) Water vapor isotherms and heat of immersion of Na/Ca-Montmorillonite systems - 2. Mixed systems. Clays Clay Minerals 27: 145-151.

[20] Dios Cancela G, Huertas FJ, Romero Taboada E, Sánchez-Rasero F, Hernández Laguna A (1997) Adsorption of Water Vapor by Homoionic Montmorillonites. Heats of Adsorption and Desorption. J. Colloid Interface Sci. 185: 343-354.

[21] van Olphen H (1965) Thermodynamics of interlayer adsorption of water in clays. I.-- Sodium vermiculite. J. Coll. Sci. 20: 822-837.

[22] Carey JW, Bish DL (1996) Equilibrium in the clinoptilolite - H2O system. Amer. Mineral. 81: 952-962.

[23] Fialips CI, Carey JW, Bish DL (2005) Hydration-dehydration behavior and thermodynamics of chabazite. Geochim. Cosmochim. Acta 69: 2293-2308.

[24] Koster van Groos AF, Guggenheim S (1984) The effect of pressure on the dehydration reaction of interlayer water in Na-montmorillonite (SWy-1). Amer. Mineral. 69: 872-879.

[25] Wu T-C, Bassett WA, Huang W-L, Guggenheim S, Koster van Groos AF (1997) Montmorillonite under high H 2 O pressures; stability of hydrate phases, rehydration hysteresis, and the effect of interlayer cations. Amer. Mineral. 82: 69-78.

[26] Koster v, Groos, A.F., Guggenheim S (1986) Dehydration of K exchanged montmorillonite at elevated temperatures and pressures. Clays Clay Minerals 34: 281-286.

[27] Petrova N, Mizota T, Fujiwara K (2001) Hydration heats of zeolites for evaluation of heat exchangers. J. Therm. Anal. Calorim. 64: 157-166.

[28] Hill TL (1949) Statistical mechanics of adsorption V. Thermodynamics and heat of adsorption. J. Chem. Phys. 17: 520-535.

[29] Hill TL (1950) Statistical Mechanics of Adsorption. IX. Adsorption Thermodynamics and Solution Thermodynamics. J. Chem. Phys.s 18: 246-256.

[30] Keren R, Shainberg I (1980) Water vapor isotherms and heat of immersion of Na/Ca-Montmorillonite systems - 3. Thermodynamics. Clays Clay Minerals 28: 204-210.

[31] Hall PL, Astill DM (1989) Adsorption of water by homoionic exchange forms of Wyoming montmorillonite (SWy-1). Clays Clay Minerals 37: 355-363.

[32] Valueva GP, Goryainov SV (1992) Chabazite during dehydration: thermochemical and Raman spectroscopy study. Russian Geologica Geophys. 33: 68-75.

[33] Xu W, Johnston CT, Parker P, Agnew SF (2000) Infrared study of water sorption on Na-, Li-, Ca- and Mg-exchanged (SWy-1 AND SAz-1) montmorillonite. Clays Clay Minerals 48: 120-131.

[34] Fu MH, Zhang ZZ, Low PF (1990) Changes in the properties of a montmorillonite-water system during the adsorption and desorption of water: hysteresis. Clays Clay Minerals 38: 485-492.

[35] Mercury L, Vieillard P, Tardy Y (2001) "Thermodynamics of Icepolymorphs and "Ice-like" water in hydrates and hydroxides. Appl. Geochem. 16: 161-181.

[36] Cox JD, Wagman DD, Medvedev VA (1989) CODATA Key values for thermodynamics. New York: Hemisphere Publishing Corporation. 271 p.

[37] Tardy Y, Mercury L, Roquin C, Vieillard P (1999) Le concept d'eau *ice-like*: hydratation-deshydratation des sels, hydroxydes, zéolites, argiles et matières organiques vivantes ou inertes. Compt. Rendus Académie Science Paris 329: 377-388.

[38] Chipera SJ, Apps JA (2001) Geochemical stability of natural zeolites. In: Bish DLM, D.W., editor. Natural Zeolites:Occurence, properties Applications. Blacksburg (Va)117-161.

[39] Glasser L, Jenkins HDB (2007) The thermodynamic solvate difference rule: Solvation parameters and their use in interpretation of the role of bound solvent in condensed-phase solvates. Inorg. Chem. 46: 9768-9778.

[40] Sidorov YI (1988) Estimating silicate standard entropies. Geochem. Intern. 80-85.

[41] Robinson GR, Haas JL (1983) Heat capacity, relative enthalpy and calorimetric entropy of silicate minerals : an empirical method of prediction. Amer. Mineral. 68: 541-443.

[42] Moiseev GK, Sestak J (1995) Some calculations methods for estimation of thermodynamical and thermochemical properties of inorganic compounds. Prog Cryst Growth Charact. 30:23-81.

[43] Burylev BP (1988) Thermodynamics of crystal hydrates of rare-earth metal chlorides. Sov. Radiochem. 29): 653-655.

[44] Burylev BP, Kritskaya EB, Moisov LP, Kostenko NB (1995) Calculation of the standard enthalpies of formation of hydrous sodium salts. Inorg. Mater. 31: 727-729.

[45] Van Hinsberg VJ, Vriend SP, Schumacher JC (2005) A new method to calculate end-member thermodynamic properties of minerals from their constituent polyhedra II: heat capacity, compressibility and thermal expansion. J. Metamorph Geol. 23: 681-693.

[46] Van Hinsberg VJ, Vriend SP, Schumacher JC (2005) A new method to calculate end-member thermodynamic properties of minerals from their constituent polyhedra I: enthalpy, entropy and molar volume. J. Metamorph. Geol. 23: 165-179.

[47] Chermak JA, Rimstidt JD (1989) Estimating the thermodynamic properties (ΔG°f and ΔH°f) of silicate minerals at 298 K from the sum of polyhedral contributions. Amer. Mineral. 74: 1023-1031.

[48] Li J, Li B, Gao S (2000) Calculation of thermodynamic properties of hydrated borates by group contribution method. Physics and Chemistry of Minerals 27: 342-346.

[49] Chen FR, Ewing RC, Clark SB (1999) The Gibbs free energies and enthalpies of formation of U6+ phases: An empirical method of prediction. Amer. Mineral. 84: 650-664.

[50] Vieillard P, Jenkins HDB (1986) Empirical relationships for estimation of enthalpies of formation of simples hydrates. Part I Hydrates of alkali- metal cations, of hydrogen and of monovalent cations. J. Chem. Res.: 444-445.

[51] Vieillard P, Jenkins HDB (1986) Empirical relationships for estimation of enthalpies of formation of simples hydrates. Part 2. Hydrates of alkaline-earth metal cations. J. Chem. Res.: 446-447.

[52] Vieillard P, Jenkins HDB (1986) Empirical relationships for estimation of enthalpies of formation of simples hydrates. Part 3. Hydrates of transition metal cations (Cr^{2+},Fe^{2+},Co^{2+},Ni^{2+},Cu^{2+},Zn^{2+},$Cd^{2+}$and UO_2^{2+}). J. Chem. Res.: 448-449.

[53] Robie RA, Waldbaum DR (1968) Thermodynamic properties of minerals and related substances at 298.15°K (25°C) and one atmosphere (1.O13 bars) pressure and at higher temperatures. U. S. Geol. Surv. Bull., 1259, 256 p.

[54] Vieillard P, Mathieu R (2009) A predictive model for the enthalpies of hydration of zeolites. . Amer. Mineral. 94: 565-577.

[55] Meier WM, Olson DH, Baerlocher C (1996) Atlas of Zeolite Structure types: 4th revised Edn. Zeolites 17: 1-230.

[56] Meier WM, Olson DH (1971) Zeolites Frameworks In: Flanigen EM, Sand LB, editors. Molecular Sieve Zeolites 155-170.

[57] Barrer RM (1982) Hydrothermal chemistry of Zeolites. Press A, editor. London New York: Harcurt Brace Jovanovich. 1-360 p.

[58] Carey JW (1995) A thermodynamic formulation of hydrous cordierite. Contrib. Miner. Petrol. 119: 155-165.

[59] Armbruster T (1993) Dehydration mechanism of clinoptilolite and heulandite: Single-crystal X-ray-study of Na-poor and K-Mg-rich clinoptilolite at 100K. Amer. Mineral. 78: 260-264.

[60] Bish DL, Carey JW (2001) Thermal behavior of natural zeolites. In: D.W. BDLM, editor. Natural zeolites:occurrence, properties Applications. Blacksburg: Mineralogical Society of America. pp. 403-452.

[61] Mathieu R, Vieillard P (2010) A predictive model for the enthalpies of formation of zeolites. . Microporous Mesoporous Mat. 132: 335-351.

[62] Carey JW, Bish DL (1997) Calorimetric measurement of the enthalpy of hydration of clinoptilolite. Clays Clay Minerals 45: 826-833.

[63] Vieillard P, Blanc P, Fialips CI, Gailhanou H, Gaboreau S (2011) Hydration thermodynamics of the SWy-1 montmorillonite saturated with alkali and alkaline-earth cations: a predictive model. Geochim. Cosmochim. Acta 75: 5664-5685.

[64] Fripiat J, Cases J, Francois M, Letellier M (1982) Thermodynamic and microdynamic behavior of water in clay suspensions and gels. J. Colloid Interface Sci. 89: 378-400.

[65] Koster van Groos AF, Guggenheim S (1987) Dehydration of a Ca- and a Mg-exchanged montmorillonite (SWy-1) at elevated pressures. Amer. Mineral. 72: 292-298.

[66] Ransom B, Helgeson HC (1994) A Chemical and Thermodynamic Model of Aluminous Dioctahedral-2/1 Layer Clay Minerals in Diagenetic Processes - Regular Solution Representation of Interlayer Dehydration in Smectite. Amer. J. Sci. 294: 449-484.

[67] Ransom BH, H.C. (1994) Estimation of the standard molal heat capacities, entropies, and volumes of 2:1 clay minerals. Geochim. Cosmochim. Acta 58: 4537-4547.

[68] Vidal O, Dubacq B (2009) Thermodynamic modelling of clay dehydration, stability and compositional evolution with temperature, pressure and H2O activity. Geochim. Cosmochim. Acta 73: 6544-6564.

[69] Dubacq B, Vidal O, De Andrade V (2010) Dehydration of dioctahedral aluminous phyllosilicates: thermodynamic modelling and implications for thermobarometric estimates. Contri. Miner. Petrol. 159: 159.

[70] Helgeson HC, Delany JM, Nesbitt HW, Bird DK (1978) Summary and critique of the thermodynamic properties of the rock-forming minerals. Am. J. Sci. 278-A: 1-229.

[71] Johnson GK, Tasker IR, Jurgens R, O'Hare PAG (1991) Thermodynamic studies of Zeolites-Clinoptilolite. J.Chem. Thermodyn. 23: 475-484.

[72] Vieillard P (2010) A predictive model for the entropies and heat capacities of zeolites. Eur. J. Mineral. 22:823-836.

[73] Marcus Y (1994) A simple empirical model describing the thermodynamics of hydration of ions of widely varying charges, sizes, and shapes. Biophys. Chem. 51: 111-127.

[74] Berman RG, Brown TH (1985) Heat capacity of minerals in the system Na2O-K2O-CaO-MgO-FeO-Fe2O3-Al2O3-SiO2-TiO2-H2O-CO2, representation, estimation and high temperature extrapolation. Contrib. Miner. Petrol. 89: 168-183.

[75] Neuhoff PS, Wang J (2007) Heat capacity of hydration. Amer. Mineral. 92: 1358-1367.

[76] Fyfe WS, Turner FJ, Verhoogen J. Metamorphic reactions and metamorphic facies. In: Memoirs GSoA, editor.1958. p. 253 p.

[77] Cruciani G (2006) Zeolites uponheating: Factors governing their thermal stability and structural changes. J. Phys. Chem. Solids 67: 1973-1994.

[78] Lassin A, Duplay J, Touret O, Tardy Y (2000) Thermodynamic treatment of hydrates. Comptes Rendus Acad. Sci. Paris SER II A 330: 533-540.

[79] Barshad I (1960) Thermodynamics of water adsorption and desorption on montmorillonite. Clays Clay Minerals: 84-101.

Chapter 14

Thermodynamics and Kinetics in the Synthesis of Monodisperse Nanoparticles

Nong-Moon Hwang, Jae-Soo Jung and Dong-Kwon Lee

Additional information is available at the end of the chapter

http://dx.doi.org/10.5772/50324

1. Introduction

The preparation of monodisperse nanoparticles with uniform size, shape and composition has been intensively pursed because of their scientific and technological interests [1-4]. The major advantage of monodisperse particles may be attributed to the uniform properties of individual particles, which makes the property of whole particles strictly controllable. They have been widely used in industries such as pharmacy, catalysts, sensors, film precursors, and information storage. The property of nanoparticles is much more sensitive to their size than that of micro-particles. For example, the florescence of monodisperse CdSe/ZnS core/shell nanoparticles depends strongly on their size [5]. The superparamagnetism also depends strongly on the size of nanoparticles [6]. The properties of these particular size nanoparticles show great potentials in the field of bio-medicals and electronics.

In general, to prepare the monodisperse nanoparticles in solution, the size had to be selected after somewhat polydisperse nanoparticles were produced. The separation procedures are very laborious and expensive because the size of nanoparticles is too small to be sorted. Moreover, the production yield of monodisperse nanoparticles decreases markedly due to the loss of nanoparticles during the separation procedures. Therefore, many efforts have been made to synthesize directly monodisperse nanoparticles without size selection procedures.

Recently, several methods have been developed successfully to synthesize gram quantity monodisperse nanoparticles directly without size selection procedures. One of them is a slow heating method developed by Hyeon et al. [6,7], which utilizes the burst nucleation followed by sustained growth of particles. Although nanoparticles have some size distribution in the nucleation stage, they became gradually monodisperse during growth. In this case, the principle for the direct synthesis of monodisperse nanoparticles could be approached by sustained growth of nuclei formed by burst nucleation with some initial size distribution [8].

The digestive ripening developed by Klabunde et al. [9-12] is another fascinating method for the direct synthesis of monodisperse nanoparticles. During digestive ripening, nanoparticles initially ranging from 2 to 40 nm were spontaneously transformed into particles with nearly uniform sizes of 4~5 nm. Lee et al. [13,14] could explain the digestive ripening process by considering the charge effect in the Gibbs-Thomson equation.

The purpose of this article is to provide the thermodynamic and kinetic basis to 'slow heating' method and digestive ripening, which are two successful processes to synthesize the monodisperse nanoparticles.

2. Evolution mechanism of monodisperse particles

To understand intuitively the underlying principle for the evolution of monodispersed particles, let's compare the growth of two spherical particles of different diamters of 1 nm and 5 nm as shown in Fig. 1(a). The problem could be simple under the assumption that all the particles grow at the same rate regardless of the size, which is valid for interface-controlled growth. The size ratio of two particles is 5, which is far from the monodisperse state. If the two particles grow by 1 nm after some time, their diameters become 2 nm and 6 nm, respectively, as shown in Fig. 1(b). Now the size ratio of the two particles is 3. If the two initial particles grow by 99 nm after an appreciable amount of time, their diameters become 100 nm and 104 nm, respectively, as shown in Fig. 1(c). Now their size ratio is 1.04, which is in the monodisperse state with the size being within 5% difference. The time evolution of the two particles in Fig. 1(a)-(c) shows that just the growth induces the monodispersity. It should be noted that the initial size difference between the two particles, which is 4 nm, is maintained throughout the growth but the size ratio continues to decrease with growth, approaching unity. This tendency to approach the monodisperse distribution by growth is often called 'focusing effect' [15].

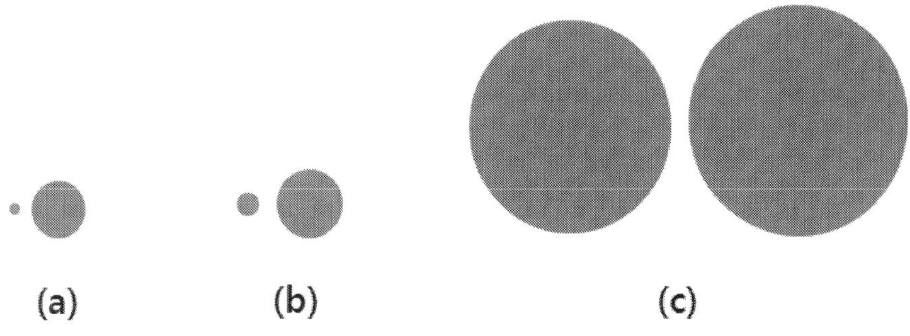

Figure 1. Initially, two spherical particles have the diameters of (a) 1 nm and 5 nm with the size ratio of 5. After growth by 1 nm, the respective particle becomes (b) 2 nm and 6 nm with the size ratio of 3. After growth by 99 nm, the respective particle becomes (c) 100 nm and 104 nm with the size ratio of 1.04, which has the monodispersity with size difference less than 5%.

If a colloidal particle grows in supersaturated solution, the solute may diffuse from the bulk liquid phase containing a uniform concentration of solute C_b to the particle surface through a diffusion layer with some concentration gradient where C_r is the solubility of the particle. In this condition the growth rate is described by

$$\frac{dr}{dt} = \frac{\frac{D}{r}\left(1+\frac{r}{\delta}\right)V_m\left(C_b - C_r\right)}{1+\frac{D}{kr}\left(1+\frac{r}{\delta}\right)} \tag{1}$$

where D is the diffusion coefficient of the solute, r is the particle radius, δ is the diffusion radius around a particle and k is the rate constant of the interface reaction of a solute at the particle interface.

If D is much larger than kr (D >> kr), the interface reaction process becomes a rate-determining step and Eq. (1) is reduced to

$$\frac{dr}{dt} = kV_m\left(C_b - C_r\right) \tag{2}$$

In this case of the interface-controlled growth, the growth rate does not depend on the particle size. If D is much smaller than kr (D << kr), however, the diffusion process of a solute becomes a rate-determining step and thus Eq. (1) is reduced to

$$\frac{dr}{dt} = \frac{DV_m}{r}(C_b - C_r) \tag{3}$$

In this case of the diffusion-controlled growth, the growth rate is inversely proportional to the radius of each particle. This means that large particles grow more slowly than small ones, which is in contrast with the interface-controlled growth, where the growth rate was the same regardless of the size. Therefore, the diffusion-controlled growth has a stronger tendency to approach the monodisperse distribution than the interface-controlled growth.

The narrowing of the size distribution was first theoretically studied in the diffusion-controlled process by Reiss [16]. The variation of the radius distribution σ^2 during growth changes with time as follows.

$$\frac{d\left(\sigma^2\right)}{dt} = 2DV_m\left(C_b - C_r\right)\left[1 - \bar{r}\overline{\left(\frac{1}{r}\right)}\right] \tag{4}$$

In Eq. (4), $\bar{r}$ and $\overline{(1/r)}$ are the mean values of r and 1/r, respectively. Since the arithmetic mean is larger than the harmonic mean, $\overline{(1/r)}$ is greater than $1/\bar{r}$. Therefore, in the supersaturated state, where $C_b > C_r$, Eq. (4) is negative. This means that the variation of the radius distribution σ^2 decreases with time as the growth continues.

nucleation is negligibly low at low supersaturation and becomes very high above certain supersaturation, varying like a step function. The supersaturation for the onset of nucleation has a special meaning and is called 'the critical supersaturation for nucleation', which is defined as the supersaturation to produce the nucleation rate of $1/cm^3 \cdot sec$.

Consequently, to inhibit the additional nucleation during growth, the supersaturation should be maintained below the critical supersaturation for nucleation. Normally, the critical supersaturation for homogeneous nucleation is high enough but that for heterogeneous nucleation is relatively low and can be easily achieved. Therefore, a special care must be taken to prepare the reactor for the synthesis of monodisperse nanoparticles free from the site for heterogeneous nucleation.

One attempt to separate growth from nucleation is the seed-mediated growth, where the preformed nanoparticles are used as seed nuclei [7, 19]. Another attempt to separate growth from nucleation is the initial burst of nucleation, where an appreciable amount of supersaturation is consumed during the burst of nucleation and the relatively low supersaturation is maintained during growth. The 'hot-injection' method [20-22] is an example of this attempt.

Rapid injection of reagents into a hot surfactant solution raises the precursor concentration above the nucleation threshold. A burst of nucleation during a short period of time partially relieves the supersaturation. If the rate of the precursor consumption by the growing colloidal nanoparticles is not exceeded by the rate of precursor addition to solution, no new nuclei would form. This process typically produces the nanoparticles with the size distribution of $10 < \sigma < 15\%$ in diameter, which are then narrowed to $< 5\%$ through additional size-selective processing.

The 'heating-up' method [6, 23-25] is a new attempt to separate growth from nucleation. This process is a batch process and very simple. Metal-oleate precursors are prepared from metal chloride and sodium oleate. If the metal-oleate precursors are heated in a high-boiling point solvent, they are thermally decomposed and produce monodisperse nanocrystals. This method proved to be advantageous for large-scale production. Park et al. [6] showed that as large as 40g of monodispese magnetite nanocrystals with a yield of >95% could be produced in a batch. The size uniformity of the nanoparticles is usually better than that by the 'hot-injection' method. Since this is a batch process, the precursors are not added during growth to supplement the supersaturation.

The supersaturation that has been consumed during growth can be indirectly estimated from the final size of the nanoparticles. The highly monodisperse nanoparticles were reported to grow as large as 22 nm in the 'heating up' process. Considering that the critical nucleus size is ~ 0.5 nm, the diameter increases by more than 20 nm, which implies that a considerable supersaturation must have been consumed. Since an appreciable amount of supersaturation is consumed in the initial nucleation process, the remaining supersaturation is usually not so high. Therefore, it is highly unlikely that the nuclei should grow as large as

10 nm by remaining supersaturation without precursors being supplied additionally. It should be reminded that in the 'hot-injection' method the additional precursors should be supplied to produce nanoparticles larger than 10 nm.

Then, a question arises as to why additional nucleation does not occur during such extensive growth. It appears that all the supersaturation, which should be consumed to produce the final size of nanoparticles, is not built up simultaneously but only a very small fraction of supersaturation, which is too small to trigger the nucleation, is maintained throughout the growth. This is an ideal situation where growth is separated from nucleation. One possibility to realize such an ideal situation is that the decomposition of the metal orleate complex into metal does not occur in the solution but occurs only at the surface of nanoparticles. In other words, the surface of nanoparticles acts as the catalytic site for decomposition of the metal orleate. This type of reaction is called 'self-catalytic'. Therefore, if a system is self catalytic, growth can be separated ideally from nucleation and monodisperse nanoparticles can be easily synthesized.

6. Digestive ripening

As mentioned earlier, the coarsening kinetics of Ostwald ripening has been well established by LSW theory [17, 18]. During Ostwald ripening, large particles with a low chemical potential grow at the expense of small particles with a high chemical potential and, as a consequence, the average size increases and the total number of particles decreases during coarsening. Eventually, only one large particle remains, which corresponds to a final equilibrium state.

However, Klabunde and co-workers reported a series of articles on the synthesis of monodisperse gold or silver nanocrystals by a digestive-ripening process of polydisperse nanocrystals. [9-12, 26, 27] Gram quantities of monodisperse gold or silver nanoparticles could be produced through digestive ripening, where colloidal particles from 2 to 40 nm are transformed to nearly monodisperse particles of 4–5 nm diameters.

Digestive ripening is the reverse process of Ostwald ripening. It is interesting and also puzzling in that small particles grow at the expense of large ones. Clearly, digestive ripening cannot be understood with the usual Gibbs–Thomson equation solely based on a curvature effect. According to the theory of diffusion-controlled coarsening [28], the growth rate of a particle with radius, r, is given by

$$\frac{dr}{dt} = \frac{D_f V_g C_o}{rRT}\left[\mu(r)^* - \mu(r)\right] \tag{5}$$

where R is the gas constant, T the absolute temperature, and D_f and C_o are the diffusivity and solubility of atoms in the solution, respectively. Eq. (5) is valid when the diffusion field does not overlap. $\mu(r)^*$ is the chemical potential of a particle of critical size, which neither grows nor shrinks at the given instant.

Since the phenomenon of digestive ripening runs counter to the decrease of interfacial free energy, a different type of free energy must be involved. The driving force for digestive ripening must compete against the reduction of the interface free energy. There are two such free energies. One is strain energy and the other is electrostatic energy. Since solid particles dispersed in liquid do not have any appreciable strain energy, the electrostatic energy is a possible candidate. In fact, Klabunde et al, who have developed the digestive ripening process, reported that their nanoparticles were negatively charged [9, 10]. If nanoparticles are electrically charged, they have an electrostatic energy inversely proportional to the radius of the particles. Since electrostatic energy increases with decreasing particle size, charged particles cannot shrink away completely. Therefore, the presence of charge can drastically change the Ostwald ripening behavior. The chemical potential change arising from the presence of charge can be treated by modifying the Gibbs-Thomson equation in consideration of the electrostatic energy.

To analyze the effect of charge on the coarsening behavior of nanoparticles, it is assumed that each particle is singly charged, electrically-conducting, and spherical with isotropic interface free energy, dispersed in a matrix phase with a dielectric constant of 1. According to this assumption, ions are regarded as the primary embryos of charged nanoparticles. It is further assumed that the charged nanoparticles do not coagulate with each other and that the atomic transfer between particles is diffusion-controlled. The Gibbs free energy of a spherical conducting particle with radius r and charge e (corresponding to the unit charge of an electron) is expressed as

$$\Delta G = 4\pi r^2 \sigma + k\frac{e^2}{2r}, \tag{6}$$

where σ is the interface free energy of the particle and k is defined by $1/(4\pi\varepsilon)$, where ε is the vacuum permittivity [29]. ke^2 is 2.3068×10^{-28} N·m^2. It should be noted that with decreasing r, the interface free energy term decreases but the electrostatic energy term increases.

From Eq. (6), the modified Gibbs-Thomson equation is derived as

$$\Delta\mu = \mu_r - \mu_o = RT ln\frac{C_r}{C_o} = V_m\left(\frac{2\sigma}{r} - k\frac{q^2}{8\pi r^4}\right), \tag{7}$$

where C_o is the solute concentration in the matrix without capillary effect. If the difference between C_r and C_o is small, the equation can be approximated as

$$ln\left(\frac{C_r}{C_o}\right) \approx \frac{C_r - C_o}{C_o} = \frac{V_m}{RT}\left(\frac{2\sigma}{r} - k\frac{q^2}{8\pi r^4}\right). \tag{8}$$

The coarsening behavior can be analyzed quantitatively by solving Eqs. (5) and (8) simultaneously under the constraint of mass conservation for a total number of particles N_p, and is expressed as

$$\sum_{n}^{N_p} 4\pi r_n^{\,2} \frac{dr_n}{dt} = 0. \tag{9}$$

Substituting Eq. (5) into Eq. (9) yields

$$\sum_{n}^{N_p} r_n D_f V_m \left(C^* - C_0 - \left(C_{r,n} - C_0\right)\right) = 0. \tag{10}$$

From Eqs. (5), (9) and (10), the following equation can be derived:

$$\frac{dr_n}{dt} = \frac{D_f V_m^2 C_0}{R T r_n} \left(\sum_{n}^{N_p} \left(\frac{r_n}{\sum_{n}^{N_p} r_n} \left(\frac{2\sigma}{r_n} - k \frac{q^2}{8\pi r_n^4} \right) \right) - \left(\frac{2\sigma}{r_n} - k \frac{q^2}{8\pi r_n^4} \right) \right). \tag{11}$$

If the growth rate of each particle is determined by Eq. (11), the new radius after dt is given by

$$r_n(t+dt) = r_n(t) + \left(\frac{dr_n}{dt} \right) dt. \tag{12}$$

Once the initial size distribution of particles is given, the time-dependent size distribution can be obtained by solving equations (11) and (12) simultaneously by iteration.

To demonstrate that this approach reproduces simple digestive ripening, one large charged nanoparticle of 15 nm radius and 600 ions of 0.2 nm radius were chosen as an initial state. A diffusivity, D_f, of 10^{-9} $m^{-1}sec^{-1}$ and a temperature of 393 K were chosen for calculation.[11] Besides, the interface energy of σ = 1 $mJ{\cdot}m^{-2}$, C_0 = 0.01 $mol{\cdot}m^{-3}$, and V_m =1.02×10^{-5} $m^3 \cdot mol^{-1}$, which is the molar volume of gold, were chosen. Here, the interface free energy of 1$mJ{\cdot}m^{-2}$ is chosen, simply because it produces monodisperse nanoparticles of the 1.5 ~ 2 nm size after digestive ripening. Experimentally, the surfactant, dodecanethiol, which is expected to diminish the interface free energy between gold and solution, appears to play a critical role in inducing digestive ripening. [30, 31] As the interface free energy decreases and the electric charge density increases, the size of finally-evolved monodisperse nanoparticles increases.

Fig. 4 shows the size evolution of charged nanoparticles with time. a1 and a2 in Fig. 4(a) represent the radius of charged embryos at 200 s and 600 s, respectively. Likewise, b1 and b2 in Fig. 4(a) represent the radius of the 15 nm particle at 200 sec and 600 sec, respectively. The size of the small charged embryos increases, whereas the size of the large charged particles decreases. Figs 4(b)-(d) show a display of the size distribution of nanoparticles at 200 s, 600 s and 900 s, respectively. Finally, the radius of every nanoparticle becomes 1.78 nm at 900 s, as shown in Fig. 4(d), which corresponds to 'F' in Fig. 4(a), producing perfect monodisperse nanoparticles. This calculation reproduces the experimentally observed digestive ripening behavior.

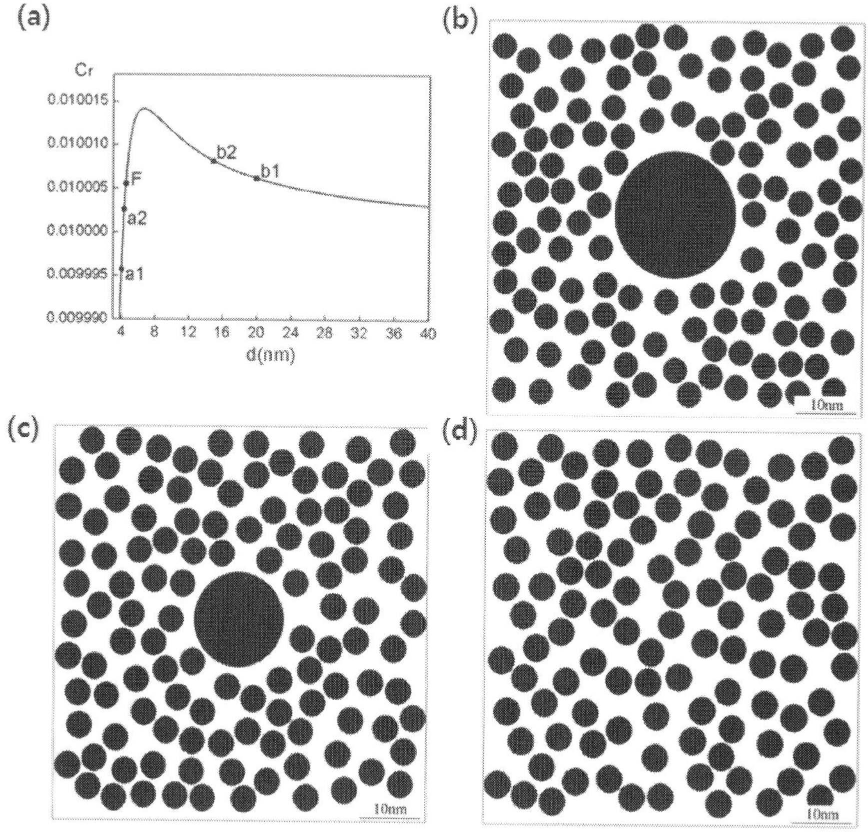

Figure 4. Coarsening behavior of one big particle of 30 nm and 300 embryos of 0.4 nm. (a) a1 and a2 indicate the size of embryos at 200 s and 500 s, respectively. Likewise b1 and b2 indicate the size of the large nanoparticle at 200 s and 500 s, respectively. Finally, the size of every nanoparticle becomes the same on F at 900s. The size distribution of nanoparticles at (b) 200 s, (c) 500 s and (d) 900 s, respectively. In (b), the size of small nanoparticles is 4 nm and that of a large nanoparticle is 19.9 nm. In (c), the size of small nanoparticles is 4.3 nm and that of a large nanoparticle is 14.9 nm. In (d), the size of every nanoparticle becomes 4.5 nm. Reprinted with permission from [13]. Copyright 2007 Elsevier.

7. Digestive ripening in a binary system

Smetana et al. [26] reported digestive ripening in a binary system, where separately prepared monodisperse Au and Ag nanoparticles with a radius of ~3.3 nm, mixed in a colloid solution, were transformed into monodisperse Au/Ag alloy nanoparticles with a radius of ~2.8 nm. The final monodisperse nanoparticles did not have a core–shell structure but had uniform composition distribution. The size of the final monodisperse nanoparticles was reduced in comparison to that of the initial unalloyed monodisperse nanoparticles, which implied that additional nucleation had occurred during digestive ripening.

To analyze the coarsening behavior of a binary system of Au and Ag nanoparticles, the chemical potential of a binary system needs to be derived. According to the phase diagram of an Au/Ag binary system [31], Au and Ag, which have the same face-centered cubic (FCC) structure and a small difference in atomic size, have a complete solid solution. Therefore, it is assumed that Au and Ag nanoparticles behave like an ideal binary solid solution, and then, the Gibbs-free energy is expressed in the whole composition range as follows [32]:

$$G = X_{Au}G_{Au} + X_{Ag}G_{Ag} + RT(X_{Au}lnX_{Au} + X_{Ag}lnX_{Ag}) \tag{13}$$

From Eq. (13), the chemical potentials of Au and Ag for the alloy are given, respectively, by

$$\begin{aligned} \mu_{Au} &= G_{Au} + RTlnX_{Au} \\ \mu_{Ag} &= G_{Ag} + RTlnX_{Ag} \end{aligned} \tag{14}$$

where G_{Au} and G_{Ag} are the molar free energies of pure Au and Ag, respectively; and X_{Au} and X_{Ag} are the mole fractions of Au and Ag, respectively. From Eqs. (7) and (14), the chemical potentials of spherically charged nanoparticles of Au and Ag with radius r can be derived as

$$\begin{aligned} \mu_{Au}(r) &= G_{Au} + RTlnX_{Au} + V_{SS}\left[\frac{2\sigma_{SS}}{r} - \frac{kz^2q^2}{8\pi r^4}\right] \\ \mu_{Ag}(r) &= G_{Ag} + RTlnX_{Ag} + V_{SS}\left[\frac{2\sigma_{SS}}{r} - \frac{kz^2q^2}{8\pi r^4}\right] \end{aligned} \tag{15}$$

where σ_{SS} and V_{SS} are the interface free energy and the molar volume of the Au/Ag solid solution, respectively. To investigate the time evolution of the coarsening behavior of the mixed Au and Ag nanoparticles, the growth rate of particles should be calculated.

Substituting Eq. (5) for dr/dt in Eq. (9) and rearranging the result yield

$$\mu(r)^* = \left[\sum_{n=1}^{N_p} r_n\mu(r_n)\right] / \sum_{n=1}^{N_p} r_n \tag{16}$$

Expressing $\mu(r)^*$ in terms of dr_i/dt and using Eq. (15) for $\mu(r)$ in Eq. (16), the growth rate of the *i*th particle by diffusion of Au and Ag atoms is obtained as

$$\frac{dr_i}{dt}=\frac{D_f V_{SS}^2 C_o}{r_i RT}\left\{-\frac{RT}{V_{SS}}lnX_{Au}-\frac{2\sigma_{SS}}{r_i}+\frac{kz^2q^2}{8\pi r_i^4}+\left(\sum_{n=1}^{N_p} r_n\right)^{-1}\sum_{n=1}^{N_p} r_n\left[\frac{RT}{V_{SS}}lnX_{Au}+\frac{2\sigma_{SS}}{r_n}-\frac{kz^2q^2}{8\pi r_n^4}\right]\right\},\quad i=1,2,\ldots \text{ for } Au$$

$$\frac{dr_i}{dt}=\frac{D_f V_{SS}^2 C_o}{r_i RT}\left\{-\frac{RT}{V_{SS}}lnX_{Ag}-\frac{2\sigma_{SS}}{r_i}+\frac{kz^2q^2}{8\pi r_i^4}+\left(\sum_{n=1}^{N_p} r_n\right)^{-1}\sum_{n=1}^{N_p} r_n\left[\frac{RT}{V_{SS}}lnX_{Ag}+\frac{2\sigma_{SS}}{r_n}-\frac{kz^2q^2}{8\pi r_n^4}\right]\right\},\quad i=1,2,\ldots \text{ for } Ag \qquad (17)$$

With Eq. (17), the radius of the *i*th particle can be calculated at a time, t +Δt, by

$$r_i(t+\Delta t)=r_i(t)+\frac{dr_i}{dt}\Delta t,\quad i=1,2,\ldots \qquad (18)$$

The analytic expression for the composition change of nanoparticles cannot be derived. Therefore, the composition change of nanoparticles should be computed from the mass change of each element, which is determined from the size change by Eq. (18) for each iteration.

To check whether these schemes reproduce the experimental result of alloy digestive ripening, a simple system of 500 Au charged nanoparticles of R_{Au} = 3.3 nm, 500 Ag charged nanoparticles of R_{Ag} = 3.3 nm and 500 charged nuclei of R_{nuclei} = 0.5 nm was considered. This system is chosen to represent the experimental condition of as-prepared Au and Ag nanoparticles mixed for digestive ripening [14]. The 500 nuclei were added because the experimental fact that the number of final nanoparticles increases after digestive ripening indicates that additional nucleation occurs. The capital letter R_i indicates an initial radius with the suffix i implying the *i*th group of particles with identical radius. The initial mole fraction of Au nanoparticles, X_{Au}, is set at 0.99 instead of 1 to avoid the infinity in calculation. In the same manner, X_{Au} of Ag nanoparticles and the size of nuclei are set at 0.01 and 0.5, respectively. The parameters used are $D_f = 0.5 \times 10^{-14}$ m²/s, $V_{SS} = 1.02 \times 10^{-5}$ m³/mol, C_o = 0.01 mol/m³, $kq^2 = 2.3068 \times 10^{-28}$ J·m, T = 393 K, σ_{SS} = 0.3 J/m², and z = 30. It should be noted that if the particles are assumed to be singly charged with z = 1, the interface energy σ_{SS} should have a much smaller value than 0.3 J/m² to reproduce the experimentally-observed digestive ripening.

For the simplification, the diffusivities of Au or Ag atoms in the nanoparticles are assumed to be high enough to be homogenized immediately because the final structure of Au/Ag alloy nanoparticles was reported not to have a core-shell structure but to have a homogeneous solid solution.

In Fig. 5, the coarsening behavior of nanoparticles with time is schematically displayed. The time evolution of the microstructure is shown in Figs. 5(a), (b), (c) and (d) at $t_1 = 0$, $t_2 = 1$, $t_3 = 2$ and $t_4 = 16$ h, respectively. Fig. 5(a) shows the initial size distribution and the composition of nanoparticles, which is represented by the gray scale from black for Au to white for Ag as shown by the gray scale bar in Fig. 5. Therefore, black, white and gray nanoparticles represent Au-rich, Ag-rich and Au/Ag alloy nanoparticles, respectively. After 1 h, the radii of both Au and Ag nanoparticles decrease to 3.07 nm and the mole fractions, X_{Au}, of Au and

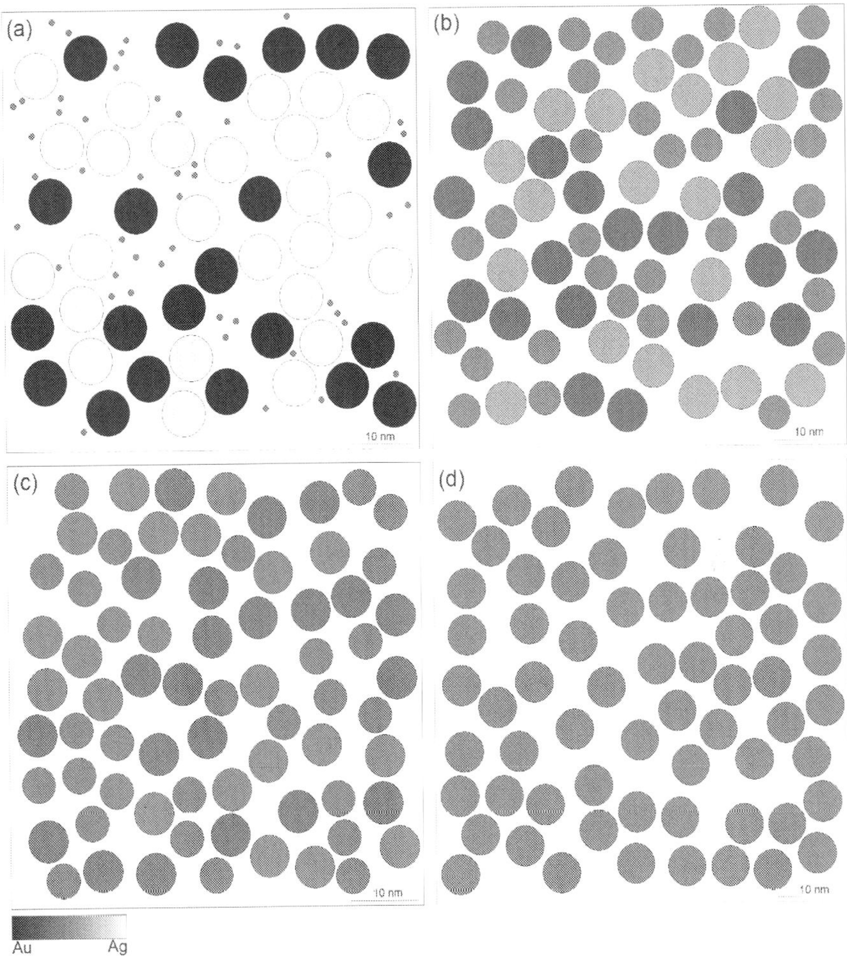

Figure 5. Display of the nanoparticles evolved through coarsening at four different times in the system of 500 Au nanoparticles, 500 Ag nanoparticles and 500 nuclei. Composition is represented by the gray scale from Au as black to Ag as white. (a) At $t_1 = 0$ h, $R_{Au} = 3.3$ nm with $X_{Au} = 0.99$, $R_{Ag} = 3.3$ nm with $X_{Au} = 0.01$ and $R_{nuclei} = 0.5$ nm with $X_{Au} = 0.5$. (b) At $t_2 = 1$ h, $r_{Au} = 3.07$ nm with $X_{Au} = 0.64$, $r_{Ag} = 3.07$ nm with $X_{Au} = 0.36$ and $r_{nuclei} = 2.4$ nm with $X_{Au} = 0.5$. (c) At $t_3 = 2$ h, $r_{Au} = 3$ nm with $X_{Au} = 0.54$, $r_{Ag} = 3$ nm with $X_{Au} = 0.46$ and $r_{nuclei} = 2.6$ nm with $X_{Au} = 0.5$. (d) At $t_4 = 16$ h, all the particles become monodisperse at 2.88 nm with the same composition of $X_{Au} = 0.5$. Reprinted with permission from [14]. Copyright 2009 Elsevier.

Ag nanoparticles become 0.64 and 0.36, respectively, as shown in Fig. 5(b). The radii of the nuclei increase to 2.4 nm with their mole fraction, X_{Au}, maintained as 0.5. After 2 h, the size distribution becomes much narrower approaching a monodisperse state and the composition becomes more homogenized as shown in Fig. 5(c). The radii of initial Au and Ag nanoparticles decrease to 3 nm, and the radii of the nuclei increase to 2.6 nm. The mole fractions of Au and Ag nanoparticles become 0.54 and 0.46, respectively. After 16 h, all the nanoparticles eventually have the same radius of 2.88 nm and the same composition of X_{Au} = 0.5, as shown in Fig. 5(d). By considering both the electrostatic energy and ideal solid solution, the coarsening behavior of the digestive ripening process, where the separated Au and Ag nanoparticles were transformed into monodisperse Au/Ag alloy nanoparticles, was successfully reproduced.

8. Conclusions

The evolution mechanism of monodisperse nanoparticles is approached thermodynamically and kinetically. Both interface-controlled and diffusion-controlled growth of particles can induce monodisperse distribution of particles if coagulation, additional nucleation and Ostwald ripening are inhibited. The diffusion-controlled growth reaches the monodisperse state at much smaller size than the interface-controlled growth. The evolution mechanism of monodisperse nanoparticles by digestive ripening is approached using the modified Gibbs-Thompson equation considering electrostatic energy. The digestive ripening behavior of both pure metal and alloy which is contrary to conventional Ostwald ripening, is well explained assuming that nanoparticles are electrically charged.

Author details

Nong-Moon Hwang and Jae-Soo Jung
Department of Materials Science and Engineering, Seoul National University, Seoul, Korea

Dong-Kwon Lee
LG Chem. Ltd, Research Park, Yoseong-gu, Daejon, Korea

Acknowledgement

This work was supported by the National Research Foundation of Korea (NRF) grant funded by the Korea government (MEST) (No. M10600000159-06J0000-15910).

9. References

[1] K. J. Klabunde, *Nanoscale Materials in Chemistry*, (Wiley-Interscience, New York, 2001).
[2] T. Sugimoto, *Monodispersed Particles*, (Elsevier, Amsterdam, 2001).
[3] G. Schmid, *Nanoparticles: From Theory to Application*, (WILEY-VCH, Weinheim, 2004).

[4] J. N. Park, S. G. Kwon, J. Joo, Y. G. Jang, and T. H. Hyeon, Angew. Chem. Int., 46, 4630 (2007).

[5] I. L. Medintz, H. T. Uyeda, E. R. Goldman, and H. Mattoussi, Nature Mater., 4, 435 (2005).

[6] J. N. Park, K. J. An, Y. S. Hwang, J. G. Park, H. J. Noh, J. Y. Kim, J. H. Park, N. M. Hwang, and T. H. Hyeon, Nature Mater., 3, 891 (2004).

[7] J. N. Park, E. W. Lee, N. M. Hwang, M. S. Kang, S. C. Kim, Y. S. Hwang, J. G. Park, H. J. Noh, J. Y. Kim, J. H. Park, and T. H. Hyeon, Angew. Chem. Int., 44, 2872 (2005).

[8] S. G. Kwon, Y. Piao, J. N. Park, S. Angappane, Y. G. Jo, N. M. Hwang, J. G. Park, and T. H. Hyeon, J. Am. Chem. Soc., 129, 12571 (2007).

[9] G. Trivino, K. J. Klabunde, and E. Dale, Langmuir, 3, 986 (1987).

[10] B. L. V. Prasad, S. I. Stoeva, C. M. Sorensen, and K. J. Klabunde, Langmuir, 18, 7515 (2002).

[11] S. Stoeva, K. J. Klabunde, C. Sorensen, and I. Dragieva, J. Am. Chem. Soc., 124, 2305 (2002).

[12] B. L. V. Prasad, S. I. Stoeva, C. M. Sorensen, and K. J. Klabunde, Chem. Mater., 15, 935 (2003).

[13] D. K. Lee, S. I. Park, J. K. Lee, and N. M. Hwang, Acta Mater., 55, 5281 (2007).

[14] D. K. Lee and N. M. Hwang, Scripta Mater., 61, 304 (2009).

[15] X. Peng, J. Wickham, A.P. Alivisatos, J. Am. Chem. Soc. 120, 5343 (1998).

[16] H. Reiss, J. Chem. Phys. 19, 482 (1951).

[17] I. M. Lifshitz, V. V. slyozov, J. Phys. Chem. Solids 19, 35 (1961).

[18] C. Wagner, Zeitschrift fuer Elektrodhemie 65, 581 (1961)

[19] N. R Jana, L. Gearheart, C. J. Murphy, Chem. Mater, 12, 2313 (2001), H. Yu, P.C. Gibbons, K. F. Kelton, W. E. Buhro, J. Am Chem, Soc. 123, 9198 (2001).

[20] C. B. Murray, D. J. Norris, M.G. Bawendi, J. Am. Chem. Soc., 155, 8706 (1993).

[21] D. V. Talapin, A. L. Rogach, A. Kornowski, M. Haase, H. Weller, Nano Lett., 1, 207 (2001).

[22] J. Hambrock, R.Becker, A. Birkner, J. Weiss, R. A. Fischer, Chem. Commun. 68 (2002), N.R. Jana, X. Peng, J. Am. Chem. Soc., 125, 14280 (2003).

[23] W. S. Seo, H. H. Jo, K. Lee, J. T. Park, Adv. Mater., 15, 795 (2003).

[24] W. W. Yu, J. C. Falkner, C. T. Yavuz, V.L. Colvin, Chem. Commun., 2306 (2004).

[25] J. Joo, H. B. Na, T. Yu, J. H. Yu, Y. W. Kim, F. Wu, J. Z. Ahang, T. Hyeon, J. Am. Chem.Soc., 125, 11100 (2003).

[26] A. B. Smetana, K. J Klabunde, C. M. Sorensen, A. A. Ponce, B. Mwale, J. Phys. Chem. B, 110, 2155 (2006).

[27] D. V. Leff, P. C Ohara, J. R Heath, W .M Gelbart, J. Phys. Chem., 99, 7036 (1995).

[28] P. G. Shewmon, Transformations in Metals, (McGraw-Hill, New York, 1969).

[29] E. A. Boucher, Nucleation (ed. A. C Zettlemoyer), (Marcel Dekker, INC, New York, 1969).

[30] I. Stalder, J. H Bilgram, J. Chem. Phys. 118, 7981 (2003).

[31] H. Okamoto, T. B. Massalski, Phase Diagrams of Binary Gold Alloys, (ASM International, Metals Park OH, 1987).

[32] D. A. Porter, K. E. Easterling, Phase Transformations in Metals and Alloys, (Chapman & Hall, London, 1992).

Chapter 15

Statistical Thermodynamics of Lattice Gas Models of Multisite Adsorption

Vasiliy Fefelov, Vitaly Gorbunov,
Alexander Myshlyavtsev and Marta Myshlyavtseva

Additional information is available at the end of the chapter

http://dx.doi.org/10.5772/51802

1. Introduction

The lattice models naturally arise in different fields of physics, chemistry and other sciences. First, it is physics of the solid state and physicochemistry of the surface. Among the many well-known lattice models the magnetic, alloys, liquid mixture, adsorption models are usually mentioned. The lattice models can be both classical and quantum. In this chapter only the classical lattice models focusing on models arising in physicochemistry of the surface will be considered. For the beginning let's give the most common formal definition of the classical lattice model.

Let there be given some finite or countable set M. Its elements will be called sites or nodes of the lattice and numbered index i. Each site is associated with the vector ($\vec{c_i}$) having k_i components. Each of the components can take a finite countable or uncountable number of values, i.e. without loss of generality, it can assume that the components of this vector take either integer or real values in some finite or infinite interval. The specific values of the $\vec{c_i}$ vector components determine the *i*-th site state. The *M* set state uniquely determines the state of all its elements. The system states can be divided into allowable and unallowable ones. Each allowable state of the $M\{\vec{c_i}\}$ set is match to the real number $E_M\{\vec{c_i}\}$ called the energy per lattice site. (Formally, unallowable state can be assigned value $E_M = +\infty$). Accepting the Boltzmann probability distribution of the system states as an additional postulate, one can receive object called a classical lattice model. All the models considered below are particular cases of the introduced generalized lattice model.

Determination of the lattice model dimension may be connected with the number of sites where can be the particle performing a random walk on the lattice in n steps. For *d*-dimensional regular lattice the number N is proportional to the volume of a box with an edge n, where $N \sim n^d$, i.e. the higher the dimension, the closer neighboring sites are located.

Following this approach 1,2,3,..., d-dimensional lattice models (for example, d-dimensional hypercubic lattices with interaction between the nearest-sites) are naturally obtained. 1,2 and 3-dimensional lattice models are of most importance for specific applications to the natural sciences. However, besides models whose dimension is equal to some natural number models which dimension defined the same way as above is fractional or infinite are of great interest. The simplest example of a model with a fractional dimension is the Ising model on the diamond-like lattice.

Concluding the general description of lattice models it is worth to introduce the concept of a homogeneous lattice model. Models properties of all their sites are identical are called homogeneous lattice models. It follows the obtained lattice model has a geometric realization possessing the property of translational invariance. As it is mentioned above, the models relating to physicochemistry of the surface are of main interest, respectively, these are primarily two-dimensional ones. It is readily to conclude that all the two-dimensional homogeneous models can be divided to three classes having in the basis the geometric realizations: the square, triangular and hexagonal lattice. Note that the lattice models having different types of sites but whose geometric realizations possess translational invariance can serve the generalization of concept of lattice model homogeneity. From the computational point of view these models are very similar to homogeneous ones. Thus, the most common definition of the classical lattice model, its dimension, homogeneity and translational invariance have been introduced.

Lattice gas model and its various generalizations are one of the most important lattice models of modern statistical physics. Despite its relative simplicity these shows fantastic variety of non-trivial physical phenomena. First, this refers to the phase transitions of various types whose study is of great interest from the standpoint of the general theory. Note that the adlayers on the monocrystal faces represent the physical realization of many interesting and important models for $d = 2$, in particular, these admitting exact solutions. The fact causes an increased interest of theorists to such systems.

Usually the lattice gas means the molecular system which differs from the continual gas following feature: molecules of such a system can occupy only such places in the space in which their centers are located at the sites of one-, two- or three-dimensional geometric lattice. The interaction energies of molecules located in different configurations respect to each other are the parameters of the model. In the general case, these parameters also depend on the relative orientation of the molecules. These interactions called lateral naturally can be divided to pairs and many-particle. Pair interactions are additive while many-particle ones are nonadditive.

It is easy to see from the standpoint of common definition of classical lattice model the lattice gas model (LGM) stated above refers to the same class of models as the Ising model. Indeed, considering the dependences of the energy of lateral interaction of molecules on their relative orientation one can obtain the lattice model in which the vector $\vec{c_i}$ is a scalar accepting two values (when all the molecules are the same type). From the above it immediately follows that the simplest LGM is isomorphic to the classical Ising model in the sense of the common definition. The latter statement is a very important because the property of one model can be obtained that of another one.

One of the simplest examples of the system for which LGM is quite adequate model is the monomolecular adsorption model. It is well known that chemisorption has localized character as in this case there occurs chemical bond between the adsorbed molecule and the substrate. For substrate being face of monocrystal chemisorbed molecules are located in well defined places of the periodic lattice formed by the surface atoms. Therefore, the simplest LGM introduced above is completely adequate model of monomolecular chemisorption on monocrystal face.

The thermodynamic Hamiltonian of the simplest LGM with one type of particles on regular lattice (for example, hypercubic) has the form:

$$H = \varepsilon \sum_{\langle n,n \rangle} n_i n_j - \mu \sum_i n_i \,, \tag{1}$$

where ε is the lateral interaction energies of the nearest neighbors; μ is the chemical potential; the occupation number n_i is equal to unity if the adsorption site is occupied and zero in the case of a vacant site. The Hamiltonian (1) for d = 2 in the first approximation describes the monomolecular (one-center) adsorption on a monocrystal face.

In some cases this description corresponds to reality but in most cases it does not [1]. It is known that even at adsorption of the most simple gases such as Ar, Kr, Xe, nitrogen, carbon monoxide, oxygen and others the adsorbate molecule size is usually larger than the distance between neighboring active centers of adsorption [2-6]. In this connection at the adsorbate molecule adsorption on one active center it simultaneously occupies one or several neighboring active centers.

The assumption of one-center adsorption becomes even more inadequate at the description of adsorption of linear and quasilinear molecules such as the simple saturated and unsaturated hydrocarbons [7,8]. In the paper complete review of the experimental results on the adsorption of simple hydrocarbons on the surface of metal monocrystals (Pt, Pd) was given and, particularly, the adsorption of such molecules was shown to occur parallel to the solid surface in connection with which multicenter nature of the adsorption becomes apparent.

In addition in this review some types of ordered structures of adsorption monolayer C_4 – C_8 alkanes experimentally observed at low temperatyres on the surface of monocrystalline platinum are discussed. The necessity of describing of the n-alkane adsorption in the framework of the models with multisite adsorption is also supported by the fact that at the interpretation of experimental adsorption isotherms by using the known multisite Langmuir model and other analytical models there exists good correlation between the model parameter k (the number of active sites occupied by adsorbate molecule) and the real number of segments in the molecule of the adsorbate [7,8,10-12]. It should be noted that the above analytical models of multisite adsorption are used to determine the specific surface of porous solids and its topography (in the case of energetically heterogeneous surfaces).

Obviously that the adsorption of more complex (in regard geometry, the chemical structure – the presence of double/triple bonds or several functional groups) of molecules is even more nontrivial [14-25]. First of all this is manifested in that complex organic molecules (cyclic hydrocarbons, aromatic systems etc.) depending on their geometry and chemical structure can form set of different ordered structures on the solid surface.

Along with the possibility of multisite adsorption that of different orientations of molecules with respect to the interface is one of the most interesting features of these systems. Indeed, over the past ten years series of experimental works devoted to the study of organic self-assembled monolayers on metal surfaces has been published. The general conclusion of these studies is that the molecule orientation in the adlayer is a function of external parameters such as concentration, pressure, temperature, electrode potential and others[14-25]. Moreover, very interesting ordered structures have been experimentally found in some of similar systems. Those are structures which simultaneously contain the molecules with different orientations in the adlayer. For example, in [18] the authors investigated behavior of the adsorption monolayer of molecules p-Sexiphenil on the Au(111) surface in ultrahigh vacuum using the method of scanning tunneling microscopy. It was shown that five various ordered structures two of which contain molecules with different orientations in the adlayer – molecules oriented parallel to the surface and tilted to the surface at an angle can be formed. In [24] phase transitions in the adlayer of acid trimezin molecules on the Au(111) surface have been investigated using the method of scanning tunneling microscopy, and one of them proved to lead to formation of the ordered structure containing the acid trimezin molecules oriented both parallel and perpendicular to the surface. Another striking example of the adsorption system in which the adsorbate molecules can have different orientations with respect to the surface is the adsorption of cyclic unsaturated hydrocarbons on the reconstructed semiconductor surface especially on the reconstructed face of the Si(001) [25-27]. This is connected to the fact that cycloaddition reaction [2+2] with formation of di-σ bond Si-C with the silicon surface atoms results from chemisorption of unsaturated organic molecules on the silicon surface. In chemisorptions of more complex hydrocarbons containing several unsaturated bonds the particle adsorbed on the surface can have several configurations depending on the number of di-σ bonds Si-C [28]. Moreover, the stable π-complex being resulted from interaction between unsaturated hydrocarbon molecule and silicon atom has been experimentally discovered recently [29,30]. In light of the above it is clear that all specified features of the behavior arising at detail studying of adsorption of simple or complex molecules can be investigated only in the framework of models taking into account the multisite character of adsorption and the possibility of different orientation of molecules both with respect to each other and with respect to the solid surface.

Practically, theoretical study of organic self-assembled adsorption monolayers or thin organic films is of great interest generally in connection to the set of possible applications in which thin organic films are used anyway [31]. The potential field of application of such systems is an organic optoelectronics, in particular, electroluminescent devices [32], photovoltaics [33], organic field-effect transistors [34]. Similar systems are used as coatings on computer hard drives to provide protection against corrosion and low friction [35].

It is difficult to overestimate the applied significance of considered systems in the field of chemistry and biochemistry where they are used as active elements of chemical and biological sensors [36], in heterogeneous catalysis [37] and as coatings for biomedical implants [38]. Polymorphism of organic thin films and the ability of molecules to constitute different crystalline forms leads to is it is very difficult to control growth and properties of such systems. Moreover, it was recently shown that the structure of the organic film significantly affects epitaxial growth of crystals on the film [39]. This fact can be used to control the morphology of nanocrystalline systems.

In all these cases location of the molecules on the surfaces is a decisive factor that determines process of growth of the film and its physical properties. Therefore, a detailed understanding of elementary phisical and chemical processes occurring in such systems is the primary motivator at the investigation of molecular self-organization on the solid surface. The development of realistic models of such systems can allow completely to control the process of self-assembly of organic and other molecules on the solid surface and to come nearer to dream of nanotechnology – to gather material possessing the necessary properties with atomic precision. Based on the above it is clear why the interest of researchers specified both practical and theoretical considerations is now shifting towards more complex lattice models some of which will be discussed in this review.

Let's consider the class of lattice models describing the so-called multisite adsorption. The simplest LGM considered above in the two-dimensional case is a model of monosite adsorption (active centers of adsorption, generally speaking, do not coincide with the surface atoms). However, as noted above at the description of many adsorption systems one is forced to abandon from idea about the monosite adsorption. In the framework of the LGM multisite adsorption is described as a system of prohibitions on certain configurations. At the same time a relatively simple lattice geometrically equivalent to the crystal lattice of the surface is persisted but the number of possible states of the site determining which part of the complex molecule is located above this site and how adsorbed molecule is oriented is increased. Further, the main results obtained in the framework of the models with multisite adsorption will be presented.

2. The models of dimer and *k*-mer adsorption

The simplest model of multisite adsorption is a dimer adsorption model. In the first approximation the dimer model described adsorption of molecules consisted of two the same atoms, for example H_2, N_2, O_2 etc. when temperature is relatively low and molecules cannot dissociate.

Statistical thermodynamics of the dimers lattice models has a long history. This is one of the earliest lattice models which take into account the own size of molecules in the frame of the lattice gas model. Apparently, the first model of the dimer has been studied in the context of the entropy of the adlayer in 1937 [40]. As it turned out, the dimer model has deep connections with the Ising model and many other important models in statistical physics. In the early sixties of the twentieth century an exact solution of the dimer model on a square

developed by Roma et al. Semi-empirical model is a combination of the exact solution for one-dimensional approximation and Guggenheim-DiMarzio approximation [65]. A new theoretical approach is significantly better than other existing approximations and allows fairly simple explaining the experimental data.

When comparing the adsorption isotherms of monomers and k-mers, it was found that in the second case the symmetry of the "particle-vacancy" is broken. The isotherms are shifted toward lower coverage with increasing coordination number of the lattice. In other words, for a given value of the chemical potential the surface coverage increases with the lattice coordination number.

As can be seen in most works devoted to the development of new analytical methods, the results are compared not only with the experimental data, but also with the results obtained by the Monte-Carlo, as the reference.

The Monte Carlo method has proved to be a very powerful tool in the study of *k*-mers adsorption. Using this method with different techniques (reweighing, finite-size scaling, and others [66,67]) many important parameters of the phase behavior for different *k*-mers adsorption systems were identified, such as the types of ordered phase structures, the points of phase transitions and critical indexes of phase transitions, etc. [68,69]. The appearance of ordered phases for the model *k*-mers in the presence of lateral interactions has its own specifics, this is due to the presence of orientation in the arrangement of the adsorbed molecules relative to each other.

In ref. [70] the authors, using the transfer-matrix method, investigated the ordered structures of the adsorption layer consisting of interacting dimers adsorbed on a square lattice. Analysis of the changes of the adlayer entropy and the surface diffusion coefficient showed that there is a finite number of ordered phases in case of repulsion lateral interactions between the nearest neighboring molecules.

Later in ref. [71] Ramirez-Pastor et al. using the Monte Carlo method have considered both attractive and repulsive interactions between adsorbed dimers on a square lattice. It was shown that in the case of attractive interactions, the phase diagram is similar to the diagram for a monoatomic gas, but the critical temperature is shifted to higher values. The most interesting case is repulsive interactions when a variety of ordered structures take place. In the case of dimers the symmetry of the "particle-vacancy", typical of monatomic particles, disturbed, that leads to the asymmetry of the adsorption isotherm with respect to the line $\theta = 0.5$, on the isotherm two steps take place. When $\theta = 0.5$ $c(2\times4)$ structure formed, which is characterized by the alternation of the adsorbed dimer and two adjacent vacancies. When the chemical potential μ increases and θ close to $\theta = 2/3$, adsorbed dimers form parallel zigzag rows (ZZ phase) [71.72]. A similar phase behavior of adsorbed layer of dimers is observed in the case of triangular and honeycomb lattices [73] (Fig. 1). In addition, the scientific group of Ramirez-Pastor, using Monte Carlo method and finite-size scaling techniques, calculated the critical exponents and critical temperatures, and calculated a phase diagram for dimers with repulsive lateral interactions on a square [71] and triangular

[74] lattices. It was shown that the system does not belong to the universality class of two-dimensional Ising model.

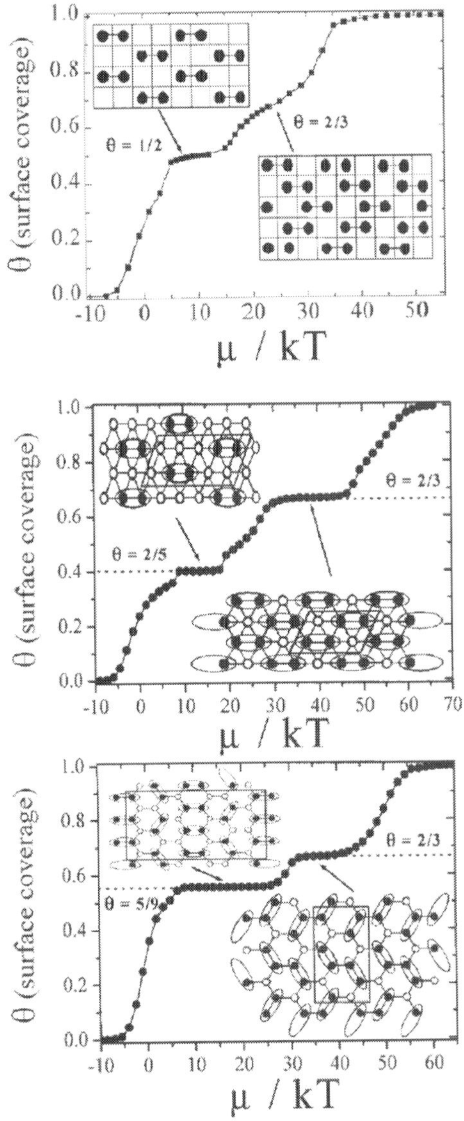

Figure 1. Adsorption isotherms of dimers on square, triangular and honeycomb lattices and corresponding ordered phases [71,73].

The study of the adsorption monolayers consisting of heteronuclear dimers showed that the phase diagram of the system greatly influenced by the quantity of energy of lateral interactions between the different types of segments (it is about two different molecules). In the study of phase diagrams of these films interesting phenomena were found. In particular, the coexistence of three phases and a variety of structural transitions, and ordered linear type structure exists even at high temperatures [75].

Similar calculations were performed for the adsorption of homonuclear dimers on heterogeneous surfaces [1,10,76-79]. In the case where the surface is represented by heterogeneous clusters of active sites of one kind and another, the approximate solution is relatively simple (the solution for multisite adsorption on a homogeneous surface can be used). In describing the adsorption of dimers on heterogeneous surfaces, created at random, the task becomes more complicated. Slightly more than twenty years ago an approximate solution was suggested for this model [1,10,79]. However, the original approach of Nitta et al. could only be applied to surfaces with a discrete distribution of adsorption energy. Later on the basis of this approximation Rudzinski and Everett [1] obtained a solution for a model with continuous distribution of adsorption energy on the surface. However, the usable area of this and other approximations [78] is limited. For example, in [77], the authors investigated an analytical approximation by Monte Carlo method to describe the adsorption of homonuclear dimers on heterogeneous surfaces, created randomly (random heterogeneous surface). The calculation shows that this approximation yields accurate results when the difference between the energies of adsorption on active sites of one and other kind is small.

There are papers devoted to the description of the first order phase transition such as "surface gas – surface liquid" [75,76,80]. The most important conclusion from all these studies is that with increasing molecular size the critical temperature shifts to higher values. On the other hand, increasing of the molecule flexibility leads to decreasing of critical temperature. A very interesting phenomenon was observed in the study of phase transition "surface gas – surfacee liquid" in the adlayer consisting of heteronuclear trimers – namely, the coexistence of three phases (there is a second phase transition "liquid – liquid"). In this case, the phase diagram is asymmetric – there are a shift of the critical density to the unit and an increasing the critical temperature.

The findings in the study of dimers adsorption on heterogeneous surfaces with a sufficient degree of accuracy can be extended to the adsorption systems of k-mers [10,78,79]. However, as was shown in ref. [81], the calculation accuracy of the approximations developed for the adsorption of dimers on heterogeneous surfaces decreases with increasing molecular size.

Summarizing the results obtained in these works, one can conclude that the phase behavior of adlayer of dimers on heterogeneous surfaces defined by the following factors: 1) the distribution of the various active sites of adsorption, and 2) the relation between the k-mer length and the size of the local heterogeneity, 3) adsorption energy on different active sites.

It should be noted about the theoretical studies of percolation threshold in systems with multisite adsorption. Information on transitions "percolate region" – "nonpercolate area" on the surface is of great interest both from the point of view of phase transitions science, as well as from the applied point of view, in particular the appearance of surface conductivity. In the case of the two-dimensional lattice gas model the percolation threshold is so minimum value of the coverage θ on an infinite surface, as infinitely large cluster of adsorbed molecules is formed. In ref. [82,83] it is shown that in addition to the dimension of the system and the lattice coordination number on the percolation threshold a size of k-mers has a significant effect. Influence of temperature on percolation in adsorption systems with flexible k-mers studied in ref. [50] by the Monte Carlo method. There was found an interesting phenomenon – a nonmonotonic behavior of the percolation threshold as function of temperature, i.e. there exists a temperature at which the percolation threshold is a minimum, and the position of the minimum does not depend on the length of k-mers.

Thus, all existing works on the k-mers adsorption, as in the case of dimers, are concentrated in four main areas: (1) study of the influence of the chemical structure of noninteracting adsorbed molecules on the behavior of their adsorption, (2) study of the influence of surface heterogeneity on the k-mers adsorption, (3) description of first order phase transition in the adlayer in the case of mutual attraction of the adsorbate – adsorbate. Only a small number of papers devoted to the study of phase transition "order-disorder", which takes places in the case of repulsive intermolecular interactions [69,70]. In particular, in ref. [69], using Monte Carlo simulation, two important characteristics of the critical properties of repulsive k-mers were identified: (1) the minimum critical temperature have on the value of k equal to 2 (dimers) and (2) for $k > 2$ the critical temperature increases monotonically with increasing k. Similar results (qualitatively) have been obtained from the analytical calculation of the mean field approximation and the principle of minimum free energy.

Additionally it should be noted that long before the above works, the study of the multisite adsorption was engaged by Soviet scientists in Temkin (1938 [84]), and Snagovski(1972 [85-87]). In ref. [84-87] the adsorption isotherms were analytically obtained in the case of multisite adsorption of two-center (k = 2), square (k = 4) and hexagonal (k = 7) complexes. Today's papers on multicenter adsorption are essentially the development, and sometimes repeating, of the works of Temkin and Snagovski.

As one can see, at present the theory of adsorption of k-mers are actively developing – adsorption isotherms are calculated, phase diagrams are constructed, percolation thresholds are determined, etc. This is due primarily to a large applied importance of such research. Already, it can be concluded that the behavior of adlayers consisting of molecules that occupy more than one active site of the surface is significantly different from the behavior of systems with monosite adsorption. However, it should be noted that in all these considered works the adsorbate molecule can adsorb only one way to the surface – is planar.

3. Multisite adsorption of orientable molecules

Today there is a small number of papers devoted to theoretical investigation of the behavior of adsorption monolayer consisting of molecules that can have a different orientation with respect to the surface [88-92]. Let's examine them in detail.

The earliest papers on theoretical study of molecular reorientation in the adsorption monolayer were carried out at MSU by Gorshtein and Lopatkin in 1971 [88,89]. They investigated one- and two-dimensional lattice models of diatomic molecules adsorption. It was assumed that the molecule can adsorb in two different ways with respect to the surface: vertically and horizontally. Each type of adsorption had its heat of adsorption, and adsorption energy of vertically oriented molecule was approximately two times smaller. The lateral interactions between adsorbed molecules were not taken into account. The authors derived an exact analytical expression for the adsorption isotherms in one-dimensional case and the approximate equation for two-dimensional lattice. It is shown that for the large values of adsorption heats at low pressures, most of the molecules adsorbed horizontally, and the number of vertically orientated molecules is very small. When coverage increasing the horizontally adsorbed molecules change the orientation, and the number of molecules adsorbed vertically grows fast. In addition, the authors had obtained expressions for isosteric heats of adsorption. Having analyzed calculated thermodynamic functions they concluded that the system exhibits two modes of adsorption: on two neighbor sites in the region of low pressure and on one site at high pressures.

The authors of [90] studied the adsorption of heteronuclear dimers (A-B) on a homogeneous surface with a mean-field approximation. In this case the dimer can be adsorbed on the surface in three different ways: horizontal adsorption with two segments at the same time and vertical adsorption with the A segment or vertical adsorption with the B segment. It was assumed that all three types of adsorption differed in the adsorption heat, and the interaction between adsorbed molecules were absent. The authors derived analytical expressions for the adsorption isotherms and isobars. In fact, the physical results obtained in this work are in qualitatively agreement with Gorshteyn and Lopatkin's ones and partially duplicate it.

There is the quite interesting model of spin-1 type, which can describe the adsorption of heteronuclear dimers on different lattices [91,92]. It is assumed in the model that all adsorbed molecules are oriented vertically to the surface only, and the energy of adsorption depends on what segment, A or B, molecule adsorb with. The authors of [91,92] considered the various sets of lateral interactions between dimers, in particular, they took into account not only interactions between nearest neighbors, but also between next-nearest neighbors. Such complex lateral interactions in the system led to the set of ordered surface structures and phase transitions. It had been shown that if the lattice is completely filled the "order-disorder" transition may occure via a continuous phase transition as well as the first-order phase transition depending on the model parameters. Moreover, the continuous phase transition is nonuniversal. It should be noted that the model does not take into account the possibility of horizontal orientation of the dimer and generally speaking this model can be attributed to the well-known Blume-Emery-Griffiths model [66].

In [93] authors proposed the general lattice gas model describing the adsorption of complex molecules. The model can be formulate by following: the molecule can be adsorbed on the surface by the k ways occupying the $m_1, m_2, \dots, m_k$ active sites located in the corresponding configurations on homogeneous or heterogeneous lattice, respectively. The simplest model of this type is the model of homonuclear dimers adsorption. In this model it is assumed that dimer may have two different orientations in the adlayer – parallel (adsorption on two active sites) or perpendicular (adsorption on one active site) to the surface. In the language of this general model homonuclear dimers adsorption model belongs to the class of models with $k=2,\ m_1=1,\ m_2=2,$ and heteronuclear dimers adsorption model [90] – to the class with $k=3,\ m_1=1,\ m_2=1,\ m_3=2.$ The model studied in [91,92] belongs to the class of $k=2,\ m_1=1,\ m_2=1.$ It should be noted that all these simple classes of models have a single representative and the set of numbers $k, m_1, m_2, \ldots, m_k$ uniquely identifies the type of model.

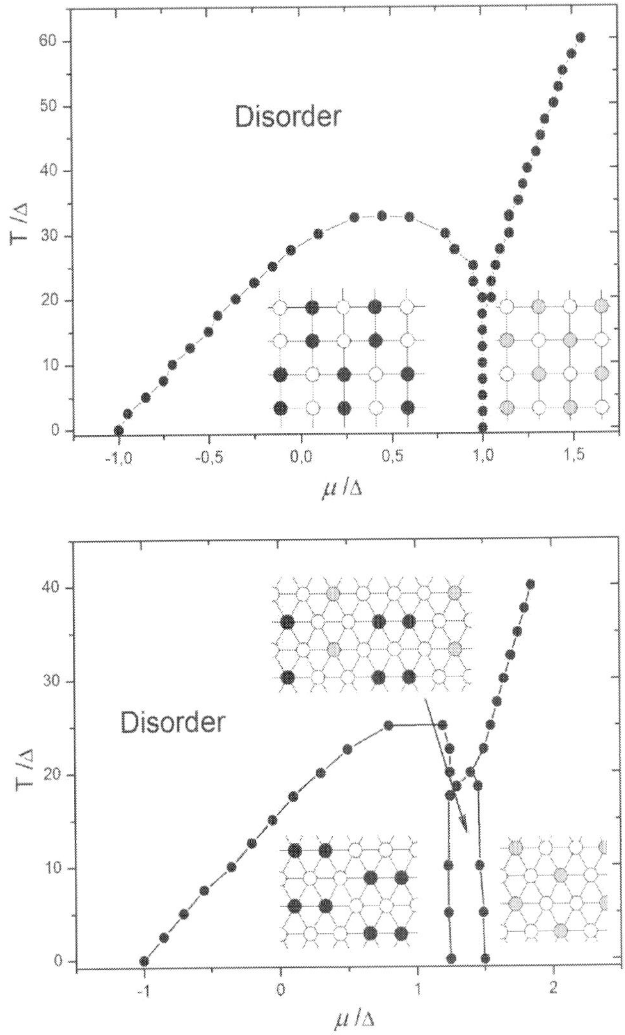

Figure 2. Phase diagrams for dimers on square and triangular lattices. Ordered structures: black circles are site occupied by dimers adsorbed on two sites; gray circles are sites occupied by dimers adsorbed on one site.

Thermodynamic equilibrium is reached by spin-flip (Glauber) dynamics [102] and diffusion relaxation (Kawasaki dynamics) [103]. To calculate thermodynamic functions successive configurations of the adlayer are generated using Metropolis transition probabilities [104] in the grand canonical ensemble.

Analysis of the ground state (T = 0K) of the model allowed to conclude that in the adlayer due to the infinitely strong repulsive interactions between nearest-neighbor molecules the set of chessboard type ordered structures forms: $c(4{\times}4)_4$, $c(3{\times}3)_{4\text{-}1}$ and $c(2{\times}2)$.The structures are schematically shown in Fig. 4. In addition, the lattice gas phase (LG) with zero coverage exists in the ground state of the system. The corresponding phase diagram of the adsorption monolayer in the ground state of the system shown in Fig. 5.

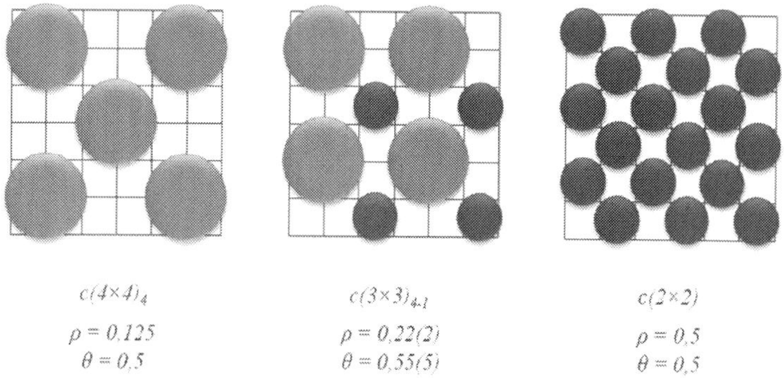

Figure 4. Ordered structure of the adlayer. Gray circle denotes a molecule adsorbed on the four active sites, and the black circle – molecule adsorbed on one active site. The structures are shown in order of their formation with the chemical potential increasing.

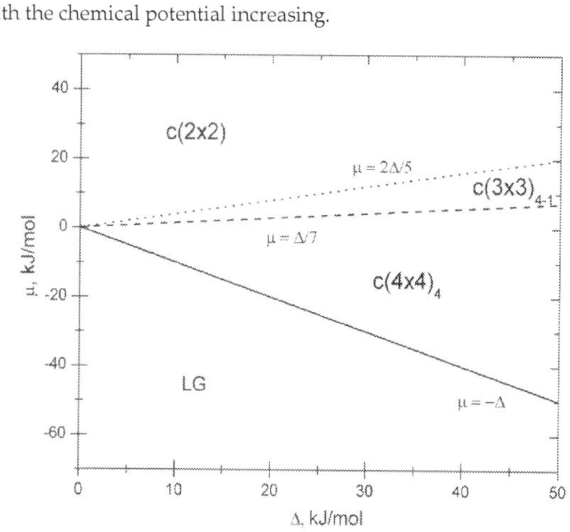

Figure 5. Phase diagram of the adlayer in the ground state. The solid line separates the stability region of the LG phase and stability region of phase $c(4{\times}4)_4$, while the dashed and dotted lines separate the stability region of phases $c(4{\times}4)_4 - c(3{\times}3)_{4\text{-}1}$ and phases $c(3{\times}3)_{4\text{-}1} - c(2{\times}2)$, respectively.

The adsorption isotherms calculated for different values of the ratio Δ/RT are shown in Fig. 6. It is seen that for sufficiently large values of Δ/RT there are three distinct plateaus on the isotherms, which correspond to thermodynamically stable phases of the adsorption monolayer at $\rho=0{,}125$, $\rho=0{,}22(2)$ и $\rho=0{,}5$, where ρ is density of the adlayer (the amount of adsorbed molecules per site). Based on the values of ρ corresponded to each horizontal plateau one can conclude that the first plateau determines the existence region of the ordered phase $c(4\times4)_4$, the second plateau – the existence region of the phase $c(3\times3)_{4\text{-}1}$, and the third – the existence region of the phase $c(2\times2)$. In addition, according to the shape of the adsorption isotherms it can be assumed that the ordered structure $c(4\times4)_4$ is formed from disordered lattice gas phase via second-order phase transition, which associated with the origin of a new symmetry element in the system. On the other hand, the phase transitions $c(4\times4)_4$ – $c(3\times3)_{4\text{-}1}$ and $c(3\times3)_{4\text{-}1}$ – $c(2\times2)$ are the first-order phase transitions which associated with sharp changing in the first derivatives of the free energy of the system, in this case it is expressed by the sharp changing in the number of adsorbed molecules in the system. It is worth to note that along with an abrupt changing in the number of adsorbed molecules in the system, phase transitions $c(4\times4)_4$ – $c(3\times3)_{4\text{-}1}$ and $c(3\times3)_{4\text{-}1}$ – $c(2\times2)$ are also attended by the reorientation of the adsorbed molecules with respect to the solid surface.

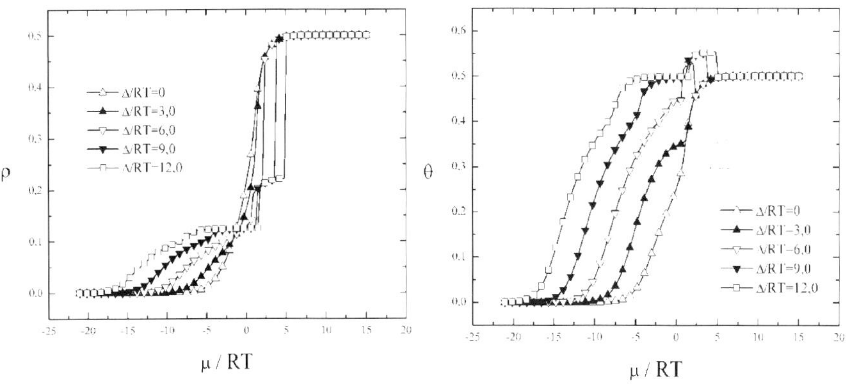

Figure 6. Adsorption isotherms (left) and the dependencies of surface coverage on reduced chemical potential (right) calculated for different values of Δ/RT.

The dependencies of surface coverage on reduced chemical potential Δ/RT demonstrate the same phase behavior of adsorbed monolayer (Fig. 6). Namely, there are three horizontal plateaus on the surface coverage curves at $\theta=0{,}5$, $\theta=0{,}55(5)$ and $\theta=0{,}5$ for sufficiently large values of the ratio Δ/RT. These values of surface coverage, as in the case of the adsorption isotherms, coincide with the values of surface coverage for the above mentioned phases $c(4\times4)_4$, $c(3\times3)_{4\text{-}1}$ and $c(2\times2)$, respectively . However, there is one significant difference between the adsorption isotherms and the dependencies of surface coverage on reduced chemical potential: the isotherm is not only coincides with the curve of surface coverage, but even out of proportion to it. Indeed, the dependencies of surface coverage on the reduced chemical potential are non-monotonic function for sufficiently large values of Δ/RT. While the adsorption isotherms in accordance with the general conditions of stability of thermodynamic systems

$$(\delta\Omega)_{Tp\mu} = \delta\left(G - \mu N\right)_{Tp\mu} = \left(\frac{\partial G(T,p,N)}{\partial N} - \mu\right)\delta N + \frac{1}{2!}\frac{\partial^2 G}{\partial N^2}\left(\delta N\right)^2 +$$

$$+\frac{1}{3!}\frac{\partial^3 G}{\partial N^3}\left(\delta N\right)^3 + \frac{1}{4!}\frac{\partial^4 G}{\partial N^4}\left(\delta N\right)^4 + \ldots + \frac{1}{m!}\frac{\partial^m G}{\partial N^m}\left(\delta N\right)^m + \sum_{n=m}^{\infty}\frac{1}{n!}\frac{\partial^n G}{\partial N^n}\left(\delta N\right)^n > 0, \quad (3)$$

$$\mu = \frac{\partial G(T,p,N)}{\partial N}, \quad \left(\frac{\partial \mu}{\partial N}\right)_T > 0,$$

are monotonically increasing functions for all values of Δ/RT, where Ω – grand potential, G – Gibbs free energy, μ – chemical potential, N – amount of adsorbed molecules, T – temperature and p – pressure in gas phase. The same effect is observed in the dimer adsorption models on square and triangular lattices. Apparently, this effect takes a place in any adsorption monolayer of molecules with different orientations with respect to the interface.

In order to uniquely identify the ordered structures appearing in the adsorption monolayer the authors of [93] calculated the curves of partial surface coverage by molecules adsorbed on one and four active sites as functions of the reduced chemical potential μ/RT (Fig. 7). It is seen the phase with coverage $\theta = 0{,}5$ formed at low values of μ/RT consists only of molecules adsorbed on four active sites, so it is obvious that it has the structure of $c(4\times4)_4$. Then, with increasing chemical potential (gas phase pressure) or decreasing in temperature of the substrate, the system undergoes the phase transition from phase $c(4\times4)_4$ to the phase with coverage $\theta=0{,}55(5)$. It is clear from Fig. 7 that this phase consists of both molecules adsorbed on four active sites and on one active site. Therefore, this phase has the structure of $c(3\times3)_{4\text{-}1}$. With further increase of the chemical potential or when the temperature decreases the system undergoes the phase transition from phase $c(3\times3)_{4\text{-}1}$ to the phase $c(2\times2)$ with $\rho=0{,}5$ and $\theta=0{,}5$ which is formed only by molecules adsorbed on one active site.

Recently, a similar phase behavior of adlayer of complex organic molecules was observed experimentally (Fig. 8). In [24] the authors investigated the behavior of the adsorption monolayer of trimesic acid on Au (111) with electrochemical scanning tunneling microscopy. It was shown that with increasing electrode potential the ordered structure of the adlayer consisting only of molecules oriented parallel to the surface changes into the ordered phase, which contains molecules adsorbed both parallel and perpendicular to the electrode surface. Further increasing in the electrode potential leads to the surface phase which is formed only by molecules oriented perpendicular to the surface.

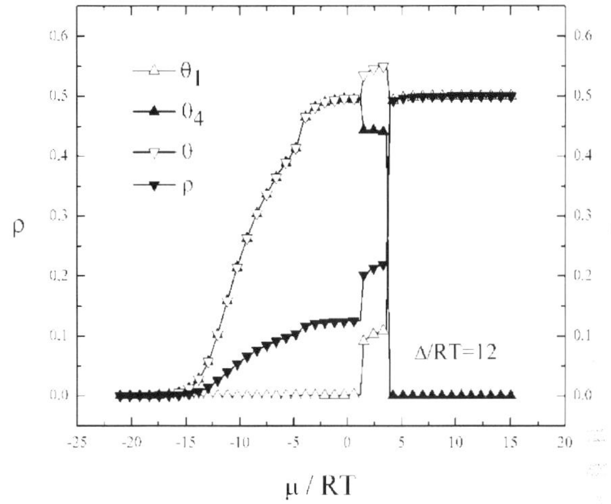

Figure 7. The curves of partial surface coverage calculated at $\Delta/RT = 12$. θ_1 – partial coverage of the surface by molecules adsorbed on one active site; θ_4 – partial coverage of the surface by molecules adsorbed on four active sites; θ – total surface coverage; ϱ – density of the adlayer (adsorption isotherm).

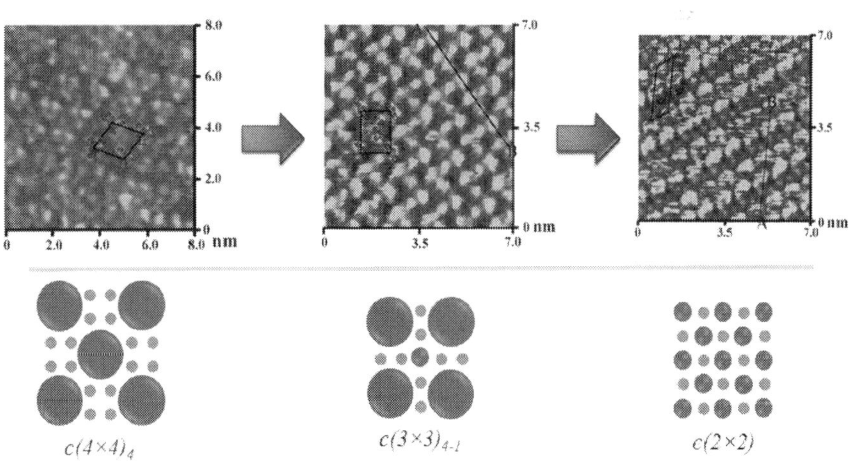

Figure 8. Ordered structure of the trimesic acid adsorption monolayer on Au (111) [21] and the corresponding structures calculated in the framework of the proposed model.

Analogous sequence of the ordered structures was also observed in other systems, in particular, in p-sexiphenyl [18] and pyridine adsorption monolayers [23] on Au (111). Thus, the phase behavior of monolayer adsorption which is realized in the model under consideration is qualitatively the same as the phase behavior of real adsorption monolayers of complex organic molecules on homogeneous surfaces. Therefore, a detailed study of the model and models similar to that is very useful for deeper understanding of the thermodynamics of self-assembled monolayers of complex organic molecules.

In the framework of the model under consideration, the modern methods of theoretical physics such as the multiple-histogram reweighting and finite-size scaling techniques (the fourth-order cumulant of the order parameter) have been used to estimate the phase diagram (T, μ) of complicated organic molecules adsorbed on the homogenous square lattice (Fig. 9) [93].

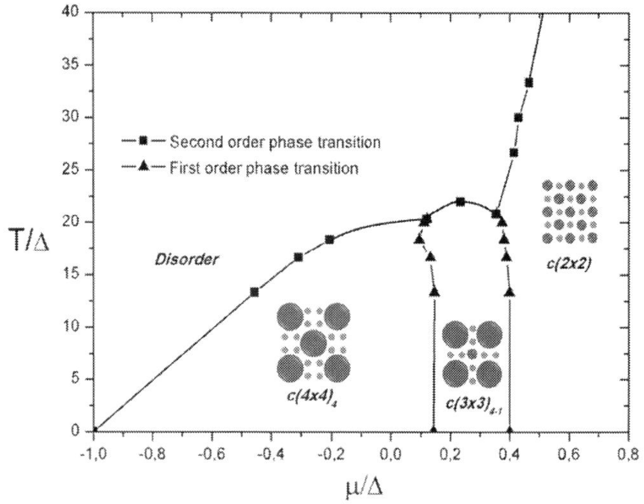

Figure 9. Phase diagram of adsorption monolayer of complex organic molecules on homogenous square lattice.

The phase diagram shown in Fig. 9 differs from the phase diagrams of dimers adsorption monolayer on square and triangular lattices that: 1) there is the phase consisting of molecules oriented in both ways (parallel and perpendicular to the solid surface) simultaneously, and 2) the critical temperature of this phase (*c*$(3{\times}3)_{4\text{-}1}$) is higher than the critical temperature of the ordered phase formed only by molecules adsorbed parallel to the surface. Indeed, the phase diagram of dimers adsorption on square lattice contains only two regions corresponding to the ordered phases due to the fact that when dimers adsorb on square lattice the ordered phase consisting of both dimers adsorbed parallel and perpendicular to the surface does not appear. In contrast, when dimers adsorb on triangular lattice that ordered phase is formed, but its critical temperature is lower than the critical temperature of the ordered structure formed only by dimers adsorbed parallel to the surface.

Thus, the numerical analysis of the thermodynamic properties of the lattice models which take into account the possibility of varying the orientation of the adsorbate molecules both with respect to each other and with respect to the solid surface showed that:

1. The phenomenon of non-monotonic changes of the surface coverage as function of the chemical potential is a general property of the systems under consideration and is not observed in models of single-site or multisite single-component gas adsorption which exclude different orientations in the adsorption monolayer.
2. There are ordered structures in the adsorption monolayer of complex organic molecules consisting of the molecules with different orientations relative to the solid surface, simultaneously.
3. The transition between the ordered structures of the adsorption monolayer occurs abruptly through the first-order phase transition.
4. The phase behavior of adsorption monolayer of complex organic molecules strongly depends on two factors: a) the geometry and chemical structure of the adsorbed molecule, and b) the geometry of the surface.

It is worth to note that the adsorption of single-component gas consisting of molecules with complex shape can exhibit the features which typical for the adsorption of multicomponent gas mixtures [90]. In particular, the model of adsorption of a gas mixture A (adsorb on m_1 active sites) and B (adsorb on m_2 active sites) is equivalent to the considered models (with k = 2) when the chemical potentials of the gas mixture components are the same, $\mu_A = \mu_B$.

4. Conclusion

Lattice gas model and its various generalizations are one of the most important models of modern statistical thermodynamics and are currently experiencing a new stage of development. On the one hand, it is related to the solution of some fundamental issues concerning the limiting behavior of two-dimensional lattice models, and, on the other hand, it is related to successful attempts to describe with the help of this model complex systems, in particular, multisite adsorption of various molecules on the solid surface.

Lattice models of multisite adsorption can take into account the number of characteristic features of complex adsorption systems, the most important of which is the possibility of the orientation varying of the adsorbate molecules, both with respect to each other and relative to the solid surface. Analysis of the thermodynamic properties of models of this type showed that despite its relative simplicity they exhibit a fantastic variety of non-trivial physical phenomena. First of all, it refers to the processes of self-assembly and phase transitions of various types, whose study is of considerable interest from the standpoint of the general thermodynamic theory. In addition, referring to the results of the experimental data, it can be argued that discussed in this chapter multisite adsorption model is firstly qualitatively, and in some cases quantitatively, reproduce the behavior of real systems and secondly have considerable predictive power. The results of study of multisite adsorption models can be summarized in following conclusions:

1. The adlayer of complex organic and inorganic molecules can form ordered structures consisting of molecules with different orientations simultaneously (both with respect to each other and relative to the solid surface).

situ scanning tunneling microscopy and atomic force microscopy // Langmuir 12 (1996) 6410.
[23] W.-B. Cai, L.-J. Wan, H. Noda, Y. Hibino, K. Ataka, M. Osawa, Orientational phase transition in a pyridine adlayer on gold(111) in aqueous solution studied by in situ infrared spectroscopy and scanning tunneling microscopy // Langmuir 14 (1998) 6992.
[24] G.-J. Su, H.-M. Zhang, L.-J. Wan, C.-L. Bai, T. Wandlowski, Potential-induced phase transition of trimesic acid adlayer on Au(111) // J. Phys. Chem. B 108 (2004) 1931.
[25] H.S. Kato, M. Wakatsuchi, M. Kawai, J. Yoshinobu, Different adsorbed states of 1,4-cyclohexadien on Si(001) controlled by substrate temperature // J. Phys. Chem. C. 111 (2007) 2557.
[26] K. Oguchi, K. Mukai, Y. Yamashita, J. Yoshinobu, Low-Temperature STM and UPS Study of Adsorption States of 1,4-Cyclohexadiene on Si(100) – c(4×2) // J. Phys. Chem. C 112 (2008) 15009.
[27] R.J. Hamers, S.K. Coulter, M.D. Ellison, Cycloaddition chemistry of organic molecules with semiconductor surfaces // Acc. Chem. Res. 33 (2000) 617.
[28] N.A. Besley, J.A. Bryan, Partial hessian vibrational analysis of organic molecules adsorbed on Si(100) // J. Phys. Chem. C, 112 (2008) 4308.
[29] M. Nagao, H. Umeyama, K. Mukai and et al., Precursor mediated cycloaddition reaction of ethylene to the Si(001)c(4×2) surface // J. Am. Chem. Soc., 126 (2004) 9922.
[30] C.-H. Chung, W.-J. Jung, I.-W. Lyo, Trapping-mediated chemisorption of ethylene on Si(001) – c(4×2) // Phys. Rev. Lett., 97 (2006) 116102.
[31] J. Seto, N. Asai, I. Fujiwara, T. Ishibashi, T. Kamei, S. Tamura, Functional organic ultra-thin films for device applications // Thin Solid Films 273 (1996) 97.
[32] T. Sano, H. Fujii, Y. Nishio, Y. Hamada, H. Takahashi, K. Shibata, Organic electroluminescent devices doped with condensed polycyclic aromatic compounds // Synthetic Metals 91 (1997) 27.
[33] L. Schmidt-Mende, A. Fechtenkötter, K. Müllen, E. Moons, R.H. Friend, J.D. MacKenzie, Self-organized discotic liquid crystals for high-efficiency organic photovoltaics // Science 293 (2001) 1119.
[34] C.D. Dimitrakopoulos and P.R.L. Malenfant, Organic thin film transistors for large area electronics // Adv. Mater. 14 (2002) 99.
[35] D. H. Kim, J. K. Kim, and P. Hwang, Anisotropic tribological properties of the coating on a magnetic recording disk // Thin Solid Films 360 (2000) 187.
[36] A.N. Shipway, E. Katz, I. Willner, Nanoparticle arrays on surfaces for electronic, optical, and sensor applications // Chem. Phys. Chem. 1 (2000) 18.
[37] M.O. Lorenzo, C.J. Baddeley, C. Muryn, R. Raval, Extended surface chirality from supramolecular assemblies of adsorbed chiral molecules // Nature 404 (2000) 376.
[38] B. Kasemo, J. Gold, Implant surfaces and interface processes // Adv. Dent. Res. 13 (1999) 8.
[39] C.A. Mitchell, L. Yu, M.D.Ward, Selective nucleation and discovery of organic polymorphs through epitaxy with single crystal substrates // J. Am. Chem. Soc. 123 (2001) 10830.

[40] R.H. Fowler, G.S. Rushbrooke, An attempt to extend the statistical theory of perfect solutions // Trans. Faraday Soc., 33 (1937) 1272-1294.
[41] M.E. Fisher, Statistical mechanics of dimers on a plane lattice // Phys. Rev., 124 (1961) 1664.
[42] P.W. Kasteleyn, The statistics of dimers on a lattice // Physica 27 (1961) 1209.
[43] H.N.V. Temperley, M.E. Fisher, Dimer problem in statistical mechanics – An exact result // Phil. Mag. 6 (1961) 1061.
[44] W.T. Lu, F.Y. Wu, Dimer statistics on a Mobius strip and the Klein bottle // Phys. Lett. A 259 (1999) 108.
[45] W.T. Lu, F.Y. Wu, Close-packed dimers on nonorientable surfaces, Phys. Lett. A 293 (2002) 235.
[46] Y. Kong, Packing dimers on (2p+1)×(2q+1) lattices // Phys. Rev. E, 73 (2006) 016101.
[47] V.S. Poghosyan, V.B. Priezzhev, P. Ruelle, Jamming probabilities for a vacancy in the dimer model // Phys. Rev. E 77 (2008) 041130.
[48] P.J. Flory, Thermodynamics of high polymer solutions // J. Chem. Phys. 10 (1942) 51.
[49] F. Roma, A.J. Ramirez-Pastor, J.L. Riccardo, Multisite occupancy adsorption: Comparative study of new different analytical approaches // Langmuir 19 (2003) 6770.
[50] G. Kondart, Influence of temperature on percolation in a simple model of flexible chains adsorption // J. Chem. Phys. 117 (2002) 6662.
[51] F. Roma, J.L. Riccardo, A.J. Ramirez-Pastor, Application of the FSTA to adsorption of linear and flexible k-mers on two-dimensional surfaces // Ind. Eng. Chem. Res. 45 (2006) 2046.
[52] A.J. Ramirez-Pastor, T.P. Eggarter, V.D. Pereyra, J.L. Riccardo, Statistical thermodynamics and transport of linear adsorbates // Phys. Rev. B 59 (1999) 11027.
[53] A.J. Ramirez-Pastor, F. Roma, A. Aligia, J.L. Riccardo, Multisite-occupancy adsorption and surface diffusion of linear adsorbates in low dimensions: rigorous results for a lattice gas model // Langmuir 16 (2000) 5100.
[54] F. Bulnes, A.J. Ramirez-Pastor, G. Zgrablich, Surface diffusion of k-mers in one-dimensional systems // Surface Science 601 (2007) 569–577.
[55] P.J. Flory, Principles of polymers chemistry, Cornell University Press, Ithaca, NY, 1953.
[56] M.L. Huggins, Some properties of solutions of long-chain compounds // J. Phys. Chem. 46 (1942) 151.
[57] M.L. Huggins, Thermodynamic properties of solutions of long-chain compounds // Ann. N.Y. Acad. Sci. 43 (1942) 1.
[58] M.L. Huggins, Theory of solutions of high polymers // J. Am. Chem. Soc. 64 (1942) 1712.
[59] E.A. DiMarzio, Statistics of orientation effects in linear polymer molecules // J. Chem. Phys. 35 (1961) 658.
[60] E.A. Guggenheim, Statistical thermodynamics of mixtures with non-zero energies of mixing // Proc. R. Soc. London A 183 (1944) 203.
[61] P.M. Centres, A.J. Ramirez-Pastor, Configurational entropy of adsorbed rigid rods: theory and Monte Carlo simulations // Physica A 388 (2009) 2001.

[62] M. Davila, F. Roma, J.L. Riccardo, A.J. Ramirez-Pastor, Quasi-chemical approximation for polyatomics: statistical thermodynamics of adsorption // Surface Science 600 (2006) 2011.

[63] A.J. Ramirez-Pastor, J.L. Riccardo, V. Pereyra, Adsorption thermodynamics with multisite Occupancy at criticality // Langmuir 16 (2000) 10167.

[64] J.L. Riccardo, F. Roma, A.J. Ramirez-Pastor, Fractional statistical theory of adsorption of polyatomics // Phys. Rev. Lett. 93 (2004) 186101.

[65] F. Roma, J. L. Riccardo, and A. J. Ramirez-Pastor, Semiempirical model for adsorption of polyatomics // Langmuir 2006, 22, 3192.

[66] D.P. Landau, K. Binder, A Guide to Monte Carlo Simulation in Statistical Physics – Cambridge: Cambridge University Press, 2000 – p.432.

[67] M.E.J. Newman, G.T. Barkema, Monte Carlo Methods in Statistical Physics – NY: Oxford University Press, 2001 – p.475.

[68] D.A. Matoz-Fernandez, D.H. Linares, A.J. Ramirez-Pastor, Critical behavior of long linear k-mers on honeycomb lattices // Physica A 387 (2008) 6513.

[69] F. Roma, A. J. Ramirez-Pastor and J. L. Riccardo, Critical behavior of repulsive linear k-mers on square lattices at half coverage: Theory and Monte Carlo simulations // Phys. Rev. B 68 (2003) 205407.

[70] A.J. Phares, F.J. Wunderlicht, D.J. Curley, Jr. D.W. Grumbine, Structural ordering of dimers on square lattice // J. Phys. A: Math. Gen. 26 (1993) 6847.

[71] A.J. Ramirez-Pastor, J.L. Riccardo, V.D. Pereyra, Monte Carlo study of dimer adsorption at monolayer on square lattices // Surf. Sci. 411 (1998) 294.

[72] F. Roma, J.L. Riccardo, A.J. Ramirez-Pastor, Critical behavior of repulsive dimers on square lattices at 2/3 monolayer coverage // Phys. Rev. B. 77 (2008) 195401.

[73] J.E. Gonzalez, A.J. Ramirez-Pastor, V.D. Pereyra, Adsorption of dimer molecules on triangular and honeycomb lattices // Langmuir 17 (2001) 6974.

[74] P.M. Pasinetti, F. Roma, J.L. Riccardo, A.J. Ramirez-Pastor, Critical behavior of repulsive linear k-mers on triangular lattices // Phys. Rev. B, 74 (2006) 155418(8).

[75] W. Rzysko, M. Borowko, Phase diagrams of heteronuclear dimers adsorbed on a square lattice // J. Chem. Phys. 117 (2002) 4526.

[76] M. Borowko and W. Rzysko, Critical Behavior of Dimers in Monolayers Adsorbed on Heterogeneous Solid Surfaces // J. Colloid Interface Sci., 244 (2001) 1.

[77] M. C. Gimenez, F. Nieto, A. J. Ramirez-Pastor, Surface order-disorder phase transitions and percolation // J. Chem. Phys. – 125 (2006) 184707.

[78] A. J. Ramirez-Pastor, M. S. Nazzarro, J. L. Riccardo, G. Zgrablich, Dimer physisorption on heterogeneous substrates // Surf. Sci. 341 (1995) 249.

[79] T. Nitta, H. Kiriama, T. Shigeta, Monte Carlo simulation study for adsorption of dimers on random heterogeneous surfaces // Langmuir 13 (1997) 903.

[80] W. Rzysko, M. Borowko, Computer simulation of phase diagrams of trimers adsorbed on a square lattice // Phys. Rev. B 67 (2003) 045403.

[81] V.F. Cabral, C.R.A. Abreu, M. Castier, F.W. Tavares, Monte Carlo simulations of the adsorption of chainlike molecules on two-dimensional heterogeneous surfaces // Langmuir 19 (2003) 1429.

[82] V. Cornette, A.J. Ramirez-Pastor, F. Nieto, Dependence ofthe percolation threshold on the size of the percolating species // Physica A 327 (2003) 71.
[83] M. Quintana, I. Kornhauser, R. López, A.J. Ramirez-Pastor, G. Zgrablich, Monte Carlo simulation of the percolation process caused by the random sequential adsorption of k-mers on heterogeneous triangular lattices // Physica A 361 (2003) 195.
[84] M.I. Temkin, The transition state in surface reactions // Rus. J. Phys. Chem. 2 (1938) 169.
[85] Yu.S. Snagovski, Adsorption and catalytic reactions of multicenter molecules. I. in Ideal adlayer equilibrium // Rus. J. Phys. Chem. 9 (1972) 2367.
[86] Yu.S. Snagovski, Adsorption and catalytic reactions of multicenter molecules. II. Reaction kinetics in ideal adlayer // Rus. J. Phys. Chem. 10 (1972) 2584.
[87] Yu.S. Snagovski, Adsorption and catalytic reactions of multicenter molecules. III. Induced heterogeneity effect // Rus. J. Phys. Chem. 10 (1972) 2589.
[88] A.B. Gorshtein, A.A. Lopatkin, The influence of reorientation of adsorbed molecules on differential heat capacity // Rus. J. Phys. Chem. 1 (1974) 177.
[89] A.B. Gorshtein, A.A. Lopatkin, Lattice models of adsorption of diatomic molecules // Theor. Exp. Chem. 9 (1975) 148.
[90] W. Engl, L. Courbin, P. Panizza, Adsorption of asymmetric rigid rods or heteronuclear diatomic molecules on homogeneous surfaces // Phys. Rev. B. 70 (2004) 165407.
[91] W. Rżysko, A. Patrykiejew, K. Binder, Phase transitions in a two-dimensional lattice gas model of orientable diatomic molecules // Phys. Rev. B 72 (2005) 165416.
[92] W. Rżysko, A. Patrykiejew, K. Binder, Phase transitions in a two-dimensional lattice gas model of orientable diatomic molecules. II. Order-disorder transitions of superantiferromagnetic and c(2x2)AF phases // Phys. Rev. B. 76 (2007) 195409.
[93] V.F. Fefelov, V.A. Gorbunov, A.V. Myshlyavtsev, M.D. Myshlyavtseva, The simplest model of adsorption of molecules with different orientations of complex organic molecules. Monte Carlo and transfer-matrix techniques // Chem. Eng. J., 154 (2009) 107.
[94] V.A. Gorbunov, A.V. Myshlyavtsev, M.D. Myshlyavtseva, V.F. Fefelov, Simulation of complex molecules adsorption with different orientation in adlayer on square lattice // Omskiy nauchnyi vestnik 2 (2007) 19 (in russian).
[95] V.A. Gorbunov, A.V. Myshlyavtsev, M.D. Myshlyavtseva, V.F. Fefelov, Simulation of unsaturated cyclic hydrocarbons adsorption on reconstructed Si(001)–2×1 // Omskiy nauchnyi vestnik 1 (2009) 215 (in russian).
[96] V.A. Gorbunov, A.V. Myshlyavtsev, M.D. Myshlyavtseva, V.F. Fefelov, Monte Carlo study of phase behavior of complex organic molecules adlayer // Omskiy nauchnyi vestnik 3 (2009) 37 (in russian).
[97] V.F. Fefelov, V.A. Gorbunov, A.V. Myshlyavtsev, M.D. Myshlyavtseva, The simplest model of adsorption of molecules with different orientations in adlayer on the stepped surface // App. Surf. Sci., 256 (2010) 5298.
[98] P. Rinaldi, F. Bulnes, A.J. Ramirez-Pastor, G. Zgrablich, Monte Carlo study of multicomponent adsorption on triangular lattices // Surf.Sci. 602 (2008) 1783.
[99] F. Bulnes, A.J. Ramirez-Pastor, G. Zgrablich, Monte Carlo simulation of adsorption of binary gas mixtures on heterogeneous surfaces // Adsorpt. Sci. Technol. 23 (2005) 81.

[100] V.A. Gorbunov, A.V. Myshlyavtsev, M.D. Myshlyavtseva, V.F. Fefelov, Simulation of dimers adsorption on triangular lattice // Izvestiya vyschyh uchebnyh zavedeniy. Chimia I chimicheskaya, 9 (2010) 66.

[101] V.F. Fefelov, V.A. Gorbunov, A.V. Myshlyavtsev, M.D. Myshlyavtseva, Model of homonuclear dimer adsorption in terms of two possible molecule orientations with respect to surface: square lattice // Phys. Rev. E. 82 (2010) 041602.

[102] R.J. Glauber, Time-dependent statistics of the Ising model // J. Math. Phys. 4 (1963) 294.

[103] K. Kawasaki, Diffusion constants near the critical point for time-dependent Ising models // Phys. Rev. 145 (1966) 224.

[104] N. Metropolis, A.W. Rosenbluth, M.N. Rosenbluth, A.W. Teller, E. Teller, Equation of state calculations by fast computing machines // J. Chem. Phys. 21 (1953) 1087.

Section 5

Non-Equilibrium Thermodynamics

Chapter 16

Influence of Simulation Parameters on the Excitable Media Behaviour – The Case of Turbulent Mixing

Adela Ionescu

Additional information is available at the end of the chapter

http://dx.doi.org/10.5772/51585

1. Introduction

Although the turbulence is often associated with fluid dynamics, it is in fact a basic feature for most systems with few or infinity freedom degrees. It can be defined as chaotic behavior of the systems with few freedom degrees and which are far from the thermodynamic equilibrium. In this area two important zones are distinguished:

- The theory of transition from laminar smooth motions to irregular motions, characteristic to turbulence;
- Characteristic studies of completely developed turbulent systems.

In hydrodynamics, the transition problem lays back to the end of last century, at the pioneering works of Osborne Reynolds and Lord Rayleigh. Since the beginnings, it was pointed out the fruitful investigation method of considering the linear stability of basic laminar flow until infinitesimal turbulences. Nonlinearity can act in the sense of stabilizing the flow and therefore the primary state is replaced with another stable motion which is considered as secondary flow. This one can be further replaced with a tertiary stable flow, and so on. It is in fact about a *bifurcations sequence*, and Couette-Taylor flow is maybe the most widespread example in this sense.

The situation becomes hard to approach if the non-linearity is acting in the sense of increasing the rate of growing the unstable linear modes. Although it was anticipated that the flows can be stable according to the linear theory, in experiments it was concluded that they are unstable. It must be noticed that Reynolds himself understood this possibility, and suggested that for the transition from laminar to turbulence for a pipe flow, "the condition must be of instability at certain size perturbations and stability at smaller perturbations".

The issue of transition in flows such Poiseuille flow until Reynolds numbers under the critical value, must be due to instabilities at finite amplitude perturbations. For these flows there was nothing relevant found concerning the eventual secondary stable motions, moreover it seems that turbulence issue directly from primary flow, at a fixed Reynolds number. These *strong turbulence problems* are quite difficult to approach and this gives evidence that still is to be added more substance to the original Reynolds' suggestion, after one hundred years of stability studies. For this quite new theory, *unifying with the classic turbulence* is far from being solving, but the recent challenges could open new research directions.

Generally, the statistical idea of a flow is represented by a map:

$$x = \Phi_t\left(X\right), X = \Phi_{t=0}\left(X\right) \tag{1}$$

That means **X** *is mapped in* **x** after a time t. In the continuum mechanics the relation (1) is named *flow,* and it is a diffeomorphism of class C^k. Moreover, the relation (1) must satisfy the following equation:

$$J = \det\left(D\left(\Phi_t\left(X\right)\right)\right) = \det\left(\frac{\partial x_i}{\partial X_j}\right) \tag{2}$$

where D denotes the derivation with respect to the reference configuration, in this case **X**. The equation (2) implies two particles, X_1 and X_2, which occupy the same position **x** at a moment. Non-topological behavior (like break up, for example) *is not allowed.*

With respect to **X** there is defined the basic measure of deformation, the *deformation gradient,* **F,** namely the equation (3):

$$\mathbf{F} = \left(\nabla_X \Phi_t\left(\mathbf{X}\right)\right)^T, \; F_{ij} = \left(\frac{\partial x_i}{\partial X_j}\right) \tag{3}$$

where ∇_X denotes differentiation with respect to **X**. According to equation (2), **F** is non singular. The basic measure for the deformation with respect to **x** is the *velocity gradient.*

After defining the basic deformation of a material filament and the corresponding relation for the area of an infinitesimal material surface, there can be defined the basic deformation measures: the *length deformation* λ and *surface deformation* η, with the following relations (Ottino, 1989):

$$\lambda = \left(\mathbf{C} : \mathbf{MM}\right)^{1/2}, \; \eta = \left(\det F\right) \cdot \left(\mathbf{C}^{-1} : \mathbf{NN}\right)^{1/2} \tag{4}$$

where **C** (=$\mathbf{F}^T \cdot \mathbf{F}$) is the *Cauchy-Green deformation tensor,* and the vectors **M,N** - the orientation versors in length and surface respectively, are defined by:

$$\mathbf{M}=\frac{d\mathbf{X}}{|d\mathbf{X}|},\ \mathbf{N}=\frac{d\mathbf{A}}{|d\mathbf{A}|} \tag{5}$$

Very often, in practice is used the scalar form of (4), namely the following relation:

$$\lambda^2=C_{ij}\cdot M_i\cdot N_j,\ \eta^2=(\det F)\cdot C_{ij}^{-1}\cdot M_i\cdot N_j \tag{6}$$

where **D** is the deformation tensor, obtained by decomposing the velocity gradient in its symmetric and non-symmetric part.

The flow $x=\Phi_t(X)$ has *a good mixing* if the mean values $D(ln\lambda)/Dt$ and $D(ln\eta)/Dt$ are not decreasing to zero, for any initial position P and any initial orientations **M** and **N**. As the above two quantities are bounded, the deformation efficiency can be naturally quantified. Thus, there is defined (Otino, 1989) the *deformation efficiency in length*, $e_\lambda = e_\lambda\ (\mathbf{X},\mathbf{M},t)$ of the material element d**X**, as follows:

$$e_\lambda=\frac{D(\ln\lambda)/Dt}{(\mathbf{D}:\mathbf{D})^{1/2}}\leq 1 \tag{7}$$

Similarly, there is defined the *deformation efficiency in surface*, , $e_\eta= e_\eta\ (\mathbf{X},\mathbf{N},t)$ of the area element d**A**: in the case of an isochoric flow (the jacobian equal 1), there exists the equation (8):

$$e_\eta=\frac{D(\ln\eta)/Dt}{(\mathbf{D}:\mathbf{D})^{1/2}}\leq 1 \tag{8}$$

The deformation tensor **F** and the associated tensors **C**, $\mathbf{C}^{-1}$, form the fundamental quantities for the analysis of deformation of infinitesimal elements. In most cases the flow $x=\Phi_t(X)$ is unknown and has to be obtained by integration from the Eulerian velocity field. If this can be done analytically, then **F** can be obtained by differentiation of the flow with respect to the material coordinates **X.** The flows of interest belong to two classes: i) flows with a special form of ∇v and ii) flows with a special form of **F**. The second class is what we are interested for, as it contains the so-called Constant Stretch History Motion – CSHM.

2. Issues on turbulent mixing. Results and challenges

The central problem exhibited in this chapter is the challenge of unifying the theory of turbulent mixing. This implies few levels: analytical, computational and experimental. The specific literature is rich in works both on analytical and experimental models. There were realized few comparative analysis, both for analytical and computational standpoint, for 2D and 3D models of different flow types.

From physical standpoint (Savulescu, 1998), the vortex produced in the installation implies four mechanisms:

- Convection (scalar transport), due to streamlines which are directed towards the ending lid near the tube walls, with the pressure source near the tube centre line;
- Turbulent diffusion due to the pressure and velocity variation;
- Stratification effects due to pressure and temperature gradients;
- Turbulent mixing due to the velocity concentration in annular structures.

The convection keeps in the transport and deposit of the powder pollutant when the graining diameter is greater than 5μm. However, the graining spectra domain under 5μm undergoes a turbulent diffusion and stratification effects which are generally out of control. For revealing the local concentration of the streamlines, 2D simplified models were tested. The multiphase 3D flows simulation is still in study.

The turbulent mixing in multiphase flow reveals the following experiment components:

- The topological limit of the multiphase flow when the pollutant enters in the aggregation state or remains fragmented, in the host fluid;
- The rheological behaviour of non-Newtonian compositions, under macroscopic effects of the air velocity.

The installation was realized in two versions: a small scale vortex tube (10-20mm diameter) and a large scale one (100-300mm). They correspond generally at two particles processing classes, although in most cases the classes can be superposed.

The first category refers to collecting and separating various powders, from gaseous emissions to ceramic powders. The parameters which must be taken into account are the graining spectra, the atmosphere nature and concentration. The vorticity concentration can be used for processing the deposit of different particles in powder or cement form by an adequate closing lid.

The second category contains the processing on small scale, including deformation and breaking up mechanisms for various particles in a host fluid. It is studied the vorticity concentration near the closing lid.

During such an experiment (Ionescu, 2002), it was processed the aquatic algae *Spirulina Platensis* in the host fluid water. After the processing, the long chains of cellular filaments were fragmented, producing isolated cell units or – sometimes – there was recorded the breaking up of some cellular membranes (with less than 100 Angstrom thickness). The initial and final observations (after the vortexation) are exhibited in figure 1 below. It has been used the non-dimensional parameter $\tau_a = \frac{t \cdot Q}{D^3}$, where t represents the time (in seconds), Q the installation debit (m^3/sec) and D the diameter (m^3). As it can be seen from the picture, the fragmentation degree starts to increase as τ_a grows. There can also be observed the algae form before and after the fragmentation.

Following the opinion of the specialists in cellular biology, this new technological method of processing the flows is more efficient than the classical centrifugation method.

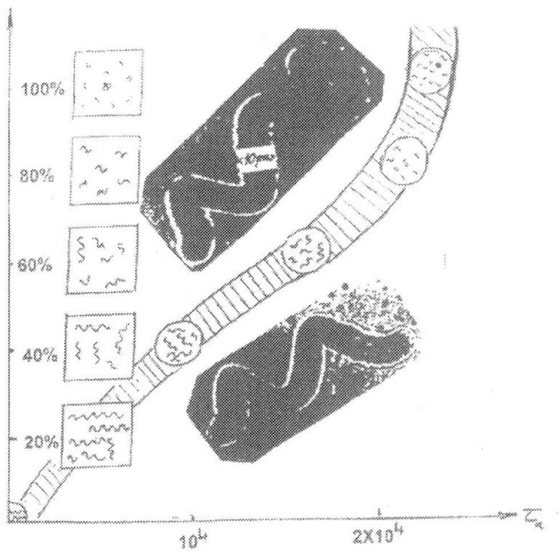

Figure 1. The fragmentation degree variation of *Spirulina Platensis*

Three specific applications were performed as fluid waste management:

- the agglomeration of short fibers (*aerodynamic spinning*);
- the retention of particles under 5μm without any material filter;
- the breakout of cell membranes of the phytoplankton from polluted waters and the providing of a cell content solution *with important bio-stimulating features.*

The mathematical modeling and analytically testing of the above experiment confirmed the experimental study. Concerning the phenomenon scale, there were taken into account medium helicoidally streamlines with approximating 10μm width. *There were not gone further to molecular level.* Also, it is important to notice that for this moment of the analytical testing of the model, there has not been studied the influence of strictly biological parameters (such as pH, the temperature, the humidity, etc).

The complex multiphase flow necessarily implies a theoretical approach for discovering the ways of optimization, developing and control of the installation. Numeric simulation of 3D multiphase flows is currently in study. In the mathematical framework, the flow complexity implies the following three stages:

- modeling the global swirling streamlines;
- local modeling of the concentrated vorticity structure;
- introducing the elements of chaotic turbulence.

The mathematical model associated to the vortex phenomena is, basically, the 3D version of the widespread isochoric two-dimensional flow (Ottino, 1989):

$$\begin{cases} \dot{x}_1 = G \cdot x_2 \\ \dot{x}_2 = K \cdot G \cdot x_1, \ -1 \langle K \langle 1, G \rangle 0 \end{cases} \tag{9}$$

Namely, the following vortex flow model is in study:

$$\begin{cases} \dot{x}_1 = G \cdot x_2 \\ \dot{x}_2 = K \cdot G \cdot x_1 \quad , -1 \langle K \langle 1, \, c = const. \\ \dot{x}_3 = c \end{cases} \tag{10}$$

In the first stage, the flow model (10) was studied from the analytical standpoint. Namely, the solution of the Cauchy problem associated to (10) was found. In order to analyze the length and surface deformations of algae filaments in certain vortex conditions, the Cauchy-Green deformation tensor was calculated. There were obtained quite complex formulas for e_λ, e_η (Ionescu 2002, 2010).

The second stage has been a computational / simulation standpoint. It has been realized a computational analysis of the length and surface deformation efficiency, in some specific vortexation conditions. With the numeric soft MAPLE11, the analysis has been two parts. Firstly, the following Cauchy problems has been solved:

$$\begin{aligned} e_\lambda(t) = 0, \ x(0) = 0 \\ e_\eta(t) = 0, \ x(0) = 0 \end{aligned} \tag{11}$$

using a *discrete* numeric calculus procedure (Abell, 2005). The output of the procedure is a listing of the form $(t_i, x(t_i)), i = 0..25$. In the second part there have been realized discrete time plots, in 25 time units, following the listings obtained in the first part. The plots represent in fact *the image of the length and surface deformations, in the established scale time.*

Very few *irrational* value sets were chosen for the length, respectively surface versors, taking into account the versor condition:

$$\sum M_i = 1, \sum N_j = 1 \tag{12}$$

The studied cases are in fact represented by the events associated to different values of length and surface orientation versors: (M_1, M_2, M_3) and (N_1, N_2, N_3) for length and surface respectively. The events were very few, about 60. Their statistical interpretation is synthesized in (Ionescu 2010), including also the two-dimensional case.

The graphical events are illustrating the analysis of the deformations for Spirulina Platensis, in 25 time units' vortexation. According to (Ionescu 2001, Ottino 1989), the algae filaments represent *Lagrange markers*. The special spiral form of the algae gives the answer to the computational context that the *surface deformation e_η is significant,* since it brings more cases of filaments' break up. It was called *rare event* the event of very sudden breaking up of the algae filaments, this corresponding from mathematical standpoint to the break out of the program (because of impossibility of maintaining the required accuracy). Thus, a very important fact happened: *the mathematical simulations matched the experimental analysis.*

Sumarising these basic stages in the behavior analysis of the turbulent mixing flow, is important to notice that the filaments breaking up is due to alternating loadings that the filament undergoes, in a space-time with *random events,* available to the break phenomena. The modeling has been the following basic stages:

- Associating known streamlines to the medium flow (that means helical flow in a cylinder);
- Determining the linear and surface deformations from continuum fluid mechanics;
- Associating a sequence of random values to the (vectorial) length and surface orientations.

In fact, there were found four types of processes, all of which were matched by the simulations

1. Processes with relative linear behavior;
2. Linear-negative processes, which correspond to alternate tasks of stretching and folding of filaments and are the most;
3. Mixing phenomena where there issue smaller or larger deviations or strong discontinuities; these concern the situations when some pieces are suddenly coming off from the whole filament, followed by the restarting of vortexation for the rest of algae;
4. Rare events – these correspond to the turbulent mixing and represent the sudden break up of spirulina filaments.

The validity of the model is confirmed by two aspects:

- the statistic increment in time of the breaking cases, according to the experiments;
- the relative singularity of the events which could produce the breaking, fact which is confirmed by the quite long duration of the experiments which led to the filaments break up; in very rare cases, the cellular membrane could be broken, and the cellular content collected.

Crossing over from 2D to 3D case, it is easy to deduce the requirement of a special analysis of the influence of parameters on the behavior of this complex mixing flow. This would include more paramater analysis types for some perturbation models in 2D and 3D case, but also another mathematical analysis types, for example spectral analysis.

2.3. New influence of simulation parameters

The analysis recently has been continued with more computational simulations, for 2D model, both in periodic and – non – periodic case, and for 3D model, too. A lot of comparisons between periodic and non- periodic case, 2D and 3D case were realized (Ionescu 2008, 2009). In the same time, the computational appliances were varied. If innitialy, the model was studied from the standpoint of mixing efficiency, in the works that come after, new appliances of the MAPLE11 soft were tested (Abell 2005, Ionescu 2011), in order to collect more statistical data for the turbulent mixing theory. In what follows, few of these appliances are described.

2.3.1. Few MAPLE11 apliances used in simulations

For the purpose of this chapter, there are presented only the recent used appliances of MAPLE11 soft. There have been used two graphical appliances of this soft, namely „DePlot" and „Phaseportrait" tools, from „DETools" package. Both of them are based on numeric methods in producing trajectories of differential equations systems. The most used numeric method is *Fehlberg fourth-fifth order Runge-Kutta* method - the so-called „rkf45" method - with degree four interpolant (Abell 2005).

DETools[DePlot] – this tool plots solutions to a system of differential equations. The calling sequences are as follows:

DEplot(deqns, vars, trange, options)

DEplot(deqns, vars, trange, inits, options)

DEplot(deqns, vars, trange, xrange, yrange, options)

DEplot(deqns, vars, trange, inits, xrange, yrange, options)

DEplot(dproc, vars, trange, number, xrange, yrange, options)

Parameters:

deqns - list or set of first order ordinary differential equations, or a single differential equation of any order;

dproc - a Maple procedure representation for first order ordinary differential equations, or a single differential equation of any order;

vars - dependent variable, or list or set of dependent variables;

trange - range of the independent variable;

number - equation of the form 'number'=integer indicating the number of differential equations when **deqns** is given as a function (dproc) instead of expressions;

inits - set or list of lists; initial conditions for solution curves;

xrange - range of the first dependent variable;

yrange - range of the second dependent variable;

options - (optional) equations of the form keyword=value;

Description:

- Given a set or list of initial conditions (see below), and a system of first order differential equations or a single higher order differential equation, **DEplot** plots solution curves, *by numerical methods*. This means that the initial conditions of the problem must be given in standard form, that is, the function values and all derivatives up to one less than the differential order of the differential equation at the same point.
- A system of two first order differential equations produces a direction field plot, provided the system is determined to be autonomous. In addition, a single first order differential equation produces a direction field (as it can always be mapped to a system of two first order autonomous differential equations). A system is determined to be autonomous when all terms and factors, other than the differential, are free of the independent variable. For systems not meeting these criteria, no direction field is produced (only solution curves are possible in such instances). There can be **only one** independent variable
- The default method of integration is ***method=rkf45***. Other methods can be specified in the optional equations. Note that because numerical methods are used to generate plots, the output is subject to the characteristics of the numerical method in use. In particular, unusual output may occur when dealing with asymptotes.
- The direction field presented consists of either a grid of arrows or a set of randomly generated arrows. In either case, the arrows are tangential to solution curves. For each grid point, the arrow centered at **(x,y)** has slope **dy/dx**. This slope is computed using **(dy/dt)/(dx/dt)**, where these two derivatives are specified in the first argument to **DEplot**. The curved arrow types (**curves** and **comet**) require additional data for the curvature of the direction field, which is computed by moving an epsilon in the direction of the slope **dy/dx**, and computing **dy/dx**, then moving an epsilon in the direction opposite the slope, and computing **dy/dx**. This data is then sufficient to draw a small portion of the direction field lines local to the point, which is then used to draw the curved arrows.
- By default, the two dependent variables are plotted, unless otherwise specified in the **scene** option.
- The **deqns** parameter can be given as a procedure, but must conform to the specification as given in **dsolve/numeric**, and the **number** option must be included before the initial conditions. In this instance, **deqns** must be of the form:

```
proc( N, ivar, Y, YP )
...
YP[1] := f1(ivar,Y);
YP[2] := f2(ivar,Y);
...
end proc
```

where **N** represents the number of first order equations, **ivar** is the independent variable, **Y** is a vector of length **N**, and **YP** is a vector of derivatives which is updated by the procedure (for the equivalent first order system), also of length **N.**

- The **inits** parameter must be specified as:

[[x(t0)=x0,y(t0)=y0], [x(t1)=x1,y(t1)=y1], ...]

[[y(t0)=y0], [y(t1)=y1], ...]

[y(t0)=y0, y(t1)=y1, ...]

where, in the above, sets can be used in place of lists, or

[[t0,x0,y0], [t1,x1,y1], ...]

{ [t0,x0,y0], [t1,x1,y1], ... }

[[t0,x0], [t1,x1], ...]

where the above is a list or set of lists, each sublist specifying one group of initial conditions.

- The **xrange** and **yrange** parameters must be specified as follows.

x(t) = x1..x2, y(t) = y1..y2 or

x = x1..x2, y = y1..y2

More details about the parameters can be found in (Abell 2005).

DETools[Phaseportrait]. This tool has the following parameters:

deqns - list or set of first order ordinary differential equations, or a single differential equation of any order;

vars - dependent variable, or list or set of dependent variables;

trange - range of the independent variable;

inits - set or list of lists; initial conditions for solution curves;

options - (optional) equations of the form keyword=value;

Description

- Given a list (or set) of initial conditions (see below), and a system of first order differential equations or a single higher order differential equation, **phaseportrait** plots solution curves, by numerical methods. This means that the initial conditions of the problem must be given in standard form, that is, the function values and all derivatives up to one less than the differential order of the differential equation at the same point.
- A system of two first order differential equations also produces a direction field plot, provided the system is determined to be autonomous. In addition, a single first order differential equation also produces a direction field (as it can always be mapped to a system of two first order autonomous differential equations). For systems not meeting these criteria, no direction field is produced (only solution curves are possible in such instances). There can be ONLY one independent variable.
- All of the properties and options available in **phaseportrait** are also found in **Deplot**.

- **inits** should be specified as

  [[x(t0)=x0,y(t0)=y0], [x(t1)=x1,y(t1)=y1], ...]

where the above is a list (or set) of lists, each sublist specifying one group of initial conditions.

2.3.2. *Comparative computational analysis*

In what follows there is presented the comparative analysis for 2D and 3D mixing flow models. There are followed two main leves of comparison:

i. The 2D (non-periodic) mixing flow, namely the differential system (9):

$$\begin{cases} \dot{x}_1 = G \cdot x_2 \\ \dot{x}_2 = K \cdot G \cdot x_1, \ -1 \langle K \langle 1, G \rangle 0 \end{cases}$$

in comparison with a „perturbed" version of it, the system (13):

$$\begin{cases} \dot{x}_1 = G \cdot x_2 + x_1 \\ \dot{x}_2 = K \cdot G \cdot x_1 + G \cdot (x_2 - x_1) \end{cases} \tag{13}$$

ii. The 3D mixing flow model (10)

$$\begin{cases} \dot{x}_1 = G \cdot x_2 \\ \dot{x}_2 = K \cdot G \cdot x_1 \quad , -1 \langle K \langle 1, c = const. \\ \dot{x}_3 = c \end{cases}$$

For all models it is used the same set of parameter values (G, KG) as used in (Ionescu 2008), containing three simulation cases:

- case1: $G = 0.25, K \cdot G = -0.035$;
- case2: $G = 0.755, K \cdot G = -0.65$;
- case3: $G = 0.85, K \cdot G = -0.25$

Each simulation case is labelled on the figure. The time units number was succesively increased, in order to better analyse the solution behavior for each model and corresponding case. The „stepsize" option for the „rkf45" numeric method, on which the „Phaseportrait" procedure is based, is implicitely assigned to 0.05. Also, it must be noticed that the above choice of parameters (one positive and another negative) is optimal for analyzing the direction field associated to the models' solutions.

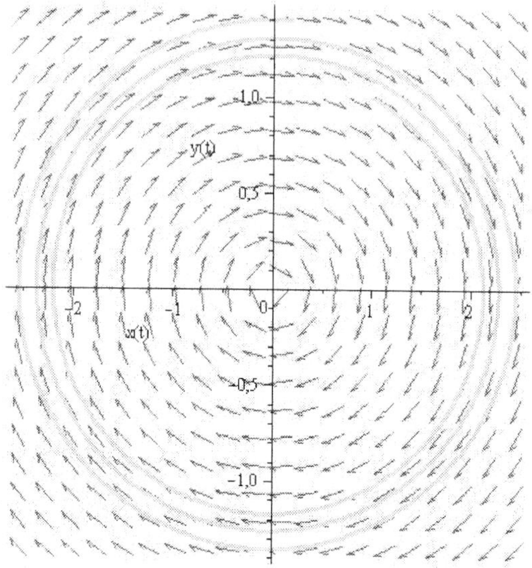

Figure 6. Case3 of simulation for the model (9). Time units number t=0.40

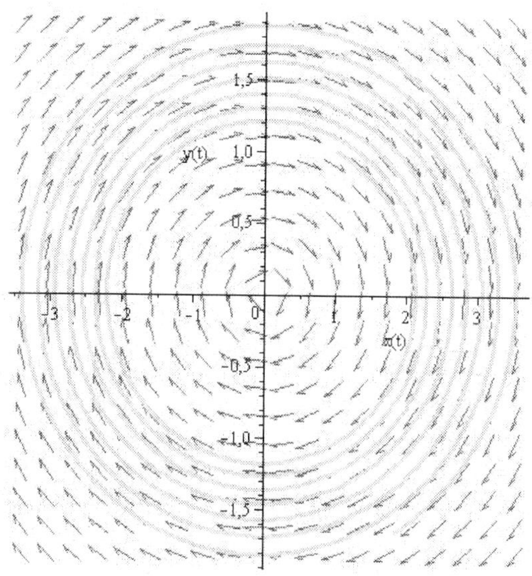

Figure 7. Case3 of simulation for the model (9) with t=0..100. The trajectory multiplies but does not draw near the origin

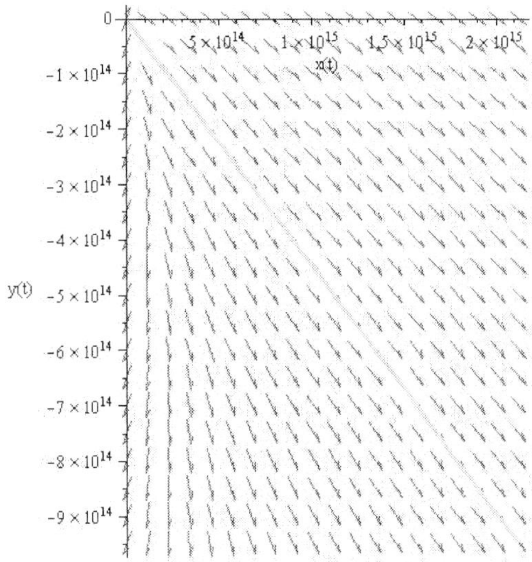

Figure 8. Case1 of simulation for the model (13). The trajectory changes, it seems to be no longer periodic, but tends to infinity

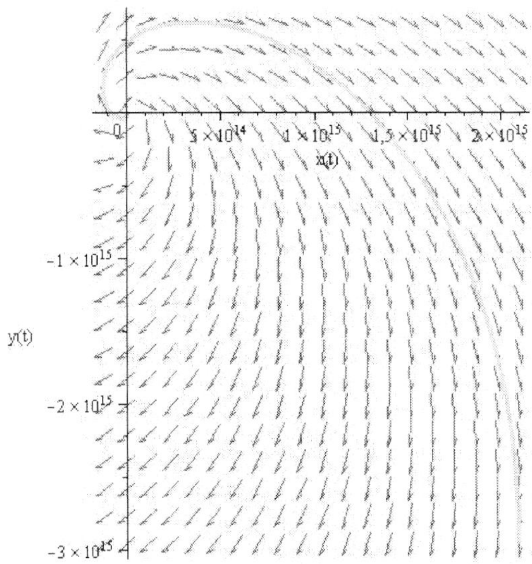

Figure 9. Case2 of simulation for the model (13). The time units number t=0..40. The trajectory changes again and becomes negative

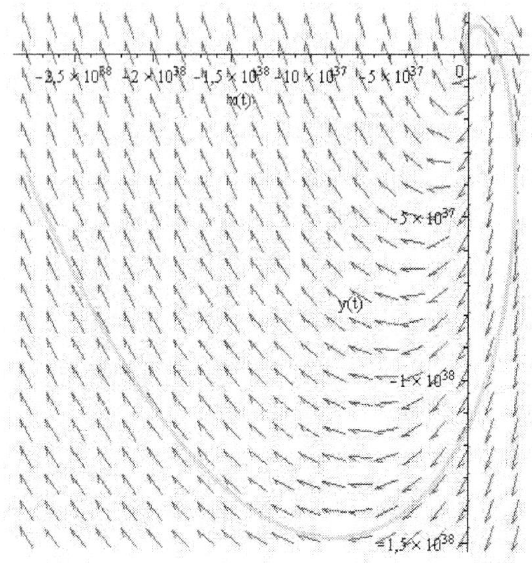

Figure 10. Case2 of simulation for the model (13), with t=0..100.

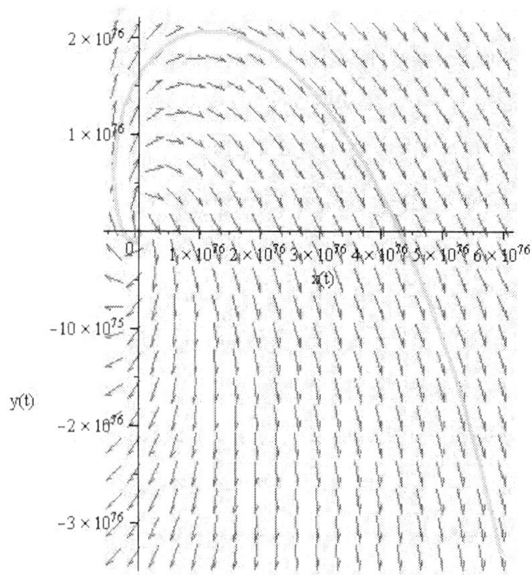

Figure 11. Case2 of simulation for the model (13) with t=0..200

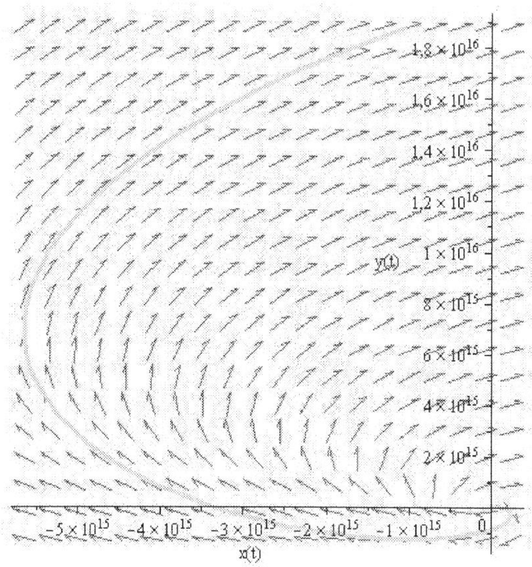

Figure 12. Case3 of simulation for the model (13), t=0..40. The trajectory becomes positive

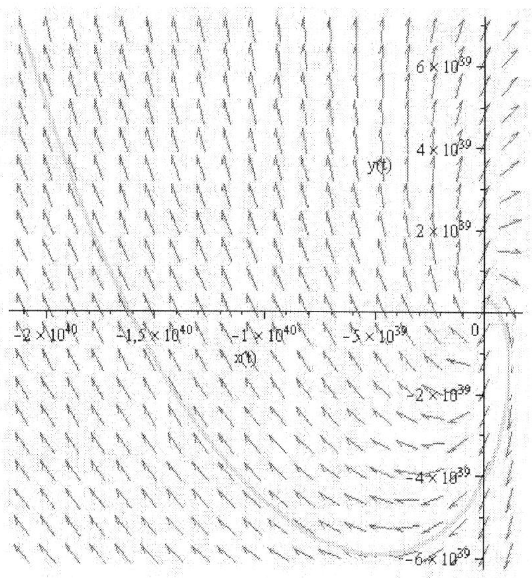

Figure 13. Case3 of simulation for the model (13) with t=0..100

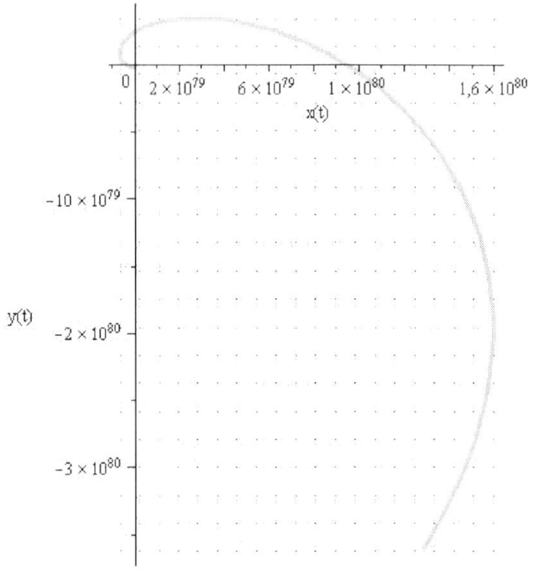

Figure 14. Case3 of simulation for the model (13) with t=0..200. The program does not show the field arrows anymore.

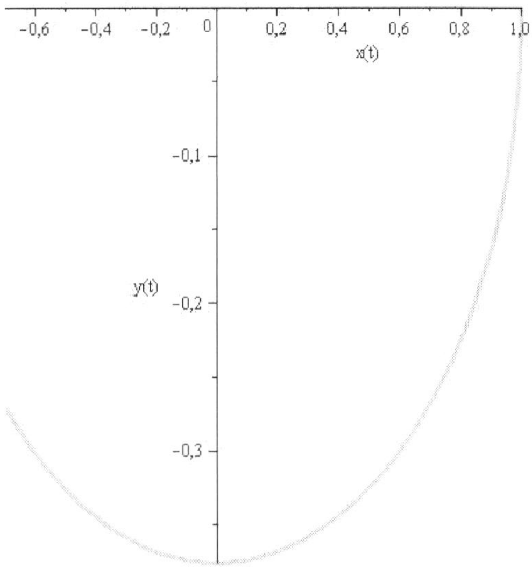

Figure 15. Case1 of simulation for the model (10). Time units number t=0..25

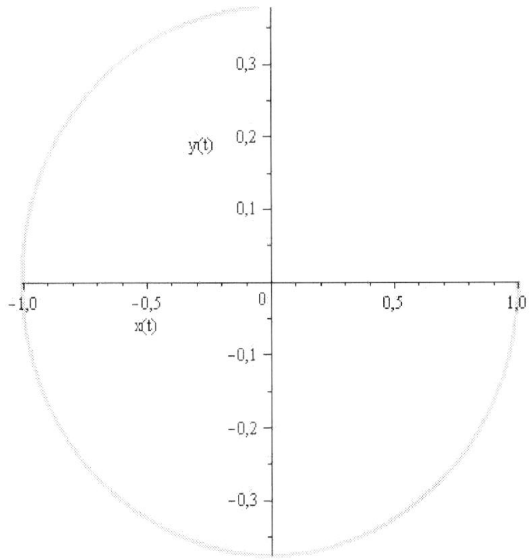

Figure 16. Case1 of simulation for the model (10) with t=0..50

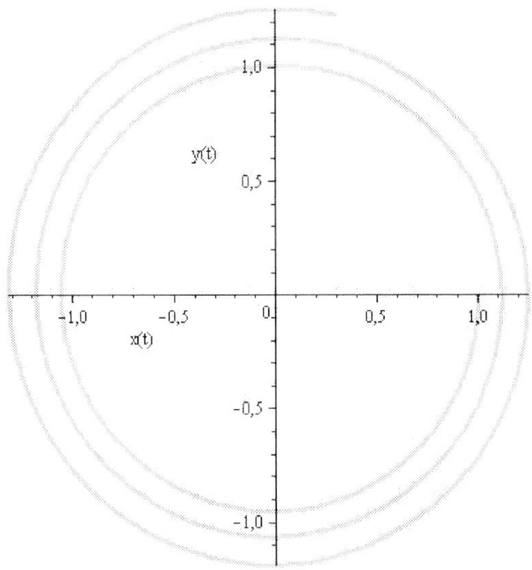

Figure 17. Case2 of simulation for the model (10), t=0..25. A periodic behaviour of the trajectory

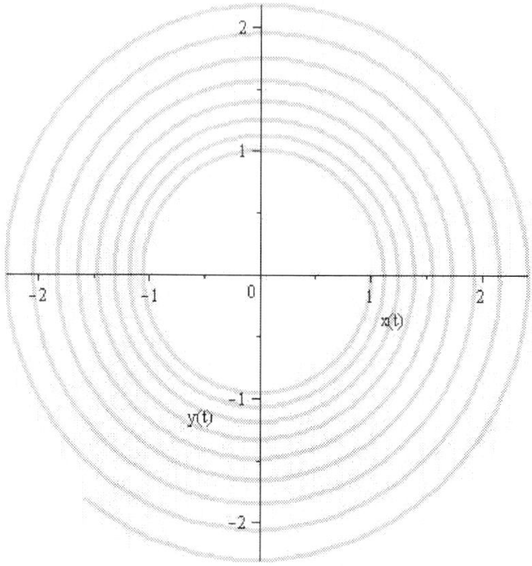

Figure 18. Case2 of simulation for the model (10), t=0..75

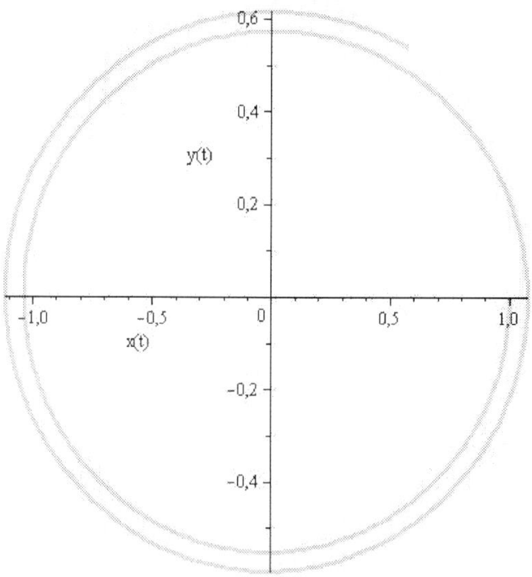

Figure 19. Case3 of simulation for the model (10), time t=0..25

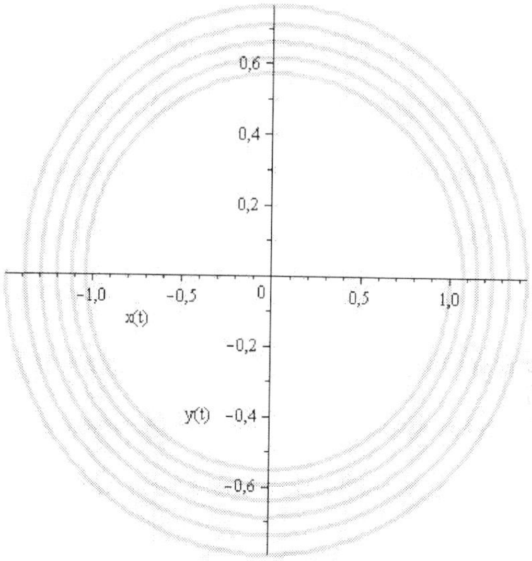

Figure 20. Case3 of simulation for the model (10), time t=0..75. The "circular loops" are fewer than in the case2, for the same simulation time.

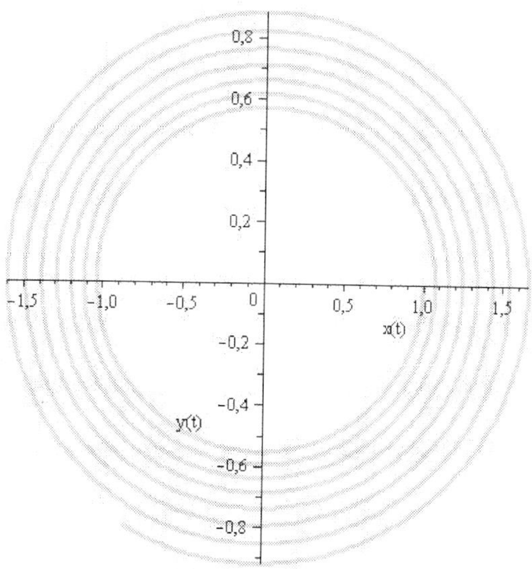

Figure 21. Case3 of simulation for the model (10) with t=0..100.

state 3 is condensed to state 4 by removing $\dot{Q}_C$ heat rate. The condensed working fluid at state 4 is then throttled by a valve and enters the evaporator at state 4′. The liquid working fluid is evaporated due to heat transfer rate $\dot{Q}_L$ from the cooling space to the working fluid (4′–5). Finally, the vaporized working fluid is absorbed by the weak solution in the absorber, and by means of $\dot{Q}_A$ heat rate release in the absorber, state 1 is reached.

Work input required by the solution pump in the system is negligible relative to the energy input to the generator and is often neglected for the purpose of analysis. Under such assumption, the equation for the first law of thermodynamics is written as:

$$\dot{Q}_H+\dot{Q}_L-\dot{Q}_C-\dot{Q}_A=0 \tag{1}$$

Absorption refrigeration systems operate between three temperature levels, if $T_A=T_C$, or four temperature levels when $T_A\neq T_C$. In this chapter, by taking $T_A=T_C$, the cycle of the working fluid consists of three irreversible isothermal and three irreversible adiabatic processes. The temperatures of the working fluid in the three isothermal processes are different from those of the external heat reservoirs so that heat is transferred under a finite temperature difference, as shown in Fig. 2 where

$$\dot{Q}_O=\dot{Q}_C+\dot{Q}_A \tag{2}$$

T_1 and T_2 are, respectively, the temperatures of the working fluid in the generator and evaporator. It is assumed that the working fluid in the condenser and absorber has the same temperature T_3 [2]. $\dot{Q}_{LC}$ is the heat leak from the heat sink to the cooled space.

The heat exchanges between the working fluid and heat reservoirs obey a linear heat transfer law, so that the heat-transfer equations in the generator, evaporator, condenser and absorber are, respectively, expressed as follows:

$$\dot{Q}_H=U_H A_H\left(T_H-T_1\right) \tag{3}$$

$$\dot{Q}_L=U_L A_L\left(T_L-T_2\right) \tag{4}$$

$$\dot{Q}_O=U_O\left(A_A+A_C\right)\left(T_3-T_O\right) \tag{5}$$

where A_H, A_L, A_C and A_A are, respectively, the heat-transfer areas of the generator, evaporator, condenser and absorber, U_H and U_L are, respectively, the overall heat-transfer coefficients of the generator and evaporator, and it is assumed that the condenser and absorber have the same overall heat-transfer coefficient U_O [2].

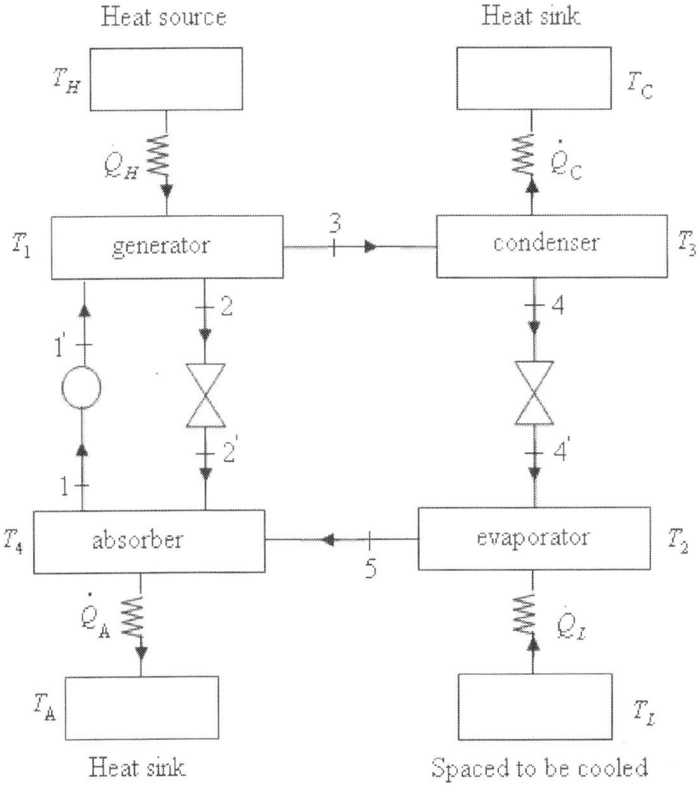

Figure 1. Schematic diagram of absorption refrigeration system [2]

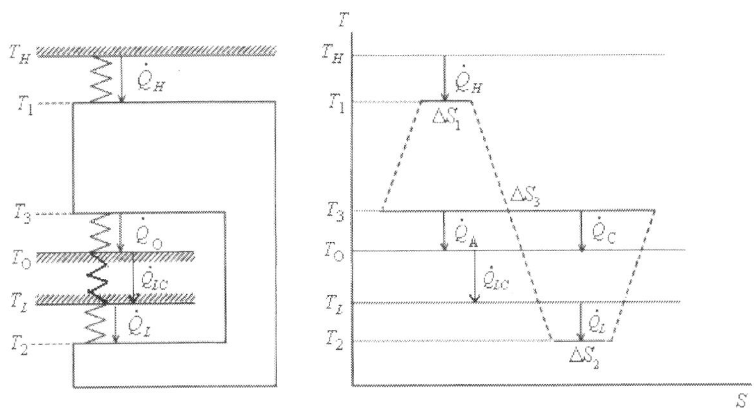

Figure 2. Considered irreversible absorption refrigeration model and its T–S diagram.

The absorption refrigeration system does not exchange heat with other external reservoirs except for the three heat reservoirs at temperatures T_H , T_L and T_O, so the total heat-transfer area between the cycle system and the external heat reservoirs is given by the relationships:

$$A = A_H + A_L + A_O \tag{6}$$

where

$$A_O = A_C + A_A. \tag{7}$$

The rate of heat leakage $\dot{Q}_{LC}$ from the heat sink at temperature T_O to the cold reservoir at temperature T_L was first provided by Bejan [3] and it is given as:

$$\dot{Q}_{LC} = K_{LC}\left(T_O - T_L\right) \tag{8}$$

where K_{LC} is the heat leak coefficient.

Real absorption refrigerators are complex devices and suffer from a series of irreversibilities. Besides the irreversibility of finite rate heat transfer which is considered in the endoreversible cycle models and the heat leak from the heat sink to the cooled space, there also exist other sources of irreversibility. The internal irreversibilities that result from friction, mass transfer and other working fluid dissipations are an another main source of irreversibility, which can decrease the coefficient of performance and the cooling load of absorption refrigerators. The total effect of the internal irreversibilities on the working fluid can be characterized in terms of entropy production. An irreversibility factor is introduced to describe these internal irreversibilities:

$$I = \frac{\Delta S_3}{\Delta S_1 + \Delta S_2} \tag{9}$$

On the basis of the second law of thermodynamics, $\Delta S_3 > \Delta S_1 + \Delta S_2$ for an internally irreversible cycle, so that $I > 1$. If the internal irreversibility is neglected, the cycle is endoreversible and so $I = 1$. The second law of thermodynamics for an irreversible three-heat-source cycle requires that:

$$\oint \frac{\delta \dot{Q}}{T} = \frac{\dot{Q}_H}{T_1} + \frac{\dot{Q}_L}{T_2} - \frac{\dot{Q}_O}{T_3} \leq 0 \tag{10}$$

From Eq. (9), the inequality in Eq. (10) is written as:

$$\frac{\dot{Q}_H}{T_1} + \frac{\dot{Q}_L}{T_2} - \frac{\dot{Q}_O}{IT_3} = 0 \tag{11}$$

The coefficient of performance of the irreversible three-heat-source absorption refrigerator is:

$$COP = \frac{\dot{Q}_L - \dot{Q}_{LC}}{\dot{Q}_H} = \frac{\dot{Q}_L}{\dot{Q}_H}\left(1 - \frac{\dot{Q}_{LC}}{\dot{Q}_L}\right) \tag{12}$$

From Eq. (6), it is expressed as:

$$A_L = \frac{A}{1 + \frac{A_H}{A_L} + \frac{A_O}{A_L}} \tag{13}$$

Using Eqs. (3)-(5), Eq. (13) is rewritten as:

$$A_L = \frac{A}{1 + \frac{\dot{Q}_H}{\dot{Q}_L}\frac{U_L(T_L - T_2)}{U_H(T_H - T_1)} + \frac{\dot{Q}_O}{\dot{Q}_L}\frac{U_L(T_L - T_2)}{U_O(T_3 - T_O)}} \tag{14}$$

Combining Eqs. (1) and (11), the following ratios are derived:

$$\frac{\dot{Q}_L}{\dot{Q}_H} = \frac{T_2(T_1 - IT_3)}{T_1(IT_3 - T_2)} \tag{15}$$

$$\frac{\dot{Q}_O}{\dot{Q}_L} = \frac{IT_3(T_1 - T_2)}{(T_1 - IT_3)T_2} \tag{16}$$

The first is the coefficient of performance of the irreversible three-heat-source absorption refrigeration cycle without heat leak losses.

Substituting Eqs. (15) and (16) into Eq. (14), the heat-transfer area of the evaporator is expressed as a function of T_1, T_2 and T_3 for a given total heat-transfer areas :

$$A_L = \frac{A}{1 + \frac{U_L T_1(IT_3 - T_2)(T_L - T_2)}{U_H(T_H - T_1)(T_1 - IT_3)T_2} + \frac{U_L IT_3(T_1 - T_2)(T_L - T_2)}{U_O(T_3 - T_O)(T_1 - IT_3)T_2}} \tag{17}$$

By investigating similar reasoning, the heat-transfer areas of the generator and of condenser and absorber are given respectively by:

$$A_H = \frac{A}{1 + \frac{U_H T_2(T_1 - IT_3)(T_H - T_1)}{U_L T_1(IT_3 - T_2)(T_L - T_2)} + \frac{U_H IT_3(T_1 - T_2)(T_H - T_1)}{U_O T_1(IT_3 - T_2)(T_3 - T_O)}} \tag{18}$$

and

$$A_O = \frac{A}{1+\dfrac{U_O T_1(IT_3-T_2)(T_3-T_O)}{U_H IT_3(T_1-T_2)(T_H-T_1)}+\dfrac{U_O T_2(T_1-IT_3)(T_3-T_O)}{U_L IT_3(T_1-T_2)(T_L-T_2)}} \tag{19}$$

Substituting Eq. (17) into Eq. (4):

$$\dot{Q}_L = \frac{A}{\dfrac{1}{U_L(T_L-T_2)}+\dfrac{T_1(IT_3-T_2)}{U_H(T_H-T_1)(T_1-IT_3)T_2}+\dfrac{IT_3(T_1-T_2)}{U_O(T_3-T_O)(T_1-IT_3)T_2}} \tag{20}$$

Combining Eqs. (8), (12), (15) and (20), the coefficient of performance of the irreversible three-heat-source refrigerator as a function of the temperatures T_1, T_2 and T_3 of the working fluid in the generator, evaporator, condenser and absorber is obtained:

$$COP = \frac{T_2(T_1-IT_3)}{T_1(IT_3-T_2)}\left\{1-\xi(T_O-T_L)\left[\frac{1}{U_L(T_L-T_2)}+\frac{T_1(IT_3-T_2)}{U_H(T_H-T_1)(T_1-IT_3)T_2}+\frac{IT_3(T_1-T_2)}{U_O(T_3-T_O)(T_1-IT_3)T_2}\right]\right\} \tag{21}$$

where the parameter

$$\xi = \frac{K_{LC}}{A} \tag{22}$$

represents the heat leakage coefficient and its dimension is w/(Km²)

The specific cooling load of the irreversible three-heat-source refrigerator is deduced as:

$$r = \frac{\dot{Q}_L-\dot{Q}_{LC}}{A} = \left[\frac{1}{U_L(T_L-T_2)}+\frac{T_1(IT_3-T_2)}{U_H(T_H-T_1)(T_1-IT_3)T_2}+\frac{IT_3(T_1-T_2)}{U_O(T_3-T_O)(T_1-IT_3)T_2}\right]^{-1} - \xi(T_O-T_L) \tag{23}$$

The specific entropy production rate of the irreversible three-heat-source absorption refrigerator is:

$$s = \frac{\dot{\sigma}}{A} = \frac{\dfrac{\dot{Q}_O-\dot{Q}_{LC}}{T_O}-\dfrac{\dot{Q}_H}{T_H}-\dfrac{\dot{Q}_L-\dot{Q}_{LC}}{T_L}}{A} \tag{24}$$

Using Eq. (1) s is rewritten as:

$$s=\left(\frac{1}{T_L}-\frac{1}{T_O}\right)\frac{\dot{Q}_{LC}}{A}+\left(\frac{1}{T_O}-\frac{1}{T_H}\right)\frac{\dot{Q}_H}{\dot{Q}_L}\frac{\dot{Q}_L}{A}+\left(\frac{1}{T_O}-\frac{1}{T_L}\right)\frac{\dot{Q}_L}{A} \tag{25}$$

or

$$s=\left(\frac{1}{T_O}-\frac{1}{T_H}\right)\frac{\dot{Q}_H}{A}+\left(\frac{1}{T_O}-\frac{1}{T_L}\right)\frac{\dot{Q}_L-\dot{Q}_{LC}}{A} \tag{26}$$

Substituting Eqs.(8), (15) and (20) into Eq. (25), the specific entropy production rate as a function of T_1, T_2 and T_3 is given by :

$$s=\left(\frac{1}{T_L}-\frac{1}{T_O}\right)\left\{\xi(T_O-T_L)-\left[1-\frac{\varepsilon_r T_1(IT_3-T_2)}{T_2(T_1-IT_3)}\right]\left[\frac{1}{U_L(T_L-T_2)}+\frac{T_1(IT_3-T_2)}{U_H(T_H-T_1)(T_1-IT_3)T_2}+\frac{IT_3(T_1-T_2)}{U_O(T_3-T_O)(T_1-IT_3)T_2}\right]^{-1}\right\} \tag{27}$$

where

$$\varepsilon_r=\frac{\left(1-\frac{T_O}{T_H}\right)}{\left(\frac{T_O}{T_L}-1\right)} \tag{28}$$

is the coefficient of performance for reversible three-heat-source refrigerator.

According to the definition of the general thermo-ecological criterion function for different heat engine models [4-9], a two-heat-source refrigerator [10, 11] and three-heat-source absorption refrigerator [2], the new thermo-ecological objective function called ecological coefficient of performance (*ECOP)* of an absorption refrigerator is defined as:

$$ECOP=\frac{\dot{Q}_L-\dot{Q}_{LC}}{T_{env}\,\sigma}=\frac{\dot{Q}_L-\dot{Q}_{LC}}{AT_{env}s} \tag{29}$$

Putting Eq.(26) into Eq. (29):

$$ECOP=\frac{1}{T_{env}\left[\frac{1}{T_O}-\frac{1}{T_H}+\left(\frac{1}{T_O}-\frac{1}{T_L}\right)\frac{1}{COP}\right]} \tag{30}$$

When Eq. (21) is put in Eq. (30), the ecological coefficient of performance of the irreversible three-heat-source absorption refrigerator as a function of T_1, T_2 and T_3 is derived as :

$$ECOP = \frac{\dfrac{1}{T_{env}\left(T_O^{-1} - T_L^{-1}\right)}}{1 - \dfrac{\varepsilon_r T_1\left(IT_3 - T_2\right)}{T_2\left(T_1 - IT_3\right)}\left\{1 - \xi\left(T_O - T_L\right)\left[\dfrac{1}{U_L\left(T_L - T_2\right)} + \dfrac{T_1\left(IT_3 - T_2\right)}{U_H\left(T_H - T_1\right)\left(T_1 - IT_3\right)T_2} + \dfrac{IT_3\left(T_1 - T_2\right)}{U_O\left(T_3 - T_O\right)\left(T_1 - IT_3\right)T_2}\right]\right\}^{-1}} \tag{31}$$

where T_{env} is the temperature in the environment conditions.

3. Performance optimization for a three-heat-source irreversible absorption refrigerator based on ECOP criterion

The ECOP function given in Eq. (31) is plotted with respect to the working fluid temperatures (T_1, T_2 and T_3) for different internal irreversibility parameters as shown in Fig. 3(a), (b) and (c). As it can be seen from the figure, there exists a specific T_1, T_2 and T_3 that maximize the ECOP function for given I and ξ values. Therefore, Eq. (31) can be maximized (or optimized) with respect to T_1, T_2 and T_3. The optimization is carried out analytically.

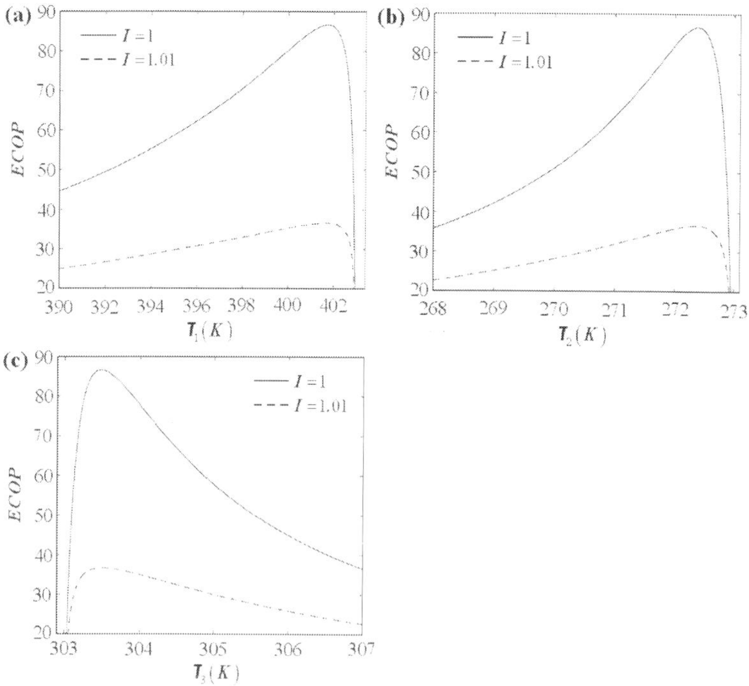

Figure 3. Variation of the ECOP objective function with respect to T_1 (a), T_2 (b) and T_3 (c) for different I values ($T_G = 403K$, $T_L = 273K$, $T_O = 303K$, $T_{env} = 290K$, $U_G = 1163\ W/m^2K$, $U_E = 2326\ W/m^2K$, $U_O = 4650\ W/m^2K$, $K_L = 1082\ W/K$, $A = 1100\ m^2$)

For the sake of convenience, let

$$x = \frac{IT_3}{T_1} \tag{32}$$

$$y = \frac{IT_3}{T_2} \tag{33}$$

$$z = IT_3 \tag{34}$$

Then Eq. (31) is rewritten as:

$$ECOP = \frac{\dfrac{1}{T_{env}\left(T_O^{-1} - T_L^{-1}\right)}}{1 - \dfrac{\varepsilon_r (y-1)}{1-x}\left\{1 - \xi(T_O - T_E)\left[\dfrac{y}{U_L(T_L y - z)} + \dfrac{x(y-1)}{U_H(T_H x - z)(1-x)} + \dfrac{y-x}{U(z-T)(1-x)}\right]\right\}^{-1}} \tag{35}$$

where

$$T = IT_O \tag{36}$$

and

$$U = \frac{U_O}{I}. \tag{37}$$

Starting from Eq. (35), the extremal conditions:

$$\frac{\partial ECOP}{\partial x} = 0 \tag{38}$$

$$\frac{\partial ECOP}{\partial y} = 0 \tag{39}$$

$$\frac{\partial ECOP}{\partial z} = 0 \tag{40}$$

give respectively:

$$\frac{1}{\xi(T_O - T_E)} - \frac{y}{U_L(T_L y - z)} - \frac{1}{U(z-T)} - \frac{z(y-1)}{U_H(T_H x - z)^2} = 0 \tag{41}$$

$$\frac{1}{\xi(T_O - T_E)} - \frac{y}{U_L(T_L y - z)} - \frac{1}{U(z-T)} - \frac{z(y-1)}{U_L(T_L y - z)^2} = 0 \qquad (42)$$

$$\frac{y(1-x)}{U_L(T_L y - z)^2} + \frac{x(y-1)}{U_H(T_H x - z)^2} - \frac{y-x}{U(z-T)^2} = 0 \qquad (43)$$

Combining Eqs (41)-(43), the following general relation is found:

$$\sqrt{U_H}(T_H x - z) = \sqrt{U_L}(T_L y - z) = \sqrt{U}(z - T) \qquad (44)$$

From Eqs (44), it is derived as:

$$x = \frac{(1+b_1)z}{T_H} - \frac{b_1 T}{T_H} \qquad (45)$$

$$y = \frac{(1+b_2)z}{T_L} - \frac{b_2 T}{T_L} \qquad (46)$$

where

$$b_1 = \sqrt{\frac{U}{U_H}} \qquad (47)$$

$$b_2 = \sqrt{\frac{U}{U_L}} \qquad (48)$$

When Eqs. (45) and (46) are substituted into Eq. (43):

$$z = T\frac{D + b_2}{1 + b_2} \qquad (49)$$

where

$$D = \frac{1 + \sqrt{d_1\left[1 - \frac{T_L(1-d_1)}{T}\right]}}{1 - d_1} \qquad (50)$$

$$d_1 = \xi \frac{\left(1+b_2\right)^2}{U}\left(\frac{T_O}{T_L} - 1\right) \tag{51}$$

Therefore Eqs. (45) and (46) are rewritten as:

$$x = \frac{T}{T_H} B_1 \left(D + B\right) \tag{52}$$

$$y = \frac{T}{T_L} D \tag{53}$$

where

$$B = \frac{b_2 - b_1}{1 + b_1} \tag{54}$$

$$B_1 = \frac{1 + b_1}{1 + b_2} \tag{55}$$

Using Eqs. (49), (52) and (53) with Eqs.(32)-(34), the corresponding optimal temperatures of the working fluid in the three isothermal processes when the ecological coefficient of performance is a maximum, are, respectively, determined by:

$$T_1^* = T_H \frac{D + b_2}{\left(1 + b_1\right)\left(D + B\right)} \tag{56}$$

$$T_2^* = T_L \frac{D + b_2}{D\left(1 + b_2\right)} \tag{57}$$

$$T_3^* = T_O \frac{D + b_2}{1 + b_2} \tag{58}$$

Substituting Eqs. (56)-(58) into Eqs. (21), (23), (27) and (31) the maximum *ECOP* function and the corresponding optimal coefficient of performance, optimal specific cooling load and optimal specific entropy generation rate are derived, respectively, as:

$$ECOP_{\max} = \frac{1}{T_{env}\left(T_O^{-1} - T_L^{-1}\right)}$$

$$\times\frac{1}{1-\frac{\varepsilon_r T_H\left(TD-T_L\right)}{\left[T_H-B_1\left(D+B\right)T\right]T_L}\left\{1-\xi\left(T_O-T_L\right)\frac{T_HD-B_1^{\,2}\left(D+B\right)\left(T_L+BDT\right)}{U^*B_1^{\,2}\left(D-1\right)\left[T_H-B_1\left(D+B\right)T\right]T_L}\right\}^{-1}} \tag{59}$$

$$COP^*=\frac{\left[T_H-B_1\left(D+B\right)T\right]T_L}{T_H\left(TD-T_L\right)}\left\{1-\xi\left(T_O-T_L\right)\frac{T_HD-B_1^{\,2}\left(D+B\right)\left(T_L+BDT\right)}{U^*B_1^{\,2}\left(D-1\right)\left[T_H-B_1\left(D+B\right)T\right]T_L}\right\} \tag{60}$$

$$r^*=\frac{U^*B_1^{\,2}\left(D-1\right)\left[T_H-B_1\left(D+B\right)T\right]T_H}{T_HD-B_1^{\,2}\left(D+B\right)\left(T_L+BDT\right)}-\xi\left(T_O-T_L\right) \tag{61}$$

$$s^*=\left(\frac{1}{T_L}-\frac{1}{T_O}\right)\left\{\xi\left(T_O-T_L\right)-\left[1-\frac{\varepsilon_r T_H\left(TD-T_L\right)}{\left[T_H-B_1\left(D+B\right)T\right]T_L}\right]\left[\frac{U^*B_1^{\,2}\left(D-1\right)\left[T_H-B_1\left(D+B\right)T\right]T_L}{T_HD-B_1^{\,2}\left(D+B\right)\left(T_L+BDT\right)}\right]\right\} \tag{62}$$

where

$$U^*=\frac{U}{\left(1+b_1\right)^2} \tag{63}$$

From Eqs. (17)-(19) and (56)-(58), it is found that , when the three-heat-source absorption refrigerator is operated in the state of maximum ecological coefficient of performance, the relations between the heat-transfer areas of the heat exchangers and the total heat-transfer area are determined by:

$$A_H^{\;*}=A\frac{b_1}{1+b_1}\frac{B_1^{\,2}\left(D+B\right)\left(TD-T_L\right)}{T_HD-B_1^{\,2}\left(D+B\right)\left(T_L+BDT\right)} \tag{64}$$

$$A_L^{\;*}=A\frac{b_2}{1+b_1}\frac{DB_1\left[T_H-B_1T\left(D+B\right)\right]}{T_HD-B_1^{\,2}\left(D+B\right)\left(T_L+BDT\right)} \tag{65}$$

$$A_O^{\;*}=A\frac{1}{1+b_1}\frac{B_1\left[T_HD-B_1T_L\left(D+B\right)\right]}{T_HD-B_1^{\,2}\left(D+B\right)\left(T_L+BDT\right)} \tag{66}$$

From Equations (64)-(66), a concise optimum relation for the distribution of the heat-transfer areas is obtained as:

$$\sqrt{U_H}A_H^{\;*}+\sqrt{U_L}A_L^{\;*}=\sqrt{U}A_O^{\;*} \tag{67}$$

Obviously, this relation is independent of the heat leak and the temperatures of the external heat reservoirs.

4. Comparison with COP criterion

In Fig.4, the variation of the normalized ECOP ($\overline{ECOP}=\dfrac{ECOP}{ECOP_{max}}$), normalized COP ($\overline{COP}=\dfrac{COP}{COP_{max}}$) and the specific cooling load (r) with respect to the specific entropy generation rate (s) are demonstrated. One interesting observation from this figure is that maximum of the ECOP and COP coincides although their functional forms are different: the coefficient of performance gives information about the necessary heat rate input in order to produce certain amount of cooling load and the ecological coefficient of performance gives information about the entropy generation rate or loss rate of availability in order to produce certain amount of cooling load. The maximum ECOP and COP conditions give the same amount of cooling load and entropy generation rate. It is also seen analytically that the performance parameters T_1^* , T_2^* , T_3^* , A_1^* , A_2^* , A_3^* , r^* , s^* and $COP^*=COP_{max}$ at the maximum ECOP and maximum COP are same. Getting the same performance at maximum ECOP and COP conditions is an expected and logical result. Since, for a certain cooling load the maximum COP results from minimum heat consumption so that minimum environmental pollution. The minimum environmental pollution is also achieved by maximizing the $ECOP$. Although the optimal performance conditions $ECOP$ and COP criteria are same, their impact on the system design performance is different. The coefficient of performance is used to evaluate the performance and the efficiency of systems. This method only takes into account the first law of thermodynamics which is concerned only with the conversion of energy, and therefore, can not show how or where irreversibilities in a system or process occur. Also, when different sources and forms of energy are involved within a system, the COP criterion of a system doesn't describe its performance from the view point of the energy quality involved. This factor is taken into account by the second law of thermodynamics characterized by the entropy production which appears in the ecological coefficient of performance criterion ($ECOP$). This aspect is of major importance today since that with the requirement of a rigorous management of our energy resources, one should have brought to be interested more and more in the second principle of thermodynamics, because degradations of energy, in other words the entropy productions, are equivalent to consumption of energy resources. For this important reason, the $ECOP$ criterion can enhance the system performance of the absorption refrigerators by reducing the irreversible losses in the system. A better understanding of the second law of thermodynamics reveals that the ecological coefficient of performance optimization is an important technique in achieving better operating conditions.

[11] Ust, Y.: Performance analysis and optimization of irreversible air refrigeration cycles based on ecological coefficient of performance criterion. Applied Thermal Engineering. 29, 47–55 (2009).

Section 6

Thermodynamics in Diverse Areas

Chapter 18

Thermodynamics of Microarray Hybridization

Raul Măluţan and Pedro Gómez Vilda

Additional information is available at the end of the chapter

http://dx.doi.org/10.5772/51624

1. Introduction

Microarrays make the use of hybridization properties of nucleic acids to monitor Deoxyribonucleic acid (DNA) or Ribonucleic acid (RNA) abundance on a genomic scale in different types of cells. The hybridization process takes place between surface-bound DNA sequences - the probes, and the DNA or RNA sequences in solution - the targets. Hybridization is the process of combining complementary, single-stranded nucleic acids into a single molecule. Nucleotides will bind to their complement under normal conditions, so two perfectly complementary strands will bind to each other readily. Conversely, due to the different geometries of the nucleotides, a single inconsistency between the two strands will prevent them from binding.

In oligonucleotide microarrays hundreds of thousands of oligonucleotides are synthesized *in situ* by means of photochemical reaction and mask technology. Probe design in these microarrays is based on complementarity to the selected gene or an expressed sequence tag (EST) reference sequence. An important component in designing an oligonucleotide array is ensuring that each probe binds to its target with high specificity.

The dynamics of the hybridization process underlying genomic expression is complex as thermodynamic factors influencing molecular interaction are still fields of important research [1] and their effects are not taken into account in the estimation of genetic expression by the algorithms currently in use.

2. State of the art

Many techniques have been developed to identify trends in the expression levels inferred from DNA microarray data, and recently the attention was devoted to methods to obtain accurate expression levels from raw data on the underlying principles of the thermodynamics and hybridization kinetics. The development of DNA chips for rapidly

Finding the location of a gene or gene product by adding specific radioactive or chemically tagged probes for the gene and detecting the location of the radioactivity or chemical on the chromosome or in the cell after hybridization is called *in-situ* hybridization.

In the same way, in microarray technology, hybridization is used in comparing mRNA abundance in two samples, or in one sample and a control. RNA from the sample and control are extracted and labeled with two different fluorescent labels, *e.g.* a red dye for the RNA from the sample population and green dye for that from the control population. Both extracts are washed over the microarray and gene sequences from the extracts hybridized to their complementary single-strand DNA molecule previously attached to the microarray. Then, to measure the abundance of the hybridized RNA, the array is excited by a laser.

In the oligonucleotide microarrays the hybridization process occurs in the same way, the only difference here is that the sequences to be laid over the chip are sequences of 25 nucleotides length, perfect complementary to same length sequence of the gene, PM – perfect match, and sequences of 25 nucleotides length, designed to correspond to PM, but having the middle base - the 13th one, changed by its complementary base, MM – mismatch, as in Figure 2. The MM probes give some estimates of the random hybridization and cross hybridization signals. One principle to be followed in the design of oligonucleotide arrays is ensuring that the probes bind to their target with high accuracy. When the two strands are completely complementary they will bind by a specific hybridization, as it can be seen in Figure 3. On the contrary if there are mismatches between the nucleotides of the strands and they bind, a process called non-specific hybridization or cross-hybridization occurs.

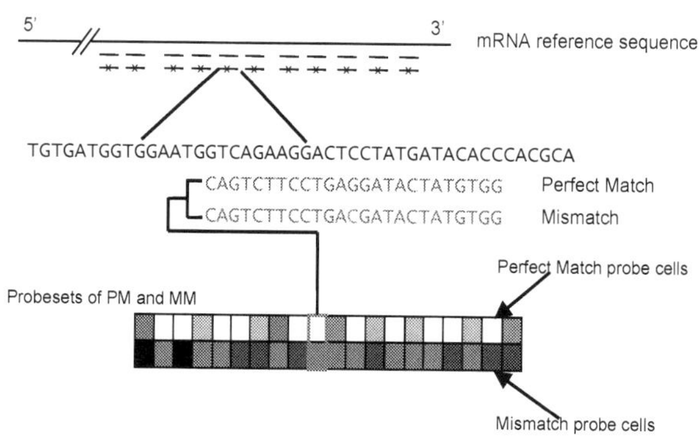

Figure 2. Perfect Match – Mismtach probeset strategy. Sequence of 25-mer length complementary to the selected part of mRNA sequence form a Perfect Match probe, while the Mismatch probe is artificially created by changing middle base with its complementary. In an oligonucleotide array a gene is represented by 11 to 20 probes. (modified from [13])

The hybridization process has been studied from point of view of interaction between base pairs, the interaction with unintended targets and also from its kinetics processes. Because in practice the DNA chips are immersed in the target solution for a relatively short time, the arrival to equilibrium is not guaranteed. Yet full analysis of the reaction kinetics requires knowledge of the equilibrium state. An understanding of the equilibrium state is also necessary to identify the relative importance of kinetic controls of the performance of the DNA microarrays. The effect of the cross-hybridization on probe intensity is predictable in the oligonucleotide microarrays, and models for avoiding this have been developed [14], [15], [16] some aspects of it going to be described in the following section.

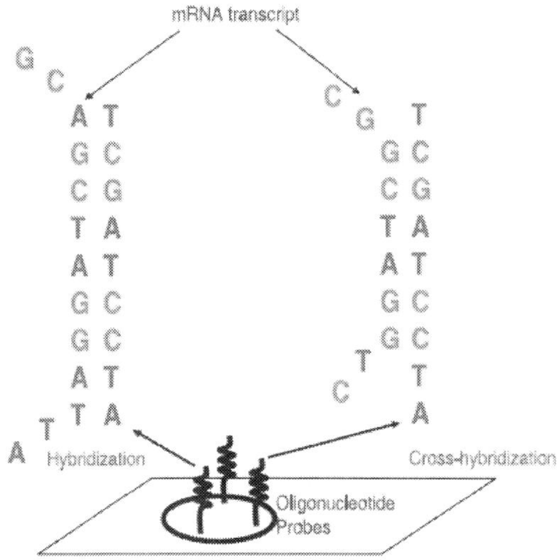

Figure 3. Cross-hybridization on a nucleotide probe. In specific hybridization the sequences are completely complementary, while in non-specific or cross hybridization the sequences contain mismatches. (from [17])

4. Technical factors affecting gene expression

4.1. Thermodynamics parameters

Black and Hartley [18] define enthalpy as the sum of the internal energy of a thermodynamic system plus the energy associated with work done by the system on the atmosphere, which is the product of the pressure times the volume, as in equation (1)

$$H = U + pV \tag{1}$$

Because enthalpy is a property, its value can be determined for a simple compressible substance once two independent, intensive thermodynamic properties of the substance are known, and the change in enthalpy is independent of the path followed between two equilibrium states

In [18] the entropy, S, was defined using the following equation:

$$dS = \frac{\delta Q}{T} \tag{2}$$

where δQ is an amount of heat introduced to the system and T is a constant absolute temperature. Since this definition involves only differences in entropy, the entropy itself is only defined up to an arbitrary additive constant.

The following models to be described use the state function parameters, enthalpy and entropy. State functions define the properties of a thermodynamic state. In a change between two thermodynamic states, the change in value of the state function is given by the symbol Δ .

The standard enthalpy change, ΔH° , is the difference in the standard enthalpies of formation between the products and the reactants. This state function is associated with changes in bonding between reactants and products. Changes in enthalpy during reactions are measured by calorimetry experiments.

The standard entropy change, ΔS° , is the difference in standard entropies between reactants and products. Entropy is a measure of the degree of order in a chemical system due to bond rotations, other molecular motions, and aggregation. The more random a system (disorder), the greater the entropy is. The larger a structure, the more degrees of freedom it has, and the greater its entropy.

4.2. Interaction between pairs

The nucleic acid duplex stability can be endangered by the interaction between the nucleotide bases. Thermodynamics for double helix formation of DNA/DNA, RNA/RNA or DNA/RNA can be estimated with nearest neighbour parameters. Enthalpy change, ΔH° , entropy change, ΔS° , free energy change, ΔG° , and melting temperature, T_m, were obtained on the basis of the nearest-neighbour model.

The nearest-neighbour model for nucleic acids, known as the NN model, assumes that the stability of a given base pair depends on the identity and orientation of neighbouring base pairs [3]. Previous studies in NN model parameters were brought forth in [15] and [19].

In the NN model, sequence dependent stability is considered in terms of nearest-neighbour doublets. In duplex DNA there are 10 such unique internal nearest-neighbour doublets. Listed in the 5′-3′ direction, these are AT/AT TA/TA AA/TT AC/GT CA/TG TC/GA CT/AG CG/CG GC/GC and GG/CC. Dimmer duplexes are represented with a slash separating strands in antiparallel orientation *e.g.* AC/TG means 5′-AC-3′ Watson–Crick base-paired with 3′-TG-5′.

The total difference in the free energy of the folded and unfolded states of a DNA duplex can be approximated at 37º, with a nearest-neighbour model:

$$\begin{aligned}\Delta G^{o}(\text{total}) &= \textstyle\sum_{i} n_{i} \Delta G^{o}(i) + \Delta G^{o}(\text{init w/term G}\cdot\text{C}) \\ &+ \Delta G^{o}(\text{init w/term A}\cdot\text{T}) + \Delta G^{o}(\text{sym})\end{aligned} \quad (3)$$

where $\Delta G^{\circ}(i)$ are the standard free-energy changes for 10 possible Watson-Crick nearest neighbours, *e.g.* $\Delta G^{o}(1) = \Delta G^{o}_{37}(\text{AA/TT}).$, $\Delta G^{o}(2) = \Delta G^{o}_{37}(\text{TA/AT}).$, n_i is the number of occurrences of each nearest neighbour, i, and $\Delta G^{o}(\text{sym})$ equals +0.43 kcal/mol if the duplex is self complementary and zero if it is not self-complementary. The total difference in the free energy at 37°, ΔG_{37}^{o}, can be computed from ΔH^{o} and ΔS^{o} parameters using the equation:

$$\Delta G_{37}^{o} = \Delta H^{o} - T\Delta S^{o} \tag{4}$$

For a specific temperature one can compute the total free energy using the values from Table 1. As described in [19] the melting temperature T_m is defined as the temperature at which half of the strands are in double helical and half are in the random-coil state. A random-coil state is a polymer conformation where the monomer subunits are oriented randomly while still being bonded to adjacent units.

For self-complementary oligonucleotides, the T_m for individual melting curves was calculated from the fitted parameters using the following equation:

$$T_m = \Delta H^{o} / \left(\Delta S^{o} + R\ln C_T\right) \tag{5}$$

where R is the general gas constant, *i.e.* 1.987cal/K mol, the C_T is the total strand concentration, and T_m is given in K. For non-self-complementary molecules, C_T in equation (5) was replaced by $C_T/4$.

Sequence	ΔH° kcal/mol	ΔS° kcal/mol
AA/TT	-7.9	-22.2
AT/TA	-7.2	-20.4
TA/AT	-7.2	-21.3
CA/GT	-8.5	-22.7
GT/CA	-8.4	-22.4
CT/GA	-7.8	-21.0
GA/CT	-8.2	-22.2
CG/GC	-10.6	-27.2
GC/CG	-9.8	-24.4
GG/CC	-8.0	-19.9
Init. w/term G•C	0.1	-2.8
Init. w/term A•T	2.3	4.1
Symmetry correction	0	-1.4

Table 1. Unified oligonucleotide ΔH° and ΔS° nearest neighbour parameters in 1M NaCl. The table shows the values of the total enthalpy and entropy for the dimmer duplexes as used in [3].

The nearest-neighbour parameters of Delcourt et al. (1991) [20], SantaLucia et al. (1996) [19], Sugimoto et al. (1996) [15] and Allawi et al. (1997) [21] were evaluated from the analysis of optical melting curves of a variety of short synthetic DNA duplexes in 1 M Na+.

The observed trend in nearest-neighbor stabilities at 37°C is GC/CG = CG/GC > GG/CC > CA/GT = GT/CA = GA/CT = CT/GA > AA/TT > AT/TA > TA/AT, as in Table 2. This trend suggests that both sequence and base composition are important determinants of DNA duplex stability. It has long been recognized that DNA stability depends of the percent G-C content.

Sequence	ΔG 37 (kcal/mol)			
	Delcourt et al.	*SantaLucia et al.*	*Sugimoto et al.*	*Allawi et al.*
AA/TT	-0.67	-1.02	-1.20	-1.00
AT/TA	0.62	-0.73	-0.90	-0.88
TA/AT	-0.70	-0.60	-0.90	-0.58
CA/GT	-1.19	-1.38	-1.70	-1.45
GT/CA	-1.28	-1.43	-1.50	-1.44
CT/GA	-1.17	-1.16	-1.50	-1.28
GA/CT	-1.12	-1.46	-1.50	-1.30
CG/GC	-1.87	-2.09	-2.80	-2.17
GC/CG	-1.85	-2.28	-2.30	-2.24
GG/CC	-1.55	-1.77	-2.10	-1.84
Average	-1.20	-1.39	-1.64	-1.42
Init. w/term G•C	NA	0.91	1.70	0.98
Init. w/term A·T	NA	1.11	1.70	1.03

Table 2. Comparison of computed NN free energy parameters at 37°C

On the other hand, the nearest neighbour ΔH° parameters from Table 1, do not follow this trend. This suggests that stacking, hydrogen bonding, and other contributions to the ΔH° present a complicated sequence dependence.

4.3. Interaction with unintended targets

As seen in previous sections the major issue in microarray oligonucleotide technology is the selection of probe sequences with high sensitivity and specificity. It has been shown [22] that the use of MM probes for assessment of non-specific binding is unreliable. Since the duplex formation in solution has been studied using the nearest neighbour model [3], [15] the microarray design in terms of probe selection has been achieved by using a model based on the previously mentioned nearest neighbour model [16]. The model of Zhang *et al.* presents some modification to the nearest neighbour model, firstly to assign different weight factors at each nucleotide position on a probe with the scope of reflecting that the binding parts of a probe may contribute differently to the stability of bindings, and secondly to take into account two different modes of binding the probes: gene specific binding, *i.e.* formation of DNA-RNA duplexes with exact complementary sequences, and non-specific binding, *i.e.* formation of duplexes with many mismatches between the probe and the attached RNA molecule. They called their model, the positional-dependent-nearest-neighbour model.

According with their method, the observed signal I_{ij} for probe i in the probe set for gene j is modelled as:

$$I_{ij} = \frac{N_j}{1+e^{E_{ij}}} + \frac{N^*}{1+e^{E^*_{ij}}} + B \tag{6}$$

where B is the background intensity, N_j is the number of expressed mRNA molecules contributing to gene specific binding, N^* represents the number of RNA molecule contributing to nonspecific binding, E and E^* are the binding energies for gene specific and respectively nonspecific binding. These energies are calculated as the weighted sum of stacking energies:

$$E_{ij} = \sum \omega_k \varepsilon\left(b_k, b_{k+1}\right) \tag{7}$$

$$E^*_{ij} = \sum \omega^*_k \varepsilon^*\left(b_k, b_{k+1}\right) \tag{8}$$

where ω_k and ω^*_k are the weight factors that depend on the position along the probe from the 5′ to 3′ end, and $\varepsilon\left(b_k, b_{k+1}\right)$ is the same as the stacking energy used in nearest neighbour model [15].

The positional-dependent-nearest-neighbour model appears to indicate that the two ends of a probe contribute less to binding stability according to their weight factors, see Figure 4. a). It also can be observed that there is a dip in the gene specific binding weight factors of MM probes around the mismatch position, probably due the mismatch which destabilizes the duplex structure. In Figure 4. b) it can be noted that stacking energies in the positional-dependent-nearest-neighbour model can give an explanation for the presence of negative probe pair signals.

This model, together with the nearest neighbour model solves the problem of binding on microarrays, but still there are factors that affect the gene expression measuring. One of them affects the process of competing adsorption and desorption of target RNA to from probe-target duplexes at the chip surface.

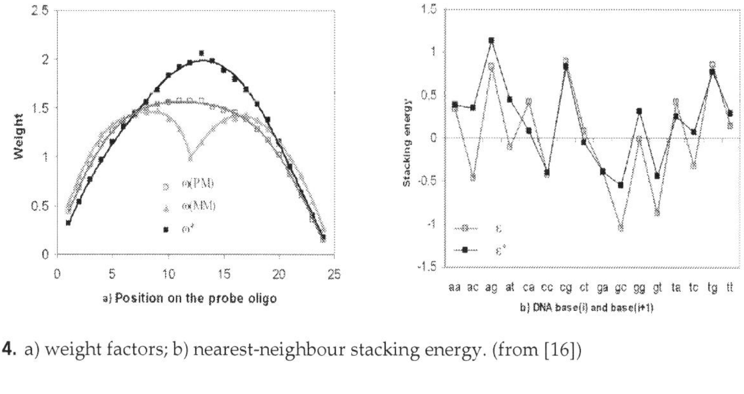

Figure 4. a) weight factors; b) nearest-neighbour stacking energy. (from [16])

4.4. Kinetic processes in hybridization thermodynamics

4.4.1. Derivation of the Langmuir isotherm

For molecules in contact with a solid surface at a fixed temperature, the Langmuir Isotherm, developed by Irving Langmuir in 1916, describes the partitioning between the gas phase and adsorbed species as a function of applied pressure.

The adsorption process between gas phase molecules, A, vacant surface sites, S, and occupied surface sites, SA, can be represented by the following chemical equation, assuming that there are a fixed number of surface sites present on the surface, as in Figure 5.

$$S + A \rightleftarrows SA \tag{9}$$

When considering adsorption isotherms it is conventional to adopt a definition of surface coverage (θ) which defines the maximum (saturation) surface coverage of a particular adsorbate on a given surface always to be unity, *i.e.* $\theta_{max} = 1$.

4.4.2. Thermodynamic derivation

An equilibrium constant k can be written in terms of the concentrations of "reactants" and "products":

$$k = \frac{[SA]}{[S][A]} \tag{10}$$

where:

[SA] is proportional to the surface coverage of adsorbed molecules, or proportional to θ;
[S] is proportional to the number of vacant sites, $(1 - \theta)$;
[A] is proportional to the pressure of gas, P.
Thus it is possible to define another equilibrium constant, b:

$$b = \frac{\theta}{(1-\theta)P} \tag{11}$$

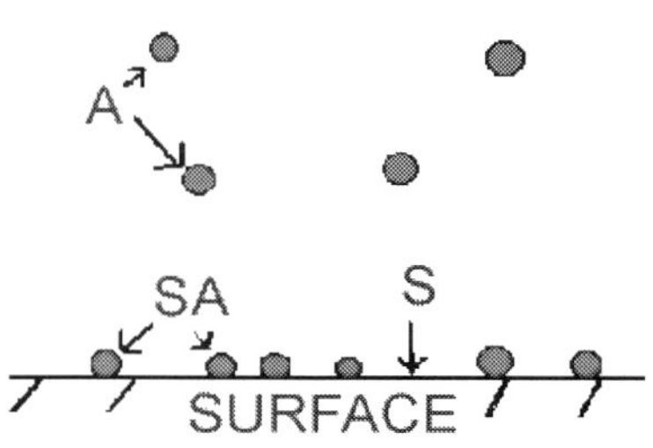

Figure 5. Absorption process.

Rearranging the equations (10) and (11) one can obtain the expression for surface coverage:

$$\theta = \frac{bP}{1+bP} \tag{12}$$

4.4.3. *Kinetic derivation*

The equilibrium that may exist between gas adsorbed on a surface and molecules in the gas phase is a dynamic state, *i.e.* the equilibrium represents a state in which the rate of adsorption of molecules onto the surface is exactly counterbalanced by the rate of desorption of molecules back into the gas phase. It should therefore be possible to derive an isotherm for the adsorption process simply by considering and equating the rates for these two processes.

The rate of adsorption will be proportional to the pressure of the gas and the number of vacant sites for adsorption. If the total number of sites on the surface is N, then the rate of change of the surface coverage due to adsorption is:

$$\frac{d\theta}{dt} = k_a pN\left(1-\theta\right) \tag{13}$$

The rate of change of the coverage due to the adsorbate leaving the surface (desorption) is proportional to the number of adsorbed species:

$$\frac{d\theta}{dt} = -k_d N\theta \tag{14}$$

In these equations, k_a and k_d are the rate constants for adsorption and desorption respectively, and p is the pressure of the adsorbate gas. At equilibrium, the coverage is independent of time and thus the adsorption and desorption rates are equal. The solution to this condition gives us a relation for θ, equation (12), where $b = k_a/k_d$. Here b is only a constant if the enthalpy of adsorption is independent of coverage.

4.4.4. *Dynamic absorption model*

Burden *et al.* [14] develop a dynamic adsorption model based on Langmuir isotherm. If x is the concentration of mRNA target and $\theta(t)$ is the fraction of sites occupied by probe-target duplex, then in the forward absorption, target mRNA attaches to probe at a rate $k_f x\left(1-\theta(t)\right)$ proportional to the concentration of specific target mRNA and the fraction $\left(1-\theta(t)\right)$ of unoccupied probes; and in the backward desorption reaction, target mRNA detaches from probes at a rate $k_b\theta(t)$ proportional to the fractions of occupied probes. The fraction of probe sites occupied by probe-target duplexes is then given by the differential equation:

$$\frac{d\theta(t)}{dt} = k_f x\left(1-\theta(t)\right) - k_b\theta(t) \tag{15}$$

For the initial condition $\theta(0)=0$, equation (15) has the following solution:

$$\theta(t)=\frac{x}{x+K}\left[1-e^{-(x+K)k_f t}\right] \tag{16}$$

where $K=k_b/k_f$.

Using equation (16) Burden *et al.* estimate the measured fluorescence intensity I, with I_0 as the background intensity at zero concentration, to be:

$$I(x,t)=I_0+\frac{bx}{x+K}\left[1-e^{-(x+K)k_f t}\right] \tag{17}$$

At equilibrium, the intensity $I(x)$ at target concentration x follows *Langmuir Isotherm* (12):

$$I(x)=I_0+\frac{bx}{x+K} \tag{18}$$

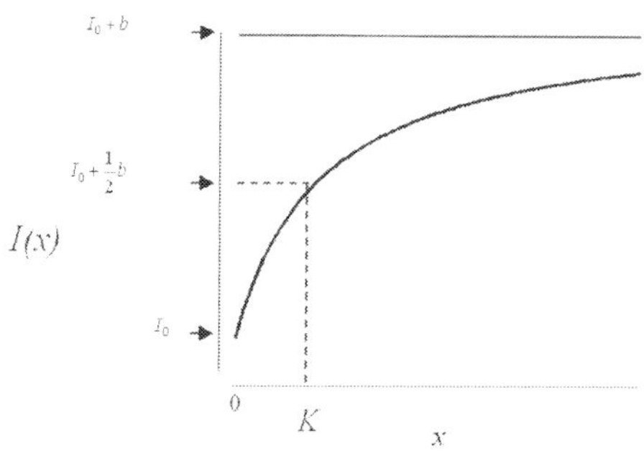

Figure 6. Hyperbolic response function for the intensity $I(x)$ according to the Langmuir isotherm.

5. Hybridization dynamics compensation

5.1. Modelling hybridization by thermodynamics

It is well known that hybridization processes may be seen under the point of view of general thermodynamic conditions [23], meaning that the hybridization probability of a given test segment will be defined by its thermodynamic conditions, *i.e.* by its hybridization temperature. Regarding this, one can state that hybridization process will respond to the dynamic equation:

$$P+T \underset{K_b}{\overset{K_f}{\rightleftarrows}} C \tag{19}$$

where P represents the number of oligonucleotides available for hybridization, T the concentration of free RNA target, C the number of bound complexes, k_f and k_b are the respective forward and backwards rate constants for the reaction. This equation has as a natural solution the following expression in the time domain:

$$C(t) = \frac{T}{T+K}\left[1-\exp(-t/\tau)\right] \tag{20}$$

where K defined as in equation (16) is an equilibrium dissociation constant, and $\tau = \frac{1}{k_f(T+K)}$ denoting a characteristic time over which the system reaches equilibrium.

Recent studies [24], [25] confirm the hypothesis that the hybridization process for the each of the probe pairs follows a time model according to the one from Figure 7. This model of evolution predicts that the probability of hybridization will be almost zero if not enough time interval is provided for the experiment to take place, and that in the limit, if enough time is allowed saturation will take place.

A practical solution to the different hybridization dynamics can be solved by using multiple regressions to convey PM-MM probe pairs to equivalent thermodynamic conditions by processing diachronic hybridization experiments [26].

The last procedure will be explained in more detail in the following paragraphs.

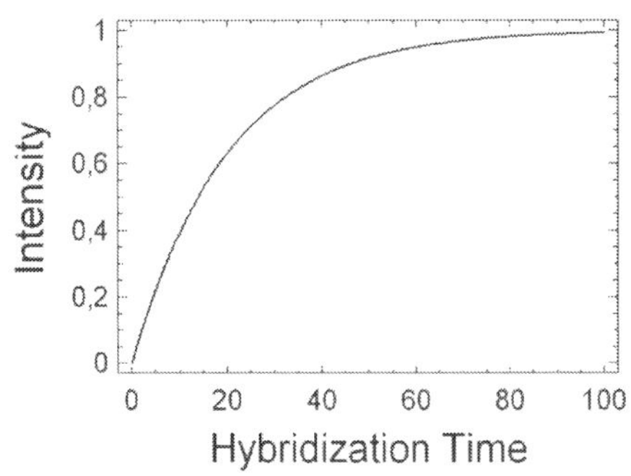

Figure 7. Theoretical model for perfect match hybridization. Intensity of perfect match versus hybridization time. (adapted from [24])

5.2. Exponential regression model

From equation (20) one can assume that a model to solve the multiple regression problem implicit in this study will have the following form:

$$y = a\left(1-e^{-bx}\right) \tag{21}$$

where a and b are parameters to be estimated adaptively using *least square fitting* and the *gradient method.*

Vertical least square fitting proceeds by finding the sum of the squares of the vertical deviations R^2 of parameters a and b:

$$R^2 = \sum_i \left[y_i - a\left(1 - e^{-bx_i}\right) \right]^2 \tag{22}$$

where:

$$\varepsilon_i = y_i - a\left(1 - e^{-bx_i}\right) \tag{23}$$

is the estimation error incurred for each component.

With this notation equation (22) will became:

$$R^2 = \sum_i \varepsilon_i^2 \tag{24}$$

The condition of R^2 to be at a minimum is that

$$\frac{\partial\left(R^2\right)}{\partial a} = 0 \tag{25}$$

$$\frac{\partial\left(R^2\right)}{\partial b} = 0 \tag{26}$$

From equations (24), (25) and (26) one will obtain:

$$\frac{\partial\left(R^2\right)}{\partial a} = \sum_i \varepsilon_i \frac{\partial \varepsilon_i}{\partial a} = -\sum_i \varepsilon_i \left(1 - e^{-bx_i}\right) = 0 \tag{27}$$

$$\frac{\partial\left(R^2\right)}{\partial b} = \sum_i \varepsilon_i \frac{\partial \varepsilon_i}{\partial b} = -\sum_i \varepsilon_i a x_i e^{-bx_i} = 0 \tag{28}$$

A solution for equations (27) and (28) can be found using the gradient method. In this case the parameters are going to be computed adaptively:

$$a_{k+1} = a_k - \beta_a \frac{\partial\left(R^2\right)}{\partial a} = a_k + \beta_a \sum_i \varepsilon_{i,k} \left(1 - e^{-b_k x_i}\right) \tag{29}$$

$$b_{k+1} = b_k - \beta_b \frac{\partial\left(R^2\right)}{\partial b} = b_k + \beta_b \sum_i \varepsilon_{i,k} a_k e^{-b_k x_i} \tag{30}$$

where $\varepsilon_{i,k}$ is defined as in equation (23) and β is a parameter used as an adjust step.

5.3. Application for experimental data

The experimental part has been complemented with artificially simulated test probes used for algorithmic validation. A diachronic database was also being produced to estimate hybridization time constants for different gene segments.

Considering these assumptions data records have been created from experimental data fitted by the above mentioned models. The time dynamics of hybridization for both probe sets and their profiles were evaluated at certain time intervals.

Firstly, the diachronic data distribution for an evolution from 0 to 30 minutes is shown in Figure 8 in both cases, for the PM probe set and the MM probe set, and in the following figures, *i.e.* Figure 9 and Figure 10, show this time evolution for 60 and 120 minutes is also shown following the model in equation (20).

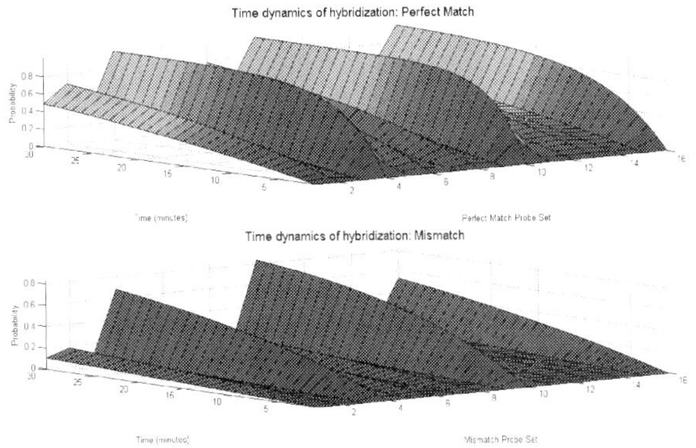

Figure 8. Time dynamics of hybridization corresponding to perfect and mismatch probes, for a maximum of 30 minutes.

The next step on data analysis was to look at the probe profiles, at certain times. Figure 11 shows the regression parameters obtained for time constants. The profiles of the perfect and mismatch were extracted for two different time values underlining the fact that if enough time is allowed to some probes, the mismatches will also hybridize completely.

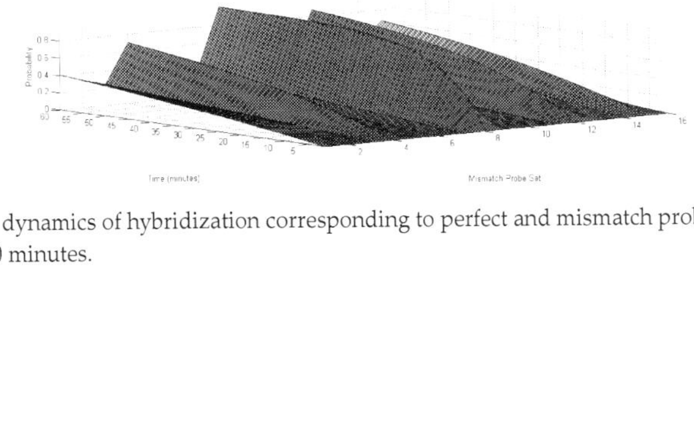

Figure 9. Time dynamics of hybridization corresponding to perfect and mismatch probes, for a maximum of 60 minutes.

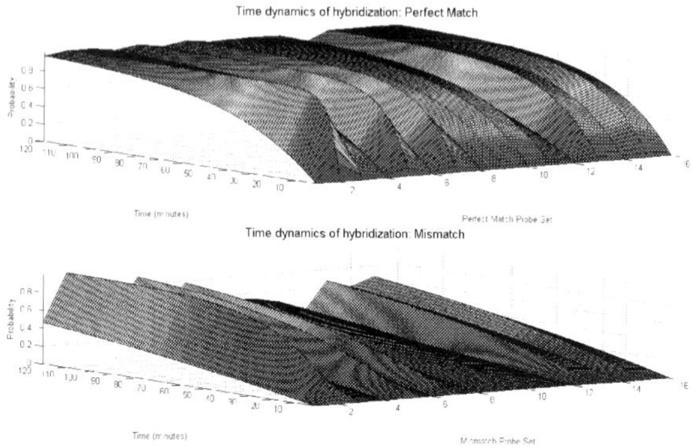

Figure 10. Time dynamics of hybridization corresponding to perfect and mismatch probes, for a maximum of 120 minutes.

Considering this and applying the regression algorithm, we observed that this algorithm searches for the matching values of expression levels of probes sets and for estimated values of perfect and mismatch probes. One of the steps of this iterative algorithm can be seen in Figure 12.

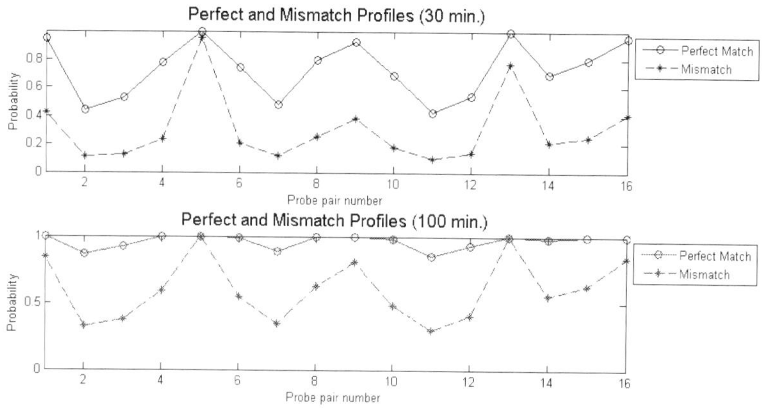

Figure 11. Profiles corresponding to perfect and mismatch probes for time constants, at 30 and 100 minutes.

Once the iterative process was complete, certain probes have reached their target. In the expression level estimation most of the perfect match probes obtained the expected values, while some of the mismatch probes did not reach their target, Figure 13. Similar results were obtained in the case of matching hybridization for time constants.

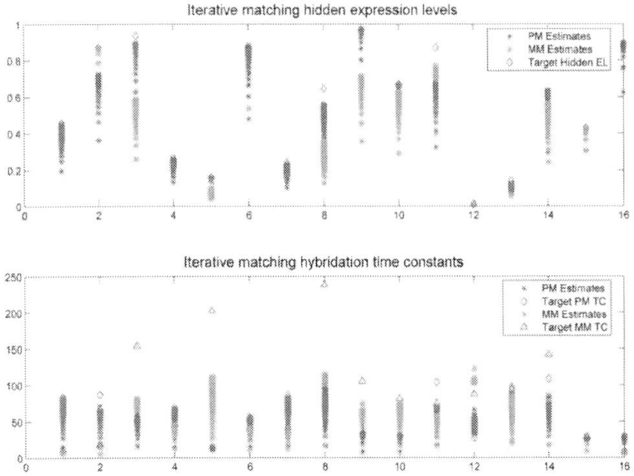

Figure 12. Top template shows the iterative matching for hidden expression levels. Bottom template shows the iterative matching for perfect and mismatch hybridization.

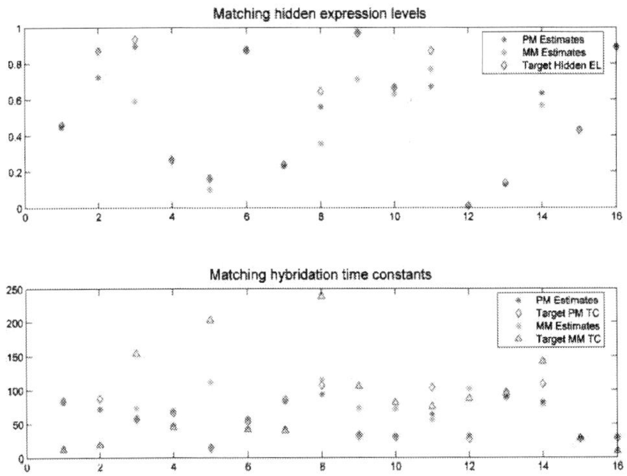

Figure 13. Results for the iterative process of matching.

6. Conclusions

The thermodynamics of oligonucleotide hybridization processes where PM-MM results do not show the expected behaviour, thus affecting to the reliability of expression estimation, was studied in this chapter and the following conclusions were emphasized:

- Modelling the hybridization process through thermodynamical principles reproduces exponential-like behaviour for each P-T segment pair.
- The hybridization process should be confined to the time interval where linear growth is granted, this is, at the beginning of the exponential curve shown in Figure 6.
- Adaptive fitting may be used to predict and regress expression levels on a specific test probe to common thermodynamic conditions. Time constants may be inferred from the regression parameters adaptively.
- The main features of the PM-MM probe sets may be reproduced from probabilistic modelling.
- It may be expected that more precise and robust estimations could be produced using this technique with diachronically expressed hybridization experiments.

Author details

Raul Măluţan*
Communications Department, Technical University of Cluj Napoca, Cluj Napoca, Romania

Pedro Gómez Vilda
DATSI, Universidad Politécnica de Madrid, Madrid, Spain

Acknowledgement

This work was supported by the project "Development and support of multidisciplinary postdoctoral programmes in major technical areas of national strategy of Research - Development - Innovation" 4D-POSTDOC, contract no. POSDRU/89/1.5/S/52603, project co-funded by the European Social Fund through Sectoral Operational Programme Human Resources Development 2007-2013.

7. References

[1] Malutan R, Gómez Vilda P, Berindan Neagoe I, Borda M (2011) Hybridization Dynamics Compensation in Microarray Experiments. Advances in Intelligent and Soft Computing. 93: 255-261

[2] Wu P, Nakano S, Sugimoto N (2002) Temperature dependence of thermodynamic properties for DNA/DNA and RNA/DNA duplex formation. European Journal of Biochemistry. 269:2821 – 2830

[3] SantaLucia Jr. J (1998) A unified view of polymer, dumbbell, and oligonucleotide DNA nearest-neighbor thermodynamics. PNAS on Biochemistry. 95:1460 – 1465

[4] Chan V, Graves D.J., McKenzie S.E. (1995) The Biophysics of DNA Hybridization with Immobilized Oligonucleotides Probes. Biophysical Journal. 69:2243 – 2255

* Corresponding Author

[5] Livshits M A, Mirzabekov A D (1996) Theoretical analysis of the kinetics of DNA hybridization with gel-immobilized oligonucleotides. Biophysical Journal. 71:2795 – 2801

[6] Dai H, Meyer M, Stepaniants S, Ziman M, Stoughton R (2002) Use of hybridization kinetics for differentiating specific from non-specific binding to oligonucleotide microarrays. Nucleic Acids Research. 30(16): e86.1 – e86.8

[7] Dorris D R, *et al.* (2003) Oligodeoxyribonucleotide probe accessibility on a three-dimensional DNA microarray surface and the effect of hybridization time on the accuracy of expression ratios. BMC Biotechnology. 3:6

[8] Binder H, Preibisch S (2005) Specific and Nonspecific Hybridization of Oligonucleotide Probes on Microarrays. Biophysical Journal. 89:337 – 352

[9] Wang J Y, Drlica K (2003) Modelling hybridization kinetics. Mathematical Bioscience. 183:37 – 47

[10] Naef F, Magnasco M O (2003) Solving the riddle of the bright mismatches: Labeling and effective binding in oligonucleotide arrays. Physical Review E., 68:011906-1 – 011906-4

[11] Wu Z, Irizarry R A (2004) Stochastic models inspired by hybridization theory for short oligonucleotide arrays. Proc. of the 8th Annual International Conference on Research in Computational Molecular Biology. 98 – 106

[12] www.accessexcellence.org/RC/VL/GG/index.html

[13] Lipshutz R L, Fodor S P A, Gingeras T R, Lockhart D J (1999) High density synthetic oligonucleotide arrays. Nature Genetics Supplement. 21:20 – 24

[14] Burden C, Pittelkow Y E, Wilson S R (2004) Statistical Analysis of Adsorption Models for Oligonucleotide Microarrays. Statistical Applications in Genetics and Molecular Biology. 3:35

[15] Sugimoto N *et al* (1996) Improved thermodynamic parameters and helix initiation factor to predict stability of DNA duplexes. Nucleic Acids Research. 24:4501 – 4505

[16] Zhang L, Miles M F, Aldape K D (2003) A model of molecular interactions on short oligonucleotide microarrays, Nature Biotechnology. 21(7):818 – 821

[17] Huang J C, Morris Q D, Hughes T R, Frey B J (2005) GenXHC: a probabilistic generative model for cross-hybridization compensation in high-density genome-wide microarray data. Bioinformatics. 21:i222 – i231

[18] Black W Z, Hartley J G (1991) Thermodynamics. Second Edition. SI Version. Harper Collins Publisher

[19] SantaLucia Jr J, Allawi H T, Seneviratne P A (1996) Improved Nearest-Neighbor Parameters for Predicting DNA Duplex Stability. Biochemistry. 35(11): 3555 – 3562

[20] Delcourt S G, Blake R D (1991) Stacking Energies in DNA. The Journal of Biological Chemistry. 266: 15160 – 15169

[21] Allawi H T, SantaLucia Jr J. Thermodynamics and NMR of Internal G•T Mismatches in DNA. Biochemistry. 36:10581 – 10594

[22] Li C, Wong W H (2001) Model-based analysis of oligonucleotide arrays: Expression index computation and outlier detection. PNAS USA. 98(1):31 – 36

[23] El Samad H, Khammash M, Petzold L, Gillespie D (2005) Stochastic Modelling of Gene Regulatory Networks. Int. Journal of Robust and Nonlinear Control. 15(15):691 – 711

[24] Dai H, Meyer M, Stepaniants S, Ziman M, Stoughton R (2002) Use of hybridization kinetics for differentiating specific from non-specific binding to oligonucleotide microarrays. Nucleic Acids Research. 30(16):e86.1 – e86.8

[25] Zhang Y, Hammer D A, Graves D J (2005) Competitive Hybridization Kinetics Reveals Unexpected Behavior Patterns. Biophysical Journal. 89:2950 – 2959

[26] Diaz F, Malutan R, Gomez P, Martinez R, Stetter B, Fe Paz M, Garcia E, Pelaez J (2006) Estimating Oligonucleotide Microarray Expression by Hybridization Process Modelling. Proc. of IEEE/NLM Life Science Systems and Applications Workshop. 1 – 2

Probing the Thermodynamics of Photosystem I by Spectroscopic and Mutagenic Methods

Xuejing Hou and Harvey J.M. Hou

Additional information is available at the end of the chapter

http://dx.doi.org/10.5772/2615

1. Introduction

Thermodynamics of a chemical reaction is a fundamental and vital issue for complete understanding of the reaction at the molecular level and involves the elucidation of the energy level of reactant and products, direction of reaction, and driving force or spontaneity of the reaction (Tadashi, 2011). Most the chemical reactions are enthalpy driven and are determined by chemical bonding energy of the reactants and products. However, some of the chemical reaction or process is entropy driven and are largely due the probability or disorder of the system during the reaction. Protein denaturation and dissolution of potassium iodide in water are such examples. In chemistry and biology, especially electron transfer reaction, the entropy changes are often assumes small and negligible. The understanding of thermodynamics of electron transfer reactions is relatively limited (Mauzerall, 2006).

To study the thermodynamics of reaction in chemistry and biology, photosynthetic reaction is an excellent model system. The photosynthesis involves multiple electron transfer reaction driven by sunlight under room temperature and neutral pH (Blankenship, 2002; Diner and Rappaport, 2002; Golbeck, 2006). The understanding of light-induced electron transfer reaction in photosynthesis will provide fundamental knowledge of chemical reactions and guide the design and fabrication in artificial photosynthetic system in address the global energy and environmental crisis in the 21st century (Lewis and Nocera, 2006). In particular the solar energy storage of solar energy using water splitting reaction mimicking photosynthesis might solved energy and pure water problems at the same time (Kanan and Nocera, 2008; Cook et al., 2010; Hou, 2010, 2011). The electron transfer reactions in photosynthesis involves four major chlorophyll binding protein complexes: Photosystem II, cytochrome b_6f, photosystem I, and ATP synthase (Figure 1). Photosystem I and photosystem II are belong to two types of different reaction centers in nature, respectively. Type I reaction centers incorporate a phylloquinone or menaquinone as secondary electron acceptor, A_1, and three tertiary iron-sulfur cluster electron acceptors, F_A, F_B, and F_X. Type II centers use two quinone acceptors: Q_A undergoes one-electron reduction, and Q_B undergoes a two-electron reduction with concomitant protonation.

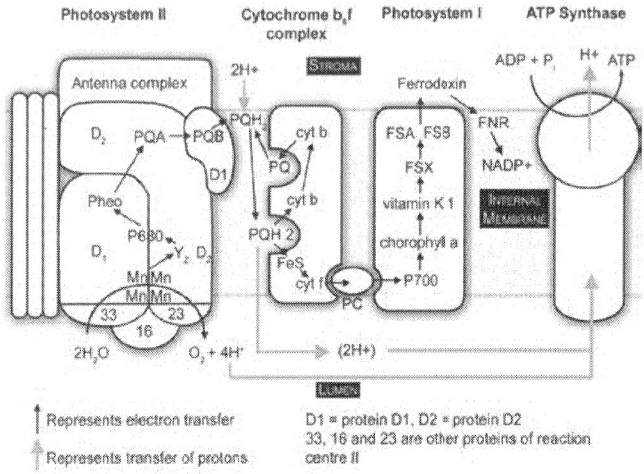

Figure 1. Four protein complexes, including photosystem II, cytochrome b_6f, photosystem I, and ATP synthase, are responsible for the electron transfer reactions in photosynthesis. (from http://www.answersingenesis.org/articles/tj/v17/n3/photosynthesis).

Photosystem I is a pigment-protein complex consisting of more than 11 polypeptides embedded in the photosynthetic membrane and catalyzes light-induced electron transfer from reduced plastocyanin (or cytochrome c_6) to oxidized ferredoxin (or flavodoxin). The electron transfer pathway and the electron transfer cofactors in photosystem I is shown as a black arrows in Figure 1. The primary electron donor is P_{700}, a pair of chlorophyll a molecules. After absorbing light photon energy, P_{700} becomes excited species P_{700}* and delivers one electron to the primary electron acceptor A_0, a chlorophyll a molecule. The reduced A_0 anion donates its electron to the secondary acceptor A_1, a phylloquione or vitamin K_1 molecule. The reduced A^1 anion transfer the electron to F_X, F_A, F_B, and finally to ferrodoxin for producing $NADPH^+$.

The three-dimensional structure of cyanobacterial PS I at 2.5 A resolution has been obtained and revealed much of the detailed orientation and binding site of electron transfer cofactors. These structural details offer a solid basis for structure and function studies at an atomic level (Figure 2). The almost complete symmetric arrangement of cofactors in PS I suggested the electron transfer might involve two electron transfer branches (A side and B side). This is different from the electron transfer mechanism in type II centers. For example, in bacterial and PS II, only one electron transfer branch (L side or D1 side) is active. The M-side (or D2 side) electron transfer is inactive and may provide protective role in the reaction center in regulating excess light energy.

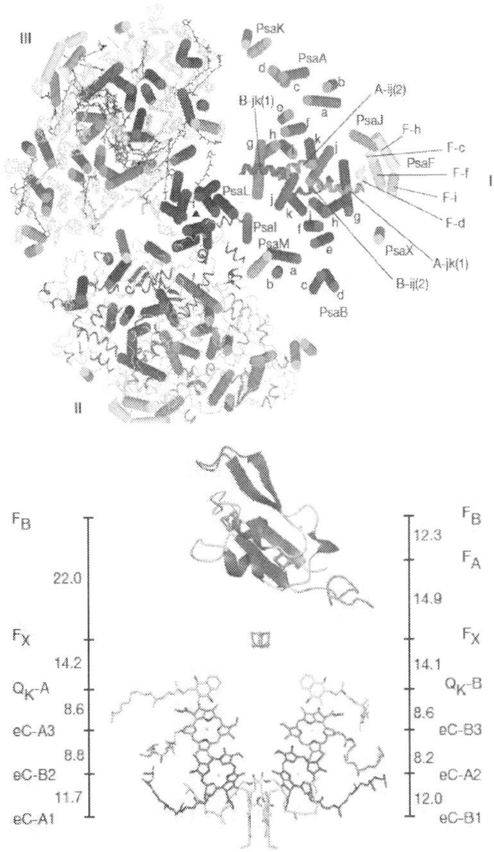

Figure 2. X-ray crystallographic structure (upper panel) and the arrangment of the cofactors (lower pahnel) in photosystem I from cyanobacterium *Synechococcus* elongatus (Jordan et al., 2001) (Reproduced with permission from Nature publishing Group).

2. Quinones in photosystem I

A quinone molecule is a perfect electron transfer cofactor due to its reversible electrochemical redox properties and plays a key role in photosynthetic electron transfer process. For example, both type I and type II reaction centers contain a quinone that operates as an intermediate electron acceptor and as a one-electron carrier. However, the local protein environment and chemical properties of the quinone in these two types of reaction centers must be different. EPR measurements revealed that there are striking difference in the binding and function of phylloquinone (A_1) in PS I and ubiquione (Q_A) in the bacterial center of R. sphaeroides (Kamlowski et al., 1998). As a type I center, PS I

In the left of the panel A in Figure 5 shows the restriction maps of the genomic regions surrounding *menA* gene in the wild type and mutant strains. A 440-bp fragment in the *menA* gene was deleted and replaced by a 1.3 kb kanamycine resistance cartridge. PCR amplification of the *menA* locus of the wild type produced the expected 1.9 kb fragment (panel A right, Figure 5). Southern blot hybridization analyses confirmed the interruption of the *menA* gene as expected. Insertional inactivation of the menB gene was also confirmed by both Southern blot hybridization and PCR amplification of menB locus from the mutant strain. The part of the menB gene was deleted and replaced with a 2-kb spectinomycin resistance cartridge (left, panel B in Figure 5). The PCR amplification of the menB locus of the wild type produced the expected fragment of 920 bp (right, panel B in Figure 5), confirming the inactivation of the menB gene.

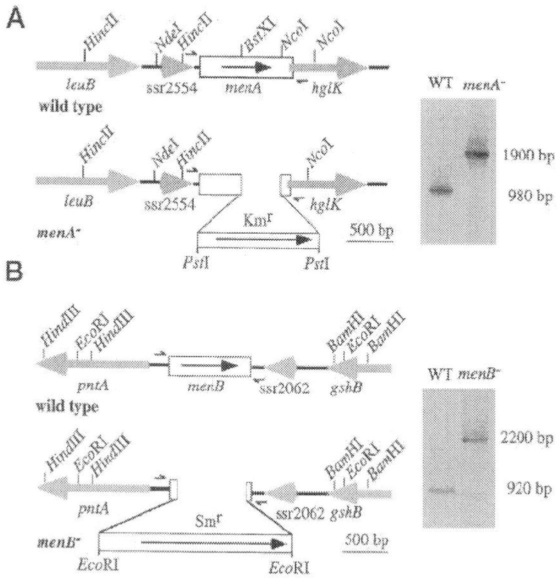

Figure 5. Construct and confirmation of *menA* and *menB* null mutants by inactivation of *menA* and *menB* genes in *Synechocystis* sp. PCC 6803 (Johnson et al., 2000) (Reproduced with permission from the American Society for Biochemistry and Molecular Biology).

3. Physiological, structural, and kinetics of *menA* and *menB* null mutants

The *menA* and *menB* null mutants are able to grow photoheterotrophically and photoautotrophically under low light conditions. The expression level of active PS I in the *menA* and *menB* mutant are 50-60% of the wild type stain. HPLC analysis indicates that the absence of phylloquinone in PS I and that the presence of high level of plastoquinone in PS I (Johnson et al., 2000). EPR, electron nuclear double resonance, and electron spin echo modulation data suggested that the orientation and distance of the foreign quinone A_P are almost identical to those of the phylloquinone A_1 in the wild type PS I. In addition, EPR measurements show that plastoquinone has been recruited into the A_1 site and functions as an efficient one electron carrier (Zybailov et al., 2000).

As shown in Figure 6, the rates of electron transfer from P_{700} to A_0 and A_1 in the mutants are similar to the wild type PS I. However, the kinetic parameter from A_1 anion to F_X is quite different in the *menA* and *menB* null mutants. The time-resolved optical studies revealed that the forward electron transfer from A_1 anion to F_X is slowed 1000-fold, to 15 and 300 ®s, compared to 20 and 200 ns in the wild type PS I (Semenov et al., 2000; Johnson et al., 2001). Based the kinetic data of electron transfer in *menA* and *menB* mutant, the redox potential of A_P in *menA* and *menB* PS I is estimated to be more oxidizing than phylloquinone so that electron transfer from A_P anion to F_X is thermodynamically unfavorable in the mutants (right panel of Figure 6).

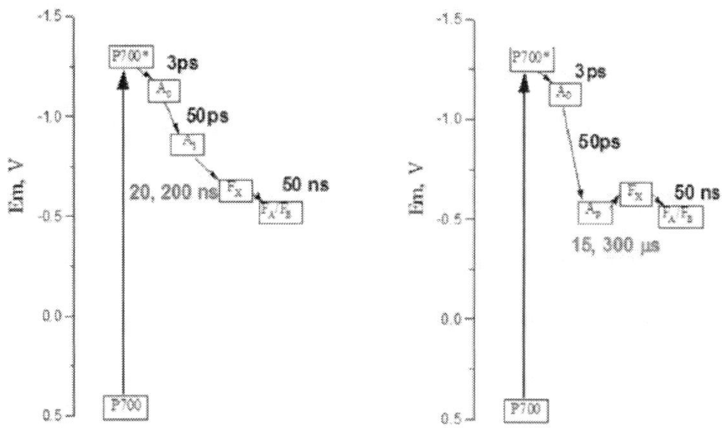

Figure 6. Kinetic data of electron transfer steps in native and *menA*/*menB* null mutants of *Synechocystis* sp. PCC 6803

4. Thermodynamics of *menA* and *menB* Null Mutants

Pulsed photoacoustic spectroscopy can provide a direct measurement of thermodynamic parameters such as volume change and enthalpy changes that accompany electron transfer reactions (Braslavsky, 1985; Carpentier et al., 1990; Arnaut et al., 1992; Small et al., 1992; Losi et al., 1995; Edens et al., 2000; Malkin, 2000; Herbert et al., 2001; Feitelson and Mauzerall, 2002; Delosme, 2003; Hou and Mauzerall, 2006; Hou and Sakmar, 2010; Hou, 2011; Hou and Mauzerall, 2011). With prior knowledge of the change in Gibbs free energy of the corresponding reactions, the apparent entropy change ($T\Delta S$) of the reaction can be calculated. This is an important parameter, knowledge of which is required to fully understand the mechanism of electron transfer, and it has been largely underreported in the literature. The ΔH, ΔV, and $T\Delta S$ of electron transfer in the photosynthetic reaction center

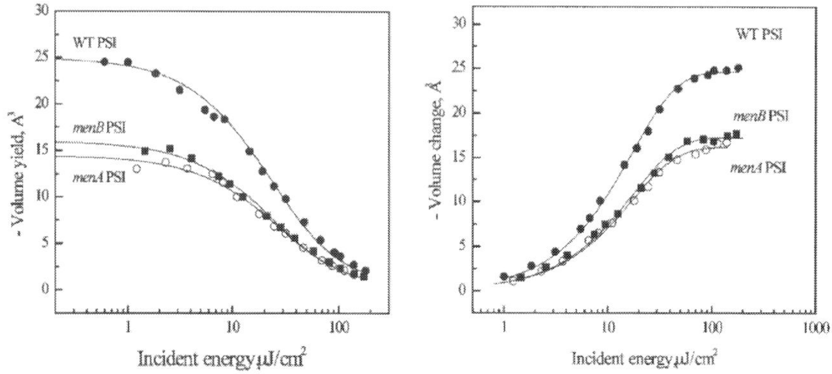

Figure 8. Molecular volume changes determined by volume yield method (left panel) and saturation curve method (right panel) on the 1-microsecond time scale (Hou et al., 2009) (Reproduced with permission from the American Chemical Society).

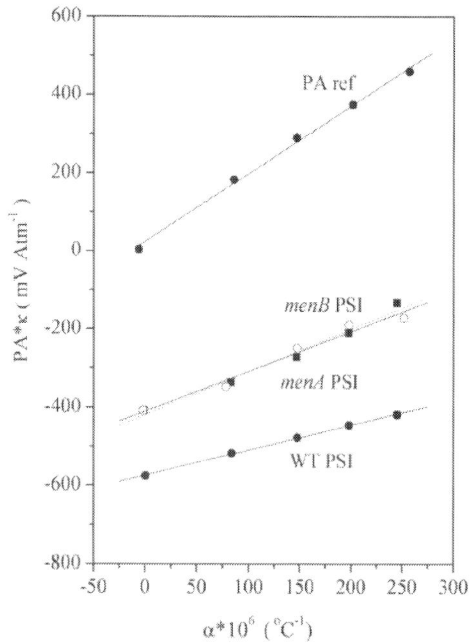

Figure 9. Enthalpy changes determined by analyzing the linear fit of the photoacoustic intensity vs. temperature (water expansivity) on the 1-microsecond time scale (Hou et al., 2009) (Reproduced with permission from the American Chemical Society).

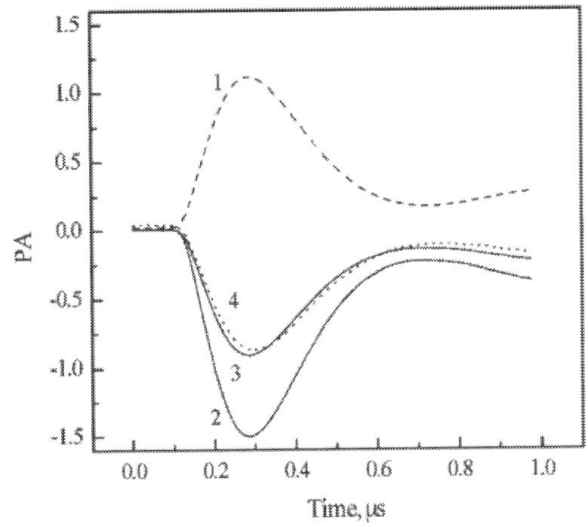

Figure 10. Photoacoustic waves of the photoacoustic reference sample (curve 1), wild-type (curve 2), *menA* (curve 3), and *menB* (curve 4) mutants of *Synechocystis* sp. PCC 6803 on the nanosecond time scale (Hou et al., 2009) (Reproduced with permission from the American Chemical Society).

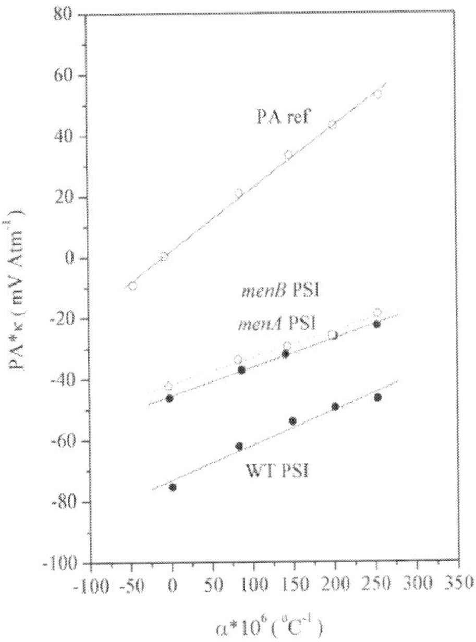

Figure 11. Enthalpy changes determined by plotting the linear fit of the photoacoustic intensity vs. temperature (water expansitivity) on the nanosecond time scale (Hou et al., 2009) (Reproduced with permission from the American Chemical Society).

To summarize the thermodynamic data, the volume changes, free energies, and enthalpy and entropy changes on *menA/B* PS I in comparison with those on the wild-type PS I are listed in Figure 12. Opened arrows are the early step forming $P_{700}{}^{+}A_1{}^{-}$ from $P_{700}{}^{*}$ for the wild-type PS I or $P7_{00}{}^{+}A_P{}^{-}$ from $P_{700}{}^{*}$ for the mutants, and solid arrows are the number of the following reaction: $P_{700}{}^{+}A_P{}^{-}$ to $P_{700}{}^{+}F_{A/B}{}^{-}$. As shown in panel A, the volume contraction of the early step of the photoreaction in the mutants (-17 Å^3) is smaller than that in the wild type (-21 Å^3). Similarly, the enthalpy change (-0.7 eV) of the early step in the mutants is smaller than that (-0.8 eV) in wild-type PS I (Figure 12B). Assuming a redox potential of -0.6 V for plastoquinone-9 in the A_1 site (21), the free energy (-0.7 eV) of this early reaction in the mutants is larger than the value (-0.6 eV) in the wild type as indicated in Figure 12C. Taking the difference of free energy and enthalpy change in the mutants, the apparent entropy change of the early step in mutants is zero. In contrast, the apparent entropy change in the wild type is calculated to be +0.2 eV. Since the apparent entropy change for the overall reaction of the generation of $P_{700}{}^{+}F_{A/B}{}^{-}$ from $P_{700}{}^{*}$ is +0.35 eV (32), it implies that the latter reaction in the mutants, i.e., the $P_{700}{}^{+}A_P{}^{-}F_{A/B}$ to $P_{700}{}^{+}A_PF_{A/B}{}^{-}$ reaction, is almost completely entropy driven ($T\Delta S$) +0.4 eV and ΔG) -0.1 eV) (Figure 12D).

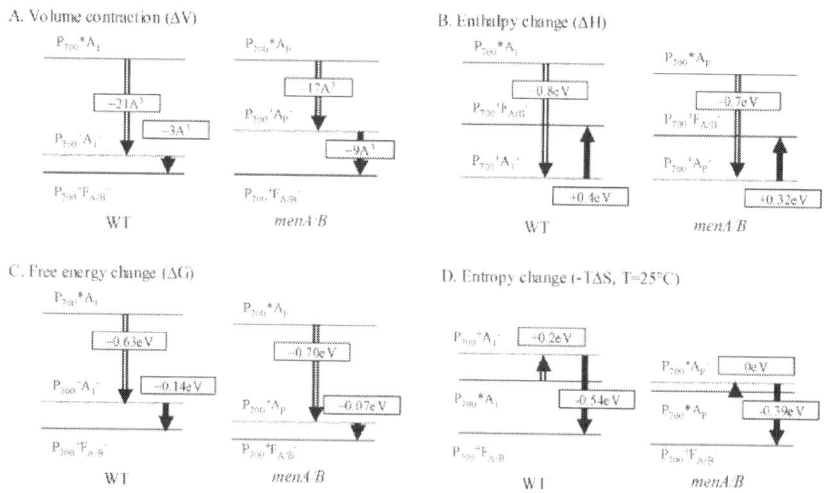

Figure 12. Thermodynamics of electron transfer reactions in photosystem I from the *menA/menB* null mutants and wild-type strains of *Synechocystis* sp. PCC 6803 (Hou et al., 2009) (Reproduced with permission from the American Chemical Society).

In the case of *menA/B* PS I, photoacoustic measurements on the microsecond time scale reveal the volume contraction to be -17 Å^3. Considering our time window of 0.1-10 μs, this value is assigned to the formation of $P_{700}{}^{+}A_P{}^{-}$ from $P_{700}{}^{*}A_P$. The intermediate step of electron transfer from $P_{700}{}^{+}A_P{}^{-}F_{A/B}$ to $P_{700}{}^{+}A_PF_{A/B}{}^{-}$ would be accompanied by a volume change of -9 Å^3, assuming the replacement of A_1 with A_P causes no change in the $F_{A/B}$ clusters. For *menA/B* PS I, the size of the benzoquinone ring in plastoquinone-9 (A_P) is smaller than the naphthoquinone ring in phylloquinone. Electron spin-echo modulation experiments showed that the distance between P_{700}^+ and A_P^- (25.3 Å) in *menA/B* PS I is the same as the distance between P_{700}^+ and A_1^- in wild-type PS I. The volume contraction of electron transfer from P_{700} to A_P is estimated to be larger (-30 Å³) than the observed one (-17 Å³). Thus a positive volume due to a protein conformational change may be possible. We offer a molecular explanation of the difference in the volume change predicted via electrostriction and that of the observed value (-17 Å³). A simple explanation would be that the quantum yield of photochemistry is lower (for example, ~0.7) in *menA/B* PS I. This, however, is unlikely because our pulse saturation data revealed a quantum yield of ~85% in PS I from both mutants, which is only sufficiently lower than the quantum yield of 96% in wild type PS I to explain less than half of the effect. Further support for a high quantum yield is that the light saturation dependence of the flavodoxin reduction rate in *menA/B* PS I is similar to that of wild-type PS I. These two arguments are consistent in showing that the smaller volume change in the *menA*/B PS I is not caused by a low quantum yield.

The smaller volume contraction may be caused by the following two factors: compressibility of protein and polarity of quinone pocket. The first factor is the effect of the foreign plastoquinone on the compressibility of the local environment of the protein. The orientation and distance of plastoquinone- 9 in the mutants are known to be similar to phylloquinone in the wild-type PS I. However, since the pocket of A_1 is adapted to phylloquinone, the smaller plastoquinone with the longer tail may not fit well into the protein. If the effect of the larger tail is to crowd the hydrophobic site, this could decrease the compressibility of the local domain and so decrease ΔV_{el}. Alternatively, the A_1 binding region in *menA/B* PS I may be more polar; i.e., it has a larger effective ε, compared to the wild-type PS I. This could be due to the small size of plastoquinone-9, allowing a water molecule to be present. This possibility also could explain the change in potential of the quinone because of hydrogen bonding to the water.

By use of the electron transfer theory and kinetic data, the redox potential of plastoquinone at the A_1 site was estimated to be -0.61 V (Hou et al., 2009). However, the error to be at least 0.1 V. The ΔG for producing $P_{700}^+A_P^-$ from P_{700}^* is then -0.71 eV. Similarly, the free energy for producing $P_{700}^+F_{A/B}$- from $P_{700}^*F_{A/B}$ is -0.77 eV (35). Thus we infer that the free energy of $P_{700}^+A_P^-F_{A/B}$ to $P_{700}^+A_PF_{A/B}^-$ reaction is -0.06±0.10 eV in the mutants. Knowing the free energy of the electron transfer step in wild-type PS I and *menA/B* PS I, the entropic contribution ($T\Delta S$) can be determined by the Gibbs relation $\Delta G = \Delta H - T\Delta S$. The enthalpy change occurring on the formation of $P_{700}^+A_P^-$ from P_{700}^* in *menA/B* PS I is -0.7 ± 0.07 eV. This is close to the estimated free energy of this reaction (-0.70 eV), and thus the apparent entropy change is close to zero (Figure 12). Considering the enthalpy change (-0.4 eV) of the overall reaction for the formation of $P_{700}^+F_{A/B}^-$ from P_{700}^* in the wild-type PS I, the electron transfer reaction from A_P^- to $F_{A/B}$ would be associated with a positive enthalpy change of +0.3 eV and thus is completely entropy driven as the free energy is zero in the mutants.

The entropy of electron transfer reactions is often assumed to be zero. However, the free energy calculated from kinetic measurements of reverse electron transfer in bacterial reaction centers shows that the free energy is time- and temperature dependent, particularly on the less than nanosecond time scale. The kinetics of these decays can only be described as "distributed", and simple analysis in terms of a single component is not trustworthy. Protein dynamics may play a key role in this electron transfer step. However, the question of whether these "relaxations" are enthalpy and/or entropy driven remains to be answered. The slow (microsecond) component observed in wild-type PS I could be such a relaxation, but only the ΔV was determined. The difference between observed enthalpies and estimated free energies as entropies highlights the problem. In addition to reaction centers of Rb. sphaeroides, similar positive entropic contribution in PS I preparations of *Synechocystis* sp. PCC 6803 were observed, but not in PS II preparations. Charge movement, but not charge separation due to proton transfer, may be the difference in PS II on the 1 μs time scale. Clear-cut and large negative entropic contribution is seen in the model system of triplet porphyrin-to-ferricyanide electron transfer in aqueous solution, where "relaxations" are too fast to be relevant.

5. Conclusion

In this chapter, thermodynamics of electron transfers in biological system can be assessed by using a combination of molecular genetics and sophisticated biophysical techniques, in particular, pulsed photoacoustic spectroscopy. Photosynthesis involves light-induced charge separation and subsequently a series of electron transfer reactions and is an ideal system for the detailed study on electron transfer mechanisms in chemistry and biology. In contrast to the susceptible and vulnerable of photosystem II complex to environment, photosystem I complex is much more stable and a perfect choice for such a study. As quinones play a central role in electron transfer reactions in both anoxygenic and oxygenic photosynthesis, the phylloquinone (A_1) of photosystem I is chosen as a probe to explore the effect and regulation of electron cofactors on kinetics and thermodynamics in vivo. The usual approach of chemical modification or replacement of phylloquinone may alter the bonding pocket of the cofactor and the interaction with its proteins.

Molecular genetic technique is utilized to block the biosynthesis of the cofactor, phylloquinone, in the cynobacterium *Synechocystis* sp. PCC 6803 and replace the phylloquinone in the A_1 site with different foreign quinones. The effect of the foreign quinones on the electron transfer is systematically studied by biophysical methodologies. Specifically, the *menA* and *menB* genes, which code for phytyl transferase and 1.4-dihydroxy-2-naphthoae synthase, respectively, are inactivated to prevent the synthesis of phylloquinone. In spite of the demonstrated absence of phylloquinone, the *menA* and *menB* null mutants grow photoheterotrophically. HPLC and EPR measurement show that plastoquinone-9 (A_P) has been recruited into the A_1 site and functions as an efficient one-electron transfer carrier. The orientation and distance between the plastoquinone and other cofactors are the same as that of the wild type. Time-resolved optical studies indicate that the forward electron transfer from A_1 anion to the iron-sulfur cluster (F_X) is slowed 1000-

fold, to 15 to 300 μs, compared to 20 and 200 ns in the wild type. Using the electron transfer theory and kinetic data, the redox potential of plastoquinone at the A_1 site was estimated to be -0.61 V, more oxidizing that phylloquinone. The electron transfer from A_P anion to F_X is thermodynamically unfavorable in the *menA* and *menB* null mutants.

Photoacoustic measurements reveal that the quantum yield of charge separation in *menA* and *menB* is slightly lower (85%) than that of the wild type. The thermodynamic parameters of the two electron transfer steps, including molecular volume changes, reaction enthalpy and apparent entropy, were obtained in *menA* and *menB* null mutants and are considerable different from the wild type. The experimental data demonstrates that the not only the kinetics but also the thermodynamics of electron transfer reaction in photosystem I are affected by the recruitment of the foreign quinone into the A_1 site. One of the most intriguing conclusions is that although free energy for the electron transfer from A_P anion to F_X in *menA* and *menB* mutant is close to zero, the apparent entropy ($T\Delta S$) for this reaction is positive, +0.4 eV. This indicates that the entropy plays a key role in the electron transfer from A_P anion to F_X. As shown in Figure 13, the electron transfer from P_{700} to A_1 is enthalpy driven; the electron transfer from A_1 to F_A/F_B is entropy driven. The driving force of electron transfer in photosynthetic systems is not only dependent on the bonding energy of the cofactor and interaction energy with the protein (enthalpy) but also dependent on the available states of interaction with the protein (entropy).

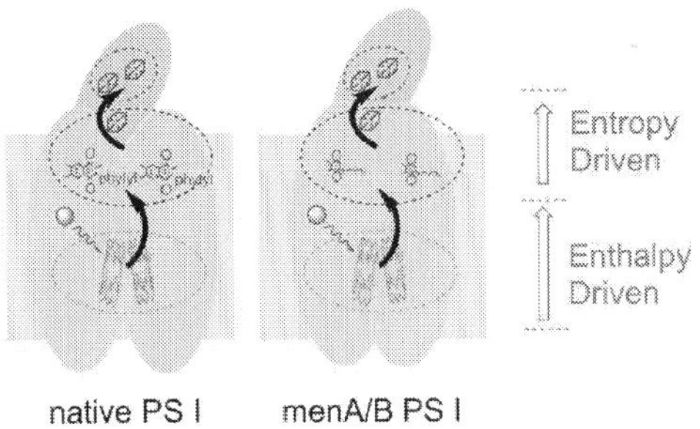

Figure 13. The enthalpy and entropy driven steps in *Synechocystis* 6803 photosystem I

Author details

Xuejing Hou
University of Massachusetts Dartmouth, USA

Harvey J.M. Hou*
University of Massachusetts Dartmouth, USA
Alabama State University, USA

* Corresponding Author

Acknowledgement

This work was supported by the Alabama State University and University of Massachusetts Dartmouth. The photoacoustic measurements were conducted in the laboratory of Professor David Mauzerall at Rockefeller University. The author thanks Professor John Golbeck and Dr. Gaozhong Shen for their collaboration and stimulating discussions on *menA/menB* project. He is also grateful to his students, Fan Zhang and Lien-Yang Chou, for data analysis and assistance.

6. References

Arnaut LG, Caldwell RA, Elbert JE, Melton LA (1992) Recent advances in photoacoustic calorimetry: theoretical basis and improvements in experimental design. Rev. Sci. Instrum. 63: 5381-5389

Blankenship RE (2002) Molecular Mechanisms of Photosynthesis. Blackwell Science

Boichenko VA, Hou J-M, Mauzerall D (2001) Thermodynamics of Electron Transfer in Oxygenic Photosynthetic Reaction Centers: Volume Change, Enthalpy, and Entropy of Electron-Transfer Reactions in the Intact Cells of the Cyanobacterium *Synechocystis* PCC 6803. Biochemistry 40: 7126-7132

Braslavsky SE (1985) Time-resolved photoacoustic and photothermal methods. Application to substances of biological interest. NATO ASI Series, Series A: Life Sciences 85: 147-158

Carpentier R, Leblanc RM, Mimeault M (1990) On the nature of the photosynthetic energy storage monitored by photoacoustic spectroscopy. Photosynth. Res. 23: 313-318

Cook TR, Dogutan DK, Reece SY, Surendranath Y, Teets TS, Nocera Daniel G (2010) Solar energy supply and storage for the legacy and nonlegacy worlds. Chem. Rev. 110: 6474-6502

Delosme R (2003) On some aspects of photosynthesis revealed by photoacoustic studies: a critical evaluation. Photosynth. Res. 76: 289-301

Delosme R, Beal D, Joliot P (1994) Photoacoustic detection of flash-induced charge separation in photosynthetic systems. Spectral dependence of the quantum yield. Biochim. Biophys. Acta, 1185: 56-64

Diner BA, Rappaport F (2002) Structure, dynamics, and energetics of the primary photochemistry of photosystem II of oxygenic photosynthesis. Annu. Rev. Plant Biol. 53: 551-580

Edens GJ, Gunner MR, Xu Q, Mauzerall D (2000) The Enthalpy and Entropy of Reaction for Formation of $P^{+}Q_{A}^{-}$ from Excited Reaction Centers of *Rhodobacter sphaeroides*. J. Am. Chem. Soc.122: 1479-1485

Feitelson J, Mauzerall D (2002) Enthalpy and Electrostriction in the Electron-Transfer Reaction between Triplet Zinc Uroporphyrin and Ferricyanide. J. Phys. Chem. B 106: 9674-9678

Golbeck JH, (2006) Photosystem I; The Light-Driven Plastocyanin: Ferredoxin Oxidoreductase. Springer

Gunner MR, Dutton PL (1989) Temperature and ΔG° dependence of the electron transfer from BPh^{-} to Q_A in reaction center protein from *Rhodobacter sphaeroides* with different quinones as Q_A. J. Am. Chem. Soc. 111: 3400-3412

Herbert SK, Han T, Vogelmann TC (2001) New applications of photoacoustics to the study of photosynthesis. Photosynth. Res. 66: 13-31

Hou HJM (2010) Structural and mechanistic aspects of Mn-oxo and Co-based compounds in water oxidation catalysis and potential application in solar fuel production. J. Integr. Plant Biol. 52: 704-711

Hou HJM (2011) Enthalpy, entropy, and voluem changes of electron transfer reactions in photosynthetic proteins, In: Application of thermodynamics to biological and material science, Mizutani Tadashi Ed., Intech, pp. 93-110.

Hou HJM (2011) Manganese-based materials inspired by photosynthesis for water-splitting. Materials 4: 1693-1704

Hou HJM, Mauzerall D (2006) The A_1-F_X to $F_{A/B}$ Step in *Synechocystis* 6803 Photosystem I Is Entropy Driven. J. Am. Chem. Soc. 128: 1580-1586

Hou HJM, Mauzerall D (2011) Listening to PS II: Enthalpy, entrhopy and volume changes. J. Photochem. Photobiol. B 104: 357-365

Hou HJM, Sakmar TP (2010) Methodology of pulsed photoacoustics and its application to probe photosystems and receptors. Sensors 10: 5642-5667

Hou HJM, Shen G, Boichenko VA, Golbeck JH, Mauzerall D (2009) Thermodynamics of Charge Separation of Photosystem I in the *menA* and *menB* Null Mutants of *Synechocystis* sp. PCC 6803 Determined by Pulsed Photoacoustics. Biochemistry 48: 1829-1837

Hou JM, Boichenko VA, Wang YC, Chitnis PR, Mauzerall D (2001) Thermodynamics of electron transfer in oxygenic photosynthetic reaction centers: a pulsed photoacoustic study of electron transfer in photosystem I reveals a similarity to bacterial reaction centers in both volume change and entropy. Biochemistry 40: 7109-7116

Johnson TW, Shen G, Zybailov B, Kolling D, Reategui R, Beauparlant S, Vassiliev IR, Bryant DA, Jones AD, Golbeck JH, Chitnis PR (2000) Recruitment of a foreign quinone into the A_1 site of photosystem I. I. Genetic and physiological characterization of phylloquinone biosynthetic pathway mutants in *Synechocystis* sp. pcc 6803. J. Biol. Chem. 275: 8523-8530

Johnson TW, Zybailov B, Jones AD, Bittl R, Zech S, Stehlik D, Golbeck JH, Chitnis PR (2001) Recruitment of a foreign quinone into the A_1 site of photosystem I. In vivo replacement of plastoquinone-9 by media-supplemented naphthoquinones in phylloquinone biosynthetic pathway mutants of *Synechocystis* sp. PCC 6803. J. Biol. Chem. 276: 39512-39521

Jordan P, Fromme P, Witt HT, Klukas O, Saenger W, Krauss N (2001) Three-dimensional structure of cyanobacterial photosystem I at 2.5 A resolution. Nature 411: 909-917

Kamlowski A, Altenberg-Greulich B, Van der Est A, Zech SG, Bittl R, Fromme P, Lubitz W, Stehlik D (1998) The Quinone Acceptor A_1 in Photosystem I: Binding Site, and Comparison to Q_A in Purple Bacteria Reaction Centers. J. Phys. Chem. B 102: 8278-8287

Kanan M, Nocera DG (2008) In Situ Formation of an Oxygen-Evolving Catalyst in Neutral Water Containing Phosphate and Co^{2+}. Science 321: 1072 - 1075

Lewis NS, Nocera DG (2006) Powering the planet: chemical challenges in solar energy utilization. Proc. Natl. Acad. Sci. U S A 103: 15729-15735

Losi A, Bedotti R, Viappiani C (1995) Time-Resolved Photoacoustics Determination of Intersystem Crossing and Singlet Oxygen Photosensitization Quantum Yields for 4,5',8-Trimethylpsoralen. J. Phys. Chem. 99: 16162-16167

Malkin S (2000) The photoacoustic effect in leaves and its applications. Probing In: Probing Photosynthesis: Mechanism, Regulation & Adaptation, M Yunus, U Pathre, and P Mohanty Eds., Taylor & Francis, pp. 484-524

Mauzerall D (2006) Thermodynamics in photosystem I. In: Photosystem I: The Light-Driven Plastocyanin: Ferredoxin Oxidoreductase, J Golbeck ed., Springer, pp. 571-581

Nagy L, Kiss V, Brumfeld V, Malkin S (2001) Thermal and structural changes of photosynthetic reaction centers characterized by photoacoustic detection with a broad frequency band hydrophone. Photochem. Photobiol. 74: 81-87

Semenov AY, Vassiliev IR, van Der Est A, Mamedov MD, Zybailov B, Shen G, Stehlik D, Diner BA, Chitnis PR, Golbeck JH (2000) Recruitment of a foreign quinone into the A_1 site of photosystem I. Altered kinetics of electron transfer in phylloquinone biosynthetic pathway mutants studied by time-resolved optical, EPR, and electrometric techniques.J. Biol. Chem. 275: 23429-23438

Semenov AY, Vassiliev IR, Van der Est A, Mamedov MD, Zybailov B, Shen G, Stehlik D, Diner BA, Chitnis PR, Golbeck JH (2000) Recruitment of a foreign quinone into the A_1 site of photosystem I. Altered kinetics of electron transfer in phylloquinone biosynthetic pathway mutants studied by time-resolved optical, EPR, and electrometric techniques. J. Biol. Chem. 275: 23429-23438

Sharma V, Hudspeth ME, Meganathan R (1996) Gene 168: 43-48

Small JR, Libertini LJ, Small EW (1992) Analysis of photoacoustic waveforms using the nonlinear least squares method. Biophys. Chem. 42: 29-48

Tadashi M (2011) Application of Thermodynamics to Biological and Materials Science. Intech

Zybailov B, van der Est A, Zech SG, Teutloff C, Johnson TW, Shen G, Bittl R, Stehlik D, Chitnis PR, Golbeck JH (2000) Recruitment of a foreign quinone into the A_1 site of photosystem I. II. Structural and functional characterization of phylloquinone biosynthetic pathway mutants by electron paramagnetic resonance and electron-nuclear double resonance spectroscopy. J. Biol. Chem. 275: 8531-8539

Chapter 20

Fuzzy Spheres Decays and Black Hole Thermodynamics

C.A.S. Silva

Additional information is available at the end of the chapter

http://dx.doi.org/10.5772/51630

1. Introduction

A black hole is, by definition, a region in spacetime in which the gravitational field is so strong that it precludes even light from escaping to infinity. A black hole is formed when a body of mass M contracts to a size less than the so called gravitational radius $r_g = 2GM/c^2$. (G is the Newton's gravitational constant, and c is the speed of light). The velocity required to leave the boundary of the black hole and move away to infinity (the escape velocity) equals the speed of light. In this way, one easily concludes that neither signals nor particles can escape from the region inside the black hole since the speed of light is the limiting propagation velocity for physical signals. From the fact that no signals can escape from a black hole, while physical objects and radiation can fall into it, the surface bounding the black hole in spacetime (called the event horizon) is a lightlike surface.

The term "black hole" was introduced by Wheeler in 1967 although the theoretical study of these objects has quite a long history. The very possibility of the existence of such objects was first discussed by Michell and Laplace within the framework of the Newtonian theory at the end of the 18th century [1–3]. In general relativity context, the problem arose within a year after the theory had been developed, i.e., after Schwarzschild (1916) obtained the first exact (spherically symmetric) solution of Einstein's equations in vacuum.

In particular, black hole developments in the last forty years have shown that black holes have thermodynamics properties like entropy and temperature, and as a consequence of the instability of the vacuum in strong gravitational fields, they are sources of quantum radiation [4–6]. String theory and loop quantum gravity, lately, showed that the origin of the black hole thermodynamics must be related with the quantum structure of the spacetime, bringing together the developments in black hole physics and the improvement of our understanding on the nature of the spacetime in quantum gravity regime [7, 8]. In this way, it is believed that black holes may play a major role in our attempts to shed some light on the quantum nature of the spacetime such as the role played by atoms in the early development of quantum mechanics.

[42] K. V. Krasnov, Gen. Rel. Grav. 30 (1998) 53.
[43] K. V. Krasnov, Class. Quant. Grav. 16 (1999) 563.
[44] A. Ashtekar, J. C. Baez and K. Krasnov, Adv. Theor. Math. Phys. 4 (2000) 1.
[45] J. Fernando Barbero G. and E.J.S. Villasenor, gr-qc/1106.3179 preprint (2011).
[46] S. Massar, R. Parentani , Nucl.Phys.B 575 (2000) 333.
[47] M. Banados, C. Teitelboim, J. Zanelli, Phys.Rev.Lett. 72 (1994) 957.
[48] Steven Carlip, Claudio Teitelboim, Class.Quant.Grav. 12 (1995) 1699.
[49] C.E. Shannon and W. Weaver, The mathematical theory of communications, University of Illinois Press, Urbana (1949).
[50] E.T. Jaynes, Phys.Rev. 106 (1957) 620.
[51] E.T. Jaynes, Phys.Rev. 108 (1957) 171.
[52] J. D. Bekenstein, Phys. Rev. D 7 (1973) 2333.
[53] S. Hod, Phys.Rev.D 61 (2000) 124016.
[54] S. Hod, Phys. Lett. A 299, (2002) 144.
[55] W. H. Zurek, Phys. Rev. Lett. 49 (1982) 1683.
[56] D.N. Page, Phys.Rev.D 13 (1976) 198.
[57] D.N. Page, Phys.Rev.D 14 (1976) 3260.
[58] D.N. Page, Phys.Rev.D 16 (1977) 2402.

Chapter 21

Bibliometric Analysis of Thermodynamic Research: A Science Citation Index Expanded-Based Analysis

Hui-Zhen Fu and Yuh-Shan Ho

Additional information is available at the end of the chapter

http://dx.doi.org/10.5772/48360

1. Introduction

Thermodynamics is the study of energy and its transformation (Holman, 1985). This subject has evolved since the beginning of the eighteenth century (Look & Saucer, 1982). Most studies of thermodynamics are primarily concerned with two forms of energy: heat and work (Holman, 1985; Keizer, 1985; Rosakis et al., 2000). Thermodynamics advanced dramatically in the 1960s and 1970s, primarily in the area of critical phenomena (Callen, 1985). In recent years, thermodynamic research has been conducted in more systems, such as corn-ethanol biofuel cycle (Patzek, 2004), lipid-peptide interactions (Seelig, 2004), DNA structural motifs (SantaLucia & Hicks, 2004), and glass-forming substances (Ngai, 2000).

Bibliometrics, as a measure of scientific performance, has been widely applied to research evaluation for a long period (Saracevi & Perk, 1973; Luukkonen, 1990; Friedberg, 2000). It has been employed to evaluate various researches, for example biology (Sainte-Marie 2010), desalination research (Tanaka & Ho, 2011), solid waste research (Fu et al., 2010), acupuncture research (Han & Ho, 2011), and global climate change (Li et al., 2011). In exited bibliometric analyses, the analyzed aspects usually covering languages (Alfaraz & Calvino, 2004; Chiu & Ho, 2007; Ferrara, 2011), annual publication outputs (Chiu & Ho, 2007; Tsay, 2008), journals (Schubert et al., 1989; Tsay, 2008), categories (Moed et al., 1995; Yamazaki, 1994), and contributing countries and institutions (Schubert et al., 1989; Chiu & Ho, 2007; Tanaka & Ho, 2011; Li et al., 2011). In recent years, author keywords which could provide a reasonably detailed picture of the article's subject (Garfield, 1990), have been quantitatively analyzed to figure out the research emphases and trends (Chiu & Ho, 2007; Fu et al., 2010; Li et al., 2011). To analyze these aspects, the common evaluation indicators number of publications and their publication share were utilized (Schubert et al., 1989; Rehn et al., 2007). Impact factor was introduced by Garfield and Sher (1963) to help select additional source journals using the recent citations received from other journals, and has been also widely used to rank and evaluate journals (Garfield, 1996; Moed, 2002). In particular, the five indicators including total, independent, collaborative, first author, and corresponding author articles have been recently used to compare the publication performance of countries and institutions in the researched of Japanese lung cancer (Ho et al., 2010), solid waste research (Fu et al., 2010), desalination research (Tanaka & Ho, 2011), acupuncture research (Han & Ho, 2011), global climate change (Li et al., 2011), and estuary pollution (Sun et al., 2012). In addition, rankings are useful indicators, especially in terms of institutions and researchers in urgent demand by politics (Weingart, 2003). These indicators including number of publication, publication share, impact factor, rankings, and five newly indicators were utilized for the evaluation of thermodynamics research.

The aim of this study was to systematically evaluate not only the publication characteristics of languages, annual publication outputs, journals and Web of Science categories, and national and inter-institutional contributors, but also the research emphases and trends by author keywords to obtain an overview of thermodynamic research during the period from 1991 to 2010.

2. Methodology

2.1. Data collection

Documents used in this study were derived from the Science Citation Index Expanded (SCI-Expanded) database of the Web of Science, the Thomson Reuters. According to Journal Citation Reports (JCR), it indexes 8,073 journals with citation references across 174 scientific disciplines in 2010. The data was collected on November 20 in 2011, and the detail data collection process is illustrated in Fig. 1. Firstly, thermodynamics related keywords including "thermodynamic", "thermodynamics", "free energy change", "enthalpy change", and "entropy change" were searched in terms of topic (including four parts: title, abstract, author keywords, and KeyWords Plus) within the publication year limitation from 1991 to 2010 based on SCI-Expanded. Total 157,140 documents were therefore found out. However, these 157,140 documents contained the documents not closely relating to thermodynamics which was searched only in terms of KeyWords Plus which providing search terms extracted from the titles of papers cited in each new article listed in *Current Contents* (Garfield, 1990). To efficiently obtain the characteristics of thermodynamics, this study only focused on the documents (50,865) with searching keywords in front page which including three parts: title, abstract, and author keywords, while the documents where searching keywords only appeared in KeyWords Plus were excluded. These documents consists of 19 document types, where articles (41,245) dominants with a percentage of 81%, followed by proceedings paper (4,860; 10%), meeting abstract (1,787; 3.5%), and review (1,407; 2.8%).

Finally, the journal articles (41,245) were extracted from the 50,865 documents for subsequent analyses.

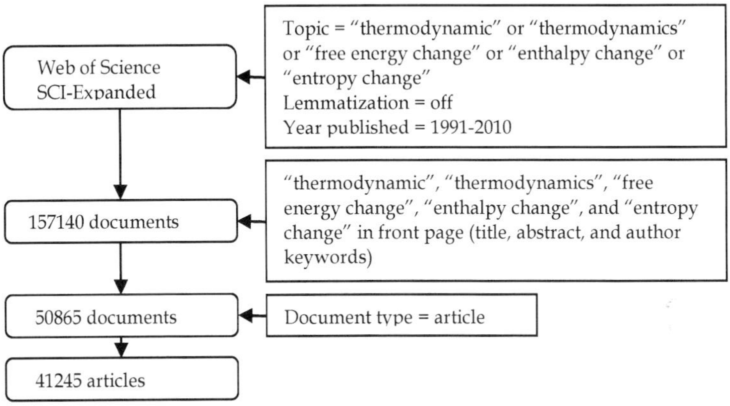

Figure 1. Data collection process for searching thermodynamic related research

2.2. Analyzing structure

The analyzing aspects were divided into five sections from different angles. To be specific, the analytic structure of analytic structure of bibliometric analysis of thermodynamic research is presented in Fig. 2. The first section dealt with the basic information covering the distribution of languages and annual production. The second section revealed the performance of productive journals and Web of Science categories. Then national contributors countries and institutional contributors institutions were analyzed in the following third and fourth sections. Finally, recent research emphases and trends were examined by the frequency of author keywords.

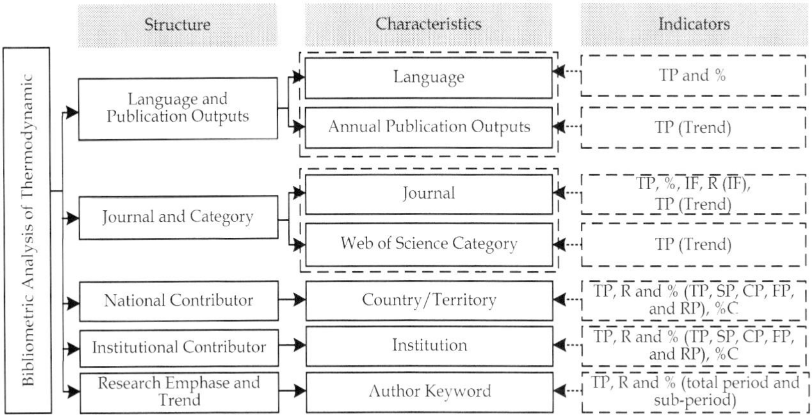

Figure 2. Analytic structure of bibliometric analysis of thermodynamic research

Journal	TP (%)	IF	Web of Science category (rankings by IF)
Russian Journal of Physical Chemistry A	827 (2.0)	0.503	physical chemistry (114/127)
Journal of Alloys and Compounds	768 (1.9)	2.134	physical chemistry (63/127); multidisciplinary materials science (50/225); metallurgy & metallurgical engineering (5/76)
Physical Review B	741 (1.8)	3.774	condensed matter physics (13/68)
Journal of Chemical Physics	716 (1.7)	2.920	atomic, molecular & chemical physics (7/33)
Journal of Chemical Thermodynamics	697 (1.7)	2.794	thermodynamics (3/51); physical chemistry (46/127)
Thermochimica Acta	652 (1.6)	1.899	analytical chemistry (33/73); physical chemistry (71/127)
Biochemistry	601 (1.5)	3.226	biochemistry & molecular biology (120/286)
Journal of Physical Chemistry B	584 (1.4)	3.603	physical chemistry (32/127)
Fluid Phase Equilibria	483 (1.2)	2.253	thermodynamics (5/51); physical chemistry (61/127); chemical engineering (24/135)
CALPHAD-Computer Coupling of Phase Diagrams and Thermochemistry	450 (1.1)	1.429	thermodynamics (16/51); physical chemistry (81/127)
Physical Review E	444 (1.1)	2.352	fluids & plasmas physics (8/31); mathematical physics (4/54)
Journal of Colloid and Interface Science	411 (1.0)	3.066	physical chemistry (40/127)
Journal of Chemical and Engineering Data	400 (1.0)	2.089	multidisciplinary chemistry (44/147); chemical engineering (28/135)
Surface Science	382 (0.93)	2.010	physical chemistry (69/127); condensed matter physics (19/68)
Journal of Solution Chemistry	360 (0.87)	1.335	physical chemistry (85/127)
Journal of Molecular Biology	339 (0.82)	4.008	biochemistry & molecular biology (86/286)
Journal of the American Chemical Society	317 (0.77)	9.019	multidisciplinary chemistry (11/147)
Physical Review D	278 (0.67)	4.964	astronomy & astrophysics (8/55); particles & fields physics (5/27)
Physical Review Letters	262 (0.64)	7.621	multidisciplinary physics (5/80)

Journal	TP (%)	IF	Web of Science category (rankings by IF)
Langmuir	242 (0.59)	4.268	multidisciplinary chemistry (24/147); physical chemistry (29/127); multidisciplinary materials science (26/225)

Table 3. Characteristics of top 20 journals with the most articles in thermodynamics

The thermodynamic research revealed a growth trend through the study period, especially in recent years. The number of articles has been climbing up from 1991 to 2010 in Fig. 3. According to the annual growth rate, it can be divided into two stages: 1991-2003 and 2004-2010. In the first stage, the growth rate was 48 articles per year from 1,412 in 1991 to 1,988 in 2003, while in the second stage the growth rate has risen four-fold to 203 articles per year from 2,169 in 2004 to 3,410 in 2010.

3.2. Journals and Web of Science categories

3.2.1. Journals

The total articles (41,245) were published in 2,434 sources among 150 Web of Science categories. Table 3 lists the top 20 journals with the greatest number of articles. The impact factors of these journals were also exhibited as reputation of a journal attached much attention by the authors and editors (van Raan, 2001). Approximately 24% of the articles resided in these 20 journals. *Russian Journal of Physical Chemistry A* (IF = 0.523) which was founded in 1930, published the most articles (827; 2.0%). Its former was *Zhurnal fizicheskoi khimii* (1930-1999) and *Russian Journal of Physical Chemistry* (1999-2006). Articles published in this journal concern chemical thermodynamics and thermochemistry (http://www.maik.ru/cgi-perl/journal.pl?name=physcha&page=main) which is closely related to the topic of this study.

Among the total institutions (1,004) contributing to *Russian Journal of Physical Chemistry A*, 831 institutions (83%) belonged to Russia while only 173 institutions (17%) belonged to other countries. Following three journals: *Journal of Alloys and Compounds* (IF = 2.134), *Physical Review B* (IF = 3.774), and *Journal of Chemical Physics* (IF = 2.920), all had the more than 700 articles. However, the journal which had the highest IF (9.019) was *Journal of the American Chemical Society* in the category of multidisciplinary chemistry which was ranked 17th. *Physical Review Letters* in multidisciplinary physics category was ranked 19th in the list and ranked 2nd in terms of IF (7.621). Furthermore, of these 2,434 sources, 1,828 (75%) sources only contained less than 10 articles; 519 (21%) sources contained 11-100 articles, 60 (2.5%) sources contained 101-200 articles; 20 (0.82%) sources contained 201-600 articles; and seven sources (0.28%) contained more than 600 articles, accounting for 11 percent of the total articles. The trends of the annual publication outputs of the top core seven journals are displayed in Fig. 4. The curves of these seven journals (*Russian Journal of Physical Chemistry A, Journal of Alloys and Compounds, Physical Review B, Journal of Chemical Physics, Journal of Chemical Thermodynamics*, and *Thermochimica Acta Biochemistry*) had been interlacing in the study period. However, after 2008, *Journal of Alloys and Compounds* which published 108 articles in 2009 and 70 articles in 2010 showed its domination in the thermodynamic field.

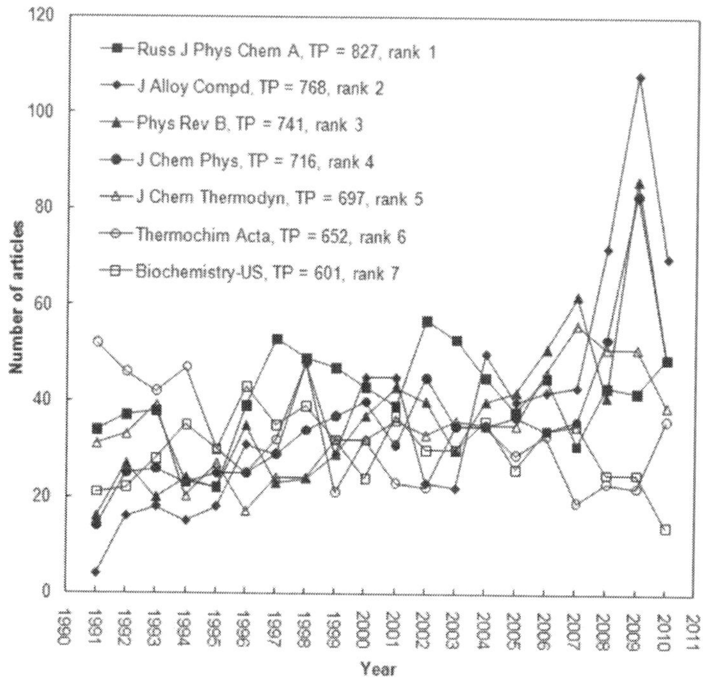

Figure 4. The trends of top seven journals with more than 600 articles

3.2.2. *Web of Science categories*

As for category analysis, excluding the 84 articles without subject category information, the remained 41,161 articles were analyzed. The most common category was physical chemistry (10,295; 25%), followed distantly by other categories. A half of the above top 20 journals are assigned to the category of physical chemistry. Within the total 150 Web of Science categories, 92 categories (61%) published articles less than 100 articles; 40 categories (27%) published 101-1000 articles; 12 categories (8.0%) published 1,001-3,000 articles; only six categories (4.0%) published more than 3,000 articles. The six core categories including physical chemistry, multidisciplinary materials science, multidisciplinary chemistry, metallurgy & metallurgical engineering, chemical engineering, and biochemistry & molecular biology, took the majority of the total articles with a great percentage of 55%. As illustrated in Fig. 5, these categories showed greater growth rates in recent years than those in 1990s. These trends were similar to the above trends of annual total publication outputs. The articles of position 1st category of physical chemistry which contained 127 journals increased from 390 in 1991 to 777 in 2010. The 2nd position category of multidisciplinary materials science which included 225 journals increased six-fold from 70 in 1991 to 423 in 2010.

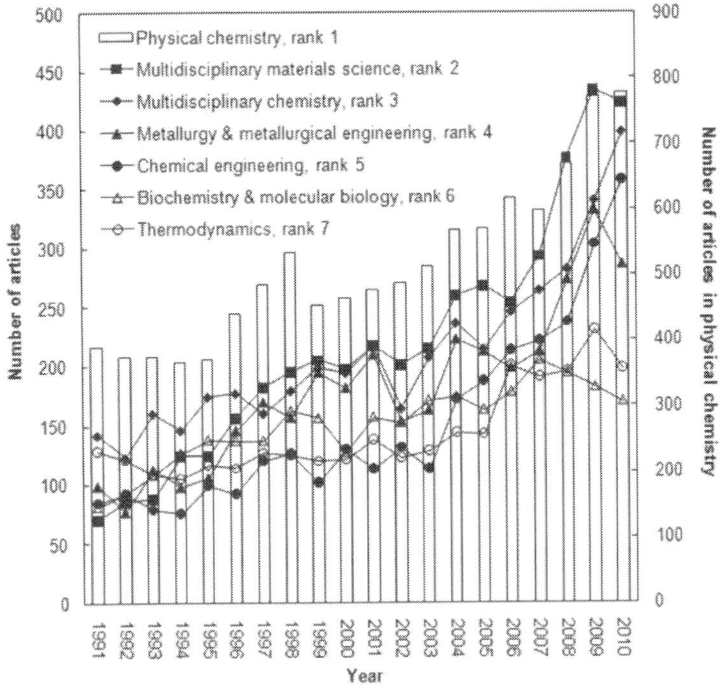

Figure 5. The trends of top seven Web of Science categories with more than 3000 articles

3.3. National contributors

3.3.1. Global geographical distribution

Excluding 263 articles without any author address information on the Web of Science, the 40,982 articles originated from 128 countries. Articles originating from England, Scotland, Northern Ireland, and Wales were reclassified as being from the United Kingdom (UK) (Chiu & Ho, 2005). Articles from Hong Kong were included in the ones from China. The geographical global distribution of thermodynamic research is shown in Fig. 6. According to their production, the 128 countries were divided into five parts. Particularly, 81 countries (63%) belonged to the first part of 1-100 articles; 36 countries (28%) belonged to the second part of 101-1,000 articles; seven countries (Japan, France, India, Italy, UK, Spain, and Canada) belonged to the third part of 1,001-3,000 articles; three countries (China, Russia, and Germany) belonged to the fourth part of 3,001-5,000 articles; and only one country (USA) which totally published 9,161 articles, belonged to the fifth part of 5,001-10,000 articles. Although there were only 11 countries which published more than 1,000 articles, these 11 countries produced 77% articles. The majority of articles originated from a small number of particular countries. Furthermore, five continents contributed to the thermodynamics research differently. Europe with 42 countries published the most articles 22,969 (56%); Asia with 42 countries published the second most articles 13,163 (32%); and America with sixteen countries published the third most articles 12,333 (30%). Africa, and Oceania two continents made much less contributions, and the quantity of them were 933 (2.3%), and 667 (1.6%), respectively. It is noticeable that Europe is taking the leading position of thermodynamic research. Asia was seeing a striking increase to catch up with Europe in terms of scientific output (Friedberg, 2000; von Bubnoff, 2005).

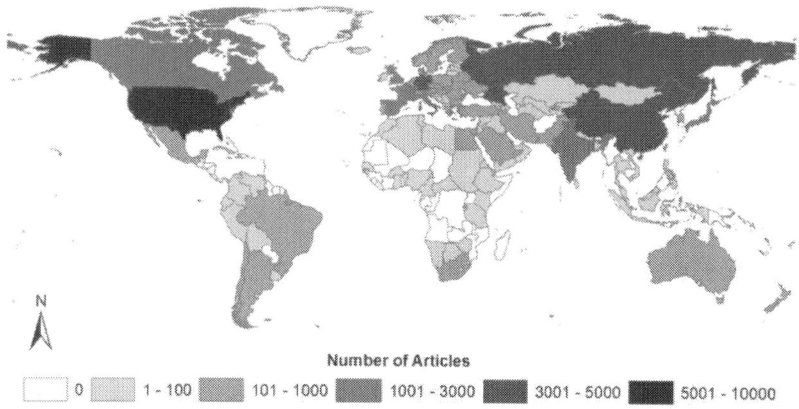

Figure 6. Global geographical distribution of thermodynamics research outputs

3.3.2. *Characteristics of top 20 countries/territories*

With respect to international collaboration, 33,234 (81%) were independent publications and 7,748 (19%) were internationally collaborative publications. The collaboration rate was a little greater than that of other areas, such as 14% biosorption technology for water treatment (Ho, 2008), 16% of desalination research (Tanaka & Ho, 2011), 16% of solid waste research (Fu et al., 2010), and 14% of acupuncture research (Han & Ho, 2011). Table 4 shows the characteristics of the top 20 productive countries. The eight major industrialized countries G8, ranked in the top 11 in Table 4, had 59% over the investigation period. Similarly, the seven major industrialized countries G7 the USA, Germany, the UK, Japan, France, Canada, and Italy accounted for a significant proportion exited in many researches (Li et al., 2009; He, 2009; Fu et al., 2010; Wang et al., 2010). The USA dominated, ranking 1st in independent and collaborative articles, as well as first author and corresponding author articles. Total 1,421 contributing institutions (13%) were affiliated with USA, followed by China (824 institutions, 7.7%), and Russia (792 institutions, 6.5%). The 2nd position China, the 7th position India, and the 19th position Iran had low percentages of collaboration less than 20%. It was not coincident that China emerged as a leading nation in science production (Zhou & Leydesdorff, 2006; Zhou & Leydesdorff, 2008). In most fields, the USA had the quantity (number of papers) lead, although the China has made dramatic strides to overtake the USA, except the biomedical field and some aspects of environmental science (Kostoff, 2008).

Country	IN	TP	TP R (%)	SP R (%)	CP R (%)	FP R (%)	RP R (%)	%C
USA	1,421	9,162	1 (22)	1 (20)	1 (33)	1 (19)	1 (18)	28
China	824	4,652	2 (11)	2 (12)	5 (10)	2 (11)	2 (11)	17
Russia	792	3,792	3 (9.3)	3 (9.1)	6 (10)	3 (8.2)	3 (8.6)	20
Germany	724	3,421	4 (8.3)	6 (5.5)	2 (21)	4 (6.3)	4 (6.2)	47
Japan	580	2,804	5 (6.8)	4 (6.3)	7 (9.0)	5 (5.8)	5 (5.8)	25
France	792	2,515	6 (6.1)	7 (4.1)	3 (15)	7 (4.6)	7 (4.5)	46
India	652	2,201	7 (5.4)	5 (5.7)	15 (3.8)	6 (5.1)	6 (5.0)	13
Italy	390	1,858	8 (4.5)	8 (3.6)	8 (8.6)	8 (3.8)	8 (3.8)	36
UK	315	1,797	9 (4.4)	9 (2.5)	4 (12)	9 (3.0)	9 (2.9)	53
Spain	210	1,487	10 (3.6)	11 (2.5)	9 (8.4)	10 (2.9)	10 (2.8)	44
Canada	192	1,446	11 (3.5)	10 (2.5)	10 (7.8)	11 (2.7)	11 (2.6)	42
Poland	171	934	12 (2.3)	12 (1.6)	11 (5.1)	12 (1.7)	12 (1.7)	42
Netherlands	145	741	13 (1.8)	18 (1.1)	13 (4.7)	17 (1.3)	19 (1.2)	50
Brazil	196	733	14 (1.8)	14 (1.5)	19 (3.1)	13 (1.5)	13 (1.5)	33
Ukraine	229	720	15 (1.8)	15 (1.4)	18 (3.1)	14 (1.4)	14 (1.4)	34
Switzerland	111	690	16 (1.7)	21 (0.93)	12 (4.9)	19 (1.2)	18 (1.2)	55
Sweden	106	645	17 (1.6)	19 (1.1)	16 (3.5)	20 (1.2)	20 (1.1)	42
South Korea	176	632	18 (1.5)	17 (1.3)	22 (2.5)	16 (1.3)	16 (1.4)	31
Iran	173	587	19 (1.4)	13 (1.5)	32 (1.0)	15 (1.4)	15 (1.4)	14
Australia	106	566	20 (1.4)	23 (0.79)	14 (3.9)	21 (1.0)	21 (1.0)	54

IN: Number of institutions.

Table 4. Characteristics of the top 20 productive countries/territories

Besides, the 9th position UK, the 13th position Netherlands, the 16th position Switzerland, and the 20th position Australia had high cooperation rates no less than 50%. It was also reported that international collaboration in both within and outside the European Communities played an increasing role for European scientific literature (Narin et al., 1991; Glänzel, 1999). South Korea and Iran had higher rankings in terms of FP and RP than that of TP, while Netherlands and Sweden had lower rankings in terms of FP and RP. Moreover, from the above languages analysis, except English, the popular languages (Russian, Chinese, French, Japanese, and German) were just the mother tongue of the top five countries China, Russia, Germany, Japan, and France except the USA. The excellent performance of these countries was consistent with the results of language analysis.

3.3.3. *Comparison of top six countries*

The comparison of top six countries with the most articles in 2010 is illustrated in Fig. 7. The USA hold the trump card based on the total number of articles in the study period, but did not have the highest growth rate. China which was ranked 1st according to the 2010 production had the highest growth rate of 32 articles per year, and surpassed the USA in 2009, following distantly by other countries. USA, India, and Russia had the lower growth rate of 10 - 13 articles per year. The following Germany and France had the growth rate of 6.3 and 5.9 articles per year. Particularly, Russia experienced a sharply increase in the first three year from 6 articles in 1991 to 174 articles in 1993, may due to its large fund for recovery of the politic reason (Goodman, 1993; Webb, 1994). The Russian annual output grew from 1980 to 1990, but fell after the dissolution of the USSR in late 1991; from 1994 there has been an inconsistent partial recovery, and by 2000 the annual output had approximately regained its 1980 value (Wilson & Markusova, 2004). It seems that the recovery time of Russian thermodynamic research was shorter.

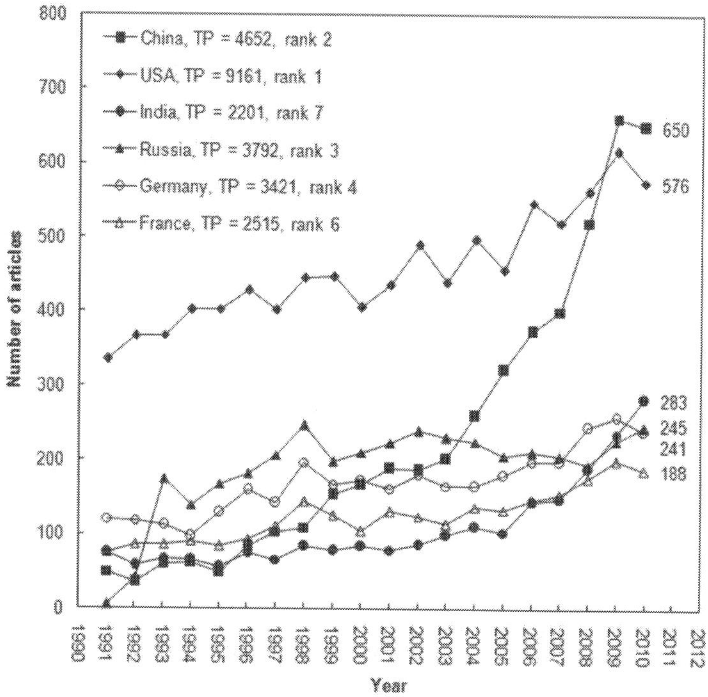

Figure 7. Trends of top six countries with most articles in 2010

3.3.4. Comparison of China and USA

In order to compare the most productive country in 1991-2010, the USA and the most active country in 2010, China, annual number of articles and citations per publication (CPP) of both were presented in Fig. 8. Total citations were collected on November 20 in 2011 from web of Science. From the above analysis, China experienced a high grow rate and got ahead of the USA in 2009 and kept on top in 2010. However, in terms of CPP, the mean number of China was 7.1 which was much less than that of the USA (23). The annual CPP of the USA showed a declined trend in 1991-2010, ranging from 53 in 1991 to 2.8 in 2010. Except the year of 1993, the annual CPP of China fluctuated ranging 6.2-11 in 1991-2006, and decreased after 2007 down to 1.9 in 2010. Both countries indicated declines as for annual CPPs, which can be attributed to that it needs time to accumulate their citations (Picknett & Davis, 1999). The outstanding performance of China's CPP in 1993 might be owned to the two highly cited articles of a series by the same authors. They were "thermodynamics of molecular recognition by cyclodextrins. 1. calorimetric titration of inclusion complexation of naphthalenesulfonates with α-, β-, and γ-cyclodextrin: enthalpy entropy compensation" (Inoue et al., 1993a) with 344 citations and "thermodynamics of molecular recognition by cyclodextrins. calorimetric titration of inclusion complexation with modified β-cyclodextrins. enthalpy-entropy compensation in host-guest complexation: from ionophore to cyclodextrin and cyclophane"(Inoue et al., 1993b) with 198 citations, respectively. Furthermore, the most frequently cited article was "a modified UNIFAC model. 2. present parameter matrix and results for different thermodynamic properties" (Gmehling et al., 1993) with 727 citations. Although this article was published by authors from Germany only, it was also assigned to China according to its address record from Web of Science, because one of the authors had permanent address of China. It is a bias of address information for articles from Web of Science. China's CPP in 1993 were still high (CPP = 20) without these highest citations articles. Likewise, the well performance of the USA's CPP in 1991 can be ascribed to the top cited article "protein folding and association: insights from the interfacial and thermodynamic

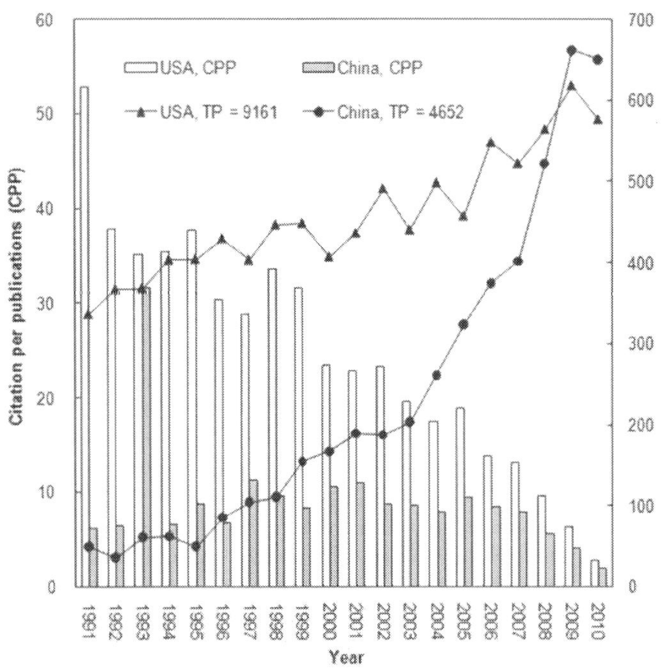

Figure 8. Comparison of USA and China by number of articles and citations per publication

Author keywords	TP	1991-2010 R (%)	1991-1995 R (%)	1996-2000 R (%)	2001-2005 R (%)	2006-2010 R (%)
thermodynamics	5,943	1 (25)	1 (31)	1 (25)	1 (23)	1 (25)
thermodynamic properties	1,577	2 (6.7)	2 (5.5)	2 (8.4)	2 (5.9)	2 (6.7)
kinetics	883	3 (3.8)	6 (2.5)	8 (2.2)	3 (3.6)	3 (4.8)
adsorption	736	5 (3.1)	11 (1.5)	10 (1.9)	6 (2.5)	4 (4.4)
thermodynamic parameters	778	4 (3.3)	3 (2.6)	3 (3.5)	4 (3.2)	5 (3.5)
thermodynamic	503	9 (2.1)	26 (1.0)	15 (1.3)	10 (2.1)	6 (2.7)
phase diagram	545	7 (2.3)	21 (1.2)	6 (2.3)	9 (2.1)	7 (2.7)
entropy	548	6 (2.3)	8 (2.0)	7 (2.2)	7 (2.5)	8 (2.4)
thermodynamic modeling	338	13 (1.4)	165 (0.27)	56 (0.50)	16 (1.3)	9 (2.2)
thermodynamic analysis	385	11 (1.6)	12 (1.5)	14 (1.7)	12 (1.3)	10 (1.8)
heat capacity	533	8 (2.3)	3 (2.6)	5 (2.7)	5 (2.8)	11 (1.7)
calorimetry	503	9 (2.1)	5 (2.5)	4 (3.0)	8 (2.2)	12 (1.7)
phase diagrams	260	17 (1.1)	26 (1.0)	31 (0.68)	25 (0.83)	13 (1.5)
CALPHAD	186	25 (0.79)	N/A	782 (0.068)	35 (0.66)	14 (1.3)
solubility	325	14 (1.4)	10 (2.0)	16 (1.3)	13 (1.3)	14 (1.3)
enthalpy	373	12 (1.6)	8 (2.0)	9 (1.9)	11 (1.7)	16 (1.3)
thermodynamic property	212	20 (0.90)	59 (0.59)	64 (0.45)	25 (0.83)	17 (1.2)
density functional theory	158	36 (0.67)	1288 (0.045)	415 (0.11)	56 (0.53)	18 (1.1)
isothermal titration calorimetry	177	28 (0.75)	1288 (0.045)	108 (0.34)	31 (0.70)	19 (1.1)
thermodynamic stability	265	16 (1.1)	38 (0.77)	16 (1.3)	13 (1.3)	20 (1.0)
isotherm	120	60 (0.51)	1288 (0.045)	565 (0.09)	231 (0.18)	21 (1.0)
equation of state	316	15 (1.3)	7 (2.4)	13 (1.7)	13 (1.3)	21 (1.0)
density	178	27 (0.76)	29 (1.0)	41 (0.59)	60 (0.51)	23 (0.94)
molecular dynamics	172	29 (0.73)	165 (0.27)	31 (0.68)	47 (0.61)	24 (0.93)
free energy	163	33 (0.69)	165 (0.27)	52 (0.52)	51 (0.59)	25 (0.92)
phase equilibria	228	19 (1.0)	32 (0.86)	25 (0.83)	18 (1.2)	26 (0.91)
biosorption	101	72 (0.43)	N/A	N/A	431 (0.11)	27 (0.89)
AB initio calculations	121	59 (0.52)	N/A	170 (0.25)	147 (0.26)	27 (0.89)
phase transition	158	36 (0.67)	52 (0.64)	58 (0.47)	60 (0.51)	29 (0.86)
nonequilibrium thermodynamics	190	24 (0.81)	23 (1.1)	27 (0.81)	39 (0.64)	30 (0.85)

N/A: Not Available

Table 6. The most frequently used 30 author keywords in 2006-2010

In particular, there were also some emerging hotspots in thermodynamics recently. Some author keywords including the 3rd "adsorption", the 9th "thermodynamic modeling", the 14th "CALPHAD", the 17th "thermodynamic property", the 18th "density functional theory", the 19th "isothermal titration calorimetry", the 20th "thermodynamic stability", the 21st "isotherm", the 25th "free energy", the 27th "biosorption", and the 28th "AB initio calculations" in the list showed the increasing trends during the study period. It is noticeable that "CALPHAD", "biosorption", and "AB initio calculations" increased rapidly from zero articles in 1991-1995 into the top 30 rankings in 2006-2010. "CALPHAD" was firstly used as the author keyword of the article entitled "compilation of 'CALPHAD' formation enthalpy data - Binary intermetallic compounds in the COST507 Gibbsian database" in 1998 (Fries & Jantzen, 1998). However, "CALPHAD method" had been used as the author keyword of the article titled "thermodynamic study on the Ag-Sb-Sn system" in 1996 (Oh et al., 1996). The first article which employed "biosorption" as author keyword was titled "determination of the equilibrium, kinetic and thermodynamic parameters of the batch biosorption of nickel(II) ions onto Chlorella vulgaris" in *Process Biochemistry* in 2002 (Aksu, 2002). "Biosorption" as a process for water treatment received increasing attention in recent years (Ho, 2008). As for "AB initio calculations", it was originally utilized as author keywords of the article "an AB initio study of structures and energetics of copper sulfide clusters" in 1996 (Dehnen et al., 1996). Similarly, "density functional theory", "isothermal titration calorimetry", and "isotherm" grew fast from the 1288th (only one article) in 1991-1995 to 18th, 19th, and 21st in 2006-2010. The article targeting at "density functional theory" could be found in 1994 and it was entitled "density-functional theory as thermodynamics" (Nagy & Parr, 1994). Thermodynamical interpretation of the density functional theory for an electronic ground state was developed (Nagy & Parr, 1994). The following article which utilized "density functional theory" as author keyword was published in 1999. For the phrase "isothermal titration calorimetry", the article which used it as author keyword appeared in 1995, and it was titled "comparative thermodynamic analyses of the Fv, Fab* and Fab fragments of anti-dansyl mouse monoclonal-antibody" (Shimba, 1995). However, it should be noticed that there was a bias that all these analysis were based on the data (57%) with author keywords information from SCI-Expanded.

4. Conclusion

Except the dominant language English, Russian, Chinese, French, Japanese and German were the most common languages. Above all, the growth process of annual thermodynamic scientific outputs can be divided into two stages, and the annual growth rate of the second stage 2004-2010 was four times of that in the first stage 1991-2003. Secondly, *Russian Journal of Physical Chemistry* was the most productive journal, and then *Journal of Alloys and Compounds and Journal of Chemical Physics*. The most common category was found to be physical chemistry. Thirdly, the global geographic distribution discovered that Europe took the leading position, following by Asia, America, Africa and Oceania. In particular, the eight major industrialized countries (G8: Canada, France, Germany, Italy, Japan, Russia, the UK, and USA) played a prominent role with more than a half contribution, especially the USA.

However, China experienced the greatest growth rate, and surpassed the USA in 2009 in terms of annual production, but needs to enhance its quality of publications. China and India had low collaboration rates while the UK, Netherlands, and Switzerland in Europe with a smaller size had high collaboration rates. In general, the thermodynamic research presented a higher international collaboration rate but a lower inter-institutional collaboration rate in comparison with other fields. Russian Academy of Sciences and Chinese Academy of Sciences took the leading position of institutions, following distantly by Tohoku University in Japan, and Centre National de la Recherche Scientifique in France. It was also revealed that the institutions with lower collaboration rates usually published greater proportion of first author articles and corresponding author articles, and vice versa. Finally, by analyzing the latest most frequently used author keywords, it was concluded that the items "thermodynamic properties", "kinetics", "adsorption", "thermodynamic parameters", "phase diagram", and "entropy" are the traditional research hotspots and will continue being the emphases in the foreseeable future. "CALPHAD", "density functional theory", "isothermal titration calorimetry", "isotherm", "biosorption", and "AB initio calculations", as the quickly increasing concernings, are the newly thermodynamic research hotspots.

Author details

Hui-Zhen Fu[1], Yuh-Shan Ho[1,2,*]

[1]*Department of Environmental Sciences, Peking University, People's Republic of China*

[2]*Trend Research Centre, Asia University, Taiwan*

5. References

Aksu, Z. (2002). Determination of the equilibrium, kinetic and thermodynamic parameters of the batch biosorption of nickel(II) ions onto Chlorella vulgaris. *Process Biochemistry*, Vol.38, No.1, (September 2002), pp. 89-99, ISSN 0032-9592

Alfaraz, P.H. & Calvino, A.M. (2004). Bibliometric study on food science and technology: Scientific production in Iberian-American countries (1991-2000). *Scientometrics*, Vol.61, No.1, pp. 89-102, ISSN 0138-9130

Astumian, R.D. (1997). Thermodynamics and kinetics of a Brownian motor. *Science*, Vol.276, No.5314, (May 1997), pp.917-922, ISSN 0036-8075

Callen, H.B. (1985). *Thermodynamics and an introduction to thermostatistics* (2nd Edition), John Wiley & Sons, pp. vii, ISBN 978-0-471-86256-7, USA

Chiu, W.T. & Ho, Y.S. (2005). Bibliometric analysis of homeopathy research during the period of 1991 to 2003. *Scientometrics*, Vol.63, No.1, (April 2005), pp. 3-23, ISSN 0138-9130

Chiu, W.T. & Ho, Y.S. (2007). Bibliometric analysis of tsunami research. *Scientometrics*, Vol.73, No.1, (October 2007), pp. 3-17, ISSN 0138-9130

Chuang, K.Y., Huang, Y.L. & Ho, Y.S. (2007). A bibliometric and citation analysis of stroke-related research in Taiwan. *Scientometrics*, Vol.72, No.2, (August 2007), pp. 201-212, ISSN 0138-9130

[*] Corresponding Author

Chuang, K.Y., Wang, M.H. & Ho, Y.S. (2011). High-impact papers presented in the subject category of water resources in the Essential Science Indicators database of the institute for scientific information. *Scientometrics*, Vol.87, No.3, (June 2011), pp. 551-562, ISSN 0138-9130

Dehnen, S., Schafer, A., Ahlrichs, R. & Fenske, D. (1996). An AB initio study of structures and energetics of copper sulfide clusters. *Chemistry-A European Journal*, Vol.2, No.4, (April 1996), pp. 429-435, ISSN 0947-6539

Ferrara, S.D., Bajanowski, T., Cecchi, R., Boscolo-Berto, R. & Viel, G. (2011). Bio-medicolegal scientific research in Europe: a comprehensive bibliometric overview. *International Journal of Legal Medicine*, Vol.125, No.3, (May 2011), pp. 393-402, ISSN 0937-9827

Friedberg, A.L. (2000). Will Europe's past be Asia's future? *Survival*, Vol.42, No.3, pp. 147-59, ISSN 0039-6338

Fries, S.G. & Jantzen, T. (1998). Compilation of 'CALPHAD' formation enthalpy data - Binary intermetallic compounds in the COST507 Gibbsian database. *Thermochimica Acta*, Vol.314, No.1-2, (April 1998), pp. 23-33, ISSN 0040-6031

Fu, H.Z., Chuang, K.Y., Wang, M.H. & Ho, Y.S. (2011). Characteristics of research in China assessed with Essential Science Indicators. *Scientometrics*, Vol.88, No.3, (September 2011), pp. 841-862, ISSN 0138-9130

Fu, H.Z., Ho, Y.S., Sui, Y.M. & Li, Z.S. (2010). A bibliometric analysis of solid waste research during the period 1993-2008. *Waste Management*, Vol.30, No.12, (December 2010), pp. 2410-2417, ISSN 0956-053X

Garfield, E. (1990). KeyWords Plus™ - ISIS breakthrough retrieval method. 1. Expanding your searching power on current-contents on diskette. *Current Contents*, Vol.32, (August 1990), pp. 5-9

Garfield, E. (1996). How can impact factors be improved? *British Medical Journal*, Vol.313, No.7054, (August 1996), pp. 411-413, ISSN 0959-535X

Garfield, E. & Sher, I.H. (1963). New factors in evaluation of scientific literature through citation indexing. *American Documentation*, Vol.14, No.3, pp. 195-201, ISSN 0096-946X

Glänzel, W., Schubert, A. & Braun, T. (2002). A relational charting approach to the world of basic research in twelve science fields at the end of the second millennium. *Scientometrics*, Vol.55, No.3, pp. 335-348, ISSN 0138-9130

Glänzel, W., Schubert, A. & Czerwon, H.J. (1999). A bibliometric analysis of international scientific cooperation of the European Union (1985-1995). *Scientometrics*, Vol.45, No.2, (June 1999), pp. 185-202, ISSN 0138-9130

Gmehling, J., Li, J.D. & Schiller, M. (1993). A modified UNIFAC model. 2. Present parameter matrix and results for different thermodynamic properties. *Industrial & Engineering Chemistry Research*, Vol.32, No.1, (January 1993), pp. 178-193, ISSN 0888-5885

Goodman, B. (1993). Noted researchers laud donation to Russian science. *Scientist*, Vol.7, No.1, (January 1993), pp. 3, ISSN 0890-3670

Han, J.S. & Ho, Y.S. (2011). Global trends and performances of acupuncture research. *Neuroscience and Biobehavioral Reviews*, Vol.35, No.3, (January 2011), pp. 680-687, ISSN 0149-7634

He, T.W. (2009). International scientific collaboration of China with the G7 countries. *Scientometrics*, Vol.80, No.3, (September 2009), pp. 571-582, ISSN 0138-9130

Ho, Y.S. (2008). Bibliometric analysis of biosorption technology in water treatment research from 1991 to 2004. *International Journal of Environment and Pollution*, Vol.34, No.1-4, pp. 1-13, ISSN 0957-4352

Ho, Y.S. & Ofomaja, A.E. (2005). Kinetics and thermodynamics of lead ion sorption on palm kernel fibre from aqueous solution. *Process Biochemistry*, Vol.40, No.11, (November 2005), pp. 3455-3461, ISSN 1359-5113

Ho, Y.S., Satoh, H. & Lin, S.Y. (2010). Japanese lung cancer research trends and performance in Science Citation Index. *Internal Medicine*, Vol.49, No.20, pp. 2219-2228, ISSN 0918-2918

Holman, J.P. (1985). *Thermodynamics* (3rd edition), McGraw-Hill, pp. 2, ISBN 978-0-070-29608-4, New York, USA

Jin, B.H. & Rousseau, R. (2005). China's quantitative expansion phase: Exponential growth but low impact. *ISSI 2005: Proceedings of the 10th International Conference of the International Society for Scientometrics and Informetrics*, pp. 362-370, ISBN 91-7140-339-6, Stockholm, Sweden, July 24-28, 2005

Inoue, Y., Hakushi, T., Liu, Y., Tong, L.H., Shen, B.J. & Jin, D.S. (1993a). Thermodynamics of molecular recognition by cyclodextrins. 1. Calorimetric titration of inclusion complexation of naphthalenesulfonates with α-, β-, and γ-cyclodextrin: enthalpy entropy compensation. *Journal of the American Chemical Society*, Vol.115, No.2, (January 1993), pp. 475-481, ISSN 0002-7863

Inoue, Y., Liu, Y., Tong, L.H., Shen, B.J. & Jin, D.S. (1993b). Thermodynamics of molecular recognition by cyclodextrins. Calorimetric titration of inclusion complexation with modified β-cyclodextrins. Enthalpy-entropy compensation in host-guest complexation: From ionophore to cyclodextrin and cyclophane. *Journal of the American Chemical Society*, Vol.115, No.23, (November 1993), pp. 10637-10644, ISSN 0002-7863

Keizer, J. (1985). Heat, work, and the thermodynamic temperature at nonequilibrium steady-states. *Journal of Chemical Physics*, Vol.82, No.6, pp. 2751-2771, ISSN 0021-9606

Kleppa, O.J. (1952). A thermodynamic study of liquid metallic solutions. 4. approximate thermodynamic data from the phase diagram for the systems copper bismuth, copper lead and copper thallium. *Journal of the American Chemical Society*, Vol.74, No.23, pp. 6047-6051, ISSN 0002-7863

Kocks, U.F., Argon, A.S. & Ashby, M.F. (1975). Thermodynamics and kinetics of slip. *Progress in Materials Science*, Vol.19, No.1-281, pp. 1-281, ISSN 0079-6425

Kostoff, R.N. (2008). Comparison of China/USA science and technology performance. *Journal of Informetrics*, Vol.2, No.4, (October 2008), pp. 354-363, ISSN 1751-1577

Li, J.F., Wang, M.H. & Ho, Y.S. (2011). Trends in research on global climate change: A Science Citation Index Expanded-based analysis. *Global and Planetary Change*, Vol.77, No.1-2, (May 2011), pp. 13-20, ISSN 0921-8181

Li, J.F., Zhang, Y.H., Wang, X.S. & Ho, Y.S. (2009). Bibliometric analysis of atmospheric simulation trends in meteorology and atmospheric science journals. *Croatica Chemica Acta*, Vol.82, No.3, (December 2009), pp. 695-705, ISSN 0011-1643

Look, D.C. & Saucer, H.J. (1982). *Thermodynamics*, Wadsworth, pp.4, ISBN 978-0-818-50491-4, California, USA

Luukkonen, T. (1990). Bibliometrics and evaluation of research performance. *Annals of Medicine*, Vol.22, No.3, (June 1990), pp. 145-150, ISSN 0785-3890

Moed, H.F. (2002). The impact-factors debate: the ISI's uses and limits. *Nature*, Vol.415, No.6873, (February 2002), pp. 731-732, ISSN 0028-0836

Moed, H.F., de Bruin, R.E. & Van Leeuwen, Th.N. (1995). New bibliometric tools for the assessment of national research performance: Database description, overview of indicators and first applications. *Scientometrics*, Vol.33, No.3, (July-August 1995), pp. 381-422, ISSN 0138-9130

Nagy, A. & Parr, R.G. (1994). Density-functional theory as thermodynamics. *Proceedings of the Indian Academy of Sciences-Chemical Sciences*, Vol.106, No.2, (April 1994), pp. 217-227, ISSN 0253-4134

Narin, F., Stevens, K. and Whitlow Edith S. (1991). Scientific co-operation in Europe and the citation of multinationally authored papers. *Scientometrics*, Vol.21, No.3, (July-August 1991), pp. 313-323, ISSN 0138-9130

Ngai, K.L. (2000). Dynamic and thermodynamic properties of glass-forming substances. *Journal of Non-Crystalline Solids*, Vol.275, No.1-2, (September 2000), pp. 7-51, ISSN 0022-3093

Nicholls, A., Sharp, K.A. & Honig, B. (1991). Protein folding and association - insights from the interfacial and thermodynamic properties of hydrocarbons. *Proteins-Structure Function and Genetics*, Vol.11, No.4, pp. 281-296, ISSN 0887-3585

Oh, C.S., Shim, J.H., Lee, B.J.A. & Lee, D.N. (1996). Thermodynamic study on the Ag-Sb-Sn system. *Journal of Alloys and Compounds*, Vol.238, No.1-2, (May 1 1996), pp. 155-166, ISSN 0925-8388

Patzek, T.W. (2004). Thermodynamics of the corn-ethanol biofuel cycle. *Critical Reviews in Plant Sciences*, Vol.23, No.6, pp. 519-567, ISSN 0735-2689

Picknett, T. & Davis, K. (1999). The 100 most-cited articles from JMB. *Journal of Molecular Biology*, Vol.293, No.2, (October 1999), pp. 173-176, ISSN 0022-2836

Rehn,C., Kronman, U. & Wadskog, D. (2007). *Bibliometric indicators-definitions and usage at Karolinska Institutet.* in Karolinska Institutet, 13.03.2012, Available from http://kib.ki.se/sites/kib.ki.se/files/Bibliometric_indicators_definitions_1.0.pdf

Rosakis, P., Rosakis, A.J., Ravichandran, G. & Hodowany, J. (2000). A thermodynamic internal variable model for the partition of plastic work into heat and stored energy in metals. *Journal of the Mechanics and Physics of Solids*, Vol.48, No.3, (March 2000), pp. 581-607, ISSN 0022-5096

Sainte-Marie, B. (2010). The first 30 years of the journal of crustacean biology - a bibliometric study. *Journal of Crustacean Biology*, Vol.30, (November 2010), pp. 541-549, ISSN 0278-0372

SantaLucia, J. & Hicks, D. (2004). The thermodynamics of DNA structural motifs. *Annual Review of Biophysics and Biomolecular Structure*, Vol.33, pp.415-440, ISSN 1056-8700

Saracevi, T. & Perk, L.J. (1973). Ascertaining activities in a subject area through bibliometric analysis. *Journal of the American Society for Information Science*, Vol.24, No.2, pp. 120-134, ISSN 0002-8231

Schubert, A., Glänzel, W. & Braun, T. (1989). Scientometric datafiles: A comprehensive set of indicators on 2649 journals and 96 countries in all major science fields and subfields 1981-1985. *Scientometrics*, Vol.16, No.1-6, (June 1989), pp. 3-478, ISSN 0138-9130

Seelig, J. (2004). Thermodynamics of lipid-peptide interactions. *Biochimica Et Biophysica Acta-Biomembranes*, Vol.1666, No.1-2, (November 2004), pp. 40-50, ISSN 0005-2736

Shimba, N., Torigoe, H. and Takahashi, H., Masudaa, K., Shimadaa, I., Arataa, Y. & Saraib, A. (1995). Comparative thermodynamic analyses of the Fv, Fab* and fab fragments of anti-dansyl mouse monoclonal antibody. *Febs Letters*, Vol.360, No.3, (March 6 1995), pp. 247-250, ISSN 0014-5793

Sun, J.S., Wang, M.H. & Ho, Y.S. (2012). A historical review and bibliometric analysis of research on estuary pollution. *Marine Pollution Bulletin*, Vol.64, No.1, (January 2012), pp. 13-21, ISSN 0025-326X

Svirbely, W.J. (1954). Thermodynamic data for the zinc indium system obtained from the phase diagram. *Journal of Physical Chemistry*, Vol.58, No.7, pp. 557-559, ISSN 0022-3654

Tanaka, H. & Ho, Y.S. (2011). Global trends and performances of desalination research. *Desalination and Water Treatment*, Vol.25, No.1-3, (January 2011), pp. 1-12, ISSN 1944-3994

Tsay, M.Y. (2008). A bibliometric analysis of hydrogen energy literature, 1965-2005. *Scientometrics*, Vol.75, No.3, (June 2008), pp. 421-438, ISSN 0138-9130

van Raan, A.F.J. (2001). Bibliometrics and Internet: Some observations and expectations. *Scientometrics*, Vol.50, No.1, (January 2001), pp. 59-63, ISSN 0138-9130

van Raan, A.F.J. (2005). For your citations only? Hot topics in bibliometric analysis. *Measurement: Interdisciplinary Research and Perspectives*, Vol.3, No.1, (January 2005), pp. 50-62, ISSN 1536-6367

von Bubnoff, A. (2005). Asia squeezes Europe's lead in science. *Nature*, Vol.436, No.7049, (July 2005), pp. 314-314, ISSN 0028-0836

Wang, M.H., Li, J.F. & Ho, Y.S. (2011). Research articles published in water resources journals: A bibliometric analysis. *Desalination and Water Treatment*, Vo.28, No.1-3, (April 2011), pp. 353-365, ISSN 1944-3994

Wang, M.H., Yu, T.C. & Ho, Y.S. (2010). A bibliometric analysis of the performance of Water Research. *Scientometrics*, Vol.84, No.3, (September 2010), pp. 813-820, ISSN 0138-9130.

Webb, T.D. (1994). Societies aid in large-scale effort to support Russian science. *Scientist*, Vol.8, No.2, (January 1994), pp. 3, ISSN 0890-3670

Weingart, P. (2003). Evaluation of research performance: The danger of numbers. *Schriften der Forschungszentrums Jülich: Vol II. Bibliometric analysis in science and research: Applications, benefits and limitations*, pp. 7-19, ISBN 3-89336-334-3, Forschungszentrums Jülich, Germany, November 5–7, 2003

Wilson, C.S. & Markusova, V.A. (2004). Changes in the scientific output of Russia from 1980 to 2000, as reflected in the Science Citation Index, in relation to national politico-economic changes. *Scientometrics*, Vol.59, No.3, pp. 345-389, ISSN 0138-9130

Yamazaki, S. (1994). Research activities in life sciences in Japan. *Scientometrics*, Vol.29, No.2, (February 1994), pp. 181-190, ISSN 0138-9130

Zhou, P. & Leydesdorff, L. (2006). The emergence of China as a leading nation in science. *Research Policy*, Vol.35, (February 2006), pp. 83-104, ISSN 0048-7333

Zhou, P. & Leydesdorff, L. (2008). China ranks second in scientific publications since 2006. *ISSI Newletter*, Vol.4, (March 2008), pp. 7-9